Boling Guo, Xiao-Feng Pang, Yu-Feng Wang, and Nan Liu

Solitons

Also of Interest

Rogue Waves. Mathematical Theory and Applications in Physics
Boling Guo, Lixin Tian, Zhenya Yan, Liming Ling, Yu-Feng Wang, 2017
ISBN 978-3-11-046942-4, e-ISBN (PDF) 978-3-11-047057-4,
e-ISBN (EPUB) 978-3-11-046969-1

Vanishing Viscosity Method. Solutions to Nonlinear Systems
Boling Guo, Dongfen Bian, Fangfang Li, Xiaoyu Xi, 2016
ISBN 978-3-11-049528-7, e-ISBN (PDF) 978-3-11-049427-3,
e-ISBN (EPUB) 978-3-11-049257-6

Stochastic PDEs and Dynamics
Boling Guo, Hongjun Gao, Xueke Pu, 2016
ISBN 978-3-11-049510-2, e-ISBN (PDF) 978-3-11-049388-7,
e-ISBN (EPUB) 978-3-11-049243-9

Nonlinear Dynamics. Non-Integrable Systems and Chaotic Dynamics
Alexander B. Borisov, Vladimir V. Zverev, 2016
ISBN 978-3-11-043938-0, e-ISBN (PDF) 978-3-11-043058-5,
e-ISBN (EPUB) 978-3-11-043067-7

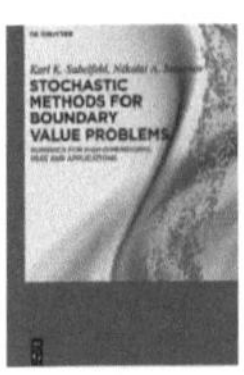

Stochastic Methods for Boundary Value Problems.
Numerics for High-dimensional PDEs and Applications
Karl K. Sabelfeld, Nikolai A. Simonov, 2016
ISBN 978-3-11-047906-5, e-ISBN (PDF) 978-3-11-047945-4,
e-ISBN (EPUB) 978-3-11-047916-4

Boling Guo, Xiao-Feng Pang, Yu-Feng Wang,
and Nan Liu

Solitons

DE GRUYTER

Mathematics Subject Classification 2010
Primary: 35Q55; Secondary: 37K40

Authors
Prof. Boling Guo
Laboratory of Computational Physics
Institute of Applied Physics and
Computational Mathematics
6 Huayuan Road
Haidian District
100088 Beijing
People's Republic of China
gbl@iapcm.ac.cn

Prof. Xiao-Feng Pang
University of Electronic Science and
Technology of China
Institute of Life Science and Technology
610054 Chengdu
People's Republic of China
Pangxf2006@aliyun.com

Prof. Yu-Feng Wang
College of Science
Minzu University of China
27 Zhongguancun South Avenue Beijing
100081 Beijing
People's Republic of China
yufeng_0617@126.com

Dr Nan Liu
Institute of Applied Physics and
Computational Mathematics
6 Huayuan Road
100088 Beijing
People's Republic of China
ln10475@163.com

ISBN 978-3-11-054924-9
e-ISBN (PDF) 978-3-11-054963-8
e-ISBN (EPUB) 978-3-11-054941-6
Set-ISBN 978-3-11-054964-5

Library of Congress Cataloging-in-Publication Data
A CIP catalog record for this book has been applied for at the Library of Congress.

Bibliographic information published by the Deutsche Nationalbibliothek
The Deutsche Nationalbibliothek lists this publication in the Deutsche Nationalbibliografie; detailed bibliographic data are available on the Internet at http://dnb.dnb.de.

Typesetting: VTeX UAB, Lithuania
Printing and binding: CPI books GmbH, Leck
Cover image: Chong Guo
∞ Printed on acid-free paper
Printed in Germany

www.degruyter.com

Contents

1 **Introduction — 1**
1.1 The background of solitons — 1
1.2 KdV equation and its soliton solutions — 3
1.3 Soliton solutions for the nonlinear Schrödinger equation and some other nonlinear evolution equations — 6
1.4 The experimental observation and applications — 9
1.5 The study of soliton theory problem — 10

2 **Inverse scattering transform — 13**
2.1 Introduction — 13
2.2 KdV equation and the associated inverse scattering transform — 13
2.3 Lax operator and the generalization of Zakharov, Shabat, AKNS — 20
2.4 A more general evolution equation (AKNS equation) — 27
2.5 Solution of the inverse scattering problems for AKNS equations — 33
2.6 Asymptotic solutions for the evolution equations ($t \to \infty$) — 43
2.7 The mathematical theory foundation of the inverse scattering transform — 53
2.8 Higher-order and multi-dimensional inverse scattering problems — 69

3 **Asymptotic behavior to initial value problems for some integrable evolution nonlinear equations — 79**
3.1 Introduction — 79
3.2 Asymptotic solutions of the KdV equation — 80
3.3 Nonlinear Schrödinger equation — 87
3.4 MKdV Equation — 96

4 **Interaction of solitons and its asymptotic properties — 105**
4.1 Interaction of solitons and its asymptotic properties as $t \to \infty$ — 105
4.2 Solution behavior for KdV equation under weak dispersion action and WKB method — 113
4.3 The soliton stability problem — 118
4.4 Water wave and wave equation under weak nonlinear action — 120

5 **Hirota method — 127**
5.1 Introduction — 127
5.2 Some properties of the D operator — 129
5.3 Solution of the bilinear differential equation — 131
5.4 Application to the sine-Gordon equation, MKdV Equation — 132
5.5 Bilinear Bäcklund transformation — 139

6 **Bäcklund transformations and the infinitely many conservation laws — 143**
6.1 Sine-Gordon equation and Bäcklund transformation — **143**
6.2 Bäcklund transformations for a class of nonlinear evolution equations — **147**
6.3 The commutativity of Bäcklund transformation for the KdV equation — **153**
6.4 Bäcklund transformations for the higher-order KdV equation and multi-dimensional sine-Gordon equation — **155**
6.5 Bäcklund transformation for the Benjamin–Ono equation — **157**
6.6 The infinitely many conservation laws for the KdV equation — **162**
6.7 Infinitely many conservation quantities for the AKNS equation — **164**
6.8 Darboux transformations — **167**

7 **Multi-dimensional solitons and their stability — 171**
7.1 Introduction — **171**
7.2 Existence problem for multi-dimensional solitons — **172**
7.3 The stability and collapse of multi-dimensional solitons — **185**

8 **Numerical computation methods for some nonlinear evolution equations — 191**
8.1 Introduction — **191**
8.2 Finite difference method and finite element method for the KdV equation — **193**
8.3 Finite difference method for the nonlinear Schrödinger equation — **197**
8.4 Numerical study of the RLW equation — **201**
8.5 Numerical study of the nonlinear Klein–Gorden equation — **203**
8.6 Numerical study of the stability problem for a class of nonlinear waves — **204**

9 **The geometric theory of solitons — 209**
9.1 Bäcklund transformation and the surface with total curvature $K = -1$ — **209**
9.2 Lie group and the nonlinear evolution equations — **212**
9.3 Prolongation structure of the nonlinear equations — **216**

10 **Global existence and blow up for the nonlinear evolution equations — 223**
10.1 Nonlinear evolution equations and the integral estimation method — **223**
10.2 Periodic initial value problem and initial value problem for the KdV equation — **224**

10.3 Periodic initial value problem for a class of nonlinear Schrödinger equations — 231
10.4 Initial value problem for the nonlinear Klein–Gordon equation — 236
10.5 RLW equation and Galerkin method — 243
10.6 Asymptotic behavior of solutions as $t \to \infty$ and the blow up problem — 250
10.7 Definite solutions problem for the Zakharov equations and some other coupling nonlinear evolution equations — 255

11 The soliton movements of elementary particles in nonlinear quantum field — 257
11.1 The elementary particles and solitons in nonlinear quantum field — 257
11.2 The movements of topological solitons in one-dimensional space — 262
11.3 The movements of topological solitons in two-dimensional space — 267
11.4 The magnetic unipolar solutions in three-dimensional spaces — 274
11.5 The topological soliton – instanton – in four-dimensional space — 280
11.6 The nontopological soliton and its properties — 283

12 The theory of soliton movement of superconductive features — 289
12.1 The macroscopic quantum effects in the superconductor and its soliton movement properties — 289
12.2 The properties of Boson condensation and spontaneous coherence of macroscopic quantum effects in the superconductor — 293
12.2.1 The nonlinear model of theoretical description for the macroscopic quantum effects — 293
12.2.2 The features of the coherent state of macroscopic quantum effects — 299
12.2.3 The boson condensed features of macroscopic quantum effects — 301
12.2.4 Differences between macroscopic quantum effects and the microscopic quantum effects and their nonlinear quantum mechanic features — 305
12.3 The soliton movements of electrons in superconductors — 306
12.3.1 The soliton features of motion of electrons in steady superconductors — 306
12.3.2 The features of soliton motion of electrons in superconductors under the action of an electromagnetic field — 309
12.3.3 The properties of soliton movement of the electrons in superconductive junctions and its relation to the macroscopic quantum effects — 312

12.4 The soliton movement features of the electrons with time dependence in the superconductor — 317
12.4.1 The dynamic equations of the superconductive electron and its soliton solutions — 317
12.4.2 The properties of soliton motion of the electrons in superconductors — 320
12.5 The transmission features of magnetic flux lines along the Josephson junctions — 321
12.5.1 The transmission equation of magnetic flux lines — 321
12.5.2 The transmission features of magnetic flux lines — 323
12.6 Conclusions — 327

13 The soliton movements in condensed state systems — 329
13.1 Soliton movements of helium atoms in superfluid systems — 329
13.1.1 The macroscopic quantum effects in helium superfluid — 329
13.1.2 The nonlinear theory of macroscopic quantum effects in a superfluid system — 331
13.1.3 The soliton motion of superfluid helium atoms in a relativistic case — 339
13.2 The soliton movement and macroscopic quantum effects in physical systems — 340
13.2.1 The macroscopic quantum effects in superfluids — 340
13.2.2 The relation between soliton movement in superconductors and superfluids — 341
13.3 The soliton excitations in the anti-ferromagnetic systems — 344
13.3.1 Dynamic features of solitons under the action of magnon–phonon interaction — 344
13.3.2 Properties of motion of solitons under the action of magnon–magnon interactions — 352
13.3.3 The nonlinear properties of magnons in different anti-ferromagnetic systems — 357

Bibliography — 361

1 Introduction

1.1 The background of solitons

In 1834, a young Scottish engineer named John Scott Russell [267] made a remarkable scientific discovery about water waves when he was conducting experiments to determine the most efficient design for canal boats. He described in his "Report on Waves" (Report of the 14th meeting of the British Association for the Advancement of Science, York, September 1844 (London 1845)): "I was observing the motion of a boat that was rapidly drawn along a narrow channel by a pair of horses, when the boat suddenly stopped – not so the mass of water in the channel which it had put in motion; it accumulated round the prow of the vessel in a state of violent agitation, then suddenly leaving it behind, rolled forward with great velocity, assuming the form of a large solitary elevation, a rounded, smooth and well-defined heap of water, which continued its course along the channel apparently without change of form or diminution of speed. I followed it on horseback, and overtook it still rolling on at a rate of some eight or nine miles an hour, preserving its original figure some thirty feet long and a foot to a foot and a half in height. Its height gradually diminished, and after a chase of one or two miles I lost it in the windings of the channel. Such, in the month of August 1834, was my first chance interview with that singular and beautiful phenomenon which I have called the Wave of Translation". He believed that this solitary motion is the steady-state solution of shallow water wave motion. Russell failed to prove this and to convince the physicists of his argument at that time; he also complained that the mathematicians failed to predict the solitary phenomenon from the known fluid motion equations. A widespread controversy among physicists about the solitary waves was caused, until 60 years later, when Korteweg and de Vries [147], assuming a long wave asymptotic and a small amplitude, established the following shallow water wave equation for movement in one direction only:

$$\frac{\partial \eta}{\partial t} = \frac{3}{2}\sqrt{\frac{g}{l}}\frac{\partial}{\partial x}\left(\frac{1}{2}\eta^2 + \frac{2}{3}\alpha\eta + \frac{1}{3}\sigma\frac{\partial^2 \eta}{\partial x^2}\right), \tag{1.1.1}$$

where η represents the elevation of the surface above the bottom, l represents the depth of the liquid, g is the gravity acceleration, and α, σ are small but arbitrary constants. They made a complete analysis of the solitary phenomenon and, finally, derived the solitary wave solution with unchanging shape, which was identical to the one described by Russell. In doing so, they proved the existence of solitary waves theoretically, while there still were unsolved questions: Is the solitary wave stable? Does the shape of the solitary wave stay invariant or change after a collision? These questions have not been answered and some people even doubt that the shape of solitary waves is destroyed when a collision occurs, since the solution for the nonlinear partial differential equation of equation (1.1.1) cannot satisfy the principle of superposition.

https://doi.org/10.1515/9783110549638-001

This argument lead to the notion of "instability" of solitary waves. Solitary waves were buried for a long time, until new discoveries were made.

There exists another question: Can solitary waves, as Russell said, appear in other physical fields than fluid dynamics? This was also an elusive problem during the beginning of the 20th century. The silence was broken by Fermi, Pasta, and Ulam [78] in the 1950s. They conducted numerical experiments (i.e., computer simulations) of a vibrating string that included a nonlinear term (quadratic in one test, cubic in another, and a piecewise linear approximation to a cubic in a third). They made a dynamical system of 64 particles with forces acting between neighbors with fixed end points. Initially, all of the energy of these oscillators is concentrated on one site. According to the classical theory, energy equipartition will happen as long as nonlinear effects exist. They thought that any weak nonlinear interaction can cause the system to transition from nonequilibrium to equilibrium. The results of their computations showed features that surprised everyone in the field. "Instead of a gradual, continuous flow of energy from the first mode to the higher modes, all the problems show an entirely different behavior." In fact, after a long time, almost all the energy returned to the original initial distribution, as shown in Figure 1.1. Finding an explanation for this phenomenon was called the famous Fermi–Pasta–Ulam (FPU) problem. Because the investigations considered the frequency space only, they did not find the solitary wave, so a solution to the problem was not found. Later, people simulated this situation by replacing the lattice with spring chains with mass. The correct answers to the

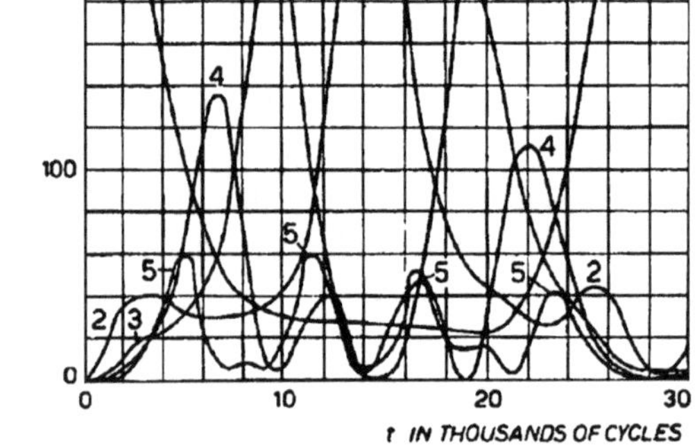

Figure 1.1: The energy quantity is plotted, with the units for energy being arbitrary. The initial form of the string was a single sine wave.

FPU problem were given by Toda, who studied the nonlinear oscillation in this mode and obtained the expected solitary wave solutions.

In 1962, numerical results made by Perring and Skyrme [245], when they investigated the elementary particles using the sine-Gordon equation, showed that such solitary waves did not disperse, i.e., solitary waves kept their original shapes and velocities invariant after the collision.

In 1965, Zabusky and Kruskal [314] investigated the interaction process of the "soliton" in a collisionless plasma in detail through numerical simulation. Their results confirmed the hypothesis that solitons pass through one another without losing their identity.

The aforementioned results and the fact that stable "solitons" were observed in several physical models in succession attracted the attention and interest of physicists and mathematicians. A complete system describing solitons was formed gradually.

So, what is the definition of a "soliton"? Generally speaking, we call the localized traveling wave solutions for nonlinear evolution equations "solitons". The adjective "localized" means that solutions for the field equation tend to zero or a certain constant at spatial infinity [258]. We name the stable solitary wave that retains its original shape and velocity after a collision "soliton". In some literature, solitons are confused with solitary waves.

In physics, the soliton is defined in terms of stable solutions for the classical field equation whose energy density $\rho(x,t)$ is finite and without dispersion, i.e.,

$$0 < H = \int \rho(x,t) d^m x < +\infty, \quad \lim_{t\to\infty} \max \rho(x,t) \neq 0, \quad \text{for certain } x,$$

where m is the space dimension. That is to say, the soliton can be regarded as a finite stable block mass with field energy without dispersion which cannot be destroyed during propagation or collisions. For a lot of nonlinear wave equations, there are four types of soliton shapes (as shown in Figure 1.2): the envelop type (bell-shaped), the swirl type (upside-down bell-shaped), the kink type, and the anti-kink type.

Based on the study of elementary particles, topological and nontopological solitons have been divided by Lee [156]. The respective definitions and investigations will be discussed in Chapter 10, where works on nontopological solitons will be presented briefly.

1.2 KdV equation and its soliton solutions

As mentioned, Korteweg and de Vries established the shallow water equation (1.1.1). After modification and simplification, we have

$$u_t + uu_x + \mu u_{xxx} = 0, \tag{1.2.1}$$

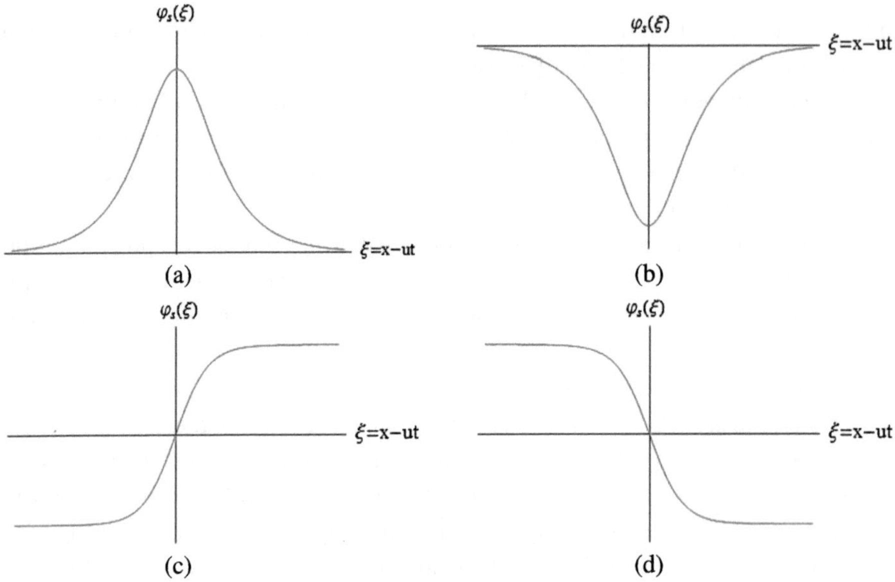

Figure 1.2: Different soliton types, where $\varphi_s(\xi)$ denotes the traveling wave solution and u is the velocity.

where μ is an arbitrary constant. If $\mu < 0$, via the transformations $u \to -u$, $x \to -x$, and $t \to -t$, equation (1.2.1) transforms to

$$u_t + uu_x - \mu u_{xxx} = 0. \tag{1.2.2}$$

Thus, we set $\mu > 0$. Equation (1.2.1) is the well-known Korteweg–de Vries (KdV) equation.

Making $u(x,t) = u(\xi)$, $\xi = x - Dt$, and $D =$ constant and integrating twice with respect to ξ, we have

$$3\mu\left(\frac{du}{d\xi}\right)^2 = -u^3 + 3Du^2 + 6Au + 6B = f(u), \tag{1.2.3}$$

where A, B are constants of integration. If and only if $f \geq 0$, equation (1.2.3) has a real solution. If $f(u)$ only has one real root, then it must be unbounded. Now, we assume that function $f(u)$ has three real roots, i.e., $f(u) = -(u - c_1)(u - c_2)(u - c_3)$, $c_1 < c_2 < c_3$. We conclude $D = \frac{1}{3}(c_1 + c_2 + c_3)$, $A = \frac{1}{6}(c_1c_2 + c_2c_3 + c_3c_1)$, $B = \frac{1}{6}c_1c_2c_3$. The general form of $f(u)$ can be expressed by curve A in Figure 1.3.

The exact solution for equation (1.2.3) reads

$$u = u(x,t) = c_2 + (c_3 - c_2)cn^2\left[\sqrt{\frac{c_3 - c_1}{12\mu}}\left\{x - \frac{1}{3}(c_1 + c_2 + c_3)t\right\}; k\right], \tag{1.2.4}$$

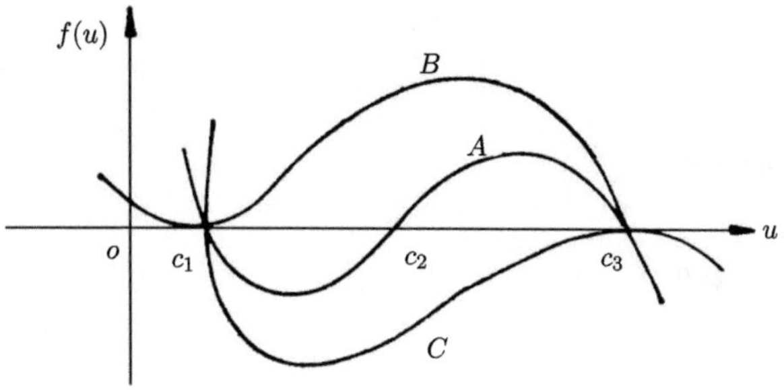

Figure 1.3: The profile of $f(u)$.

where $k^2 = (c_3 - c_2)/(c_3 - c_1)$. Equation (1.2.4) is usually called a cnoidal wave, whose period is $T_p = 4K\sqrt{\frac{3\mu}{c_3-c_1}}$, since the real period of function cn is $2K$, where K is the Legendre elliptic integral.

Under $K = 0$, $\mathrm{cn}(\xi, 0) = \cos\xi$, the oscillation solution for equation (1.2.3) reads

$$u = \bar{c} + a\cos\left[2\sqrt{\frac{c_3 - c_1}{12\mu}}\left\{x - \frac{1}{3}(c_1 + c_2 + c_3)t\right\}\right], \tag{1.2.5}$$

where $\bar{c} = \frac{c_2+c_3}{2}$, $a = \frac{c_3-c_2}{2}$.

The case where $K = 1$, $\mathrm{cn}(\xi, 1) = \mathrm{sech}\,\xi$ corresponds to curve B in Figure 1.3, whose period becomes infinite as $c_2 \to c_1$, i.e., the soliton solution for equation (1.2.1) is

$$u = c_1 + (c_3 - c_1)\,\mathrm{sech}^2\left[\sqrt{\frac{c_3 - c_1}{12\mu}}\left\{x - \frac{1}{3}(2c_1 + c_3)t\right\}\right]. \tag{1.2.6}$$

If $c_1 = u_\infty$, $c_3 - c_1 = a$, equation (1.2.6) transforms to

$$u = u_\infty + a\,\mathrm{sech}^2\left[\sqrt{\frac{a}{12\mu}}\left\{x - \left(u_\infty + \frac{a}{3}\right)t\right\}\right], \tag{1.2.7}$$

where u_∞ is the homogeneous state at infinity and a denotes the soliton amplitude. From solution (1.2.7), we see that the velocity which corresponds to the homogeneous state is proportional to the amplitude, while the width is inversely proportional to the square root of the amplitude. The amplitude is independent of the homogeneous state. If $u_\infty = 0$, $\mu = 1$, we get

$$u(x,t) = 3D\,\mathrm{sech}^2\sqrt{\frac{D}{2}}(x - Dt), \tag{1.2.8}$$

as shown in Figure 1.4.

It is known that a large range of wave equations with weak nonlinear effects can be summed up as KdV equations assuming a long wave asymptotic and a small and finite amplitude, such as (1) magnetohydrodynamics in cold plasmas, (2) motion in

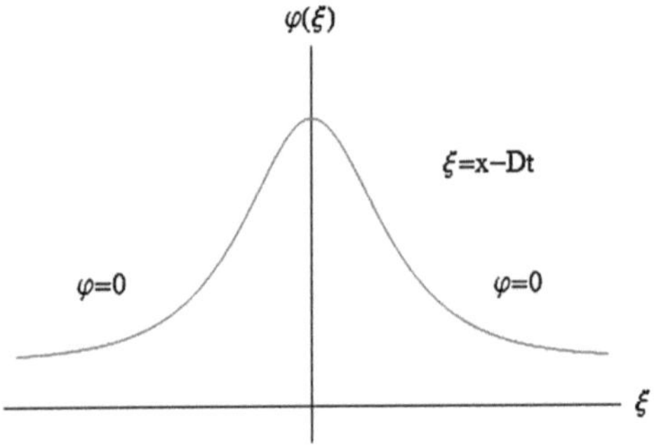

Figure 1.4: The amplitude of $u(\xi)$.

nonharmonic lattice, (3) ion acoustic waves in plasmas, (4) longitudinal dispersion fluctuations in an elastic bar, (5) pressure wave motion in a mixture state of liquid and gas, (6) the rotation of a fluid at the bottom of a tube, (7) thermal excitation of the phonon wave packet in nonlinear lattices at low temperature, etc.

1.3 Soliton solutions for the nonlinear Schrödinger equation and some other nonlinear evolution equations

The cubic nonlinear Schrödinger equation

$$iu_t + u_{xx} + v|u|^2u = 0, \tag{1.3.1}$$

or a more generalized form

$$u_t - ru_{xx} = \chi u - \beta|u|^2u, \tag{1.3.2}$$

where $\beta = \beta_0 + i\beta_1$, $r = r_0 + ir_1$, $i = \sqrt{-1}$, and $\beta_0, \beta_1, r_0, r_1, \chi, v$ are real constants, and other equations of this type have been found in many physical areas. For example, in beam flow of nonlinear optics,

$$2ik\frac{\partial\Psi}{\partial x} + \nabla_\perp^2\Psi + \frac{n_2}{n_0}k^2|\Psi|^2\Psi = 0, \tag{1.3.3}$$

where $\nabla_\perp^2 = \frac{\partial^2}{\partial r^2} + \frac{m}{r}\frac{\partial}{\partial r}$. The case where $m = 0$ denotes the plane, while $m = 1$ results in cylindrical symmetry; $\Psi = ae^{ik\theta}$, $\theta = kx - wt + ks(x, r)$, k represents the wave number, and $n = \frac{c_0k}{w} = n_0 + \frac{1}{2}n_2a^2$. For the flow in two dimensions, we have

$$\frac{n_2}{n_0}k^2|\Psi|^2\Psi = -2ki\Psi_x - \Psi_{xx} - \Psi_{yy}. \tag{1.3.4}$$

In addition, the nonlinear Schrödinger equation can also be used to describe the Langmuir wave in plasmas, self-modulation of one-dimensional monochromatic waves, self-focusing of two-dimensional stationary plane waves, the motion of a superconducting electron pair in an electromagnetic field, etc.

Taking account of the traveling wave solution for equation (1.3.1), we set

$$u(x,t) = e^{irx-ist}v(\xi), \quad \xi = x - Dt,$$

where r, s are undetermined, v is a real function, and D = constant. Substituting the above u into equation (1.3.1), we get the ordinary differential equation

$$v'' + i(2r - D)v' + (s - r^2)v + \nu|v|^2 v = 0. \tag{1.3.5}$$

Choosing $r = \frac{D}{2}$, $s = \frac{D^2}{4} - \alpha$, and $(\alpha > 0)$ and omitting v', we have

$$v'' - \alpha v - \nu v^3 = 0. \tag{1.3.6}$$

Integrating once, we get

$$v'^2 = A + \alpha v^2 - \frac{\nu}{2}v^4. \tag{1.3.7}$$

For the special case where $\nu > 0$, $A = 0$,

$$v(x,t) = \left(\frac{2\alpha}{\nu}\right)^{\frac{1}{2}} \operatorname{sech} \alpha(x - Dt). \tag{1.3.8}$$

It is obvious that $|u|^2 \propto \operatorname{sech}^2 \alpha(x - Dt)$ and $v(x,t)$ is called the envelop soliton. Next, we consider a more general solution for equation (1.3.1). We have

$$u(x,t) = \Phi(x,t)e^{i\theta(x,t)}, \tag{1.3.9}$$

where the real function Φ stands for the envelop wave and θ denotes the carrier wave. Substituting equation (1.3.9) into (1.3.1) and separating the real and imaginative parts, we have

$$\begin{aligned} &\Phi_{xx} - \Phi\theta_t - \Phi\theta_x^2 - \nu\Phi^3 = 0, \quad \nu > 0, \\ &\Phi\theta_{xx} + 2\Phi_x\theta_x + \Phi_t = 0. \end{aligned} \tag{1.3.10}$$

Supposing $\theta = \theta(x - D_1 t)$ and $\Phi = \Phi(x - D_2 t)$, equations (1.3.10) become

$$\Phi_{xx} + D_1\Phi\theta_x - \Phi(\theta_x)^2 + \nu\Phi^3 = 0, \tag{1.3.11}$$

$$\Phi\theta_{xx} + 2\Phi_x\theta_x - D_2\Phi_x = 0. \tag{1.3.12}$$

Fixing t as a constant variable in equation (1.3.12) and integrating with respect to x, we get

$$\Phi^2(2\theta_x - D_2) = \varphi(t). \tag{1.3.13}$$

Note that $\theta_x = \frac{D_2}{2}$ when $\varphi(t) = 0$, so equation (1.3.11) becomes

$$\int_{\Phi_0}^{\Phi} \frac{d\Phi}{\sqrt{p(\Phi)}} = x - D_2 t, \tag{1.3.14}$$

where

$$p(\Phi) = -\frac{\nu}{2}\Phi^4 + \frac{1}{4}(D_2^2 - 2D_1D_2)\Phi^2 + C.$$

If we take $C = 0$, $D_2^2 - 2D_1D_2 > 0$, then $\Phi = 0$ is the double root for $p(\Phi) = 0$ and the last two roots are $\Phi = \pm\Phi_0$ with $\Phi_0 = \sqrt{\frac{D_2^2 - 2D_1D_2}{2\nu}}$. In this case,

$$\Phi = \Phi_0 \text{sech}\left[\sqrt{\frac{\nu}{2}}\Phi_0(x - D_2 t)\right],$$

as shown in Figure 1.5.

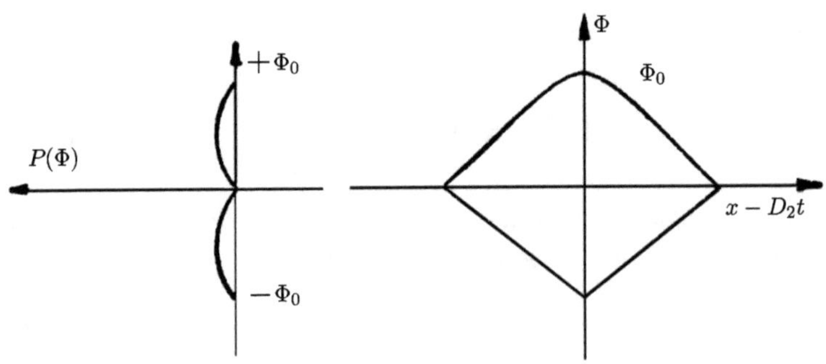

Figure 1.5: The profile of Φ with $C = 0$.

The traveling wave solution will not be obtained in the case where $p(\Phi) < 0$. If $C \neq 0$, under the conditions $[\frac{1}{4}(D_2^2 - 2D_1D_2)]^2 + 2\nu C \geq 0$ or $C > -\frac{1}{8\nu}(\frac{D_2^2}{2} - 2D_1D_2)$ and $C < 0$, the single roots $\pm\Phi_1$, $\pm\Phi_2$ will be obtained for $p(\Phi) = 0$, which can be found in Figure 1.6.

The term Φ can be expressed in the form of the elliptic function

$$\Phi = \Phi_1\left[1 - \left\{\left(1 - \frac{\Phi_1^2}{\Phi_2^2}\right)\text{sn}^2\left[\frac{\sqrt{\nu}}{2}(x - D_2 t)\right]\right\}\right]^{-\frac{1}{2}},$$

where the elliptic function sn obeys mod $r = 1 - \frac{\Phi_1^2}{\Phi_2^2}$.

The well-known sine-Gordon equation

$$u_{tt} - u_{xx} + \sin u = 0 \tag{1.3.15}$$

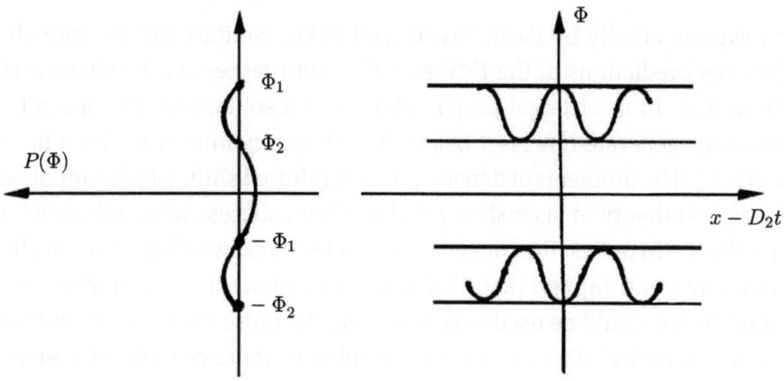

Figure 1.6: The profile of Φ with $C < 0$.

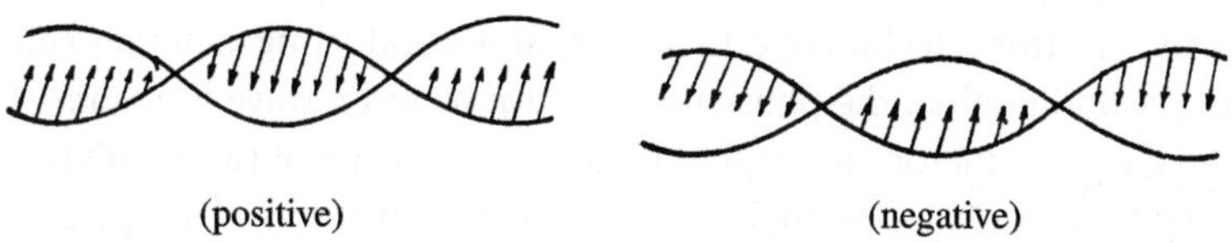

Figure 1.7: Kink and anti-kink solutions.

has the soliton solution

$$u = 4\mathrm{tg}^{-1} \pm \{\pm(1 - D^2)^{-\frac{1}{2}}(x - Dt)\}. \tag{1.3.16}$$

Kink and anti-kink types will be obtained based on the signs of the inside and outside brace (as shown in Figure 1.7).

If the positive signs are taken for both the inside and the outside brace, solution (1.3.16) respects the kink type from $\Phi = 0$ ($x = -\infty$) to $\Phi = 2\pi$ ($x = +\infty$). Otherwise, if both negative signs are taken, the kink type will be obtained from $\Phi = -2\pi$ ($x = -\infty$) to $\Phi = 0$ ($x = +\infty$). The anti-kink type will be displayed under the different signs chosen for the inside and the outside brace.

A lot of nonlinear evolution equations have soliton solutions, such as the nonlinear Klein–Gordon equation, the Toda lattice equation, the Heisenberg ferromagnetic chain equation, the nonlinear electronic filtering equation, the Boussinesq equation, the Hirota equation, the Born–Infeld equation, etc.

1.4 The experimental observation and applications

The existence of solitons was first described by Russell in water waves. Apart from the numerical calculations, experimental observations of solitons have been found constantly. In the 1970s [132], the formation and propagation of ion-acoustic solitons were observed experimentally by Ikezi, Taylor, and Baker. Solitary pulses were observed to follow the predictions of the KdV equation with respect to the shape and velocity of the soliton. In laser target propagation, vortex solitary waves caused by collapse and solitons generated by laser beams focusing in nonlinear medium have also been found [122]. The problems of density pits and infrared shift, which cannot be explained by classical theory in laser shooting, have been successfully explained by virtue of soliton theory. In optics, the concept of optical solitons was first theoretically proposed by Hasegawa and Tappert [123, 124]. They pointed out that the nonlinearity of the index of refraction could be used to compensate the pulse broadening effect of dispersion in low-loss optical fibers. Later, Bell telephone laboratories in New Jersey reported narrowing and splitting of 7-ps-duration pulses from a mode-locked color-center laser by a 700-m, single-mode silica-glass fiber [181].

The superconducting Josephson effect belongs to the most important subjects in modern physics and electronic technology. In the two superconducting materials constituting the Josephson junction, the phase difference φ between wave functions of a Cooper pair satisfies the sine-Gordon equation [304]. The fusion of two relativistic 2π-solitons of the same polarity into a single 4π-soliton has been observed in a parallel array of a Josephson junction [247].

In recent years, more aspects of solitons have been observed, in more fields. The first reported observation of soliton explosions in a passively mode-locked fiber laser was described by Runge et al. [266]. They reported the identification of clear explosion signatures in measurements of shot-to-shot spectra of a Yb-doped mode-locked fiber laser that is operating in a transition regime between stable and noise-like emission. Soliton dynamics in charge-density waves on a quasi-one-dimensional metallic surface have also been directly observed [183]. In silicon photonic crystals, soliton-effect pulse compression of picosecond pulses in silicon, despite two-photon absorption and free carriers, has been demonstrated [26].

More experimental observations will undoubtedly promote the theoretical work of soliton theory.

1.5 The study of soliton theory problem

Since soliton phenomena have some common characteristics and have been observed in many nonlinear physical problems, physicists hope to use soliton theory to discuss the motion of matter under the action of nonlinearity in plasma physics, elementary particle physics, and Bose–Einstein condensations. From a mathematical point of view, soliton solutions have been solved for a certain kind of nonlinear evolution equations. They share important characteristics, such as Bäcklund transformation, infinitely many conservation laws, and the notions that they can be solved by the inverse scattering transform, they are completely integrable, etc. There are several applications of soliton phenomena in mathematical methods: the inverse scattering transform, which is based on the boundary problems for ordinary differential equations, and the Gelfand–Levitan–Marchenko integral equation have been extended by Lax, Zakharov, and Shabat and Ablowitz, Kaup, Newell, and Suger to deal with a large variety of solvable nonlinear evolution equations; certain function transformations, like Bäcklund transformation, Darboux transformation, and the Hirota method, are efficient ways to obtain soliton solutions; the extended structure method by the aid of the exterior differential form and the Lie group is also a powerful tool. In addition, numerical simulations have been developed to discuss the stability and interaction of solitons.

Deift and Zhou [60, 62] introduced a new and general approach, named steepest descent method, to analyze the asymptotics of oscillatory Riemann–Hilbert problems. Such problems arise, in particular, when evaluating the lone-time behavior of nonlinear wave equations solvable by the inverse scattering transform. Fokas [81, 80] presented the solution for the initial boundary value problem of the nonlinear Schrödinger equation in terms of the solution of a matrix Riemann–Hilbert problem, formulated in the complex k-plane. The Riemann–Hilbert approach in integrable systems has also been developed for a lot of mathematical and physical areas, such as random matrices and orthogonal polynomials [64].

2 Inverse scattering transform

2.1 Introduction

The inverse scattering transform was first introduced by Gardner, Greene, Kruskal, and Miura during the procedure of solving the initial value problem for KdV equations [88]. After the generalization of Lax, Zakharov, and Shabat and Ablowitz, Kaup, Newell, and Suger (AKNS), the inverse scattering transform has been developed into a general and important method for solving a large range of nonlinear evolution equations, including higher-dimensional and coupled ones. The main advantage of this method is the possibility to obtain exact solutions through solving a combination of several linear equations instead of the complex nonlinear equations [153]. In this chapter, we will introduce the fundamental concepts, some results, and unsolvable problems of the inverse scattering transform.

2.2 KdV equation and the associated inverse scattering transform

By the aid of the Hopf–Cole transformation

$$u = -2\alpha \frac{w_x}{w}, \tag{2.2.1}$$

the Burgers equation

$$u_t + uu_x - \alpha u_{xx} = 0, \quad \alpha > 0, \tag{2.2.2}$$

can be transformed into the linear heat conduction equation

$$w_t = \alpha w_{xx}, \tag{2.2.3}$$

which has the solution

$$u(x,t) = \frac{\int_{-\infty}^{\infty} \frac{x-\xi}{t} \exp[-\frac{(x-\xi)^2}{4\alpha t} - \frac{1}{2\alpha}\int_0^{\xi} u_0(\xi')d\xi']d\xi}{\int_{-\infty}^{\infty} \exp[-\frac{(x-\xi)^2}{4\alpha t} - \frac{1}{2\alpha}\int_0^{\xi} u_0(\xi')d\xi']d\xi}, \tag{2.2.4}$$

where $u_0(x)$ is the initial value and $u|_{t=0} = u_0(x)$. When $\alpha \to 0$, solution (2.2.4) is proved to be the generalized solution for the quasi-linear hyperbolic equation

$$u_t + uu_x = 0. \tag{2.2.5}$$

The question is whether there exists a similar transformation like equation (2.2.2) for the KdV equation. Consider the KdV equation of the following form:

$$u_t - 6uu_x + u_{xxx} = 0. \tag{2.2.6}$$

https://doi.org/10.1515/9783110549638-002

If we take $u(x,t)$ as a known function, $u = v^2 + v_x + \lambda$ can be seen as a Riccati equation for the unknown function $v(x,t)$. When $v = \varphi_x/\varphi$, we get the one-dimensional Schrödinger equation

$$\varphi_{xx} - (u - \lambda)\varphi = 0, \tag{2.2.7}$$

where φ is a wave function, u denotes the potential, and λ corresponds to the energy spectrum. We notice that u is not merely dependent on x, but also on t. Therefore, φ and λ are related to t as well. If the solution u for the KdV equation is smooth, bounded, and decaying to zero at $|x| \to \infty$, then equation (2.2.7) exists with finite discrete spectra $\lambda_m = -k_m^2$ $(m = 1, 2, \ldots, N)$ for $\lambda < 0$ and continuous spectra $\lambda = k^2$ $(-\infty < k < \infty$, k being a real constant) for $\lambda > 0$. For a fixed t, we define the solution for scattering problems of equation (2.2.7), where $\lambda > 0$ satisfies the boundary conditions

$$\begin{cases} \varphi(x,k,t) \sim e^{-ikx} + b(k,t)e^{ikx}, & x \to +\infty, \\ \varphi(x,k,t) \sim a(k,t)e^{-ikx}, & x \to -\infty, \end{cases} \tag{2.2.8}$$

and the solution for scattering problems of equation (2.2.7), where $\lambda < 0$ satisfies the boundary conditions

$$\begin{cases} \varphi_m(x,k_m(t),t) \sim c_m(k_m(t),t)e^{-k_m x}, & x \to +\infty, \\ \varphi_m(x,k_m(t),t) \sim e^{k_m x}, & x \to -\infty, \end{cases} \tag{2.2.9}$$

where $b(k,t)$ is the reflection coefficient, $a(k,t)$ is the transmission coefficient, and c_m represents the decaying factor. This satisfies

$$\int_{-\infty}^{\infty} \varphi_m^2 dx = 1, \quad |a|^2 + |b|^2 = 1, \tag{2.2.10}$$

which can be found in Figure 2.1.

In quantum mechanics, the scattering problem for the Schrödinger equation is the following. Given potential u, we set the scattering data k_m, c_m, $a(k)$, and $b(k)$ and the

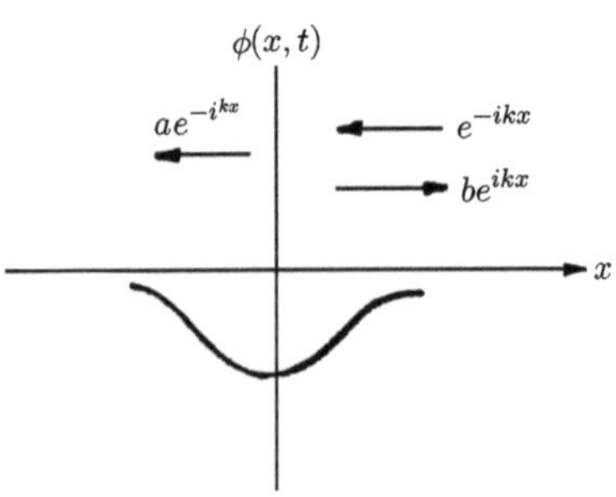

Figure 2.1: The scattering data.

wave function φ at infinity. The inverse scattering problem is to solve the potential u by the given scattering data k_m, c_m, $a(k)$, and $b(k)$ and the wave function φ at infinity. The potential u will be solved by

$$u(x,t) = -2\frac{d}{dx}K(x,x,t), \tag{2.2.11}$$

where K satisfies the Gelfand–Levitan–Marchenko integral equation

$$\begin{aligned} &K(x,y,t) + B(x+y,t) + \int_x^{\infty} B(y+z,t)K(x,z,t)dz = 0, \quad y > x, \\ &K(x,z,t) \longrightarrow 0, \quad z \to \infty, \end{aligned} \tag{2.2.12}$$

where the kernel $B(x,t)$ is given by

$$B(x,t) = \sum_{m=1}^{N} c_m^2(t)e^{-k_m x} + \frac{1}{2\pi}\int_{-\infty}^{\infty} b(k,t)e^{ikx}dk, \tag{2.2.13}$$

where $\sum$ corresponds to the discrete spectrum and $\int$ corresponds to the continuous spectrum. We see that the inverse scattering problem cannot be successfully solved from the above expressions. The evolution of u is decided by K for equation (2.2.11), while K satisfies the Gelfand–Levitan–Marchenko integral equation (2.2.12), which is solved by the scattering data of kernel $B(x,t)$. However, the scattering data were determined by u. To break the endless loop and solve the potential u for the KdV equation, we notice the following important relationship between a KdV equation and the corresponding Schrödinger equation.

Theorem 2.2.1. *Taking into account the Schrödinger equation (2.2.7), we have*

$$\varphi_{xx} - (u-\lambda)\varphi = 0, \quad -\infty < x < +\infty,$$

where the discrete eigenvalues $\lambda_1, \lambda_2, \ldots, \lambda_N$ are constants if $u(x,t)$ is the solution for the KdV equation and decays to zero as $|x| \to \infty$.

Proof. Inserting $u = \frac{\varphi_{xx}}{\varphi} + \lambda$ into the KdV equation (2.2.6) and multiplying by φ^2, we have

$$\lambda_t\varphi^2 + [\varphi R_x - \varphi_x R]_x = 0, \tag{2.2.14}$$

where

$$R \equiv \varphi_t + \varphi_{xxx} - 3(u+\lambda)\varphi_x.$$

The eigenfunction and its derivatives which correspond to λ_n tend to zero at $|x| \to \infty$. Integrating equation (2.2.14) with respect to x, we obtain

$$\lambda_{nt}\int_{-\infty}^{\infty} \varphi_n^2 dx = 0.$$

Because $\int_{-\infty}^{\infty} \varphi_n^2 dx = 1$, we have $\lambda_{nt} = 0$, i.e., $\lambda_n =$ constant. The proof is completed. □

If λ_n is constant, this means that the discrete eigenvalues and the scattering data of the inverse scattering problem of the Schrödinger equation can be solved through the initial value $u_0(x)$ of the KdV equation. By virtue of $\lambda_t = 0$, equation (2.2.14) transforms into

$$\varphi R_{xx} - R\varphi_{xx} = 0,$$

which is equal to $R_{xx} - (u - \lambda)R = 0$. Because it is of the same form as equation (2.2.7), R can be written as the linear combination of the eigenfunction of equation (2.2.7), i.e.,

$$R \equiv \varphi_t + \varphi_{xxx} - 3(u + \lambda)\varphi_x = C\varphi + D\phi, \tag{2.2.15}$$

where C and D are related to t and ϕ and φ are the linear independent solutions for equation (2.2.7). If we take $\phi = \varphi \int_0^x \frac{dx}{\varphi^2}$, we have Theorem 2.2.2.

Theorem 2.2.2. *Under the condition of Theorem 2.2.1, the scattering data for scattering problem (2.2.7) can be expressed as*

$$\begin{cases} c_n(t) = c_n(0)e^{4k_n^3 t}, \\ b(k,t) = b(k,0)e^{8ik^3 t}, \\ a(k,t) = a(k,0), \end{cases} \tag{2.2.16}$$

where $c_n(0), b(k,0), a(k,0)$ are determined by the initial value $u_0(x)$ of the KdV equation.

Proof. For the discrete spectrum, where φ_n is the eigenfunction and $\phi_n = \varphi_n \int_0^x \frac{dx}{\varphi_n^2}$, we deduce that ϕ_n is exponentially unbounded as $x \to +\infty$. Therefore, $D(t) = 0$ can be obtained from equation (2.2.15). Multiplying equation (2.2.15) by φ_n and integrating over an infinite interval, we get

$$\int_{-\infty}^{\infty} \left(\frac{1}{2}\varphi_n^2\right)_t dx + \int_{-\infty}^{\infty} \left(\varphi_n \varphi_{n,xx} - \frac{3}{2}\varphi_{n,x}^2 - 3\lambda\varphi_n^2\right)_x dx = c \int_{-\infty}^{\infty} \varphi_n^2 dx.$$

Because $\int_{-\infty}^{\infty} \varphi_n^2 dx = 1$ and due to the boundary condition, the integrations on the left side are equal to zero, so $c(t) \equiv 0$. Considering $\varphi \sim c_n(t)e^{-k_n x}$ at $x \to +\infty$ and $u \to 0$ $(x \to +\infty)$, the following relationship will be obtained from equation (2.2.15):

$$c_n'(t) - 4k_n^3 c_n(t) = 0,$$

i.e., $c_n(t) \equiv c_n(0)e^{4k_n^3 t}$.

For the continuous spectrum, λ is independent of t and φ satisfies equation (2.2.15). Taking advantage of the steady radiation condition of the plane wave, i.e.,

$$\varphi \sim a(k,t)e^{-ikx}, \quad x \to -\infty,$$

and inserting it into equation (2.2.15), we have

$$(a_t + ik^3 a + 3k^3 a)e^{-ikx} = Ca(k,t)e^{-ikx} + \frac{D}{a}e^{-ikx}\int_0^x e^{2ikx}dx,$$

$$a_t + 4ik^3 a = Ca + \frac{D}{a}\int_0^x e^{2ikx}dx.$$

Furthermore, we deduce

$$D = 0, \quad a_t + (4ik^3 - C)a = 0.$$

Substituting $\varphi \sim e^{-ikx} + b(k,t)e^{ikx}$ $(x \to +\infty)$ into equation (2.2.15) and collecting the coefficient of linearly independent functions $e^{\pm ikx}$ to zero, we get

$$C = 4ik^3, \quad b_t - 8ik^3 b = 0, \quad b(k,t) = b(k,0)e^{8ik^3 t}.$$

It is easy to see that $a_t = 0$, which means $a(k,t) = a(k,0)$. The proof is completed. □

Theorems 2.2.1 and 2.2.2 provide the procedures for solving the initial problem for the KdV equation by virtue of the Schrödinger equation and the following scattering problem:

$$\begin{cases} u_t - 6uu_x + u_{xxx} = 0, & -\infty < x < \infty,\ t > 0, \\ u(x,0) = u_0(x). \end{cases}$$

As a first step, we solve the eigenvalue problem

$$\varphi_{xx} - [u_0(x) - \lambda]\varphi = 0, \tag{2.2.17}$$

where the scattering data at $t = 0$, i.e., $k_n, c_n(0), b(k,0)$, can be given. The evolution of $c_n(t)$ and $b(k,t)$ can be obtained from equation (2.2.16), which gives

$$\begin{aligned} B(x+y,t) &= \sum_{n=1}^{N} c_n^2(t)e^{-k_n(x+y)} + \frac{1}{2\pi}\int_{-\infty}^{\infty} b(k,t)e^{ik(x+y)}dk \\ &= \sum_{n=1}^{N} c_n^2(0)e^{8k_n^3 t - k_n(x+y)} + \frac{1}{2\pi}\int_{-\infty}^{\infty} b(k,0)e^{i[8k^3 t + k(x+y)]}dk. \end{aligned}$$

As a second step, $K(x,y,t)$ will be determined using the Gelfand–Levitan–Marchenko integral equation

$$K(x,y,t) + B(x+y,t) + \int_x^{\infty} B(y+z,t)K(x,z,t)dz = 0, \quad y > x.$$

We deduce

$$u(x,t) = -2\frac{d}{dx}K(x,x;t).$$

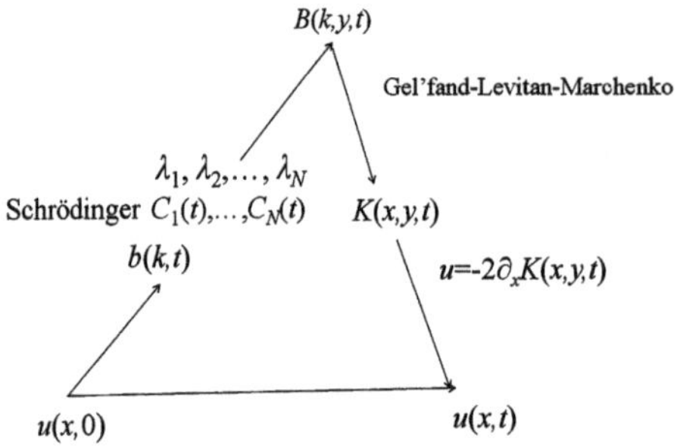

Figure 2.2: The procedures of initial problems for the KdV equation.

Following the above procedures (as seen as in Figure 2.2), the initial-value problem for a nonlinear equation (KdV equation) has been transformed into the problem of solving two linear equations. One is the Sturm–Liouville problem of the second-order ordinary differential equation. The other is solving a linear integral equation. Next, we take two examples to explain this.

If we take $u_0(x) = -2\operatorname{sech}^2 x$, the corresponding eigenvalue problem for equation (2.2.17) can be solved exactly by a hypergeometric function. The normalized constants which correspond to the discrete eigenvalue $k_1 = 1$ are $c_1(0) = \sqrt{2}$ and $b(k,0) = 0$, so we get $b(k,t) = 0$ for $t \geq 0$. The related Gelfand–Levitan–Marchenko integral equation can be expressed as

$$K(x,y,t) + 2e^{8t-x-y} + 2e^{8t-y}\int_x^{\infty} K(x,z,t)e^{-z}dz = 0.$$

Assuming that $K(x,y,t)$ is a separable variable and inserting $K(x,y,t) = L(x,t)e^{-y}$ into the above integral equation, we get

$$L(x,t) + 2e^{8t-x} + 2e^{8t}L(x,t)\int_x^{\infty} e^{-2z}dz = 0,$$

$$L(x,t) = \frac{-2e^x}{1+e^{2x-8t}}, \quad K(x,y,t) = \frac{-2e^{x-y}}{1+e^{8x-8t}}.$$

It is easy to verify that $K(x,y,t)$ is indeed the unique solution for the Gelfand–Levitan–Marchenko integral equation. Therefore, the exact solution for the initial problem of the KdV equation is written

$$u(x,t) = \frac{8e^{2x-8t}}{(1+e^{2x-8t})^2} = -2\operatorname{sech}^2(x-4t).$$

If we choose $u(x,0) = u_0(x) = -6\operatorname{sech}^2 x$, there are two different eigenvalues, $k_1 = 2$ and $k_2 = 1$. If $b(k,0) = 0$, the solution for the KdV equation is

$$u(x,t) = -12\frac{3+4\cosh(2x-8t)+\cosh(4x-64t)}{[3\cosh(x-28t)+\cosh(3x-36t)]^2}.$$

In the following part, we aim to get the N-soliton solution via the inverse scattering transform with reflection coefficient $b(k,t) = 0$. We only take the discrete spectrum into account. The related Gelfand–Levitan–Marchenko integral equation is

$$K(x,y,t) + \sum_{m=1}^{N} c_m^2(t)e^{-k_m(x+y)} + \sum_{m=1}^{N} c_m^2 e^{-k_m y} \int_x^{\infty} e^{-k_m t} K(x,z,t)dz = 0, \qquad (2.2.18)$$

where $c_m = c_m(t) = c_m(0)e^{4k_m^3 t}$ and $k_m > 0$. Assume $K(x,y,t)$ in the form of

$$K(x,y,t) = -\sum_{m=1}^{N} c_m \varphi_m(x)e^{-k_m y}, \qquad (2.2.19)$$

where φ_m is an undetermined function and c_m is the normalization factor. Substituting equation (2.2.19) into equation (2.2.18) and making the coefficient of $e^{k_m y}$ equal to zero, we derive the following linear algebra equation of $\varphi_m(x)$:

$$\varphi_m(x) + \sum_{m=1}^{N} c_m c_n \frac{e^{-(k_m+k_n)x}}{k_m + k_n} \varphi_n(x) = c_m e^{-k_m x}, \quad m = 1,2,\ldots,N. \qquad (2.2.20)$$

We denote the matrices

$$I \equiv (\delta_{mn}), \quad C \equiv \left(c_m c_n \frac{e^{-(k_m+k_n)x}}{k_m + k_n} \right), \quad \varphi = (\varphi_1, \varphi_2, \ldots, \varphi_N)^T,$$
$$E = (c_1 e^{-k_1 x}, c_2 e^{-k_2 x}, \ldots, c_N e^{-k_N x})^T.$$

Therefore, we rewrite equation (2.2.20) as

$$(I + C)\varphi = E. \qquad (2.2.21)$$

To make sure that equation (2.2.20) is solvable for φ, we need to prove C is positive definite. In fact,

$$\sum_{m=1}^{N}\sum_{n=1}^{N} p_m p_n c_m c_n \frac{e^{-(k_m+k_n)x}}{k_m + k_n} = \int_x^{\infty} \left[\sum_{m=1}^{N} p_m c_m e^{-k_m z} \right]^2 dz > 0,$$

which means $I + C$ is positive definite. The unique solution for equation (2.2.20) can be obtained using the Cramer rule. Set Q_{mn} as the algebraic cofactor of element a_{mn} of matrix $I + C$ and expand it at the nth line, to obtain

$$\triangle \equiv \det(I + C) = \sum_m \left(\delta_{mn} + c_m c_n \frac{e^{-(k_m+k_n)x}}{k_m + k_n} \right) Q_{mn},$$
$$\varphi_n(x) = \triangle^{-1} \sum_m c_m e^{-k_m x} Q_{mn}.$$

From equation (2.2.19), $y = x$, we have

$$K(x,y,t) = -\sum c_n \varphi_n(x) e^{-k_n x}$$
$$= -\triangle^{-1} \sum_m \sum_n c_m c_n e^{-(k_m+k_n)x} Q_{mn}$$
$$= \triangle^{-1} \frac{d}{dx} \triangle .$$

The potential $u(x,t)$ for the KdV equation without reflection coefficient can be given as

$$u(x,t) = -2\frac{d^2}{dx^2} \log \det(I + C). \tag{2.2.22}$$

Owing to the symmetrical structure of K and B in the Gelfand–Levitan–Marchenko integral equation, there are two ways to solve K and B. One way is to first decide B based on the scattering data and then solve K. The other way is to solve B by virtue of the Gelfand–Levitan–Marchenko integral equation and then K, which satisfies the linear hyperbolic equation. This is another method to solve the inverse scattering transform problem, as shown in Figure 2.3.

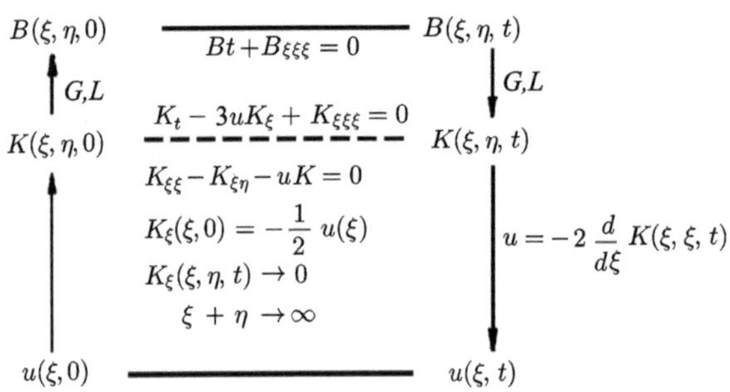

Figure 2.3: The inverse scattering transform procedure for KdV equation.

2.3 Lax operator and the generalization of Zakharov, Shabat, AKNS

Consider the general nonlinear evolution equation

$$u_t = K(u), \tag{2.3.1}$$

where $K(u)$ denotes the nonlinear operator defined in a certain proper function space. We find two linear operators, L and B [153], which are dependent on the solution u for equation (2.3.1), satisfying the following Lax equation:

$$iL_t = BL - LB = [B,L], \quad i = \sqrt{-1}, \tag{2.3.2}$$

where B is the self-adjoint operator. From equation (2.3.2), we derive the eigenfunction φ which corresponds to the operator L and eigenvalue E, i.e.,

$$L\varphi = E\varphi. \tag{2.3.3}$$

If the time evolution of φ satisfies

$$i\varphi_t = B\varphi, \tag{2.3.4}$$

we can confirm that E is independent of t. In fact, we differentiate equation (2.3.3) with respect to t and we get

$$\begin{aligned} i\left[\varphi\frac{dE}{dt} + E\frac{d\varphi}{dt}\right] &= i\left[L\varphi_t + \frac{\partial L}{\partial t}\varphi\right] \\ &= iL\varphi_t + [BL - LB]\varphi \\ &= L(i\varphi_t - B\varphi) + EB\varphi. \end{aligned}$$

Because of equation (2.3.4), $i\varphi\frac{dE}{dt} = 0$ will be derived.

To solve the initial problem of equation (2.3.1), we put forward the following steps, as seen as in Figure 2.4:

(i) The eigenvalue problem. Solve the scattering quantities (eigenvalue, reflection and diffusion coefficients, etc.) through the given initial value $u(x,0)$ at $t = 0$.
(ii) Time evolution of scattering data. Based on equation (2.3.4), consider the asymptotic solution of B at $|x| \to \infty$ and compute the evolution of the scattering data.
(iii) The inverse problem. Construct $u(x,t)$ from the Gelfand–Levitan–Marchenko integral equation.

Take the following KdV equation as an example:

$$u_t - 6uu_x + u_{xxx} = 0, \tag{2.3.5}$$

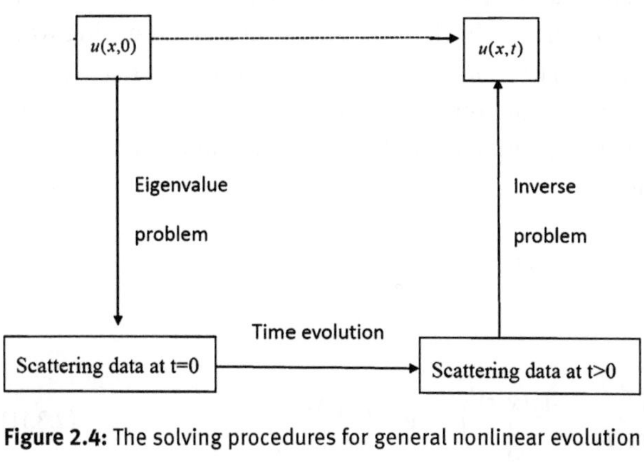

Figure 2.4: The solving procedures for general nonlinear evolution equations.

with the initial condition

$$u(x,0) = u_0(x). \tag{2.3.6}$$

Denote the second-order differential operator hierarchy $L(t)$ as

$$L(t) \equiv -\frac{\partial^2}{\partial x^2} + u(x,t), \tag{2.3.7}$$

where $u(x,t)$ is the solution for equation (2.3.5), $u(\cdot,t) \in L_2$.

The eigenvalue problem for operator L reads

$$-\varphi_{xx} + u(x,t)\varphi = E\varphi. \tag{2.3.8}$$

As mentioned in Section 2.2, for a given $u_0(x)$, we can solve the scattering quantities $(k_n, c_n, n = 1,2,\ldots,N; a(k), b(k), 0 \le k^2 < \infty)$.

Make the self-adjoint operator B as

$$B \equiv -4i\frac{\partial^3}{\partial x^3} + 3i\left(u\frac{\partial}{\partial x} + \frac{\partial}{\partial x}\cdot u\right), \tag{2.3.9}$$

where $u(x,t)$ satisfies the KdV equation (2.3.5). The operators L and B satisfy equation (2.3.2). Based on equation (2.3.4), the boundary condition of φ at infinity, and the notion that $u(x,t) \to 0$ ($|x| \to \infty$), time evolution of the scattering quantities can easily be obtained, which follows from the results of Theorem 2.2.2. We solve the Gelfand–Levitan–Marchenko integral equation and $u(x,t)$ for the initial value problem of the KdV equation will be given.

We should point out that not all the nonlinear equations (2.3.1) can be solved by the inverse scattering transform. The main difficulty is that the proper operators L and B for equation (2.3.2) cannot be found easily. Even though a certain B is found for a given L, B is trivial probably. For example, for $L = \frac{\partial^2}{\partial x^2} + u(x,t)$, choosing $B = i\frac{\partial}{\partial x}$, we have $L_t = u_t$, $[B,L] = i[D,u] = iu_x$, $iL_t = [B,L] \Rightarrow u_t = u_x$, which is a traveling wave equation with explicit solution $u = f(x-t)$. Although some confusion exists, a large variety of nontrivial nonlinear wave equations can be solved by the inverse scattering transform. Next, we state the generalization of Zakharov, Shabat, and AKNS.

Consider the linear problem

$$Lv = \zeta v, \tag{2.3.10}$$

where

$$L \equiv \begin{pmatrix} i\frac{d}{dx} & -iq(x,t) \\ ir(x,t) & -\frac{d}{dx} \end{pmatrix}, \quad v \equiv \begin{pmatrix} v_1(x,t) \\ v_2(x,t) \end{pmatrix}, \tag{2.3.11}$$

$q(x,t)$ and $r(x,t)$ are the differential functions, and ζ is constant. B is assumed in the form of

$$B = \begin{pmatrix} a(x,t;\zeta) & b(x,t;\zeta) \\ c(x,t;\zeta) & -a(x,t;\zeta) \end{pmatrix}, \tag{2.3.12}$$

where a, b, c are all the undetermined functions. The time part of the Lax equation satisfies

$$i\frac{dv}{dt} = Bv. \tag{2.3.13}$$

We rewrite Lax equations (2.3.10) and (2.3.13) as

$$\begin{cases} iv_{1,x} - iqv_2 = \zeta v_1, \\ irv_1 - iv_{2,x} = \zeta v_2 \end{cases} \tag{2.3.14}$$

and

$$\begin{cases} iv_{1,t} = av_1 + bv_2, \\ iv_{2,t} = cv_1 - av_2. \end{cases} \tag{2.3.15}$$

We differentiate equation (2.3.14) with respect to t and differentiate equation (2.3.15) with respect to x, to get

$$\begin{cases} iv_{1,xt} - iq_t v_2 - iqv_{2,t} = \zeta v_{1,t}, \\ ir_t v_1 + irv_{1,t} - iv_{2,xt} = \zeta v_{2,t} \end{cases} \tag{2.3.16}$$

and

$$\begin{cases} iv_{1,xt} = a_x v_1 + av_{1,x} + b_x v_2 + bv_{2,x}, \\ iv_{2,xt} = c_x v_1 + cv_{1,x} - a_x v_2 - av_{2,x}. \end{cases} \tag{2.3.17}$$

Omitting $v_{1,xt}$ and $v_{2,xt}$ in equations (2.3.16) and (2.3.17), we have

$$\begin{cases} a_x v_1 + av_{1,x} + b_x v_2 + bv_{2,x} = \zeta v_{1,t} + iq_t v_2 + iqv_{2,t}, \\ c_x v_1 + cv_{1,x} - a_x v_2 - av_{2,x} = ir_t v_1 + irv_{1,t} - \zeta v_{2,t}. \end{cases} \tag{2.3.18}$$

Insert equations (2.3.14), (2.3.15) into equation (2.3.18) and omit $v_{1,x}$, $v_{2,x}$, $v_{1,t}$, $v_{2,t}$, to obtain

$$\begin{aligned} &a_x v_1 + a(-i\zeta v_1 + qv_2) + b_x v_2 + b(i\zeta v_2 + rv_1) \\ &\quad = -i\zeta(av_1 + bv_2) + iq_t v_2 + q(cv_1 - av_2), \end{aligned} \tag{2.3.19}$$

$$\begin{aligned} &c_x v_1 + c(-i\zeta v_1 + qv_2) - a_x v_2 - a(i\zeta v_2 + rv_1) \\ &\quad = ir_t v_1 + r(av_1 + bv_2) + i\zeta(cv_1 - av_2). \end{aligned} \tag{2.3.20}$$

Collecting the coefficients of v_1 and v_2 in equations (2.3.19) and (2.3.20), respectively, we have

$$\begin{aligned} v_1: &\quad a_x + br = qc, \\ &\quad c_x - 2i\zeta c = ir_t + 2ar, \\ v_2: &\quad b_x + 2i\zeta b = iq_t - 2aq, \\ &\quad cq - a_x = br \end{aligned}$$

and we get the following equations, which (a, b, c, r, q) satisfy:

$$\frac{\partial a}{\partial x} = qc - rb, \tag{2.3.21}$$

$$\frac{\partial b}{\partial x} + 2i\zeta b = i\frac{\partial q}{\partial t} - 2aq, \tag{2.3.22}$$

$$\frac{\partial c}{\partial x} - 2i\zeta c = i\frac{\partial r}{\partial t} + 2ar. \tag{2.3.23}$$

In order to maintain unity, we denote

$$\frac{dv}{dt} = Mv, \quad M = \begin{pmatrix} A & B \\ C & -A \end{pmatrix}.$$

Thus, equations (2.3.21)–(2.3.23) can be rewritten as

$$A_x = qC - rB, \tag{2.3.24}$$

$$B_x + 2i\zeta B = q_t - 2Aq, \tag{2.3.25}$$

$$C_x - 2i\zeta C = r_t + 2Ar, \tag{2.3.26}$$

where $A \equiv A(x, t; \zeta)$, $B \equiv B(x, t; \zeta)$, $C \equiv C(x, t; \zeta)$, $r = r(x, t)$, and $q = q(x, t)$.

Equations (2.3.14)–(2.3.15) and (2.3.24)–(2.3.26) compose the foundation of the inverse scattering transform. For given initial values $r(x, 0)$ and $q(x, 0)$, equations (2.3.14) are used to decide the discrete eigenvalues (which are time-invariant) and the asymptotic behavior at $|x| \to \infty$ of all the eigenfunctions $v_1(x, 0; \zeta)$, $v_2(x, 0; \zeta)$ at the initial time. If a certain set of (r_t, q_t, r, q) is given, (A, B, C) will in principle be solved from equations (2.3.24)–(2.3.26). As a next step, time evolution of the asymptotic behavior of eigenfunctions v_1, v_2 at $|x| \to \infty$ will be computed from equations (2.3.15). This information is sufficient to enable us to reconstruct the potentials $r(x, t)$, $q(x, t)$ at later times. Certainly, we cannot carry this out simply, since r, q are unknown. However, the above procedures provide us the ways to solve the exact solution for nonlinear evolution equations via the inverse scattering transform.

First, we find some special solutions for equations (2.3.24)–(2.3.26), based on which a general kind of nonlinear evolution equations will be obtained. We make the following assumptions of A, B, C:

$$A = \sum_{n=0}^{N} A^{(n)}\zeta^n, \quad B = \sum_{n=0}^{N} B^{(n)}\zeta^n, \quad C = \sum_{n=0}^{N} C^{(n)}\zeta^n. \tag{2.3.27}$$

Collecting the same powers of ζ^n in equations (2.3.24)–(2.3.26), it is not difficult to see $A^{(N)} = a_N$ (a_N is independent of x and can depend on t), $B^{(N)} = C^{(N)} = 0$. $B^{(N-1)}$ and $C^{(N-1)}$ are derived from equations (2.3.25) and (2.3.26), while $A^{(N-1)}$ can be obtained from equation (2.3.24). Reiterating the process, all series of $A^{(n)}$, $B^{(n)}$, $C^{(n)}$ will be derived. Specially, the last two equations for $\zeta^{(0)}$ are

$$\begin{cases} q_t = 2A^{(0)}q + B_x^{(0)}, \\ r_t = -2A^{(0)}r + C_x^{(0)}. \end{cases} \tag{2.3.28}$$

Taking $N = 3$ as an example, we have

$$\begin{aligned} A &= A^{(0)} + A^{(1)}\zeta + A^{(2)}\zeta^2 + a_3\zeta^3, \\ B &= B^{(0)} + B^{(1)}\zeta + B^{(2)}\zeta^2, \\ C &= C^{(0)} + C^{(1)}\zeta + C^{(2)}\zeta^2. \end{aligned}$$

Inserting them into equations (2.3.24)–(2.3.26), we have

$$\begin{aligned} &A_x^{(0)} = qC^{(0)} - rB^{(0)}, \\ &A_x^{(1)} = qC^{(1)} - rB^{(1)}, \\ &A_x^{(2)} = qC^{(2)} - rB^{(2)}, \\ &B_x^{(1)} + 2iB^{(0)} = -2A^{(1)}q, \\ &B_x^{(2)} + 2iB^{(1)} = -2A^{(2)}q, \\ &C_x^{(1)} - 2iC^{(0)} = 2A^{(1)}r, \\ &C_x^{(2)} - 2iC^{(1)} = 2A^{(2)}r, \\ &A^{(2)} = a_2, \quad B^{(2)} = ia_3q, \quad C^{(2)} = ia_3r. \end{aligned}$$

Solving the above equations, the coefficients of A, B, C are obtained. A, B, C can be expressed as

$$\begin{cases} A = a_3\zeta^3 + a_2\zeta^2 + \left(\frac{1}{2}a_3qr + a_1\right)\zeta + \frac{1}{2}a_2qr - \frac{i}{4}a_3(qr_x - q_xr) + a_0, \\ B = ia_3q\zeta^2 + \left(ia_2q - \frac{1}{2}a_3q_x\right)\zeta + ia_1q + \frac{i}{2}a_3q^2r - \frac{1}{2}a_2q_x - \frac{i}{4}a_3q_{xx}, \\ C = ia_3r\zeta^2 + \left(ia_2r + \frac{1}{2}a_3r_x\right)\zeta + ia_1r + \frac{i}{2}a_3qr^2 + \frac{1}{2}a_2r_x - \frac{i}{4}a_3r_{xx}. \end{cases} \tag{2.3.29}$$

The corresponding evolution equations, i.e., equations (2.3.28), transform to

$$\begin{aligned} 0 &= q_t + \frac{i}{4}a_3(q_{xxx} - 6qrq_x) + \frac{1}{2}a_2(q_{xx} - 2q^2r) - ia_1q_x - 2a_0q, \\ 0 &= r_t + \frac{i}{4}a_3(r_{xxx} - 6qrr_x) - \frac{1}{2}a_2(r_{xx} - 2qr^2) - ia_1r_x + 2a_0r, \end{aligned}$$

which include some special cases:

(i) If $a_0 = a_1 = a_2 = 0$, $a_3 = -4i$;

(a) if $r = -1$ (KdV equation),

$$q_t + 6qq_x + q_{xxx} = 0, \tag{2.3.30}$$

(b) if $r = \mp q$ (MKdV equation),

$$q_t \pm 6q^2 q_x + q_{xxx} = 0. \tag{2.3.31}$$

(ii) If $a_0 = a_1 = a_3 = 0$, $a_2 = -2i$, $r = \mp q^*$ (nonlinear Schrödinger equation),

$$q_t - iq_{xx} \mp 2iq^2 q^* = 0. \tag{2.3.32}$$

For the KdV equation (2.3.30), the scattering problem (2.3.14) reduces to the Schrödinger equation

$$v_{2,xx} + (\zeta^2 + q(x,t))v_2 = 0. \tag{2.3.33}$$

As is well known, for a real $q(x,t)$, ζ^2 is real. The corresponding discrete eigenvalues are located at the imaginary ζ-axis and are related to the stable solitons. Generally speaking, the discrete eigenvalues which correspond to the localized pulses in the solution $q(x,t)$ are complex.

In a similar way, we can expand A, B, C in the negative powers of ζ. For instance, for

$$A(x,t;\zeta) = \frac{a(x,t)}{\zeta},$$
$$B(x,t;\zeta) = \frac{b(x,t)}{\zeta},$$
$$C(x,t;\zeta) = \frac{c(x,t)}{\zeta},$$

we obtain

$$a_x = \frac{i}{2}(qr)_t, \quad q_{xt} = -4iaq, \quad r_{xt} = -4iar. \tag{2.3.34}$$

We list several special and important cases:

(i) If $a = \frac{i}{4}\cos u$, $b = c = \frac{i}{4}\sin u$, $r = -q = \frac{1}{2}u_x$ (sine-Gordon equation),

$$u_{xt} = \sin u. \tag{2.3.35}$$

(ii) If $a = \frac{i}{4}\cosh u$, $-b = c = \frac{i}{4}\sinh u$, $r = q = \frac{1}{2}u_x$ (sinh-Gordon equation),

$$u_{xt} = \sinh u. \tag{2.3.36}$$

2.4 A more general evolution equation (AKNS equation)

The discussion in the former section suggests the question whether the evolution equations can only be solved by the inverse scattering transform in the finite power series expansions of ζ. In this section, we will show that a wider class of evolution equations indeed exist [6].

Assume that A, B, C satisfy the boundary conditions

$$\begin{cases} A(x,t;\zeta) \to A_0(\zeta), \\ B(x,t;\zeta) \to 0, \\ C(x,t;\zeta) \to 0, \quad |x| \to \infty. \end{cases} \tag{2.4.1}$$

For the case where A, B, C take on different values on the right $x \to +\infty$ and left $x \to -\infty$, the results can be found in [151].

In order to deduce the necessary integral conditions, we shall formally solve equations (2.3.24)–(2.3.26), which can easily be given in terms of a specific solution of equation (2.3.14). Therefore, we first examine the fundamental solutions for the eigenvalue problem of equation (2.3.14).

Assuming $q(x,t), r(x,t) \to 0$ as $|x| \to \infty$, for real ζ, we define the linearly independent solutions for equation (2.3.14), which satisfies the following asymptotic values:

$$\begin{cases} \varphi \to \begin{pmatrix} 1 \\ 0 \end{pmatrix} e^{-i\zeta x}, \quad x \to -\infty, \\ \bar{\varphi} \to \begin{pmatrix} 0 \\ -1 \end{pmatrix} e^{i\zeta x}, \quad x \to -\infty, \end{cases} \tag{2.4.2}$$

$$\begin{cases} \psi \to \begin{pmatrix} 0 \\ 1 \end{pmatrix} e^{i\zeta x}, \quad x \to +\infty, \\ \bar{\psi} \to \begin{pmatrix} 1 \\ 0 \end{pmatrix} e^{-i\zeta x}, \quad x \to +\infty. \end{cases} \tag{2.4.3}$$

It is customary to let the scattering data $a(\zeta,t), b(\zeta,t), \bar{a}(\zeta,t), \bar{b}(\zeta,t)$ be the coefficients relating to the following two sets of linearly independent solutions:

$$\varphi = a\bar{\psi} + b\psi \to \begin{pmatrix} ae^{-i\zeta x} \\ be^{i\zeta x} \end{pmatrix}, \quad x \to +\infty, \tag{2.4.4}$$

$$\bar{\varphi} = \bar{b}\bar{\psi} - \bar{a}\psi \to \begin{pmatrix} \bar{b}e^{-i\zeta x} \\ -\bar{a}e^{i\zeta x} \end{pmatrix}, \quad x \to +\infty. \tag{2.4.5}$$

The coefficients $a(\zeta,t), b(\zeta,t), \bar{a}(\zeta,t), \bar{b}(\zeta,t)$ are given by the Wronskian determinant of φ, $\bar{\varphi}$, ψ, $\bar{\psi}$, i.e.,

$$\begin{cases} a = W(\varphi, \psi), \\ b = -W(\varphi, \bar{\psi}), \end{cases} \qquad \begin{cases} \bar{a} = W(\bar{\varphi}, \bar{\psi}), \\ \bar{b} = W(\bar{\varphi}, \psi), \end{cases} \tag{2.4.6}$$

where $W(u, v) = u_1 v_2 - u_2 v_1$, $W(\bar{\psi}, \psi) = 1$. Since $W(\varphi, \bar{\varphi}) = -1$, we have

$$\begin{aligned} a\bar{a} + b\bar{b} &= W(\bar{\varphi}, \psi)W(\bar{\varphi}, \bar{\psi}) - W(\varphi, \bar{\psi})W(\bar{\varphi}, \psi) \\ &= (\varphi_1\psi_2 - \varphi_2\psi_1)(\bar{\varphi}_1\bar{\psi}_2 - \bar{\varphi}_2\bar{\psi}_1) - (\varphi_1\bar{\psi}_2 - \varphi_2\bar{\psi}_1)(\bar{\varphi}_1\psi_2 - \bar{\varphi}_2\psi_1) \\ &= -\varphi_1\bar{\varphi}_2(\psi_1\bar{\psi}_2 - \bar{\psi}_1\psi_2) + \bar{\varphi}_1\varphi_2(\bar{\psi}_1\psi_2 - \psi_1\bar{\psi}_2) \\ &= \bar{\varphi}_1\varphi_2 - \varphi_1\bar{\varphi}_2 = 1. \end{aligned}$$

We will show in Section 2.5 that $a(\zeta, t)$ can be analytically extended into the upper half plane ($\operatorname{Im}\zeta > 0$); $\bar{a}(\zeta, t)$ can be extended into the lower half plane ($\operatorname{Im}\zeta < 0$). The discrete eigenvalues $\{\zeta_k\}_{k=1}^{N}$ of equation (2.3.14) in the upper half plane ($\operatorname{Im}\zeta > 0$) are given by the zeros of $a(\zeta, t)$, where $\varphi(\zeta_k, t) = b_k(t)\psi(\zeta_k, t)$. Similarly, the zeros of $\bar{a}(\zeta, t)$ in the lower half plane ($\operatorname{Im}\zeta < 0$) are also eigenvalues. At these zeros,

$$\bar{\varphi}_k(\bar{\zeta}_k, t) = \bar{b}_k(t)\bar{\psi}_k(\bar{\zeta}_k, t).$$

Because of our choice of normalization in equation (2.4.2), without loss of generality, we assume that $B, C \to 0$ $(x \to -\infty)$ in equations (2.3.24)–(2.3.36). From equation (2.3.24), we find that $A(x, t; \zeta) \to$ constant $(x \to -\infty)$ and it is convenient to set

$$\lim_{x\to-\infty} A(x, t; \zeta) = A_-(\zeta), \tag{2.4.7}$$

where $A_-(\zeta)$ is an arbitrary function of ζ. Since $\varphi e^{A_- t}$ and $\varphi e^{-A_- t}$ satisfy equation (2.3.24), we have

$$\varphi_t = \begin{pmatrix} A - A_- & B \\ C & -A - A_- \end{pmatrix} \varphi, \tag{2.4.8}$$

$$\bar{\varphi}_t = \begin{pmatrix} A + A_- & B \\ C & -A + A_- \end{pmatrix} \bar{\varphi}. \tag{2.4.9}$$

We check the asymptotic behavior at $x \to +\infty$, i.e.,

$$\begin{pmatrix} a_t e^{-i\zeta x} \\ b_t e^{i\zeta x} \end{pmatrix} = \begin{pmatrix} A_+ - A_- & \lim_{x\to+\infty} B \\ \lim_{x\to+\infty} C & -A_+ - A_- \end{pmatrix} \begin{pmatrix} a e^{-i\zeta x} \\ b e^{i\zeta x} \end{pmatrix},$$

from which the time evolutions of the scattering data are given by

$$\begin{cases} a_t = (A_+ - A_-)a + B_+ b, \\ b_t = C_+ a - (A_+ + A_-)b, \end{cases} \tag{2.4.10}$$

$$\begin{cases} \bar{a}_t = -(A_+ - A_-)\bar{a} - C_+\bar{b}, \\ \bar{b}_t = -B_+a + (A_+ + A_-)\bar{b}, \end{cases} \tag{2.4.11}$$

where $A_+ = \lim_{x\to+\infty} A$, $B_+ = \lim_{x\to+\infty} Be^{2i\zeta x}$, $C_+ = \lim_{x\to+\infty} Ce^{-2i\zeta x}$. Under the special case $A_+ = A_-$, $B_+ = C_+ = 0$, equations (2.4.10) and (2.4.11) transform into

$$\begin{cases} a(\zeta,t) = a(\zeta,0), \\ b(\zeta,t) = b(\zeta,0)e^{-2A_-(\zeta)t}, \\ \bar{a}(\zeta,t) = \bar{a}(\zeta,0), \\ \bar{b}(\zeta,t) = \bar{b}(\zeta,0)e^{2A_-(\zeta)t}. \end{cases} \tag{2.4.12}$$

To solve equation (2.4.12), it is necessary to find the general solution for equations (2.3.24)–(2.3.26) and determine A_+, B_+, and C_+. Assuming $I(u,v)$ in terms of the bilinear form, we have

$$I(u,v) = \int_{-\infty}^{\infty} (-q_t u_2 v_2 + r_t u_1 v_1)dx. \tag{2.4.13}$$

A_+, B_+, and C_+ can be written as

$$\begin{cases} A_+ = -I(\psi,\bar{\psi}) + A_-(a\bar{a} - b\bar{b}), \\ B_+ = -I(\psi,\psi) + 2a\bar{b}A_-, \\ C_+ = I(\bar{\psi},\bar{\psi}) + 2\bar{a}bA_-. \end{cases} \tag{2.4.14}$$

The inverse relation to equations (2.4.4) and (2.4.5) reads

$$\begin{cases} \psi = -a\varphi + \bar{b}\bar{\varphi}, \\ \bar{\psi} = b\bar{\varphi} + \bar{a}\varphi. \end{cases} \tag{2.4.15}$$

Inserting equation (2.4.15) into equation (2.4.14) and substituting A_+, B_+, C_+ into equations (2.4.10) and (2.4.11), we have

$$\begin{cases} a_t = -I(\varphi,\psi), \\ b_t = I(\varphi,\psi), \end{cases} \tag{2.4.16}$$

$$\begin{cases} \bar{a}_t = -I(\bar{\varphi},\bar{\psi}), \\ \bar{b}_t = -I(\bar{\varphi},\psi). \end{cases} \tag{2.4.17}$$

Thus, the time evolution of scattering data from equation (2.4.16) can be expressed as

$$\begin{aligned} \left(\frac{b}{a}\right)_t &= \frac{b_t a - a_t b}{a^2} = \frac{b}{a}\cdot\frac{1}{ab}[b_t a - a_t b] \\ &= \frac{b}{a}\cdot\frac{1}{ab}[I(\varphi, b\bar{\varphi} + \bar{a}\varphi)a - I(\varphi, -a\varphi + \bar{b}\bar{\varphi})b] \end{aligned}$$

$$= \frac{b}{a}\frac{I(\varphi,\varphi)}{ab}.$$

Similarly,

$$\begin{cases} \left(\frac{\bar{b}}{\bar{a}}\right)_t = \frac{\bar{b}}{\bar{a}}\frac{I(\bar{\varphi},\varphi)}{\bar{a}\bar{b}}, \\ \left(\frac{\bar{b}}{a}\right)_t = \frac{\bar{b}}{a}\frac{I(\psi,\psi)}{a\bar{b}}, \\ \left(\frac{b}{\bar{a}}\right)_t = \frac{b}{\bar{a}}\frac{I(\bar{\psi},\bar{\psi})}{\bar{a}b}. \end{cases} \tag{2.4.18}$$

To this point, we have made no assumptions about q and r, except for the fairly weak condition that the integrals $I(u,v)$ are defined. In principle, for any q_t and r_t, we should be able to compute the time evolution of the scattering data from one time step to the next and determine $q(x,t)$ and $r(x,t)$ at later times from equation (2.4.18).

At present, we focus our attention on the analytic expressions for the evolution equation. For arbitrary complex functions $\Omega(\zeta)$ and $\bar{\Omega}(\zeta)$, if we choose

$$I(\psi,\psi) = 2\Omega(\zeta)a\bar{b}, \tag{2.4.19}$$

$$I(\bar{\psi},\bar{\psi}) = -2\bar{\Omega}(\zeta)\bar{a}b, \tag{2.4.20}$$

equation (2.4.18) can be linearized. Equation (2.4.19) may be written as

$$\int_{-\infty}^{\infty} [(r_t + 2\Omega(\zeta)r)\psi_1^2 + (-q_t + 2\Omega(\zeta)q)\psi_2^2]dx = 0. \tag{2.4.21}$$

In fact, we notice that

$$\begin{aligned} I(\psi,\psi) &= \int_{-\infty}^{\infty} (-q_t\psi_2^2 + r_t\psi_1^2)dx, \\ -\psi_1\psi_2|_{-\infty}^{\infty} &= -\psi_1\psi_2|_{-\infty} \\ &= (-a\varphi_1 + \bar{b}\bar{\varphi}_1)(-a\varphi_2 + \bar{b}\bar{\varphi}_2)|_{-\infty} \\ &= a\bar{b}. \end{aligned}$$

On the other hand, from equation (2.3.14), we obtain

$$-\psi_1\psi_2|_{-\infty}^{\infty} = -\int_{-\infty}^{\infty} \frac{d}{dx}(\psi_1\psi_2)dx = -\int_{-\infty}^{\infty} (q\psi_2^2 + r\psi_1^2)dx.$$

Therefore, equation (2.4.21) can be derived. As can be verified from equation (2.3.14) that the vector $\Psi = (\psi_1^2, \psi_2^2)^T$ satisfies

$$L\Psi = \zeta\Psi, \tag{2.4.22}$$

where T means the transpose,

$$L = \frac{1}{2i}\begin{pmatrix} -\frac{\partial}{\partial x} - 2q\int_x^{\infty}\cdot r(y)dy & -2q\int_x^{\infty}\cdot q(y)dy \\ 2r\int_x^{\infty}\cdot r(y)dy & \frac{\partial}{\partial x} + 2r\int_x^{\infty}\cdot q(y)dy \end{pmatrix}. \tag{2.4.23}$$

If we define $u = (r, q)^T$ and $\sigma_3 = \begin{pmatrix} 1 & 0 \\ 0 & -1 \end{pmatrix}$, we rewrite equation (2.4.21) as

$$\int_{-\infty}^{\infty} [\sigma_3 u_t + 2u\Omega(\zeta)]\Psi dx = 0. \tag{2.4.24}$$

If $\Omega(\zeta)$ is an entire function, we conclude

$$\Omega(\zeta)\Psi = \Omega(L)\Psi. \tag{2.4.25}$$

Defining the adjoint operator L^+ as

$$L^+ = \frac{1}{2i}\begin{pmatrix} \frac{\partial}{\partial x} - 2r\int_{-\infty}^{x} \cdot q(y)dy & 2r\int_{-\infty}^{x} \cdot r(y)dy \\ -2q\int_{-\infty}^{x} \cdot q(y)dy & -\frac{\partial}{\partial x} + 2q\int_{-\infty}^{x} \cdot r(y)dy \end{pmatrix}, \tag{2.4.26}$$

equation (2.4.24) transforms into

$$\int_{-\infty}^{\infty} [\sigma_3 u_t + 2\Omega(L^+)u]\Psi dx = 0. \tag{2.4.27}$$

Similarly, equation (2.4.20) gives

$$\int_{-\infty}^{\infty} [\sigma_3 u_t + 2\bar{\Omega}(L^+)u]\bar{\Psi} dx = 0, \quad \bar{\Psi} = (\bar{\psi}_1^2, \bar{\psi}_2^2)^T. \tag{2.4.28}$$

In the special case of $\Omega = \bar{\Omega}$, in order to satisfy equations (2.4.27)–(2.4.28), it is sufficient to denote

$$\sigma_3 u_t + 2\Omega(L^+)u = 0. \tag{2.4.29}$$

Without loss of generality, taking $\Omega(\zeta) = A_-(\zeta)$, we have

$$\sigma_3 u_t + 2A_-(L^+)u = 0. \tag{2.4.30}$$

Equation (2.4.30) is the nonlinear evolution equation whose linearized dispersion relation is defined by $A_-(\zeta)$ and which can be solved by the inverse scattering transform. For instance,

$$\begin{aligned} &\Omega(\zeta) = A_-(\zeta) = -2i\zeta^2, \\ &\sigma_3 u_t = \begin{pmatrix} r_t \\ -q_t \end{pmatrix}, \quad L^+ u = \frac{1}{2i}\begin{pmatrix} r_x \\ q_x \end{pmatrix}, \\ &2[-2i(L^+)^2 u] = i\begin{pmatrix} r_{xx} - 2qr^2 \\ q_{xx} - 2q^2 r \end{pmatrix}, \end{aligned} \tag{2.4.31}$$

which yields

$$\begin{pmatrix} r_t \\ -q_t \end{pmatrix} + i\begin{pmatrix} r_{xx} - 2qr^2 \\ q_{xx} - 2q^2 r \end{pmatrix} = 0, \tag{2.4.32}$$

which is exactly equal to equation (2.3.32) with $r = \mp q^*$.

It also follows that

$$I(\psi,\bar{\psi}) = -2A_-(\zeta)b\bar{b}, \tag{2.4.33}$$

$$I(\varphi,\varphi) = -2A_-(\zeta)ab, \tag{2.4.34}$$

$$I(\bar{\varphi},\bar{\varphi}) = 2A_-(\zeta)\bar{a}\bar{b}. \tag{2.4.35}$$

From equation (2.4.14), we note that

$$A_+ = A_-, \quad B_+ = C_+ = 0. \tag{2.4.36}$$

Therefore, the inverse problem can be solved by scattering data. In the theory of conservation law, we show that equation (2.4.36) leads to the existence of an infinite sequence of integrated densities $\{c_n\}_{n=1}^{\infty}$, which stand for motion constants. The first three are

$$c_1 = \int qr dx, \quad c_2 = \frac{1}{2}\int (rq_x - r_x q)dx, \quad c_3 = \int (q_x r_x + q^2 r^2)dx. \tag{2.4.37}$$

We also note that, when $q \to 0$ as $|x| \to \infty$ and $r = -1$, equation (2.3.14) is associated with the eigenvalue problem of the Schrödinger equation

$$v_{2,xx} + (\zeta^2 + q)v_2 = 0 \tag{2.4.38}$$

and the related evolution equation is

$$q_t + \hat{c}(4L_s^+)q_x = 0, \tag{2.4.39}$$

where

$$L_s^+ = -\frac{1}{4}\frac{\partial^2}{\partial x^2} - q + \frac{1}{2}q_x \int_x^{\infty} \cdot dy, \tag{2.4.40}$$

$\hat{c}(k^2) = \omega/k$, and ω is the dispersion relation of the linearized equation. It is easy to verify that $\omega = -k^3$ yields

$$q_t + q_{xxx} + 6qq_x = 0. \tag{2.4.41}$$

We extend these ideas by allowing the dispersion relation $\Omega(\zeta)$ to be a ratio of entire functions, i.e., $\Omega(\zeta) = \Omega_1(\zeta)/\Omega_2(\zeta)$. Then the analog to equation (2.4.30) is

$$\Omega_2(L^+)\sigma_3 u_t + 2\Omega_1(L^+)u = 0. \tag{2.4.42}$$

If we take $\Omega = i\alpha/2(\zeta - \zeta_1)$, equation (2.4.42) becomes

$$\frac{1}{2i}\begin{pmatrix} r_{xt} - 2r\int_{-\infty}^{r}(qr)_t dy - 2i\hat{\zeta}_1 r_t \\ q_{xt} - 2q\int_{-\infty}^{r}(qr)_t dy + 2i\hat{\zeta}_1 q_t \end{pmatrix} = -i\alpha\begin{pmatrix} r \\ q \end{pmatrix}. \tag{2.4.43}$$

When $\hat{c}(k^2) = \frac{1}{1+k^2}$, an equation will be derived from equation (2.4.39), i.e.,

$$q_t - q_{xxt} - 4qq_t + 2q_x \int_x^{\infty} q_t dy + q_x = 0, \tag{2.4.44}$$

which reduces to the KdV equation under the long wave approximation and small amplitude assumption.

A further extension of the analysis is possible and an even wider class of evolution equations can be obtained. The inverse scattering transform is still solvable if we do not choose $\Omega(t) = \bar{\Omega}(t)$, even in the cases where there are no motion invariants.

2.5 Solution of the inverse scattering problems for AKNS equations

In this section, we make a comprehensive study of the solvability of the inverse scattering transform problem. For the eigenvalue problem which is not self-adjoint, we still can derive the Marchenko equation. First, we provide a discussion of the analytical properties of the scattering data.

(1) Analytical properties of the scattering data.

For the eigenvalue problem given by equation (2.3.14) on the interval $-\infty < x < +\infty$, we assume q and r vanish sufficiently rapidly to zero as $|x| \to \infty$, so in these limits, the right-hand side in equation (2.3.14) can be neglected. Set $\varphi, \bar{\varphi}, \psi, \bar{\psi}$ to be the Jost functions of equation (2.3.14), satisfying the boundary conditions (2.4.2)–(2.4.3). φ and $\bar{\varphi}$ are linearly independent, as are ψ and $\bar{\psi}$. Therefore, for real ζ, we have

$$\varphi(\zeta, x) = a(\zeta)\bar{\psi}(\zeta, x) + b(\zeta)\psi(\zeta, x), \tag{2.5.1}$$

$$\bar{\varphi}(\zeta, x) = -\bar{a}(\zeta)\psi(\zeta, x) + \bar{b}(\zeta)\bar{\psi}(\zeta, x), \tag{2.5.2}$$

which define a, $\bar{a}$, b, $\bar{b}$. From equation (2.3.14), if $u(\zeta, x)$ and $v(\zeta, x)$ are solutions for equation (2.3.14), then we have

$$\frac{dW(u,v)}{dx} = 0, \tag{2.5.3}$$

where

$$W(u,v) \equiv u_1(\zeta, x)v_2(\zeta, x) - u_2(\zeta, x)v_1(\zeta, x). \tag{2.5.4}$$

In fact, equation (2.5.3) will be derived from

$$\begin{aligned}&[u_{1,x} + i\zeta u_1 = qu_2] \cdot v_2 + u_1[v_{2,x} - i\zeta v_2 = rv_1]\\ &-[u_{2,x} - i\zeta u_2 = ru_1] \cdot v_1 - u_2[v_{1,x} + i\zeta v_1 = qv_2] = 0.\end{aligned}$$

The relations (2.4.6) and $W(\bar{\varphi}, \varphi) = 1$ imply

$$\bar{a}(\zeta)a(\zeta) + \bar{b}(\zeta)b(\zeta) = 1. \tag{2.5.5}$$

The inverse of equations (2.5.1)–(2.5.2) are expressed as

$$\psi(\zeta, x) = -a(\zeta)\bar{\varphi}(\zeta, x) + \bar{b}(\zeta)\varphi(\zeta, x), \tag{2.5.6}$$

$$\bar{\psi}(\zeta, x) = \bar{a}(\zeta)\varphi(\zeta, x) + b(\zeta)\bar{\varphi}(\zeta, x). \tag{2.5.7}$$

By virtue of $\varphi_{2,x} - i\zeta\varphi_2 = r\varphi_1$ and the boundary condition $\varphi_2 \to 0$ $(x \to -\infty)$, we have

$$e^{i\zeta x}\varphi_2(x) = \int_{-\infty}^{x} e^{2i\zeta(x-y)} r(y) e^{i\zeta y}\varphi(y)dy. \tag{2.5.8}$$

Substituting equation (2.5.8) into $\varphi_{1,x} + i\zeta\varphi_1 = q\varphi_2$ and considering $\varphi_1 \to e^{-i\zeta x}$ $(x \to -\infty)$, we obtain

$$e^{i\zeta x}\varphi_1(x) = 1 + \int_{-\infty}^{x} M(\zeta, x, y) e^{i\zeta y}\varphi_1(y)dy, \tag{2.5.9}$$

where

$$M(\zeta, x, y) \equiv r(y) \int_{y}^{x} e^{2i\zeta(x-y)} q(z)dz. \tag{2.5.10}$$

Under suitable conditions, we extend φ into the upper half of the ζ-plane ($\zeta = \xi + i\eta$, $\eta > 0$). To see this, let

$$R_n(x) \equiv \int_{-\infty}^{x} |y^n||r(y)|dy, \tag{2.5.11}$$

$$Q_n(x) \equiv \int_{-\infty}^{x} |y^n||q(y)|dy, \tag{2.5.12}$$

where we assume r and q to vanish sufficiently rapidly as $x \to -\infty$ for at least some of these integrals to exist when $n > 0$. For $\eta \geq 0$, we have

$$\begin{aligned} |e^{i\zeta x}\varphi_1(x)| &\leq 1 + \int_{-\infty}^{x} |q(z)|dz \int_{-\infty}^{z} |r(y)||e^{i\zeta y}\varphi_1(y)|dy \\ &= 1 + \int_{-\infty}^{x} Q_0'(z)dz \int_{-\infty}^{z} R_0'(y)|e^{i\zeta y}\varphi_1(y)|dy \\ &\leq 1 + R_0(x)Q_0(x) + \frac{[R_0(x)Q_0(x)]^2}{(2!)^2} + \frac{[R_0(x)Q_0(x)]^3}{(3!)^2} + \cdots \end{aligned}$$

or

$$|e^{i\zeta x}\varphi_1(x)| \leq I_0(S(x)), \tag{2.5.13}$$

where $S(x) = 2(R_0(x)Q_0(x))^{1/2}$ and $I_0(S(x))$ is the zero-order Bessel function with imaginary argument. By equations (2.5.1) and (2.5.4), we know $\varphi_1 e^{i\zeta x} \to a(\zeta)$ as $x \to +\infty$. Thus, we conclude that $a(\zeta)$ is bounded in the upper half of the ζ-plane ($\eta \geq 0$) if $R_0(\infty)$ and $Q_0(\infty)$ are finite. Returning to equation (2.4.6), we see that the Neumann series solution of

$$\begin{aligned} e^{i\zeta x}\varphi_1(x) &= 1 + \int_{-\infty}^{x} M(\zeta,x,y)dy \\ &\quad + \int_{-\infty}^{x} M(\zeta,x,y)dy \int_{-\infty}^{y} M(\zeta,y,z)dz + \cdots \end{aligned} \tag{2.5.14}$$

is absolutely convergent in the upper half plane. Furthermore, one may differentiate equation (2.5.14) with respect to ζ and find that $e^{i\zeta x}\varphi_1(x)$ is analytical if $\eta > 0$. To be analytic for $\eta = 0$, it is easy to see that simply requiring $R_0(\infty)$ and $Q_0(\infty)$ to be finite is not sufficient. For instance, since ζ occurs in an exponential in equation (2.5.10), differentiation will give a $(z - y)$-term. To ensure the first differential to exist at $\eta = 0$, q and r must vanish faster than x^{-2} as $x \to \pm\infty$. Upon doing the same for $\bar{\varphi}$, ψ, $\bar{\psi}$ as for φ, we have the following theorem.

Theorem 2.5.1. *Under the conditions*

$$R_0(\infty) < \infty, \quad Q_0(\infty) < \infty, \tag{2.5.15}$$

$e^{i\zeta x}\varphi(\zeta,x)$, $e^{-i\zeta x}\psi(\zeta,x)$ are analytic functions of ζ ($\eta > 0$), while $e^{-i\zeta x}\bar{\varphi}(\zeta,x)$, $e^{i\zeta x}\bar{\psi}(\zeta,x)$ are analytic functions of ζ ($\eta < 0$). In addition, the above four functions are bounded when $\eta = 0$. Furthermore, for a given integer n satisfying

$$R_l(\infty) < \infty, \quad Q_l(\infty) < \infty, \quad l = 0,1,2,\ldots,n, \tag{2.5.16}$$

these four functions are also n-fold differentiable at $\eta = 0$ with respect to ζ. If equation (2.5.16) is true for all n, then the range of analyticity will include the real ζ-axis ($\eta = 0$).

As a corollary, we have the following results from equation (2.4.6).

Corollary 2.5.2. *When equation (2.5.15) is satisfied, $a(\zeta)$ is an analytic function of ζ for $\eta > 0$ and $\bar{a}(\zeta)$ is an analytic function of ζ for $\eta < 0$. If equation (2.5.16) is satisfied for all n, both $a(\zeta)$ and $\bar{a}(\zeta)$ are analytic when $\eta = 0$.*

When the more stringent conditions are placed on r and q, one can prove the following theorem.

Theorem 2.5.3. *If there exist the finite and positive constants $\hat{R}$, $\hat{Q}$, and K, which satisfy*

$$|r(x)| \leq \hat{R}e^{-2K|x|}, \quad |q(x)| \leq \hat{Q}e^{-2K|x|}, \tag{2.5.17}$$

for all x, then $e^{i\zeta x}\varphi(\zeta,x), e^{-i\zeta x}\psi(\zeta,x)$ are analytic functions of ζ when $\eta > -K$, while $e^{-i\zeta x}\bar{\varphi}(\zeta,x), e^{i\zeta x}\bar{\psi}(\zeta,x)$ are analytic functions of ζ when $\eta < +K$.

Immediately following from equation (2.4.6), we have the following corollary.

Corollary 2.5.4. *If equation (2.5.17) is satisfied, $a(\zeta)$ is an analytic function of ζ when $\eta > -K$, $\bar{a}(\zeta)$ is an analytic function of ζ when $\eta < +K$, and both $b(\zeta)$ and $\bar{b}(\zeta)$ are analytic when $-K < \eta < +K$.*

We note that, when q and r are on compact support, K in equation (2.5.17) can be chosen as large as desired, so we also have the following corollary.

Corollary 2.5.5. *Equation (2.5.16) is true when r and q are on compact support. Then $e^{i\zeta x}\varphi(\zeta,x)$, $e^{-i\zeta x}\bar{\varphi}(\zeta,x), e^{-i\zeta x}\psi(\zeta,x)$, and $e^{i\zeta x}\bar{\psi}(\zeta,x)$ are entire functions of ζ. Therefore, $a(\zeta)$, $\bar{a}(\zeta)$, $b(\zeta)$, $\bar{b}(\zeta)$ are also entire functions of ζ.*

We return to equations (2.5.8)–(2.5.10) in the upper half plane of ζ. When $|\zeta| \to -\infty$, we get the asymptotic series

$$\varphi_1 e^{i\zeta x} \to 1 - \frac{1}{2i\zeta}\int_{-\infty}^{x} r(y)q(y)dy + O\left(\frac{1}{\zeta^2}\right), \tag{2.5.18}$$

$$\varphi_2 e^{i\zeta x} \to -\frac{1}{2i\zeta}r(x) + O\left(\frac{1}{\zeta^2}\right) \tag{2.5.19}$$

and, similarly,

$$\psi_1 e^{-i\zeta x} \to \frac{1}{2i\zeta}q(x) + O\left(\frac{1}{\zeta^2}\right), \tag{2.5.20}$$

$$\psi_2 e^{-i\zeta x} \to 1 - \frac{1}{2i\zeta}\int_{x}^{\infty} r(y)q(y)dy + O\left(\frac{1}{\zeta^2}\right), \tag{2.5.21}$$

while, for ζ in the lower half plane, the asymptotic series as $|\zeta| \to \infty$ are

$$\bar{\varphi}_1 e^{-i\zeta x} \to -\frac{1}{2i\zeta}q(x) + O\left(\frac{1}{\zeta^2}\right), \tag{2.5.22}$$

$$\bar{\varphi}_2 e^{-i\zeta x} \to -1 - \frac{1}{2i\zeta}\int_{-\infty}^{x} q(y)r(y)dy + O\left(\frac{1}{\zeta^2}\right), \tag{2.5.23}$$

$$\bar{\psi}_1 e^{i\zeta x} \to 1 + \frac{1}{2i\zeta}\int_{x}^{\infty} q(y)r(y)dy + O\left(\frac{1}{\zeta^2}\right), \tag{2.5.24}$$

$$\bar{\psi}_2 e^{i\zeta x} \to -\frac{1}{2i\zeta}r(x) + O\left(\frac{1}{\zeta^2}\right). \tag{2.5.25}$$

Thus, in each respective half plane as $|\zeta| \to \infty$, we have

$$a(\zeta) \to 1 - \frac{1}{2i\zeta}\int_{-\infty}^{\infty} q(y)r(y)dy + O\left(\frac{1}{\zeta^2}\right), \tag{2.5.26}$$

$$\bar{a}(\zeta) \to 1 + \frac{1}{2i\zeta}\int_{-\infty}^{\infty} q(y)r(y)dy + O\left(\frac{1}{\zeta^2}\right). \tag{2.5.27}$$

When $a(\zeta)$ has zero point ζ_k $(k = 0, 1, 2, \dots, N)$ in the upper half plane $(\eta > 0)$, with N being a finite number, at $\zeta = \zeta_k$, we have

$$\varphi = b_k\psi, \tag{2.5.28}$$

where b_k is the proportional factor. In the case where r and q are on compact support, $b_k \equiv b(\zeta_k)$. In addition, $\bar{a}(\zeta)$ possesses the zero point $\bar{\zeta}_k$ $(k = 0, 1, 2, \dots, \bar{N})$ in the lower half plane. At $\zeta = \bar{\zeta}_k$, we have

$$\bar{\varphi} = \bar{b}_k\psi. \tag{2.5.29}$$

If r and q are on compact support, $\bar{b}(k) \equiv \bar{b}(\bar{\zeta}_k)$. N and $\bar{N}$ are finite numbers.

Unlike the Schrödinger equation, whose zero points must be simple as a consequence of being self-adjoint, the above eigenvalue problem may have zeros of any order for a and $\bar{a}$. However, these cases can be analyzed as the limit of the case where all zeros are simple. For example, a double zero of $a(\zeta)$ at ζ_1 is simply obtained when letting a have two simple zeros at ζ_1 and ζ_2 and then $\zeta_2 \to \zeta_1$.

Whenever r is linearly related to q or q^*, simplifications occur. First consider the case

$$r = \alpha q, \tag{2.5.30}$$

where α is any nonzero, finite, complex constant. In this case, we have

$$\bar{\psi}(\zeta, x) = S\psi(-\zeta, x), \tag{2.5.31}$$

$$\bar{\varphi}(\zeta, x) = -\frac{1}{\alpha}S\varphi(-\zeta, x), \tag{2.5.32}$$

where

$$S = \begin{pmatrix} 0 & 1 \\ \alpha & 0 \end{pmatrix}. \tag{2.5.33}$$

Consequently,

$$\bar{a}(\zeta) = a(-\zeta), \tag{2.5.34}$$

$$\bar{b}(\zeta) = -\frac{1}{\alpha}b(-\zeta). \tag{2.5.35}$$

The zeros of a and $\bar{a}$ are paired such that

$$\bar{N} = N, \tag{2.5.36}$$

$$\bar{\zeta}_k = -\zeta_k, \quad k = 1, 2, \dots, N, \tag{2.5.37}$$

$$\bar{b}_k = -\frac{1}{\alpha} b_k. \tag{2.5.38}$$

In the case where

$$r = \alpha q^*, \tag{2.5.39}$$

where α is a nonzero, finite, real constant, we have

$$\bar{\psi}(\zeta, x) = S\psi^*(\zeta^*, x), \tag{2.5.40}$$

$$\bar{\varphi}(\zeta, x) = -\frac{1}{\alpha} S\varphi^*(\zeta^*, x), \tag{2.5.41}$$

which gives

$$\bar{a}(\zeta) = a^*(\zeta^*), \tag{2.5.42}$$

$$\bar{b}(\zeta) = -\frac{1}{\alpha} b^*(\zeta^*). \tag{2.5.43}$$

Similarly, the zeros of a and $\bar{a}$ are paired, but in a different manner, i.e.,

$$\bar{N} = N, \tag{2.5.44}$$

$$\bar{\zeta}_k = \zeta_k, \quad k = 1, 2, \ldots, N, \tag{2.5.45}$$

$$\bar{b}_k = -\frac{1}{\alpha} b_k^*. \tag{2.5.46}$$

If equations (2.5.30) and (2.5.39) both hold, which means r and r^* are both proportional to q, then ζ_k is either purely imaginary or $-\zeta_k^*$ is also another eigenvalue.

(2) The inverse scattering transform.

First, we will obtain the integral representations for the four Jost functions defined by equation (2.3.14), from which we will obtain the inverse equations of the Marchenko type. For simplicity, we assume r and q to be on compact support so that the solutions for equation (2.3.14) and the scattering data will be entire functions of ζ. We define the contour C to be the contour in the complex ζ-plane, starting from $\zeta = -\infty + i0^+$, passing over all zeros of $a(\zeta)$, and ending at $\zeta = +\infty + i0^+$. Similarly, we define $\bar{C}$ as the contour starting from $\zeta = -\infty + i0^-$, passing under all zeros of $\bar{a}(\zeta)$, and ending at $\zeta = +\infty + i0^-$.

Consider the integral

$$\int_C \frac{d\zeta'}{a(\zeta')} \frac{\varphi(\zeta', x)}{\zeta' - \zeta} e^{i\zeta x} \tag{2.5.47}$$

with ζ under C. From equations (2.5.18), (2.5.19), and (2.5.26), we find the value of the above integral to be $-i\pi \binom{1}{0}$. Through equation (2.5.1) and closing the contour for the integral containing $\bar{\psi}$ in the lower ζ-plane, from equations (2.5.24)–(2.5.25), we deduce

$$\bar{\psi}(\zeta, x) e^{i\zeta x} = \begin{pmatrix} 1 \\ 0 \end{pmatrix} + \frac{1}{2\pi i} \int_C \frac{d\zeta'}{\zeta' - \zeta} \frac{b(\zeta')}{a(\zeta')} \psi(\zeta', x) e^{i\zeta' x}, \tag{2.5.48}$$

for ζ below C. Similarly, considering the integral

$$\int_{\bar{C}} \frac{d\zeta'}{\zeta'-\zeta} \frac{\bar{\varphi}(\zeta',x)}{a(\zeta')} e^{-i\zeta' x}, \tag{2.5.49}$$

for ζ above $\bar{C}$, we obtain

$$\psi(\zeta,x)e^{-i\zeta x} = \begin{pmatrix} 0 \\ 1 \end{pmatrix} + \frac{1}{2\pi i} \int_{\bar{C}} \frac{d\zeta'}{\zeta'-\zeta} \frac{\bar{b}(\zeta')}{\bar{a}(\zeta')} \bar{\psi}(\zeta',x)e^{-i\zeta' x}. \tag{2.5.50}$$

Likewise, replacing $\varphi e^{i\zeta x}$ and $\bar{\varphi} e^{-i\zeta x}$ by $\psi e^{-i\zeta x}$ and $\bar{\psi} e^{i\zeta x}$, respectively, in the above contour integrals, we get

$$\bar{\varphi}(\zeta,x)e^{-i\zeta x} = -\begin{pmatrix} 0 \\ 1 \end{pmatrix} - \frac{1}{2\pi i} \int_{C} \frac{d\zeta'}{\zeta'-\zeta} \frac{\bar{b}(\zeta')}{a(\zeta')} \psi(\zeta',x)e^{-i\zeta' x}, \tag{2.5.51}$$

$$\varphi(\zeta,x)e^{i\zeta x} = \begin{pmatrix} 1 \\ 0 \end{pmatrix} - \frac{1}{2\pi i} \int_{\bar{C}} \frac{d\zeta'}{\zeta'-\zeta} \frac{b(\zeta')}{\bar{a}(\zeta')} \bar{\psi}(\zeta',x)e^{i\zeta' x}, \tag{2.5.52}$$

where ζ lies between the contours C and $\bar{C}$.

Assume that φ, $\bar{\varphi}$, ψ, $\bar{\psi}$ can be represented as

$$\psi(\zeta,x) = \begin{pmatrix} 0 \\ 1 \end{pmatrix} e^{i\zeta x} + \int_x^{\infty} K(x,s)e^{i\zeta s}ds, \tag{2.5.53}$$

$$\bar{\psi}(\zeta,x) = \begin{pmatrix} 1 \\ 0 \end{pmatrix} e^{-i\zeta x} + \int_x^{\infty} \bar{K}(x,s)e^{-i\zeta s}ds, \tag{2.5.54}$$

$$\varphi(\zeta,x) = \begin{pmatrix} 1 \\ 0 \end{pmatrix} e^{-i\zeta x} - \int_{-\infty}^{x} L(x,s)e^{-i\zeta s}ds, \tag{2.5.55}$$

$$\bar{\varphi}(\zeta,x) = -\begin{pmatrix} 0 \\ 1 \end{pmatrix} e^{-i\zeta x} - \int_{-\infty}^{x} \bar{L}(x,s)e^{i\zeta s}ds, \tag{2.5.56}$$

where K, $\bar{K}$, L, $\bar{L}$ are column vectors. Inserting the above expressions into equations (2.5.48) and (2.5.50)–(2.5.52) and taking the Fourier transformation, we will get the following Marchenko type:

$$\bar{K}(x,y) + \begin{pmatrix} 0 \\ 1 \end{pmatrix} F(x+y) + \int_x^{\infty} K(x,s)F(s+y)ds = 0, \quad y > x, \tag{2.5.57}$$

$$K(x,y) - \begin{pmatrix} 1 \\ 0 \end{pmatrix} \bar{F}(x+y) - \int_x^{\infty} \bar{K}(x,s)\bar{F}(s+y)ds = 0, \quad y > x, \tag{2.5.58}$$

$$\bar{L}(x,y) + \begin{pmatrix} 1 \\ 0 \end{pmatrix} G(x+y) - \int_{-\infty}^{x} L(x,s)G(s+y)ds = 0, \quad x > y, \tag{2.5.59}$$

$$L(x,y) + \begin{pmatrix} 0 \\ 1 \end{pmatrix} \bar{G}(x+y) + \int_{-\infty}^{x} \bar{L}(x,s)\bar{G}(s+y)ds = 0, \quad x > y, \tag{2.5.60}$$

where

$$F(z) = \frac{1}{2\pi}\int_C \frac{b(\zeta)}{a(\zeta)} e^{i\zeta z}\, d\zeta, \tag{2.5.61}$$

$$\bar{F}(z) = \frac{1}{2\pi}\int_{\bar{C}} \frac{\bar{b}(\zeta)}{\bar{a}(\zeta)} e^{-i\zeta z}\, d\zeta, \tag{2.5.62}$$

$$G(z) = \frac{1}{2\pi}\int_C \frac{\bar{b}(\zeta)}{a(\zeta)} e^{-i\zeta z}\, d\zeta, \tag{2.5.63}$$

$$\bar{G}(z) = \frac{1}{2\pi}\int_{\bar{C}} \frac{b(\zeta)}{\bar{a}(\zeta)} e^{i\zeta z}\, d\zeta. \tag{2.5.64}$$

As the next step, we will prove the existence and uniqueness of the integral kernels $K(x,s), \bar{K}(x,s), L(x,s), \bar{L}(x,s)$ for equations (2.5.53)–(2.5.56). Substituting equation (2.5.53) into

$$\begin{cases} \psi_{1,x} + i\zeta\psi_1 = q(x)\psi_2, \\ \psi_{2,x} - i\zeta\psi_2 = r(x)\psi_1, \end{cases} \tag{2.5.65}$$

we have

$$\begin{aligned} &\int_x^\infty e^{i\zeta s}[(\partial x - \partial s)K_1(x,s) - q(x)K_2(x,s)]ds \\ &\quad - [q(x) + 2K_1(x,s)]e^{i\zeta x} + \lim_{s\to\infty}[K_1(x,s)e^{i\zeta s}] = 0, \end{aligned} \tag{2.5.66}$$

$$\begin{aligned} &\int_x^\infty e^{i\zeta s}[(\partial x + \partial s)K_2(x,s) - r(x)K_1(x,s)]ds \\ &\quad - \lim_{s\to\infty}[K_2(x,s)e^{i\zeta s}] = 0. \end{aligned} \tag{2.5.67}$$

It is necessary and sufficient to obtain the following relations:

$$(\partial_x - \partial_s)K_1(x,s) - q(x)K_2(x,s) = 0, \tag{2.5.68}$$

$$(\partial_x - \partial_s)K_2(x,s) - r(x)K_1(x,s) = 0, \tag{2.5.69}$$

which satisfy

$$K_1(x,x) = -\frac{1}{2}q(x), \tag{2.5.70}$$

$$\lim_{s\to\infty} K(x,s) = 0. \tag{2.5.71}$$

In order to ensure the existence of the solutions for equations (2.5.68)–(2.5.69) under the conditions (2.5.70)–(2.5.71), we introduce the coordinates

$$\mu = \frac{1}{2}(x+s), \quad \nu = \frac{1}{2}(x-s). \tag{2.5.72}$$

Upon transforming to these coordinates, equations (2.5.68)–(2.5.71) become

$$\partial_\nu K_1(\mu,\nu) - q(\mu+\nu)K_2(\mu,\nu) = 0, \tag{2.5.73}$$

$$\partial_\mu K_2(\mu,\nu) - r(\mu+\nu)K_1(\mu,\nu) = 0, \tag{2.5.74}$$

$$K_1(\mu,0) = -\frac{1}{2}q(\mu), \tag{2.5.75}$$

$$\lim_{\mu-\nu\to\infty} K(\mu,\nu) = 0. \tag{2.5.76}$$

From the theory of characteristics, the solution exists and is unique. Similarly, we can prove that $\bar{K}$, L, $\bar{L}$ exist and are unique.

Finally, we consider the existence and uniqueness of the solution of Marchenko equations (2.5.57)–(2.5.60) under the following restrictions:

$$r(x) = -q^*(x) \tag{2.5.77}$$

or

$$r(x) = q^*(x) \tag{2.5.78}$$

and

$$Q(\infty) = \int_{-\infty}^{\infty} |q|dx < 0.523. \tag{2.5.79}$$

Neither of these restrictions is necessary. Requirements which are both necessary and sufficient have not yet been determined.

Taking account of the homogeneous equations corresponding to equations (2.5.57)–(2.5.58) $(y > x)$, we have

$$\begin{cases} \varphi_1(y) + \int_x^{\infty} \varphi_2(s)F(s+y)ds = 0, \\ \varphi_2(y) - \int_x^{\infty} \varphi_1(s)\bar{F}(s+y)ds = 0. \end{cases} \tag{2.5.80}$$

Suppose $\varphi(y) = \binom{\varphi_1}{\varphi_2}$ is a solution for equation (2.5.80) which vanishes when $y < x$. By the Fredholm alternatives, it is sufficient to show that $\varphi(y) \equiv 0$. We multiply equation (2.5.80) by φ_1^* and φ_2^*, respectively, and integrate in y, via

$$\int_x^{\infty} |\varphi_i(y)|^2 dy = \int_{-\infty}^{\infty} |\varphi_i(y)|^2 dy, \quad i = 1, 2,$$

to obtain

$$\int_{-\infty}^{\infty} \left\{ |\varphi_1|^2 + |\varphi_2|^2 + \int_{-\infty}^{\infty} [\varphi_2(s)\varphi_1^*(y)F(s+y) - \varphi_1(s)\varphi_2^*(y)\bar{F}(s+y)]ds \right\} dy = 0. \tag{2.5.81}$$

We consider two special cases. First, when $r = -q^*$, $\bar{F}(s+y) = F^*(s+y)$ will be deduced from equations (2.5.40)–(2.5.47) ($\alpha = -1$). Hence, equation (2.5.81) becomes

$$\int_{-\infty}^{\infty}\left\{|\varphi_1|^2 + |\varphi_2|^2 + 2i\,\mathrm{Im}\int_{-\infty}^{\infty}\varphi_1^*(y)\varphi_2(s)F(s+y)ds\right\}dy = 0. \tag{2.5.82}$$

The real and imaginary parts must be zero, so we know that $\varphi(y) = 0$ and the solution for equations (2.5.57)–(2.5.58) exists and is unique. Second, if $r(x) = q^*(x)$, the problem is formally self-adjoint, the spectrum lies on the real axis, and $\bar{F}(s+y) = -F^*(s+y)$. In this case, equation (2.5.81) transforms into

$$\int_{-\infty}^{\infty}\left\{|\varphi_1|^2 + |\varphi_2|^2 + 2\,\mathrm{Re}\int_{-\infty}^{\infty}\varphi_1^*(y)\varphi_2(s)F(s+y)ds\right\}dy = 0. \tag{2.5.83}$$

If we require

$$|a(\zeta)| > 0, \quad (\eta \geq 0), \tag{2.5.84}$$

there is no discrete eigenvalue on the real axis, so

$$F(z) = \frac{1}{2\pi}\int_{-\infty}^{\infty}\frac{b(\zeta)}{a(\zeta)}e^{i\zeta z}d\zeta. \tag{2.5.85}$$

The Fourier transform of $\varphi_j(y)$ is

$$\hat{\varphi}_j(\xi) = \int_{-\infty}^{\infty}\varphi_j(y)e^{-i\xi y}dy, \tag{2.5.86}$$

which satisfies Parseval's relation

$$\int_{-\infty}^{\infty}|\varphi_j|^2dy = \frac{1}{2\pi}\int_{-\infty}^{\infty}|\hat{\varphi}_j|^2d\xi. \tag{2.5.87}$$

Substituting equations (2.5.85)–(2.5.87) into equation (2.5.82) and reversing the order of integration, we have

$$\int_{-\infty}^{\infty}\left\{|\hat{\varphi}_1(-\xi)|^2 + |\hat{\varphi}_2^*(\xi)|^2 + 2\,\mathrm{Re}\left[\frac{b(\xi)}{a(\xi)}\hat{\varphi}_1(-\xi)\hat{\varphi}_2^*(\xi)\right]\right\}d\xi = 0. \tag{2.5.88}$$

If

$$\left|\frac{b(\xi)}{a(\xi)}\right| < 1, \tag{2.5.89}$$

then

$$\left|2\,\mathrm{Re}\left[\frac{b(\xi)}{a(\xi)}\hat{\varphi}_1(-\xi)\hat{\varphi}_2(\xi)\right]\right| \leq 2|\hat{\varphi}_1(-\xi)||\hat{\varphi}_2^*(\xi)| \leq |\hat{\varphi}_1|^2 + |\hat{\varphi}_2|^2.$$

Hence the only solution for equation (2.5.88) is $\varphi \equiv 0$, which implies the existence and uniqueness of the solution for equations (2.5.57) and (2.5.58). In the case where $r = q^*$, due to the relations $\bar{a}a + \bar{b}b = 1$, $\bar{a} = a^*$, and $\bar{b} = b^*$, equation (2.5.89) can be rewritten as

$$|a|^2 > \frac{1}{2}, \tag{2.5.90}$$

which is more stringent than equation (2.5.84). The above condition is satisfied if

$$|a(\zeta) - 1| < 1 - \frac{1}{\sqrt{2}}. \tag{2.5.91}$$

From equation (2.5.13), we deduce

$$|a(\zeta) - 1| \leq I_0(2Q(\infty)) - 1 < 1 - \frac{1}{\sqrt{2}}.$$

Therefore, the condition $Q(\infty) < 0.523$ in equation (2.5.79) is sufficient.

2.6 Asymptotic solutions for the evolution equations ($t \to \infty$)

In previous sections, we explained the method of inverse scattering transforms and pointed out that a class of nonlinear evolution equations can be solved as initial value problems by this method. In this section, in order to determine the asymptotic behavior of the solution for equation (2.4.30), we need to solve the following integral equations ($y > x$):

$$\begin{cases} K(x,y;t) - \begin{pmatrix} 1 \\ 0 \end{pmatrix} \bar{F}(x+y;t) - \int_x^{\infty} \bar{K}(x,s;t)\bar{F}(s+y;t)ds = 0, \\ \bar{K}(x,y;t) + \begin{pmatrix} 0 \\ 1 \end{pmatrix} F(x+y;t) + \int_x^{\infty} K(x,s;t)F(s+y;t)ds = 0. \end{cases} \tag{2.6.1}$$

The asymptotic solution will be shown to be similar to that of the KdV equation, although there are some important differences. In the following, we will discuss separately the contribution to the solution from the discrete spectrum, the continuous spectrum, and their combination. In addition, we will make the estimate on the discrete spectrum.

1. The discrete spectrum

Firstly, we consider the solvability of equations (2.6.1). An important difference between the scattering problem (2.3.14) and the eigenvalue problem (2.3.33) is that equation (2.6.1) does not necessarily have a solution and the solution for the evolution equation will become unbounded after a finite amount of time. We explain this with an example. Let $q(x,0)$, $r(x,0)$ be the smooth initial data which satisfy equation (2.5.15). In addition, the spectrum of $q(x,0)$, $r(x,0)$ consists of two kinds of discrete eigenvalues, i.e., ζ ($\operatorname{Im}\zeta > 0$) and $\bar{\zeta}$ ($\operatorname{Im}\bar{\zeta} < 0$). Then

$$\begin{cases} F(z,t) = -ice^{i\zeta z - 2A_0(\zeta)t}, \\ \bar{F}(z,t) = i\bar{c}e^{-i\bar{\zeta}z + 2A_0(\bar{\zeta})t}, \end{cases} \tag{2.6.2}$$

where c and $\bar{c}$ are constants and $A_0(\zeta)$ is related to the linear dispersion relation. In this case, the integral kernel is degenerate. From the relations

$$\begin{cases} K_1(x,x;t) = -\frac{1}{2}q(x,t), \\ K_2(x,x;t) = \frac{1}{2}\int_x^\infty q(x,t)r(x,t)dx, \\ \bar{K}_2(x,x;t) = \frac{1}{2}r(x,t), \end{cases} \tag{2.6.3}$$

we have

$$\begin{cases} q(x,t) = -\frac{2i\bar{c}e^{2A_0(\bar{\zeta})t - 2i\bar{\zeta}x}}{D(x,t)}, \\ r(x,t) = -\frac{2ice^{-2A_0(\zeta)t + 2i\zeta x}}{D(x,t)}, \\ \int_x^\infty q(x,t)r(x,t)dx = \frac{2ic\bar{c}e^{2(A_0(\bar{\zeta}) - A_0(\zeta))t + 2i(\zeta - \bar{\zeta})x}}{(\zeta - \bar{\zeta})D(x,t)}, \end{cases} \tag{2.6.4}$$

with

$$D(x,t) = 1 - \frac{c\bar{c}}{(\zeta - \bar{\zeta})^2}e^{2(A_0(\bar{\zeta}) - A_0(\zeta))t + 2i(\zeta - \bar{\zeta})x}. \tag{2.6.5}$$

The problem is that, if $A_0(\zeta)$, $q(x,0)$, $r(x,0)$ are unrestricted, $D(x,t) = 0$ locates at a certain countable set of points $(x,0)$. At these points, the homogeneous integral equations corresponding to equation (2.6.1) have infinitely many solutions and equation (2.6.1) has no solution. None of these points (x,t) occurs at $t = 0$, since $q(x,0)$, $r(x,0)$ are smooth and decay rapidly as $|x| \to \infty$. However, after a finite time, both $q(x,t)$ and $r(x,t)$ become unbounded at the particular location x of $D(x,t) = 0$. Thus, it is possible for q, r to satisfy equation (2.5.15) initially and to evolve with time in accordance with equation (2.4.30). At a particular location at x, q, r will burst. This kind of "bursting" solitons will not occur in the KdV equation, while the existence of "bursting" solitons represents a major difference between the scattering problem (2.3.14) and the eigenvalue problem (2.3.33). For the eigenvalue problem (2.3.33), whose solution satisfies the constraints

$$\int_{-\infty}^\infty (1 + |x|)|u|dx < \infty, \quad (t = 0), \tag{2.6.6}$$

the solution satisfies the above conditions at any future time for $t > 0$. The occurrence of the "burst" in a physical problem would require a reexamination of the assumptions.

However, if we set

$$r(x,t) = \alpha q^*(x,t), \quad \alpha \text{ being real constant,} \tag{2.6.7}$$

a "burst" will never happen. We deduce $D(x,t) \neq 0$, since the first conserved density $\int_{-\infty}^{\infty} qr dx$ is time-invariant and $\int_{x}^{\infty} qr dx$ is bounded. The global solution for the evolution equation can be obtained from equation (2.6.4).

We note that equation (2.6.7) includes two special cases, $\alpha = 1$, $\alpha = -1$, in which a unique solution for equation (2.6.1) is known to exist, while the necessary and sufficient conditions for the existence of the solution for equation (2.6.1) in a general case have not been confirmed. For simplicity, we assume that a unique solution for equation (2.6.1) exists. The solution for equation (2.6.4) can be written as

$$q(x,t) = i\bar{c}e^{-i\varphi} \operatorname{sech} \theta, \tag{2.6.8}$$

with

$$\begin{aligned} \varphi &= i(A_0(\zeta) + A_0(\bar{\zeta}))t + (\bar{\zeta} + \zeta)x - iy, \\ \theta &= (A_0(\zeta) - A_0(\bar{\zeta}))t + i(\zeta - \bar{\zeta})x + y, \\ e^{2y} &= -\frac{c\bar{c}}{(\zeta - \bar{\zeta})^2}. \end{aligned}$$

Solution (2.6.8) is the basic soliton solution with speed

$$V = \operatorname{Re}\left\{\frac{A_0(\bar{\zeta}) - A_0(\zeta)}{-i(\zeta - \bar{\zeta})}\right\}, \tag{2.6.9}$$

with an amplitude proportional to $(\zeta - \bar{\zeta})$ and a wavelength proportional to $1/(\zeta - \bar{\zeta})$. These waves are basically nonlinear. We take two examples to illustrate this. For the Zakharov–Shabat problem

$$q_t - iq_{xx} - 2iq^2q^* = 0, \quad q|_{t=0} = q_0(x),$$

where $A_0(\zeta) = -2i\zeta^2$, $r = -q^*$, $\bar{c} = c^*$, and $\bar{\zeta} = \zeta^* = \xi - i\eta$, the solution reads

$$q(x,t) = 2\eta e^{[-4i(\xi^2-\eta^2)t-2i\xi x+i\varphi]} \operatorname{sech}[2\eta(x - x_0) + 8\eta\xi t]. \tag{2.6.10}$$

The above soliton is an envelope of oscillating waves, with the amplitude and wavelength depending on η, which can keep the invariant profiles during the propagation with velocity 4ξ. For the sine-Gordon equation

$$u_{xt} = \sin u,$$

the physical variables are

$$X = x + t, \quad T = x - t, \quad u = -\int_{-\infty}^{x} 2q dz,$$

with $A_0(\zeta) = \frac{i}{4\zeta}$, $r = -q$, $\bar{\zeta} = -\zeta = -i\eta$, and $\bar{c} = -c$. The soliton solution reads

$$u(X,T) = 4\tan^{-1}\left\{\exp\left[\left(c\eta + \frac{1}{4\eta}\right)(X - X_0) + \left(\eta - \frac{1}{4\eta}\right)T\right]\right\}, \tag{2.6.11}$$

which is called a kink solution.

For any problems solved by equation (2.3.14), as long as the spectrum is purely discrete and the integral kernel is degenerate, equation (2.6.1) can always be solved. Of course, the conditions for the existence of solution for equation (2.6.1) must be satisfied. The N-soliton solutions for some evolution equations have been obtained in this way. For large time t, N-solitons propagate with different velocities and the asymptotic solutions are the well-separated N-waves in the form of equation (2.6.8). The separation process has been discussed in detail for the case where $r = -q^*$ by Zakharov and Shabat [317]. They pointed out that the asymptotic effect on the soliton interaction is just a phase shift.

What we know from equation (2.6.9) is that solitons corresponding to a locus of eigenvalues share the same speed. Instead of separating each other as $t \to \infty$, a multi-soliton structure forms, which cannot occur in the solution for the KdV equation. In the Zakharov–Shabat problem, the locus which is defined by $\mathrm{Re}(\zeta) = \xi_0$ has been analyzed. For the sine-Gordon equation, the locus is given by $|\zeta| = c_0$ and the solution is given as

$$u(X,T) = 4\tan^{-1}\left[\frac{\eta\cos\{\xi(\eta(T - T_0) - (4 - v)X)\}}{\xi\cosh\{\eta(v(X - X_0) - (4 - v)T)\}}\right],$$

where $v = 2 + (1/2|\zeta|^2)$. We have mentioned that the nonlinear evolution equation which was generated from an arbitrary ratio of entire functions, $A_0(\zeta)$, can be solved by the inverse scattering transform. If $A_0(\zeta)$ has any poles, we can see from equation (2.6.9), when the eigenvalues are near a pole of $A_0(\zeta)$, the corresponding solitons move with extraordinary speed. This can occur in the sine-Gordon equation ($A_0(\zeta) + i/4\zeta$), where these high speeds are close to the speed of light. Of course, these high speeds will have other physical interpretations, but their existence should always be significant.

2. The continuous spectrum

In this section, we consider the contribution from the continuous spectrum to the asymptotic solution for the evolution equation. We start with the simplest possible case, in which the initial data satisfy

$$R(\infty)Q(\infty) = \int_{-\infty}^{\infty} |r| dx \int_{-\infty}^{\infty} |q| dx < 0.817 \tag{2.6.12}$$

and

$$R(\infty)Q(\infty) < 0.383. \tag{2.6.13}$$

Equation (2.6.12) guarantees that there are no discrete eigenvalues, while equation (2.6.13) guarantees the validity of the methods of the inverse scattering transform. The scattering data give

$$F(x,t) = \frac{1}{2\pi}\int_{-\infty}^{\infty} \frac{b(k)}{a(k)} e^{i(kx+2iA_0(k)t)} dk, \tag{2.6.14}$$

$$\bar{F}(x,t) = \frac{1}{2\pi}\int_{-\infty}^{\infty} \frac{\bar{b}(k)}{\bar{a}(k)} e^{-i(kx+2iA_0(k)t)} dk. \tag{2.6.15}$$

As $t \to \infty$, the dominant wave numbers in F and $\bar{F}$ at a particular location x are those whose phase is stationary. We have

$$\chi'(k) = \frac{x}{t} + 2iA_0'(k) = 0, \tag{2.6.16}$$

$$\chi(k) \approx \chi(k_0) + (k-k_0)^2\chi''(k_0). \tag{2.6.17}$$

As a specific example, under $A_0(\zeta) = -2i\zeta^2$, the evolution equations are

$$\begin{cases} iq_t + q_{xx} - 2(qr)q = 0, \\ ir_t - r_{xx} + 2(qr)r = 0. \end{cases} \tag{2.6.18}$$

Then equation (2.6.16) becomes

$$\frac{x}{t} = -8k_0 \tag{2.6.19}$$

and the deformed path passes through (k_0) at an angle of $\frac{\pi}{4}$ for F and $-\frac{\pi}{4}$ for $\bar{F}$, which make the same sign as $\chi''(k_0)$. Asymptotically, as $t \to \infty$, for $\frac{x}{t}$ being fixed, we have

$$\begin{cases} F(x,t) = \dfrac{1}{4\sqrt{\pi t}} \dfrac{b(-\frac{x}{8t})}{a(-\frac{x}{8t})} \exp\left[-\dfrac{i}{16}\left(\dfrac{x}{t}\right)^2 t + \dfrac{\pi}{4}i\right] + O(t^{-\frac{3}{2}}), \\ \bar{F}(x,t) = \dfrac{1}{4\sqrt{\pi t}} \dfrac{\bar{b}(-\frac{x}{8t})}{\bar{a}(-\frac{x}{8t})} \exp\left[\dfrac{i}{16}\left(\dfrac{x}{t}\right)^2 t - \dfrac{\pi}{4}i\right] + O(t^{-\frac{3}{2}}). \end{cases} \tag{2.6.20}$$

The integral equations (2.6.1) can be combined into

$$K_1(x,y;t) - \bar{F}(x+y;t) + \int\int_x^{\infty} K_1(x,z;t)F(z+s;t)\bar{F}(s+y;t)dzds = 0. \tag{2.6.21}$$

There is a similar equation for $\bar{K}_2(x,y;t)$. We seek an approximation solution for equation (2.6.21) in the form of

$$K_1(X,Y;t) = \frac{1}{4\sqrt{\pi t}} f(X,Y) \exp\left[\frac{i}{16}(X+Y)^2 t - \frac{\pi}{4}i\right] + \cdots \tag{2.6.22}$$

with $X = \frac{x}{t}$, $Y = \frac{y}{t}$. Substituting equations (2.6.20) and (2.6.22) into equation (2.6.21) and computing the integral at the stationary point, we have

$$f(X,Y) = \frac{\frac{\bar{b}}{\bar{a}}(-\frac{X+Y}{8})}{1 - \alpha\frac{\bar{b}}{\bar{a}}(-\frac{X+Y}{8})\frac{b}{a}(-\frac{X+Y}{8})}, \tag{2.6.23}$$

where

$$\alpha = \begin{cases} \frac{1}{2}, & \text{if } X \neq Y, \\ \frac{1}{4}, & \text{if } X = Y. \end{cases}$$

Because $q(x,t) = 2K_1(x,x;t)$, we get

$$q(x,t) \sim -\frac{1}{2\sqrt{\pi t}} \frac{\frac{\bar{b}}{\bar{a}}(-\frac{x}{4t})}{1 - \frac{1}{4}\frac{\bar{b}}{\bar{a}}(-\frac{x}{4t})\frac{b}{a}(-\frac{x}{4t})} \exp\left[\frac{i}{4}\left(\frac{x}{t}\right)^2 t - \frac{\pi}{4}i\right], \tag{2.6.24}$$

$$r(x,t) \sim \frac{1}{2\sqrt{\pi t}} \frac{\frac{b}{a}(-\frac{x}{4t})}{1 - \frac{1}{4}\frac{\bar{b}}{\bar{a}}(-\frac{x}{4t})\frac{b}{a}(-\frac{x}{4t})} \exp\left[-\frac{i}{4}\left(\frac{x}{t}\right)^2 t + \frac{\pi}{4}i\right]. \tag{2.6.25}$$

Condition (2.6.13) ensures that the denominator of equation (2.6.23) does not vanish. For the KdV equation, the solution corresponding to equations (2.6.24) and (2.6.25) is not a uniformly valid asymptotic approximation and it is essential to seek the similarity solution. The asymptotic approximation solution for the KdV equation

$$u_t + 6uu_x + u_{xxx} = 0 \tag{2.6.26}$$

can be written as

$$u(x,t) \sim \frac{r_0(\frac{i}{2}\sqrt{\frac{x}{3t}})(\frac{x}{3t})^{\frac{1}{4}} e^{-2(\frac{x}{3t})^{\frac{3}{2}} t}}{2\sqrt{3\pi t}} \left[1 + O\left(\frac{1}{t}\right)\right], \tag{2.6.27}$$

where $r_0(k)$ is the initial reflection coefficient. The similarity solution for equation (2.6.26) reads

$$u = \frac{1}{(3t)^{2/3}}\left[f(\eta) - \frac{1}{(3t)^{1/3}} f_1(\eta) + \frac{1}{(3t)^{2/3}} f_2(\eta) + \cdots\right], \tag{2.6.28}$$

where $f(\eta)$ satisfies the following nonlinear equation:

$$f''' + 6ff' - (2f + \eta f') = 0. \tag{2.6.29}$$

All the other $f_k(\eta)$ satisfy linear equations, with $\eta = \frac{x}{(3t)^{1/3}} = O(1)$. If $|r_0(0)| > 1$, $f(\eta)$ has the second-order pole and is unbounded at finite locations. For $|r_0(0)| < 1$, $f(\eta)$ is oscillating at $\eta \to -\infty$, which is in the form of

$$f(\eta) = 2d(-\eta)^{\frac{1}{4}} \cos\theta - 2d^2(-\eta)^{-\frac{1}{2}}(1 - \cos 2\theta) + O((\eta)^{-\frac{5}{4}}), \tag{2.6.30}$$

where

$$\theta = \frac{2}{3}(-\eta)^{\frac{3}{2}} - 3d^2 \ln(-\eta) + \theta_0 + O((-\eta)^{-\frac{3}{2}})$$

and d and θ_0 are the constants depending on $r_0(0)$. In the case where $|r_0(0)| = 1, f(\eta)$ is in the asymptotic form ($\eta \to -\infty$), i.e.,

$$f(\eta) = \frac{1}{2}\eta - \frac{1}{2}(-2\eta)^{-\frac{1}{2}} + \frac{1}{2}(-2\eta)^{-2} - \frac{5}{2}(-2\eta)^{-\frac{7}{2}} + O((-2\eta)^{-5}). \tag{2.6.31}$$

In this problem, equations (2.6.24) and (2.6.25) are uniformly valid, but we still expect the similarity solution to play a role in the asymptotic development of the solution. The similarity solution is written as

$$q(x,t) = Q_0 t^{-\frac{1}{2}} \exp\left(\frac{i}{4}\frac{x^2}{t} + 2iQ_0R_0 \log t\right), \tag{2.6.32}$$

$$r(x,t) = R_0 t^{-\frac{1}{2}} \exp\left(-\frac{i}{4}\frac{x^2}{t} - 2iQ_0R_0 \log t\right), \tag{2.6.33}$$

where Q_0 and R_0 are constants. Equations (2.6.32) and (2.6.33) will be matched to equations (2.6.24) and (2.6.25). In fact, q or r grows unboundedly as $t \to \infty$ if $|\mathrm{Im}(Q_0, R_0)| > \frac{1}{4}$. This behavior reflects an inherent instability in equation (2.6.18). In the regions where the spatial curvatures (q_{xx}, r_{xx}) are small, equation (2.6.18) is approximated by the simplified equations

$$\begin{cases} iq_t - 2(qr)q = 0, \\ ir_t + 2(qr)r = 0. \end{cases} \tag{2.6.34}$$

When (qr) is a constant and $\mathrm{Im}(qr) \neq 0$, q or r grows exponentially. However, equation (2.6.13) ensures that this instability does not occur and the solution behaves well.

Therefore, if the initial conditions satisfy equations (2.6.12) and (2.6.13), the solution for equation (2.6.18) can be approximated by equations (2.6.24) and (2.6.25). If the initial data are "small", i.e.,

$$R(\infty)Q(\infty) = \int_{-\infty}^{\infty} |r| dx \int_{-\infty}^{\infty} |q| dx \ll 1,$$

and the nonlinear terms in the evolution equation are unimportant, we expect its solution can be well approximated by the solution for the linearized problem.

3. Estimates of the discrete spectrum

The outstanding feature of nonlinear evolution equations which can be solved by the inverse scattering transform is that their solutions achieve comparatively simple asymptotic states as $t \to \pm\infty$. The contribution from the continuous spectrum decays and the dominant asymptotic solution is determined by the discrete spectrum of the scattering problem at $t = 0$. In this section, we derive some simple bounds for the discrete eigenvalues of equation (2.3.14), for example for the location of the zero points of $a(\zeta)$ and $\bar{a}(\zeta)$.

As we know, if $q(x)$ and $r(x)$ are related, the zeros of $\bar{a}(\zeta)$ can be deduced from the zeros of $a(\zeta)$, while, if $q(x)$ and $r(x)$ are independent, the zeros of $\bar{a}(\zeta)$ should be recomputed. We assume that

$$\begin{cases} \int_{-\infty}^{\infty} |x|^n |q(x)| dx < \infty, \\ \int_{-\infty}^{\infty} |x|^n |r(x)| dx < \infty, \end{cases} \tag{2.6.35}$$

for all n, so we conclude that $a(\zeta)$ and $\bar{a}(\zeta)$ are analytic on the whole plane, including the real axis $\operatorname{Im}\zeta = 0$. For simplicity, we write

$$R = \int_{-\infty}^{\infty} |r| dx, \quad Q = \int_{-\infty}^{\infty} |q| dx \tag{2.6.36}$$

and we make the following analysis:

(1) $a(\zeta)$ has only finitely many zeros on $\operatorname{Im}(\zeta) \geq 0$. As has been pointed previously, equation (2.6.35) guarantees that $a(\zeta)$ is analytic for $\operatorname{Im}(\zeta) \geq 0$ and $a(\zeta) \to 1$ as $|\zeta| \to \infty$. It follows that the zeros of $a(\zeta)$ are all isolated and lie in a bounded region. Therefore, $a(\zeta)$ has at most a finite number of zeros here.

(2) $a(\zeta)$ can have zeros on $\operatorname{Im}(\zeta) = 0$. A soliton cannot occur since there is no square-integrable eigenfunction.

(3) Let N be the number of zeros of $a(\zeta)$ with $\operatorname{Im}(\zeta) > 0$, including the nonsimple multiplicity zeros. Assume $|\zeta_0|$ to be the radius of a circle which contains all the zeros of $a(\zeta)$. Making $\xi_+ > |\zeta_0|$ and $\xi_- < -|\zeta_0|$, as $|\xi_\pm| \to \infty$, we have

$$\frac{1}{2\pi}\{\arg(a(\xi_+)) - \arg(a(\xi_-))\} \longrightarrow N. \tag{2.6.37}$$

(4) As aforementioned, if r is proportional to q or q^*, the zeros of $\bar{a}(\zeta)$ are paired with the zeros of $a(\zeta)$. In addition, if $r(x)$ and $q(x)$ are real, the zeros of $a(\zeta)$ itself occur in pairs. This pair of eigenvalues is associated with a special solution, named "breather" or 0π-pulse, which behaves differently from the usual soliton.

(5) If $r(x) = +q^*(x)$, the eigenvalue problem (2.3.14) is self-adjoint. There are no eigenvalues with $\operatorname{Im}(\zeta) > 0$.

(6) For arbitrary r and q, $a(\zeta)$ has no zeros for $\operatorname{Im}(\zeta) \geq 0$ under the condition

$$RQ = \int_{-\infty}^{\infty} |r| dx \int_{-\infty}^{\infty} |q| dx < 0.817, \tag{2.6.38}$$

or more precisely,

$$I_0(2\sqrt{RQ}) < 2. \tag{2.6.39}$$

Equation (2.6.39) implies that, for $\mathrm{Im}(\zeta) \geq 0$,

$$|a(\zeta) - 1| < 1. \tag{2.6.40}$$

In fact, from equations (2.5.8), (2.5.9), and (2.5.13) and

$$\lim_{x\to\infty} \varphi_1(x)e^{i\zeta x} = a(\zeta),$$

we have

$$\begin{aligned} a(\zeta) - 1 &= \int_{-\infty}^{\infty} r(z) \int_{z}^{\infty} q(y)e^{2i\zeta(y-z)} dy(\varphi_1 e^{i\zeta z})dz, \\ |a(\zeta) - 1| &\leq \int_{-\infty}^{\infty} |r(z)| \int_{z}^{\infty} |q(y)|dy|\varphi_1 e^{i\zeta z}|dz, \\ &\leq I_0(2\sqrt{RQ}) - 1. \end{aligned} \tag{2.6.41}$$

Therefore, to prove equation (2.6.40), we only need to set

$$I_0(2\sqrt{RQ}) < 2.$$

(7) The asymptotic method requires not only equations (2.6.38) or (2.6.39), but also

$$\left|\frac{b(\xi)\bar{b}(\xi)}{a(\xi)\bar{a}(\xi)}\right| < 2, \tag{2.6.42}$$

for all real ξ. Thus, we deduce

$$RQ < 0.383.$$

Because $a\bar{a} + b\bar{b} = 1$, equation (2.6.42) can be rewritten as

$$|1 - a(\xi)\bar{a}(\xi)| < 2|a(\xi)\bar{a}(\xi)|.$$

If we let $a(\xi)\bar{a}(\xi) = \alpha + i\beta$, the above relation becomes

$$\left(\alpha + \frac{1}{3}\right)^2 + \beta^2 > \left(\frac{2}{3}\right)^2.$$

Thus we require $|a\bar{a}| > \frac{1}{3}$ and $|a| > \frac{1}{\sqrt{3}}$, $|\bar{a}| > \frac{1}{\sqrt{3}}$.

Finally, we require

$$|a(\xi) - 1| < 1 - \frac{1}{\sqrt{3}}, \quad |\bar{a}(\xi) - 1| < 1 - \frac{1}{\sqrt{3}}$$

and the condition which satisfies both requirements for all real ξ is

$$I_0(2\sqrt{RQ}) < 2 - \frac{1}{\sqrt{3}},$$

i.e.,

$$RQ < 0.383.$$

(8) Approximations of the largest eigenvalues ζ_0 require some additional smoothness of $q(x)$ and $r(x)$ and the first three upper bounds on ζ have been derived.

(a) Let $q(x)$ be continuously differentiable for x and set

$$q'_m \equiv \max_x |q'(x)|, \quad A = \int_{-\infty}^{\infty} |qr|dx, \quad B = I_0(2\sqrt{RQ}). \tag{2.6.43}$$

If

$$|\zeta| > \frac{B}{4}\left[A + \left\{A^2 + \frac{4Rq'_m}{B}\right\}^{\frac{1}{2}}\right] = \zeta_0, \tag{2.6.44}$$

then $a(\zeta) \neq 0$. Thus, all the discrete eigenvalues must lie within a circle whose radius satisfies $|\zeta| \leq \zeta_0$, with ζ_0 determined by equation (2.6.44).

We give the proof that equation (2.6.44) implies $|a(\zeta) - 1| < 1$. From equation (2.6.41), we deduce

$$\begin{aligned} I &= \int_z^{\infty} q(y)e^{2i\zeta(y-z)}dy = \int_0^{\infty} q(z+p)e^{2i\zeta p}dp \\ &= q(z)\int_0^{\infty} e^{2i\zeta p}dp + \int_0^{\infty} q'(z+m)pe^{2i\zeta p}dp, \end{aligned}$$

where $0 < m < p$. We have

$$|I| \leq \frac{|q(z)|}{2|\zeta|} + \frac{q'_m}{4\eta^2}, \tag{2.6.45}$$

where $\eta = \operatorname{Im}\zeta$. Substituting equations (2.5.13) and (2.6.45) into equation (2.6.41), we require

$$I_0(2\sqrt{RQ})\left\{\frac{1}{2|\zeta|}\int_{-\infty}^{\infty} |rq|dx + \frac{1}{4\eta^2}R_{\infty}q'_m\right\} < 1.$$

Using $|\zeta|^2 \geq \eta^2$ and equation (2.6.43) yields

$$|\zeta|^2 > B\left\{\frac{A|\zeta|}{2} + \frac{Rq'_m}{4}\right\},$$

from which equation (2.6.44) follows.

(b) If $q(x) \in c^2$, a better bound than equation (2.6.44) will be obtained. Setting $q''_m = \max_x |q''(x)|$ and with ζ_1 satisfying

$$I_0(2\sqrt{RQ})\left\{\frac{1}{2|\zeta_1|}\int_{-\infty}^{\infty} |rq|dx + \frac{1}{4|\zeta_1|^2}\int_{-\infty}^{\infty} |rq'|dx + \frac{1}{8\eta_1^3}Rq''_m\right\} < 1, \tag{2.6.46}$$

$a(\zeta) \neq 0$ will be derived as $|\zeta| > |\zeta_1|$.

(c) Note that the quantities obtained in equations (2.6.45) and (2.6.46) are related to the polynomial conserved densities of integral rank. Therefore, another bound on ζ_0 will be obtained by virtue of conservation laws directly. If, for some $\zeta_0 > 0$,

$$\sum \frac{|c_n|}{|2\zeta_0|^n} < \infty, \tag{2.6.47}$$

then $a(\zeta) \neq 0$ for $|\zeta| > |\zeta_0|$.

2.7 The mathematical theory foundation of the inverse scattering transform

In the former sections, the sketch of procedures of solving problems by the inverse scattering transform was introduced. However, it is just a process of obtaining the formal solution, where the strictness of mathematics should be examined. For example, for the existence of solution of the eigenvalue problem for the one-dimensional Schrödinger equation in quantum mechanics, could the potential $q(x)$ be determined uniquely by the bounded state and the reflection coefficient? What conditions should a scattering matrix satisfy to make sure $q(x) \in L_2^1$ (denoting $L_2^1 : \{p(x) : \int_{-\infty}^{\infty} |p(x)|(1 + x^2)dx < \infty\}$)? Under what condition does the unique solution for the Gelfand–Levitan–Marchenko integral equation exist? Especially, if we intend to seek the solution for a differential equation via the inverse scattering transform, we need to examine and prove the differentiability of the reflection coefficient, the differentiability of the integral equation's solution, the differentiability of the functions which are constructed by the inverse scattering transform, and whether the functions satisfy differential equations. We call all above questions, which should be answered by theoretical mathematics, "the mathematical theory foundation of the inverse scattering transform". In this part, we only sketch some important results and give a taste of some proofs. More details can be found in [59].

Lemma 2.7.1. *For each k,* $\operatorname{Im} k \geq 0$, *the integral equation*

$$m(x,k) = 1 + \int_{-\infty}^{\infty} D_k(t-x)q(t)m(t,k)dt$$

has a solution $m(x,k)$, *which uniquely solves the Schrödinger equation*

$$m'' + 2ikm' = q(x)m$$

with

$$D_k(y) \equiv \int_0^y e^{2ikt}dt = \frac{1}{2ik}(e^{2iky} - 1),$$

where the boundary condition $m(x,k) \to 1$ as $x \to +\infty$. In addition, $'$ means the derivative respective to x and $m(x,k)$ satisfies $\overline{m(x,k)} = m(x,-\bar{k})$ and obeys the following estimates:

(i)

$$|m(x,k)-1| \le e^{\eta(x)/|k|}\frac{\eta(x)}{|k|} \le e^{\text{constant}/|k|}\frac{\text{constant}}{|k|}, \qquad k \neq 0.$$

(ii)

$$\begin{aligned}|m(x,k)-1| &\le K\frac{(1+\max(-x,0))\int_x^\infty (1+|t|)|q(t)|dt}{1+|k|}\\ &\le K_1\frac{(1+\max(-x,0))}{1+|k|}.\end{aligned} \tag{2.7.1}$$

(iii)

$$\begin{aligned}|m'(x,k)| = \left|\frac{dm(x,k)}{dx}\right| &\le K_2\frac{\int_x^\infty (1+|t|)|q(t)|dt}{1+|k|}\\ &\le \frac{K_3}{1+|k|}, \qquad -\infty < x < \infty.\end{aligned}$$

(iv)

$$|m'(x,k)| \le K_4\frac{\int_x^\infty |q(t)|dt}{1+|k|}, \qquad 0 \le x < \infty,$$

where $\eta(x) = \int_x^\infty |q(t)|dt$ and constants K and k_j depend on

$$\int_{-\infty}^{\infty} (1+|x|^j)|q(x)|dx, \quad j = 0,1,2.$$

For each x, $m(x,k)$ is analytic in $\operatorname{Im} k > 0$ and continuous in $\operatorname{Im} k \ge 0$. In particular, by (ii), $m(x,k) - 1 \in H^{2+}$, where H^{2+} respects the Hardy space of the function $h(k)$, $H^{2+} = \{h(k) \in L^2(-\infty,\infty), \operatorname{supp}\hat{h} \in (-\infty,0)\}$, and

$$\hat{h} = \frac{1}{\pi}\int_{-\infty}^{\infty} e^{2iky}h(k)dk.$$

Finally, $k\dot{m}(x,k)$ is continuous everywhere for all $\operatorname{Im} k \ge 0$, $k \neq 0$, with $\dot{m}(x,k) = \frac{d}{dk}m(x,k)$. If $q(x) \in L_2^1$, then $\dot{m}(x,k)$ also exists and is continuous at $k = 0$, so we have the estimate.

(v)

$$|\dot{m}(x,k)| \le \text{constant}(1+x^2), \quad \forall \operatorname{Im} k \ge 0,\ q \in L_2^1.$$

Proof. The iterates of the Volterra integral equation always converge. We have

$$m(x,k) = 1 + \sum_{n=1}^{\infty} g_n(x,k),$$

where

$$g_n(x,k) = \int_{x\le x_1\le\cdots\le x_n} D_k(x_1-x)\cdots D_k(x_n-x_{n-1})q(x_1)\cdots q(x_n)dx_1\cdots dx_n,$$

$$|g_n(x,k)| \le \int_{x\le x_1\le\cdots\le x_n} \frac{1}{|k|^n}|q(x_1)|\cdots|q(x_n)|dx_1\cdots dx_n = \frac{1}{|k|^n}\frac{(\int_x^{\infty}|g(t)|dt)^n}{n!}.$$

Note that $|D_k(y)| \le \frac{1}{|k|}$, $\operatorname{Im} k \ge 0$ has been used, so the proof of (i) is given.
Alternatively,

$$\begin{aligned}
g_n(x,k) &\le \int_{x\le x_1\le\cdots\le x_n} (x_1-x)(x_2-x_1)\cdots(x_n-x_{n-1})|q(x_1)|\cdots|q(x_n)|dx_1\cdots dx_n,\\
&\le \int_{x\le x_1\le\cdots\le x_n} (x_1-x)(x_2-x)\cdots(x_n-x)|q(x_1)|\cdots|q(x_n)|dx_1\cdots dx_n,\\
&= \frac{(\int_x^{\infty}(t-x)|q(t)|dt)^n}{n!},
\end{aligned}$$

where $|D_k(y)| \le y$, $k \ge 0$, and $y \ge 0$, so we have

$$|m(x,k)-1| \le e^{\gamma(x)}\gamma(x),$$

where

$$\gamma(x) = \int_x^{\infty}(t-x)|q(t)|dt.$$

We know that

$$\begin{aligned}
|m(x,k)| &\le 1 + \int_x^{\infty}(t-x)|q(t)||m(t,k)|dt\\
&= 1 + \int_x^{\infty} t|q(t)||m(t,k)|dt + \int_x^{\infty}(-x)|q(t)||m(t,k)|dt\\
&\le 1 + \int_0^{\infty} t|q(t)||m(t,k)|dt + \int_x^{\infty}(-x)|q(t)||m(t,k)|dt.
\end{aligned}$$

Note that the second inequality holds for x both positive and negative. Also, we have

$$1 + \int_0^{\infty} t|q(t)||m(t,k)|dt \le 1 + (1+e^{\gamma(0)}\gamma(0))\int_0^{\infty} t|q(t)|dt = K < \infty.$$

Setting $M(x,k) = m(x,k)/K(1+|x|)$, $p(x) = (1+|x|)|q(x)| \in L^1$, we get

$$|M(x,k)| \le 1 + \int_x^{\infty} p(t)|M(t,k)|dt,$$

which can be solved by iteration as above to obtain

$$|M(x,k)| \le \exp\left\{\int_x^\infty (1+|t|)|q(t)|dt\right\} \le K_1 < \infty,$$

i.e.,

$$|m(x,k)| \le K_2(1+|x|).$$

As above,

$$\begin{aligned}|m-1| &\le \int_0^\infty t|q(t)||m(t,k)|dt + \int_x^\infty (-x)|q(t)||m(t,k)|dt\\ &\le e^{\gamma(0)}\gamma(0)\int_0^\infty t|q(t)|dt + (-x)K_2\int_x^\infty (1+|t|)|q(t)|dt.\end{aligned}$$

For $x \le 0$, we have

$$|m-1| \le K_3(1+|x|)\int_x^\infty (1+|t|)|q(t)|dt,$$

while, for $x \ge 0$, we have

$$|m-1| \le e^{\gamma(0)}\gamma(x) \le e^{\gamma(0)}\int_x^\infty t|q(t)|dt.$$

Combining with (i), (ii) will be obtained. The estimates (iii) and (iv) will be followed by substituting (ii) in the following equation:

$$m'(x,k) = -\int_x^\infty e^{2ik(t-x)}q(t)m(t,k)dt.$$

A direct calculation implies that m solves the Schrödinger equation uniquely with $m \to 1$ as $x \to +\infty$. The locally uniform convergence of the series for m proves the analyticity in $\operatorname{Im} k > 0$ and the continuity in $\operatorname{Im} k \ge 0$.

Next, we consider the estimate of $\dot{m}(x,k)$. We have

$$\dot{m}(x,k) = \int_x^\infty D_k(t-x)q(t)\dot{m}(t,k)dt + \int_x^\infty \dot{D}_k(t-x)q(t)m(t,k)dt. \tag{2.7.2}$$

For $q \in L_1^1$, via the inequality

$$|k\dot{D}_k(t-x)| = \int_0^{t-x} u\left[\frac{\partial}{\partial u}e^{2iku}\right]du \le 2|t-x|,$$

we have

$$\begin{aligned}\left|\int_x^\infty k\dot{D}_k(t-x)q(t)m(t,k)dt\right| &\le K(1+\max(-x,0))\int_x^\infty (t-x)|q(t)|dt\\ &\le K(x) < \infty.\end{aligned}$$

We see that $\dot{m}(x,k)$ exists ($k \neq 0$, $\operatorname{Im} k \geq 0$) and $k\dot{m}(x,k)$ is continuous even as $k \to 0$. In fact, $\lim_{k\to 0} k\dot{m}(x,k) = 0$. For $q \in L_2^1$, through

$$|\dot{D}_k(t-x)| \leq \left|\int_0^{t-x} 2iue^{2iku}du\right| \leq (t-x)^2,$$

we have

$$\left|\int_x^\infty \dot{D}_k(t-x)q(t)m(t,k)dt\right| \leq \int_x^\infty (t-x)^2|q(t)||m(t,k)|dt.$$

Assuming $x < 0$, we get

$$\begin{aligned}
&\int_x^\infty t^2|q(t)||m(t,k)|dt \\
&= \int_0^\infty t^2|q(t)||m(t,k)|dt + \int_x^0 t^2|q(t)||m(t,k)|dt \\
&\leq \int_0^\infty t^2|q(t)||m(t,k)|dt + x^2\int_x^0 |q(t)||m(t,k)|dt \\
&\leq \text{constant}(1+x^2),
\end{aligned}$$

where the final step is derived from $|m(t,k)| \leq K(1+\max(-t,0))$. If $x \geq 0$, then

$$\int_x^\infty t^2|q(t)||m(t,k)|dt \leq K\int_0^\infty t^2|q(t)|dt.$$

Therefore, for all x,

$$\int_x^\infty t^2|q(t)||m(t,k)|dt \leq K(1 - x\max(-x,0)).$$

Now supposing $x \geq 0$, we have

$$\int_x^\infty (t-x)^2|q(t)||m(t,k)|dt \leq \int_x^\infty t^2|q(t)||m(t,k)|dt \leq K.$$

If $x \leq 0$,

$$\begin{aligned}
&\int_x^\infty (t-x)^2|q(t)||m(t,k)|dt \\
&\leq 2\int_x^\infty t^2|q(t)||m(t,k)|dt + 2x^2\int_x^\infty |q(t)||m(t,k)|dt \\
&\leq 2K(1+x^2) + 2x^2K\int_{-\infty}^\infty |q(t)|(1+|t|)dt \\
&\leq K_1(1+x^2).
\end{aligned}$$

Therefore,

$$\int_x^\infty (t-x)^2|q(t)||m(t,k)|dt \leq K_2(1 - x\max(-x,0)),$$

so that

$$|\dot{m}(x,k)| \le K_2(1 - x\max(-x,0)) + \int_x^\infty (t-x)|q(t)||\dot{m}(t,k)|dt.$$

After iterating, we have

$$|\dot{m}(x,k)| \le K_2(1 - x\max(-x,0))e^{\gamma(x)}.$$

This bound ensures that the iterates of $\dot{m}(x,k)$ in equation (2.7.2) converge uniformly in k, which makes it clear that $\dot{m}(x,k)$ exists and is continuous everywhere in $\operatorname{Im} k \ge 0$, including $k = 0$.

Finally, for any x,

$$\begin{aligned}
|\dot{m}(x,k)| &\le K_2(1+x^2) + \int_x^\infty t|q(t)||\dot{m}(t,k)|dt + (-x)\int_x^\infty |q(t)||\dot{m}(t,k)|dt \\
&\le K_2(1+x^2) + \int_0^\infty t|q(t)||\dot{m}(t,k)|dt + |x|\int_x^\infty |q(t)||\dot{m}(t,k)|dt \\
&\le K_2(1+x^2) + K_2 e^{\gamma(0)}(1+x^2)\int_0^\infty t|q(t)|dt + |x|\int_x^\infty |q(t)||\dot{m}(t,k)|dt,
\end{aligned}$$

i.e.,

$$h(x,k) \le 1 + \int_x^\infty (1+t^2)|q(t)||h(t,k)|dt,$$

with $h = \frac{|\dot{m}(x,k)|}{K_3(1+x^2)}$. Iterating this, we have

$$|h(x,k)| \le \exp\left\{\int_x^\infty (1+t^2)|q(t)|dt\right\},$$

i.e.,

$$|\dot{m}(x,k)| \le K_4(1+x^2),$$

which proves (v) and the lemma. □

In the following, the special characters of zeros of $m(x,k)$ in $\operatorname{Im} k \ge 0$ will be shown.

Lemma 2.7.2. *For any x, $m(x,k)$ has a finite number of zeros in* $\operatorname{Im} k \ge 0$*, all of which are simple and lie on the imaginary k-axis. If $k = i\beta$ ($\beta > 0$) is a zero for $m(x,k)$, then $k^2 = -\beta^2$ is a nondegenerate eigenvalue of the operator $H \equiv -\frac{d^2}{dy^2} + q(y)$ acting in L^2 ($x < y < \infty$) with a Dirichlet boundary condition at $y = x$. For any x, $m(x,k)$ has no zeros for real k except possibly at $k = 0$. If $m(x,0) = 0$, we say that the Dirichlet operator $-\frac{d^2}{dy^2} + q(y)$ on L^2 ($x < y < \infty$) has a virtual level. $k^2 = 0$ is not an eigenvalue of the operator.*

By Lemma 2.7.1, $m - 1 \in H^{2+}$ and

$$m(x,k) = 1 + \int_0^\infty B(x,y)e^{2iky}dy,$$

where $B(x,y) \in L^2$ $(0 < y < \infty)$ for each x. B has many properties which are listed in the following.

Lemma 2.7.3. *The integral equation*

$$B(x,y) = \int_{x+y}^\infty q(t)dt + \int_0^y dz \int_{x+y-z}^\infty dtq(t)B(t,z), \quad y \geq 0,$$

has a real and unique solution $B(x,y)$*, which satisfies*

$$B(x,y) \leq e^{\gamma(x)}\eta(x+y).$$

Especially, $B(x,y) \in L^1 \cap L^\infty$ $(0 < y < \infty)$ *with*

$$\|B(x,\cdot)\|_\infty \leq e^{\gamma(x)}\eta(x), \quad \|B(x,\cdot)\|_1 \leq e^{\gamma(x)}\gamma(x).$$

$B(x,y)$ *is absolutely continuous in* x *and* y *and satisfies*

$$\left|\frac{\partial}{\partial x}B(x,y) + q(x+y)\right| \leq e^{\gamma(x)}\eta(x+y)\eta(x),$$

$$\left|\frac{\partial}{\partial y}B(x,y) + q(x+y)\right| \leq 2e^{\gamma(x)}\eta(x+y)\eta(x).$$

$B(x,y)$ *solves the wave equation*

$$\frac{\partial^2}{\partial x \partial y}B(x,y) - \frac{\partial^2}{\partial x^2}B(x,y) + q(x)B(x,y) = 0, \quad y \geq 0,$$

with boundary condition $-\frac{\partial B(x,0^+)}{\partial x} = -\frac{\partial B(x,0^+)}{\partial y} = q(x)$.

$m(x,k) = 1 + \int_0^\infty B(x,y)e^{2iky}dy$ *is the Jost function in Lemma 2.7.1.*

Proof. The equation is solved by modifying the iteration of Agranovich–Marchenko as follows:

$$B(x,y) = \sum_{n=0}^\infty K_n(x,y),$$

$$K_0(x,y) = \int_{x+y}^\infty q(t)dt,$$

$$K_{n+1}(x,y) = \int_0^y dz \int_{x+y-z}^\infty q(t)K_n(t,z)dt, \quad n = 0,1,2,\ldots.$$

We now show that

$$|K_n(x,y)| \le \frac{\gamma^n(x)}{n!}\eta(x+y), \quad n \ge 0. \tag{2.7.3}$$

Assuming that equation (2.7.3) is valid for n (it is obviously true for $n=0$), we have

$$\begin{aligned}
|K_{n+1}(x,y)| &\le \int_0^y dz \int_{x+y-z}^{\infty} |q(t)|\eta(t+z)\frac{\gamma^n(t)}{n!}dt \\
&\le \eta(x+y)\int_0^y dz \int_{x+y-z}^{\infty} |q(t)|\frac{\gamma^n(t)}{n!}dt \\
&= \eta(x+y)\left(\int_x^{x+y} |q(t)|\frac{\gamma^n(t)}{n!}\left(\int_{x+y-t}^{y} dz\right)dt + \int_{x+y}^{\infty} |q(t)|\frac{\gamma^n(t)}{n!}\left(\int_0^y dz\right)dt\right) \\
&= \eta(x+y)\left(\int_x^{x+y} |q(t)|\frac{\gamma^n(t)}{n!}(t-x)dt + \int_{x+y}^{\infty} |q(t)|\frac{\gamma^n(t)}{n!}ydt\right) \\
&\le \eta(x+y)\int_x^{\infty} |q(t)|(t-x)\frac{\gamma^n(t)}{n!}dt \\
&\le \eta(x+y)\int_x^{\infty} |q(t)|(t-x)\frac{(\int_t^{\infty}(u-x)|q(u)|du)^n}{n!}dt \\
&= \eta(x+y)\frac{\gamma^{n+1}(x)}{(n+1)!}.
\end{aligned}$$

The proof of induction is completed. We obtain $|B(x,y)| \le e^{\gamma(x)}\eta(x+y)$. Especially, $\|B(x,\cdot)\|_\infty \le e^{\gamma(x)}\eta(x)$ and

$$\|B(x,\cdot)\|_1 \le e^{\gamma(x)} \int_0^{\infty} \eta(x+y)dy = e^{\gamma(x)}\gamma(x).$$

It is obvious that B is absolutely continuous in x and y. A quick calculation shows that B solves the wave equation. In addition,

$$\begin{aligned}
&\left|\frac{\partial}{\partial x}B(x,y) + q(x+y)\right| \\
&= \left|-\int_0^y q(x+y-z)B(x+y-z,z)dz\right| \\
&\le \int_0^y |q(x+y-z)|e^{\gamma(x+y-z)}\eta(x+y)dz \\
&\le e^{\gamma(x)}\eta(x+y)\eta(x).
\end{aligned}$$

The calculation for $\frac{\partial B(x,y)}{\partial y}$ is similar. Finally, if we define

$$m(x,k) = 1 + \int_0^{\infty} B(x,y)e^{2iky}dy,$$

the above estimates imply that $m'(x,k)$ exists and

$$m'(x,k) = \int_0^{\infty}\left[\frac{\partial}{\partial x}B(x,y)\right]e^{2iky}dy$$

$$
\begin{aligned}
&= \int_0^\infty \left[\frac{\partial}{\partial x}B(x,y) - \frac{\partial}{\partial y}B(x,y)\right]e^{2iky}dy + \int_0^\infty \left[\frac{\partial}{\partial y}B(x,y)\right]e^{2iky}dy \\
&= -\int_0^\infty \left[\int_x^\infty q(t)B(t,y)dt\right]e^{2iky}dy - B(x,0) - 2ik\int_0^\infty B(x,y)e^{2iky}dy.
\end{aligned}
$$

This in turn implies that $m''(x,k)$ exists almost everywhere and we have

$$m'' + 2ikm' = qm,$$

where $|m(x,k)-1| \le \|B(x,\cdot)\|_1 \le e^{\gamma(x)}\gamma(x) \to 0$ as $(x \to +\infty)$. Therefore, m is the unique Jost function of Lemma 2.7.1. The proof is completed. □

Set $m_1(x,k)$, $m_2(x,k)$ as the Jost functions of Lemma 2.7.1. Let $f_1(x,k) \equiv e^{ikx}m_1(x,k)$, $f_2(x,k) \equiv e^{-ikx}m_2(x,k)$, so $f_1(x,k)$ and $f_2(x,k)$ solve the Schrödinger equation

$$-f_j'' + qf_j = k^2 f_j, \quad j = 1,2,$$

where $f_1 \sim e^{ikx}$ as $x \to +\infty$ and $f_2 \sim e^{-ikx}$ as $x \to -\infty$. Now $f_1(x,k)$ and $f_1(x,-k)$ are two independent solutions for real $k \neq 0$, since the Wronskian

$$
\begin{aligned}
[f_1(x,k), f_1(x,-k)] &\equiv f_1'(x,k)f_1(x,-k) - f_1(x,k)f_1'(x,-k) = \text{constant} \\
&= \lim_{x\to+\infty}(e^{ikx}(ik)e^{-ikx} - e^{ikx}(-ik)e^{-ikx} + o(1)) \\
&= 2ik \neq 0.
\end{aligned}
$$

Similarly, $[f_2(x,k), f_2(x,-k)] = -2ik \neq 0$.

It follows that there are unique transmission coefficients $T_1(k)$, $T_2(k)$ and reflection coefficients $R_1(k)$, $R_2(k)$ satisfying

$$
\begin{aligned}
f_2(x,k) &= \frac{R_1(k)}{T_1(k)}f_1(x,k) + \frac{1}{T_1(k)}f_1(x,-k), \\
f_1(x,k) &= \frac{R_2(k)}{T_2(k)}f_2(x,k) + \frac{1}{T_2(k)}f_2(x,-k),
\end{aligned}
$$

for real $k \neq 0$. For m_1 and m_2, the relations are

$$
\begin{aligned}
T_1(k)m_2(x,k) &= R_1(k)e^{2ikx}m_1(x,k) + m_1(x,-k), \\
T_2(k)m_1(x,k) &= R_2(k)e^{-2ikx}m_2(x,k) + m_2(x,-k).
\end{aligned}
$$

We define the scattering matrix

$$S(k) = \begin{pmatrix} T_1(k) & R_2(k) \\ R_1(k) & T_2(k) \end{pmatrix}, \quad k \neq 0.$$

Then

$$\frac{1}{T_1(k)} = \frac{1}{2ik}[f_1(x,k), f_2(x,k)] = \frac{1}{T_2(k)},$$

$$\frac{R_1(k)}{T_1(k)} = \frac{1}{2ik}[f_2(x,k), f_1(x,k)],$$
$$\frac{R_2(k)}{T_2(k)} = \frac{1}{2ik}[f_2(x,-k), f_1(x,k)],$$

from which we see that

$$T_1(k) = T_2(k) = T(k),$$
$$R_1(k)T_2(-k) + R_2(-k)T_1(k) = 0$$

and

$$\overline{T(k)} = T(-k), \quad \overline{R_1(k)} = R_1(-k), \quad \overline{R_2(k)} = R_2(-k).$$

Inserting the one algebraic relation into the other, we get

$$|T(k)|^2 + |R_1(k)|^2 = 1 = |T(k)|^2 + |R_2(k)|^2.$$

Here, we conclude that $S(k)$ is a unitary matrix for each real $k \neq 0$. Using the properties of $m_1(x,k)$ in Lemma 2.7.1, we have

$$\begin{aligned} m_1(x,k) &= 1 + \int_{-\infty}^{\infty} \left(\frac{e^{2ik(t-x)} - 1}{2ik} \right) q(t) m_1(t,k) dt \\ &= e^{-2ik} \left(\frac{1}{2ik} \int_{-\infty}^{\infty} e^{2ikt} q(t) m_1(t,k) dt \right) \\ &\quad + \left(1 - \frac{1}{2ik} \int_{-\infty}^{\infty} q(t) m_1(t,k) dt \right) + o(1). \end{aligned}$$

On the other hand,

$$\begin{aligned} m_1(x,k) &= \frac{R_2(k)}{T(k)} e^{-2ikx} m_2(x,k) + \frac{1}{T(k)} m_2(x,-k) \\ &= e^{-2ikx} \frac{R_2(k)}{T(k)} + \frac{1}{T(k)} + o(1). \end{aligned}$$

The integral representations for the scattering coefficients are obtained as follows:

$$\frac{R_2(k)}{T(k)} = \frac{1}{2ik} \int_{-\infty}^{\infty} e^{2ikx} q(t) m_1(t,k) dt,$$
$$\frac{1}{T(k)} = 1 - \frac{1}{2ik} \int_{-\infty}^{\infty} q(t) m_1(t,k) dt.$$

The main properties of scattering matrix S can be displayed by the following theorem.

Theorem 2.7.4. *Let $q(x)$ be a real potential in L_1^1. Then*

$$S(k) = \begin{pmatrix} T_1(k) & R_2(k) \\ R_1(k) & T_2(k) \end{pmatrix}$$

is continuous for all real $k \neq 0$, while $S(k)$ is also continuous at $k = 0$ if $q(x) \in L_2^2$. $S(k)$ has the following properties:

(I) *Symmetry. We have*

$$T_1(k) = T_2(k) \equiv T(k).$$

(II) *Unitarity. We have*

$$\begin{aligned}&T(k)\overline{R_2(k)} + R_1(k)\overline{T(k)} = 0,\\ &|T(k)|^2 + |R_1(k)|^2 = 1 = |T(k)|^2 + |R_2(k)|^2,\end{aligned}$$

so that

$$|T(R)|, \quad |R_j(k)| \le 1, \quad j = 1, 2.$$

(III) *Analyticity.* $T(k)$ *is meromorphic in* $\operatorname{Im} K > 0$ *with a finite number of simple poles* $i\beta_1, \dots, i\beta_n$ *on the imaginary axis, while* $\beta_j > 0$*, whose residues are*

$$i\left(\int_{-\infty}^{\infty} f_1(x, i\beta_j) f_2(x, i\beta_j)dx\right)^{-1}, \quad j = 1, 2, \dots, n.$$

The numbers $-\beta_1^2, \dots, -\beta_n^2$ *are the simple eigenvalues of operator* H. $T(k)$ *is continuous in* $\operatorname{Im} k \ge 0$, $k \ne 0$, $i\beta_1, \dots, i\beta_n$*. If* $q(x) \in L_2^1$*, then* $T(k)$ *is continuous in* $\operatorname{Im} k \ge 0$, $k \ne i\beta_1, \dots, i\beta_n$.

(IV) *Asymptotics. We have:*

(i)

$$T(k) = 1 + O\left(\frac{1}{k}\right) \quad \text{as } |k| \to \infty, \ \operatorname{Im} k \ge 0.$$

(ii)

$$R_j(k) = O\left(\frac{1}{k}\right), \quad j = 1, 2, \quad \text{as } |k| \to \infty, \ k \text{ is real.}$$

Moreover, if $q(x)$ *has N-order derivatives which are in* L^1 $(-\infty < x < \infty)$*, then* $R_j(k) = O(1/k^{N+1})$ *as* $|k| \to \infty$*, k is real.*

(iii) *If H has no eigenvalues, then*

$$T(k) - 1 \in H^{2+}, \quad |T(k)| \le 1, \quad \text{everywhere in } \operatorname{Im} k \ge 0.$$

(V) *Rate at* $k = 0$*. We have*

$$|T(k)| > 0 \quad \text{for all } \operatorname{Im} k \ge 0, \ k \ne 0, \ |k| \le \text{constant}|T(k)|, \ \text{as } k \to 0.$$

If $q(x) \in L_2^2$*, there are two possibilities:*

(i) $0 < \text{constant} \le |T(k)|$ *and hence* $R_j(k) \le \text{constant} < 1, j = 1, 2,$ *or*

(ii) $T(k) = \alpha k + o(k)$, $\alpha \ne 0$*, as* $k \to 0$, $\operatorname{Im} k \ge 0$*, and* $1 + R_j(k) = \alpha_j k + o(k)$, $j = 1, 2$*, as* $k \to 0$*, k is real.*

(VI) *Realness. We have*

$$\overline{T(k)} = T(-k), \quad \overline{R_j(k)} = R_j(-k), \quad j = 1, 2.$$

Theorem 2.7.5.

$$\frac{R(k)}{T(k)} = \frac{1}{2ik}\int_{-\infty}^{\infty} e^{-2ikt}\Pi_1(t)dt,$$
$$\frac{1}{T} = 1 - \frac{1}{2ik}\int_{-\infty}^{\infty} q(t)dt - \frac{1}{2ik}\int_{0}^{\infty} \Pi_2(t)e^{2ikt}dt,$$

where

$$|\Pi_1(y)| \le |q(y)| + KL(y) \in L^1 \ (-\infty < y < \infty),$$
$$|\Pi_2(y)| \le K\left(\int_{y/2}^{\infty} |q(t)|dt + \int_{-\infty}^{y/2} |q(t)|dt\right) \in L^1 \ (0 < y < \infty),$$

with

$$L(y) = \begin{cases} \int_y^{\infty} |q(t)|dt, & y \ge 0, \\ \int_{-\infty}^{y} |q(t)|dt, & y < 0. \end{cases}$$

Theorem 2.7.6. *The asymptotics for $T(k), m_1(x,k), m_2(x,k)$ are displayed as follows:*
(i) *If $q(x) \in L_1^1$,*

$$m_1(x,k) = 1 + \frac{1}{2ik}\int_{x}^{\infty} (e^{2ik(t-x)} - 1)q(t)dt + \frac{1}{2(2ik)^2}\left(\int_{x}^{\infty} q(t)dt\right)^2 + o\left(\frac{1}{k^2}\right),$$
$$m_2(x,k) = 1 + \frac{1}{2ik}\int_{-\infty}^{x} (e^{2ik(x-t)} - 1)q(t)dt + \frac{1}{2(2ik)^2}\left(\int_{x}^{\infty} q(t)dt\right)^2 + o\left(\frac{1}{k^2}\right),$$
$$T(k) = 1 + \frac{1}{2ik}\int_{-\infty}^{\infty} q(t)dt + \frac{1}{2(2ik)^2}\left(\int_{-\infty}^{\infty} q(t)dt\right)^2 + o\left(\frac{1}{k^2}\right).$$

(ii) *If $q(x) \in L_1^1$, $q'(x) \in L^1$,*

$$m_1(x,k) = 1 - \frac{1}{2ik}\int_{x}^{\infty} q(t)dt + \frac{1}{2(2ik)^2}\left(\int_{x}^{\infty} q(t)dt\right)^2 - \frac{q(x)}{(2ik)^2} + o\left(\frac{1}{k^2}\right),$$
$$m_2(x,k) = 1 - \frac{1}{2ik}\int_{-\infty}^{x} q(t)dt + \frac{1}{2(2ik)^2}\left(\int_{-\infty}^{x} q(t)dt\right)^2 - \frac{q(x)}{(2ik)^2} + o\left(\frac{1}{k^2}\right).$$

The proof is straightforward.

Theorem 2.7.7 will display some important expressions and estimates of potential $q(x)$ and scattering data.

Theorem 2.7.7. *Assume $q \in L_1^1$ with bounded states $-\beta_n^2 < \cdots < -\beta_1^2$, norming constants $c_j, j = 1, 2, \ldots, n$, and the reflection coefficient R satisfies:*

(i)

$$q(x) = \lim_{a\to\infty} \frac{2i}{\pi}\int_{-a}^{a} kR(k)e^{2ikx}m^2(x,k)dk + \sum_{j=1}^{n}(2c_j\exp(-2\beta_j x))'m^2(x,i\beta_j)$$
$$= \lim_{b\to\infty}\frac{1}{b}\int_0^b da\left(\frac{2i}{\pi}\int_{-a}^{a} kR(k)e^{2ikx}m^2(x,k)dk\right)$$
$$+\sum_{j=1}^{n}(2c_j\exp(-i\beta_j x))'m^2(x,i\beta_j),$$

where the convergence of the Cesàro means is almost everywhere.

(ii)

$$q(x) = F'(x) + 2\int_0^\infty F'(x+t)B(x,t)dt + \int_0^\infty F'(x+t)(B_x * B_x)(t)dt$$
$$+\sum_{j=1}^{n}(2c_j\exp(-2\beta_j x))'m_1^2(x,i\beta_j)$$
$$= \Omega'(x) + 2\int_0^\infty \Omega'(x+t)B(x,t)dt + \int_0^\infty \Omega'(x+t)(B_x * B_x)(t)dt,$$

where

$$\Omega(t) = F(t) + \sum_{j=1}^{n} 2c_j\exp(-2\beta_j t).$$

(iii) *If* $kR(k)\in L^1$,

$$q(x) = \frac{2i}{\pi}\int_{-\infty}^{\infty} kR(k)e^{2ikx}m^2(x,k)dk + \sum_{j=1}^{n}(2c_j\exp(-2\beta_j x))'m^2(x,i\beta_j).$$

Theorem 2.7.8. *Let* $q(x)\in L_1^1$ *and*

$$\Omega(y) = F(y) + \sum_{j=1}^{n} 2c_j\exp(-2\beta_j y).$$

$\Omega(y)$ *and* $F(y)$ *are absolutely continuous with*

$$|q(x) - \Omega'(x)| \le K_1(x)\left(\int_x^\infty |q(t)|dt\right)^2,$$

where $K_1(x)$ *is nonincreasing,*

$$\int_a^\infty |F'(t)|(1+|t|)dt \le K_2(a) < \infty,$$
$$\int_a^\infty |F(t)|dt \le K_3(a) < \infty,$$

for all x *and* a.

Theorem 2.7.9. *The necessity and sufficiency of the conditions for the matrix*

$$\begin{pmatrix} T_1(k) & R_2(k) \\ R_1(k) & T_2(k) \end{pmatrix}, \quad -\infty < k < \infty,$$

being the scattering matrix of real potential $q(x) \in L_2^1$, *are:*

(i) *Symmetry. We have* $T_1(k) = T_2(k) = T(k)$.

(ii) *Unitarity. We have*

$$|T(k)|^2 + |R_1(k)|^2 = |T(k)|^2 + |R_2(k)|^2 = 1,$$
$$R_1(k)\overline{T(k)} + \overline{R_2(k)}T(k) = 0.$$

(iii) *Analyticity.* $T(k)$ *is analytic in the open upper half plane and continuous down to the axis.*

(iv) *Asymptotics. We have*

$$T(k) = 1 + O\left(\frac{1}{|k|}\right), \quad \operatorname{Im} k \geq 0,$$

$$R_i(k) = O\left(\frac{1}{|k|}\right), \quad k \text{ is real}, \ i = 1, 2.$$

(v) *Rate at* $k = 0$. $|T(k)| > 0$, $\operatorname{Im} k \geq 0$, $k \neq 0$ *and either*
   (1) $0 < c < |T(k)|$ *for all* $\operatorname{Im} k \geq 0$, *or*
   (2) $T(k) = T(0)k + o(k)$, $T(0) \neq 0$, $\operatorname{Im} k \geq 0$, $1 + R_1(k) = \rho_i k + o(k)$, $i = 1, 2$, k *is real.*

(vi) *Reality. We have*

$$T_j(k) = \overline{T_j(-k)}, \quad R_j(k) = \overline{R_j(-k)}, \quad j = 1, 2.$$

(vii) *We have*

$$F_j(y) = \frac{1}{\pi} \int_{-\infty}^{\infty} R_j(k) e^{2iky} dk, \quad j = 1, 2,$$

which are absolutely continuous with

$$\int_a^{\infty} |F_1'(t)|(1 + t^2) dt < \infty,$$
$$\int_{-\infty}^{a} |F_2'(t)|(1 + t^2) dt \leq c(a) < \infty,$$

for all $-\infty < a < \infty$.

For the KdV equation, the soliton solution can be constructed via the inverse scattering transform. In addition, the smoothness and decay of the solution corresponding to the initial function as $|x| \to \infty$ will be obtained.

Consider the initial value problem of the KdV equation

$$u_t - 6uu_x + u_{xxx} = 0,$$
$$u|_{t=0} = U(x).$$

Let the initial function $U(x)$ satisfy:
(i) $U(x) \in c^s(R)$, $s > 3$,
(ii) $U^{s+1}(x)$ is piecewise continuous,
(iii) $U^j(x) = o(|x|^{-N})$ for certain $N > 0, j \leq s+1$.

We solve $B_\pm(x,y,t)$ from the Marchenko equation

$$B_\pm(x,y,t) \pm \int_0^{\pm\infty} \Omega_\pm(x+y+z,t)B_\pm(x,z,t)dz + \Omega_\pm(x+y,t) = 0, \tag{2.7.4}$$

where

$$\Omega_\pm(x,t) = F_\pm(x,t) + 2\sum_{j=0} c_j(t)e^{\pm 2\eta_j x},$$
$$F_\pm(x,t) = \frac{1}{\pi}\int_{-\infty}^{\infty} R_\pm(\xi,t)d\xi = e^{\pm 2i\xi t}d\xi.$$

Therefore, let

$$u(x,t) = -B_+^{1,0,0}(x,0,t) = B_-^{1,0,0}(x,0,t)$$

be the solution for the initial value problem of the KdV equation, with

$$B^{(j,k,l)}(x,y,t) = \partial_x^j\partial_y^k\partial_t^l B = \left(\frac{\partial}{\partial x}\right)^j\left(\frac{\partial}{\partial y}\right)^k\left(\frac{\partial}{\partial t}\right)^l B.$$

The existence and smoothness of the solution for the initial value problem will be given by the following two theorems.

Theorem 2.7.10. *For fixed x, t,*

$$+\int_x^{+\infty} (1+|s|)|\Omega_\pm^{1,0}(s,t)|ds < \infty, \quad \textit{for } x \in R, \tag{2.7.5a}$$
$$-\int_x^{+\infty} (1+|s|)|\Omega_\pm^{1,0}(s,t)|ds < \infty, \quad \textit{for } x \in R, \tag{2.7.5b}$$

where either equation (2.7.5a) or (2.7.5b) will be satisfied. The corresponding equation (2.7.4) has the solution $B_\pm(x,y,t)$, which satisfies

$$\pm\int_x^{\infty} (1+|x|)|B_\pm^{1,0,0}(x,0,t)|dx < \infty, \quad \textit{for } x \in R.$$

Theorem 2.7.11. *For fixed t, conditions*

$$\pm \int_x^{+\infty} (1+s^2)|\Omega_\pm(s,t)|ds < \infty, \quad \text{for } x \in R \tag{2.7.6}$$

and

$$\begin{aligned} R_+(\xi) &= -1 + A\xi + o(\xi) \quad \text{if } \lim_{\xi\to 0} \xi a_+(\xi) \neq 0, \\ T(\xi) &= \alpha\xi + o(\xi), \quad \alpha \neq 0, \text{ as } \xi \to 0 \end{aligned} \tag{2.7.7}$$

are satisfied, so the solution for equation (2.7.4) fulfills

$$-B_+^{1,0,0}(x,0,t) = B_-^{1,0,0}(x,0,t).$$

Moreover, if u is defined as

$$u(x,t) = -B_+^{1,0,0}(x,0,t) = B_-^{1,0,0}(x,0,t),$$

then the related Schrödinger equation

$$L_u\psi \equiv \psi_{xx} + u(x,t)\psi = \zeta^2\psi$$

has the scattering data (2.4.12).

Theorems 2.7.10 and 2.7.11 can be verified via the integral representation of reflection coefficients $R_\pm(\xi)$ about the initial function $U(x)$. From Theorems 2.7.10 and 2.7.11, we obtain the solution $u(x,t)$ for the initial value problem of the KdV equation. $u(x,t)$ satisfies

$$\int_{-\infty}^{\infty} (1+|x|^2)|u(x,t)|dx < \infty, \quad \text{for all } t \in R.$$

We have the following theorem.

Theorem 2.7.12. (a) *If $j + 3l \leq 2[N] - 6 - \mu$, then solution $u(x,t)$ for the initial value problem exists. In addition, $u^{(j,l)}(x,t)$ exists at $t \neq 0$.*

(b) $u^{(j,0)}(x,t) \to U^{(j)}(x)$ *as* $t \to 0, j = 0,1,2.$

(c) *For* $t > 0$,

$$u^{(j,0)}(x,t) = \begin{cases} O(|x|^{[\frac{j}{2}]+3-[N]+\frac{\mu}{2}}), & x \to +\infty, j \leq 2[N] - 6 - \mu, \\ O(|x|^{-\frac{1}{2}(4-j)-\delta}), & x \to -\infty, j \leq 2. \end{cases}$$

(d) *For* $t < 0$,

$$u^{(j,0)}(x,t) = \begin{cases} O(|x|^{-\frac{1}{2}(4-j)-\delta}), & x \to +\infty, j \leq 2, \\ O(|x|^{[\frac{j}{2}]+3-[N]+\frac{\mu}{2}}), & x \to -\infty, j \leq 2[N] - 6 - \mu, \end{cases}$$

where $\delta = \frac{1}{16}$, $N > 6 + \frac{\mu}{2}$,

$$\mu = \begin{cases} 0, & \textit{Jost functions } f_\pm(x,0) \textit{ are linearly dependent,} \\ 2, & \textit{Jost functions } f_\pm(x,0) \textit{ are linearly independent.} \end{cases}$$

From Theorem 2.7.12, we see that, if $t < 0$, u decays fast as $x \to +\infty$ and slow as $x \to -\infty$.

2.8 Higher-order and multi-dimensional inverse scattering problems

In Section 2.3, we considered the second-order inverse scattering problem, which can be written in the form of the following matrix:

$$V = \begin{pmatrix} V_1 \\ V_2 \end{pmatrix},$$

$$V_x = i\zeta \begin{pmatrix} -1 & 0 \\ 0 & 1 \end{pmatrix} V + \begin{pmatrix} 0 & q \\ r & 0 \end{pmatrix} V, \tag{2.8.1}$$

$$V_t = QV, \quad Q = \begin{pmatrix} A & B \\ C & -A \end{pmatrix}. \tag{2.8.2}$$

Now we consider the higher-order inverse scattering problem,

$$V = \begin{pmatrix} V_1 \\ \vdots \\ V_n \end{pmatrix},$$

$$\begin{cases} V_x = i\zeta DV + NV, \\ V_t = QV, \end{cases} \tag{2.8.3}$$

where $D = (d_j\delta_{ij})$, $N_{ii} = 0$, and d_i is a constant. By virtue of $V_{xt} = V_{tx}$ and $\zeta_t = 0$, we have

$$Q_x = N_t + i\zeta(DQ - QD) + (NQ - QN), \tag{2.8.4}$$

or

$$Q_x = N_t + i\zeta[D, Q] + [N, Q]. \tag{2.8.5}$$

We aim to find Q to satisfy equation (2.8.5). Under equation (2.8.5), the two equations of (2.8.3) are compatible, from which we can derive the nonlinear evolution equation. Expand Q as

$$Q = Q^{(1)}\zeta + Q^{(0)}. \tag{2.8.6}$$

Substituting the above expansion into equation (2.8.5), we have

$$Q_x^{(1)}\zeta + Q_x^{(0)} = N_t + [N, Q^{(0)}] + i\zeta\{[D, Q^{(0)}] + [N, Q^{(0)}]\} + i\zeta^2[D, Q^{(1)}].$$

Collecting the coefficient of ζ^2, we have

$$i[D, Q^{(1)}] = 0,$$

or

$$\sum_k (D_{ik}Q_{kj}^{(1)} - Q_{ik}^{(1)}D_{kj}) = 0.$$

From $D_{ik} = \delta_{ik}d_i$, we get

$$(d_i - d_j)Q_{ij}^{(1)} = 0,$$

so

$$Q_{ij}^{(1)} = q_i\delta_{ij}. \tag{2.8.7}$$

Setting q_i = constant and comparing the coefficient of ζ, we have

$$Q_x^{(1)} = i[D, Q^{(0)}] + [N, Q^{(1)}].$$

Therefore,

$$\begin{aligned}
&Q_{ijx}^{(1)} = i\sum_k (D_{ik}Q_{kj}^{(0)} - Q_{ik}^{(0)}D_{kj}) + \sum_k (N_{ik}Q_{kj}^{(1)} - Q_{ik}^{(1)}N_{kj}),\\
&i(d_i - d_j)Q_{ij}^{(0)} + (q_j - q_i)N_{ij} = 0,
\end{aligned}$$

or

$$Q_{ij}^{(0)} = \frac{q_i - q_j}{i(d_i - d_j)}N_{ij}, \quad i \neq j, \tag{2.8.8}$$

and $Q_{ij}^{(0)} = 0$ if $i = j$.

Definition. Assuming $a_{ij} = \frac{1}{i}\frac{q_i - q_j}{d_i - d_j} = a_{ji}$, we have

$$Q_{ij}^{(0)} = a_{ij}N_{ij}, \quad i \neq j. \tag{2.8.9}$$

Comparing the coefficient of $\zeta^{(0)}$, because $Q_x^{(0)} = N_t + [N, Q^{(0)}]$, we get

$$a_{ij}N_{ij,x} = N_{ij,t} + \sum_k (N_{ik}a_{kj}N_{kj} - a_{ik}N_{ik}N_{kj}).$$

Therefore, $N(N-1)$ numbers of evolution equations are obtained. We have

$$N_{ij,t} = a_{ij}N_{ij,x} + \sum_k (a_{ik} - a_{kj})N_{ik}N_{kj}, \tag{2.8.10}$$

or

$$N_{ij,t} = a_{ij}N_{ij,x} + \sum_{k\neq i,j} (a_{ik} - a_{kj})N_{ik}N_{kj}.$$

Making $N_{ij} = \sigma_{ij}N_{ji}^* \ (i > j)$, equation (2.8.10) and its conjugation are compatible if $\sigma_{ik}\sigma_{kj} = -\sigma_{ij} \ (i > k > j)$ and a_{ij} is real. Thus, the number of equations will decrease. In fact,

$$N_{ij,t}^* = a_{ji}N_{ji,x}^* + \sum_{k\neq j,i} (a_{ij} - a_{ki})N_{jk}^*N_{ki}^*.$$

Multiplying by σ_{ij}, because $a_{ij} = a_{ji}$, we get

$$N_{ij,t} = a_{ij}N_{ij,x} + \sum_{k\neq j,i} (a_{jk} - a_{ki})\sigma_{ik}N_{jk}^*N_{ki}^*.$$

Considering $\sigma_{ij} = -\sigma_{ik}\sigma_{kj}$, we have

$$N_{ij,t} = a_{ij}N_{ij,x} + \sum_{k\neq i,j} (a_{ik} - a_{kj})N_{ik}N_{kj}.$$

In another form,

$$\begin{aligned} N_{ij,t} = a_{ij}N_{ij,x} &+ \sum_{k>j>i} (a_{ik} - a_{kj})N_{ik}\sigma_{kj}N_{jk}^* \\ &+ \sum_{j>k>i} N_{ij}N_{jk}(a_{ik} - a_{kj}) + \sum_{j>i>k} \sigma_{ik}N_{ki}^*N_{kj}(a_{ik} - a_{kj}). \end{aligned}$$

Example 2.8.1. If $n = 3$,

$$N = \begin{pmatrix} 0 & N_{12} & N_{13} \\ \sigma_{21}N_{12}^* & 0 & N_{23} \\ \sigma_{31}N_{13}^* & \sigma_{32}N_{23}^* & 0 \end{pmatrix} = \begin{pmatrix} 0 & A_1 & A_2 \\ \sigma_{21}A_1^* & 0 & A_3 \\ \sigma_{31}A_2^* & \sigma_{32}A_3^* & 0 \end{pmatrix}.$$

Denoting $a_{12} = V_1$, $a_{13} = V_2$, $a_{23} = V_3$, using equation (2.8.10), we derive

$$\begin{cases} A_{1,t} = V_1A_{1,x} + \sigma_{32}(V_2 - V_3)A_2A_3^*, \\ A_{2,t} = V_2A_{2,x} + (V_1 - V_3)A_1A_3, \\ A_{3,t} = V_3A_{3,x} + \sigma_{21}(V_1 - V_2)A_1^*A_2, \end{cases} \tag{2.8.11}$$

where $-\sigma_{31} = \sigma_{21}\sigma_{32}$. Equations (2.8.11) are named three-wave equations. For simplicity, introducing $A_1 = ia_1u_1, A_2 = -ia_2u_2, A_3 = ia_3u_3$, and

$$a_1^2 = \frac{q^2}{(V_1 - V_3)(V_1 - V_2)}, \quad a_2^2 = \frac{q^2}{(V_2 - V_3)(V_1 - V_2)}, \quad a_3^2 = \frac{a_1q}{a_2(V_2 - V_3)},$$

equations (2.8.11) become

$$\begin{cases} u_{1,t} = V_1 u_{1,x} + \sigma_{32} iqu_2 u_3^*, \\ u_{2,t} = V_2 u_{2,x} - iqu_1 u_3, \\ u_{3,t} = V_3 u_{3,x} + \sigma_{21} iqu_1^* u_2. \end{cases} \tag{2.8.12}$$

We expand Q to ζ^2 and obtain

$$Q = Q^{(2)}\zeta^2 + Q^{(1)}\zeta + Q^{(0)}.$$

We insert this into

$$Q_x = N_t + i\zeta[D,Q] + [N,Q]$$

and the following relations will be obtained:

$$\begin{aligned} &\beta_{ij} N_{ij,xx} + \varepsilon_{ij} N_{ij,x} - \sum_{k\neq i,j} \gamma_{ijk}(N_{ik}N_{kj})_x \\ &= N_{ij,t} + \sum_{k\neq i,j} (\varepsilon_{kj} - \varepsilon_{ik}) N_{ik} N_{kj} + N_{ij}\Big\{ 2\beta_{ij} N_{ik} N_{kj} + \sum_{k\neq i,j} (\beta_{kj} + \gamma_{ikj}) N_{jk} N_{kj} \\ &\quad - (\beta_{ki} + \gamma_{kji}) N_{ik} N_{ki} \Big\} + \sum_{k\neq i,j} (\beta_{kj} N_{ik} N_{kj,x} - \beta_{ik} N_{ki} N_{ik,x}) \\ &\quad + \sum_{k\neq i,j} \sum_{m\neq i,j} (\gamma_{ikm} N_{kj} N_{im} N_{mk} - \gamma_{kjm} N_{ik} N_{km} N_{mj}), \end{aligned} \tag{2.8.13}$$

with

$$\begin{aligned} a_{ij} &= \frac{q_i^{(2)} - q_j^{(2)}}{i(d_i - d_j)} = a_{ji}, \\ \beta_{ij} &= \frac{d_{ij}}{i(d_i - d_j)} = -\beta_{ji}, \\ \gamma_{ijk} &= \frac{a_{kj} - a_{ik}}{i(d_i - d_j)} = \gamma_{jik} = \gamma_{kij}, \\ \varepsilon_{ij} &= \frac{q_i^{(1)} - q_j^{(1)}}{i(d_i - d_j)} = \varepsilon_{ji}. \end{aligned}$$

In fact, collecting the coefficients of $\zeta^3, \zeta^2, \zeta^1, \zeta^0$, we have

$$\begin{aligned} &i[D, Q^{(2)}] = 0, \quad Q_x^{(1)} = i[D, Q^{(0)}] + [N, Q^{(1)}], \\ &Q_x^{(2)} = i[D, Q^{(1)}] + [N, Q^{(2)}], \\ &Q_x^{(0)} = N_t + [N, Q^{(0)}], \quad Q_{ik}^{(2)} = q_i^{(2)} \delta_{ik}, \quad D_{ik} = d_i \delta_{ik}. \end{aligned}$$

From $Q_x^{(2)} = i[D, Q^{(1)}] + [N, Q^{(2)}]$, i.e.,

$$i\sum_k (D_{ik}Q_{kj}^{(1)} - Q_{ik}^{(1)}D_{kj}) + \sum_k (N_{ik}Q_{kj}^{(2)} - Q_{ik}^{(2)}N_{kj}) = 0,$$

we have

$$Q_{ij}^{(1)} = \frac{q_i^{(2)} - q_j^{(2)}}{i(d_i - d_j)} N_{ij} = a_{ij}N_{ij}.$$

From

$$\begin{aligned} &Q_x^{(1)} = i[D, Q^{(0)}] + [N, Q^{(1)}], \\ &a_{ij}N_{ij,x} = i(d_i - d_j)Q_{ij}^{(0)} + \sum_{k\neq i,j} (a_{kj} - a_{ik})N_{ik}N_{kj} + (Q_{ji}^{(1)} - Q_{ii}^{(1)})N_{ij}, \end{aligned}$$

we derive

$$\begin{aligned} &Q_{ij}^{(0)} = \frac{a_{ij}}{i(d_i - d_j)} N_{ij,x} + \sum_{k\neq i,j} \frac{(a_{ik} - a_{kj})}{i(d_i - d_j)} N_{ik}N_{kj} - \frac{Q_{ij}^{(1)} - Q_{ii}^{(1)}}{i(d_i - d_j)}, \\ &N_{ij} = \beta_{ij}N_{ij,x} - \sum_{k\neq i,j} \gamma_{ijk}N_{ik}N_{kj} + \varepsilon_{ij}N_{ij}. \end{aligned}$$

Substituting the above into $Q_x^{(0)} = N_t + [N, Q^{(0)}]$, equation (2.8.13) will be obtained.

Example 2.8.2. For the Boussinesq equation

$$W_{tt} - W_{xx} - 6(W^2)_{xx} + W_{xxxx} = 0, \tag{2.8.14}$$

$$N = \begin{pmatrix} 0 & 0 & 1 \\ N_{21} & 0 & (1 + W_3)N_{31} \\ N_{31} & 1 & 0 \end{pmatrix}, \tag{2.8.15}$$

where $W_3 = e^{-2\pi i/3}$, whose eigenvalue problem is

$$\Psi_{xxx} + (\lambda + Q_1)\Psi + Q_2\Psi_x = 0, \tag{2.8.16}$$

with $Q_1 = N_{31,x} + N_{21}$, $Q_2 = (2 + W_3)N_{31}$.

Example 2.8.3. If we take

$$N = \begin{pmatrix} 0 & A & iB \\ 0 & 0 & A^* \\ -i & 0 & 0 \end{pmatrix}, \tag{2.8.17}$$

the equations which describe the interaction between the infragravity and capillary waves in shallow water will be obtained. We have

$$\begin{cases} iA_t + \lambda A_{xx} = AB, \\ B_t = -\alpha(|A|^2)_x. \end{cases} \tag{2.8.18}$$

Next, we consider the inverse scattering transform for higher dimensions. The eigenvalue problems for the y-variable added are

$$\frac{\partial v}{\partial x}(x,y,t) = i\zeta d(y)v(x,y,t) + \int_{-\infty}^{\infty} N(x,y,z;t)v(x,z,t)dz, \tag{2.8.19}$$

$$\frac{\partial v}{\partial t}(x,y,t) = \int_{-\infty}^{\infty} Q(x,y,z;t)v(x,y,z;t)dz. \tag{2.8.20}$$

Because $v_{xt} = v_{tx}$ and $\zeta_t = 0$, we have

$$\begin{aligned} Q_x(x,y,z;t) &= N_t(x,y,z;t) + i(d(y) - d(z))Q(x,y,z;t) \\ &\quad + \int_{-\infty}^{\infty} [Q(x,z',z;t)N(x,y,z';t) - N(x,z',z;t)Q(x,y,z';t)]dz'. \end{aligned} \tag{2.8.21}$$

Expanding Q as $Q = Q^{(1)} + \zeta Q^{(0)}$, the following integral-differential equation will be derived:

$$\begin{aligned} N_t(x,y,z;t) &= \alpha(y,z)N_x(x,y,z;t) \\ &\quad + \int_{-\infty}^{\infty} [\alpha(y,z') - \alpha(y,z)]N(x,y,z';t)N(x,z',z;t)dz', \end{aligned} \tag{2.8.22}$$

with $\alpha(y,z) = [c(z)-c(y)]/i[d(z)-d(y)] = \alpha(z,y)$. The symmetry condition $N(x,y,z;t) = \sigma(y,z)N^*(x,z,y;t)$ $(y > z)$ will be satisfied if $\sigma(y,z')\sigma(z',z) = -\sigma(y,z)$, $y > z' > z$. Now, we take account of the two-dimensional case. We have

$$\begin{cases} V_x = i\zeta DV + NV + BV_y, \\ V_t = QV + CV_y. \end{cases} \tag{2.8.23}$$

We set $\zeta_t = 0$, so, as B, D, C are constants,

$$\begin{aligned} V_{xt} &= i\zeta D[QV + CV_y] + N_tV + N(QV + CV_y) + B(Q_yV + QV_y + CV_{yy}), \\ V_{tx} &= Q_xV + Q[i\zeta DV + NV + BV_y] + C[i\zeta DV_y + N_yV + NV_y + BV_{yy}]. \end{aligned}$$

By virtue of $V_{xt} = V_{tx}$, collecting the coefficients of V, V_y, V_{yy}, we get

$$V_{yy}: \quad [C,B] = 0, \tag{2.8.24}$$

$$V_y: \quad i\zeta[C,D] + [Q,B] + [C,N] = 0, \tag{2.8.25}$$

$$V: \quad i\zeta[Q,D] + [Q,N] + Q_x + CN_y - BQ_y = N_t. \tag{2.8.26}$$

We take the simplest case, so

$$C = c_i\delta_{ij}, \quad B = b_i\delta_{ij}, \quad D = d_i\delta_{ij}, \quad N_{ii} = 0,$$

where a_i, b_i, d_i are constants and

$$\sum_k (C_{ik}B_{kj} - B_{ik}C_{kj}) = 0 = c_ib_i - c_ib_i, \quad \text{from equation (2.8.24)},$$

$$i\zeta[C, D] + \sum_k (Q_{ik}B_{kj} - B_{ik}Q_{kj}) + \sum_k (c_{ik}N_{kj} - N_{ik}c_{kj}) = 0, \quad \text{from equation (2.8.25)}.$$

We deduce

$$Q_{ij} = \frac{c_i - c_j}{b_i - b_j}N_{ij}, \quad i \neq j,$$
$$Q_{ii} = q_i, \quad q_i \text{ is constant.}$$

Following the definition

$$a_{ij} = \frac{c_i - c_j}{b_i - b_j} = a_{ji},$$

we get

$$Q_{ij} = a_{ij}N_{ij}, \quad i \neq j.$$

From equation (2.8.26), we deduce

$$i\zeta \sum_k (Q_{ik}D_{kj} - D_{ik}Q_{kj}) + \sum (Q_{ik}N_{kj} - N_{ik}Q_{kj}) + Q_{ij,x} + \sum (C_{ik}N_{kj,y} - B_{ik}Q_{kj,y}) = N_{ij,t}. \tag{2.8.27}$$

For $i \neq j$, this becomes

$$i\zeta[a_{ij}N_{ij}(d_j - d_i)] + (q_i - q_j)N_{ij} + \sum_{k\neq i,j} (a_{ik} - a_{kj})N_{ik}N_{kj} + a_{ij}N_{ij,x} + c_iN_{ij,x} - b_ia_{ij}N_{ij,y} = N_{ij,t}. \tag{2.8.28}$$

For $i = j$, it is satisfied logically. Notice that there exists ζ in equation (2.8.28), while $q_i = q_i(\zeta)$ is chosen. Since q_i is undetermined, we take

$$q_i - q_j = i\zeta a_{ij}(d_i - d_j).$$

Equation (2.8.28) develops into

$$N_{ij,t} = a_{ij}N_{ij,x} + \beta_{ij}N_{ij,y} + \sum_{k\neq i,j} (a_{ik} - a_{kj})N_{ik}N_{kj}, \tag{2.8.29}$$

where the group velocities in the y- and x-directions are written, respectively, as

$$\beta_{ij} = c_i - b_i a_{ij} = \frac{b_j c_j - c_i b_j}{b_i - b_j},$$
$$a_{ij} = \frac{c_i - c_j}{b_i - b_j}.$$

$N_{ij} = \sigma_{ij} N_{ji}^*$ are compatible if $\sigma_{ij} = -\sigma_{ik}\sigma_{kj}$ $(i > k > j)$.

Example 2.8.4. Take $N_{12} = A_1$, $N_{13} = A_2$, $N_{23} = A_3$ in the following three-wave equation:

$$\begin{cases} A_{1,t} = a_{12}A_{1,x} + \beta_{12}A_{1,y} + a_{32}(a_{13} - a_{23})A_2A_3^*, \\ A_{2,t} = a_{13}A_{2,x} + \beta_{13}A_{2,y} + (a_{12} - a_{23})A_1A_3, \\ A_{3,t} = a_{23}A_{3,x} + \beta_{23}A_{3,y} + \sigma_{21}(a_{12} - a_{23})A_1^*A_3. \end{cases} \tag{2.8.30}$$

If we take

$$V_t = QV + c_1V_y + c_2V_{yy} \tag{2.8.31}$$

in equation (2.8.23) and we set B, c_1, c_2 to be diagonal constant matrices, then

$$N = \begin{pmatrix} 0 & A \\ \pm A^* & 0 \end{pmatrix} \tag{2.8.32}$$

and the evolution equations will be

$$\begin{cases} iA_t + A_{xx} + A_{yy} + (Q_1 - Q_2)A = 0, \\ Q_{1,x} + k_1Q_{1,y} = \mp[(AA^*)_x - k_1(AA^*)_y], \\ Q_{2,x} + k_2Q_{2,y} = \pm[(AA^*)_x - k_2(AA^*)_y], \end{cases} \tag{2.8.33}$$

with

$$k_1 = \frac{ib_1}{\sqrt{b_1b_2}}, \quad k_2 = \frac{ib_2}{\sqrt{b_1b_2}}.$$

The higher-dimensional nonlinear Schrödinger equation is included in equation (2.8.33). We have

$$iA_t + \nabla^2A + kA^2A^* = 0.$$

If it is independent of y, equation (2.8.33) reduces to

$$iA_t + A_{xx} \mp 2A^2A^* = 0,$$

which is the classic nonlinear Schrödinger equation. Similarly, if the x-coordinate is independent, it becomes

$$iA_t + A_{yy} \pm 2A^2A^* = 0.$$

For the two-dimensional case, the KdV equation

$$u_{xt} + 6(uu_x)_x + u_{xxxx} + 3b^2u_{yy} = 0,$$

whose corresponding eigenvalue problem is

$$v_{xx} + (\lambda + u)v + bv_{yy} = 0,$$

can be obtained from

$$B = \begin{pmatrix} 0 & 0 \\ -b & 0 \end{pmatrix}, \quad N = \begin{pmatrix} 0 & 1 \\ -u & 0 \end{pmatrix}, \quad D = \begin{pmatrix} 1 & 0 \\ 0 & -1 \end{pmatrix}.$$

3 Asymptotic behavior to initial value problems for some integrable evolution nonlinear equations

3.1 Introduction

It is well known that the asymptotic behavior of solutions for nonlinear integrable evolution equations has been studied for a long time. Significant and interesting work on the long-time behavior of nonlinear wave equations solvable by the inverse scattering method was first carried out by Shabat [278], Manakov [173], and Ablowitz and Newell [4] in 1973. The decisive step was taken in 1976 when Zakharov and Manakov [316] were able to write down precise formulas, depending explicitly on initial data for the leading asymptotics of the nonlinear Schrödinger (NLS) equation in the physically interesting region $x = O(t)$. A complete description of the leading asymptotics of the solution of the Cauchy problem for the Korteweg–de Vries (KdV) equation, with connection formulas between different asymptotic regions, was presented by Ablowitz and Segur [5], but without precise information on the phase. In a later development [277], they used a modification of the method of [316] to derive the leading asymptotics for the solution of the modified KdV (MKdV), KdV, and sine-Gordon equations, including full information on the phase. The asymptotic formulas of the Zakharov–Manakov type were rigorously justified and extended to all orders by Buslaev and Sukhanov [41] in the case of the KdV equation and by Novokshenov [191] in the case of the NLS equation. Also, both Novokshenov [192, 193] and Sukhanov [286–288] extended the method to other equations.

The method of Zakharov and Manakov, pursued rigorously in [191–193], involves an ansatz for the asymptotic form of the solution and utilizes techniques that are somewhat removed from the classical framework of Riemann–Hilbert (RH) problems. In 1981, Its [133] returned to a method first proposed in 1973 by Manakov [173], which was tied more closely to standard methods for the inverse problem. In [133], the RH problem was conjugated, up to small errors which decay as $t \to \infty$, by an appropriate parametrix, to a simpler RH problem, which in turn was solved explicitly by techniques from the theory of isomonodromic deformations. This technique provides a viable and, in principle, rigorous approach to the question of long-time asymptotics for a wide class of nonlinear wave equations [134]. Finally we note that, in [40], Buslaev derived asymptotic formulas for the KdV equation from an exact determinant formula for the solution of the inverse problem.

What emerges from the developments in [133] is the following. In realizing one's hope for a nonlinear stationary phase or steepest descent method, the classical analysis of an oscillatory integral at the points of stationary phase must be replaced by the analysis of an explicitly solvable RH problem localized at the points of stationary phase.

https://doi.org/10.1515/9783110549638-003

Recently, Deift and Zhou [60] developed a steepest descent method for oscillatory RH problems. The method is computationally systematic and yields, with rigorous error estimates, the long-time asymptotics of a general class of integrable systems. A key ingredient in the method is to deform the given RH problem to an equivalent RH problem on an augmented contour adapted to the directions of steepest descent of the associated phase factor ($e^{8iz^3t+2izx}$ for MKdV). The jump matrix $v_{x,t}$ for the deformed RH problem converges in $L^1 \cap L^2 \cap L^\infty(dz)$ to the identity as $t \to \infty$ away from any neighborhood of the stationary phase points ($\pm z_0 = \pm\sqrt{\frac{-x}{12t}}$ for MKdV). The problem then reduces to an RH problem with nontrivial jumps only in a small neighborhood of the stationary phase points. After scaling at each stationary point, one again obtains an RH problem of the isomonodromy type, which can be solved explicitly as in [133] above.

In this chapter, the elegant approach to the calculation of the asymptotic solutions proposed by Ablowitz and Segur [5] is presented for the case of the KdV equation. Moreover, we illustrate the nonlinear steepest descent method of Deift and Zhou [60] by calculating the long-time asymptotics of the defocusing NLS and the MKdV equation.

3.2 Asymptotic solutions of the KdV equation

The discoveries of solitons by Zabusky and Kruskal [314] and of the inverse scattering transform by Gardner, Greene, Kruskal, and Miura [88] have made a substantial impact on mathematical physics. The basic ideas of their work, which they used to study the KdV equation

$$u_t + 6uu_x + u_{xxx} = 0, \tag{3.2.1}$$

have been shown to apply to a wide class of physically relevant problems. In particular, in this section, we consider the asymptotic solution of the KdV equation (3.2.1).

We begin by reviewing certain aspects of the inverse scattering transform. It is assumed that $u(x, 0)$, the initial data for (3.2.1), is infinitely differentiable and vanishes rapidly, along with its derivatives, as $|x| \to \infty$. Further, the spectrum of $u(x, 0)$, as a potential in

$$\psi_{xx} + [k^2 + u(x)]\psi = 0, \tag{3.2.2}$$

is assumed to be purely continuous (here it suffices that $u(x, 0) \leq 0$). The reflection coefficient $r(k)$ and the transmission coefficient $[a(k)]^{-1}$ are defined for real k by requiring that the solution of (3.2.2) satisfies

$$\begin{aligned} \psi(x; k) &\sim [a(k)]^{-1} e^{-ikx}, \qquad && x \to -\infty, \\ &\sim e^{-ikx} + r(k)e^{ikx}, && x \to +\infty. \end{aligned} \tag{3.2.3}$$

We can show that

$$\begin{aligned}&\left|[a(k)]^{-1}\right|^2+|r(k)|^2=1, \quad |r(k)|<1, \text{ for } k\neq 0,\\ &\text{if } |r(0)|=1, \qquad \text{then } r(0)=-1,\\ &|r(k)|=O(k^{-1}), \quad \text{as } k\to\infty,\ r(-k)=r^*(k).\end{aligned} \tag{3.2.4}$$

The time-dependent scattering data are

$$\begin{aligned}r(k;t)&=r(k)e^{8ik^3t},\\ a(k;t)&=a(k).\end{aligned} \tag{3.2.5}$$

We define

$$B(x,t)=\frac{1}{2\pi}\int_{-\infty}^{\infty} r(k)e^{ikx+8ik^3t}dk. \tag{3.2.6}$$

$B(2x,t)$ satisfies the linearized KdV equation. The solution of (3.2.1) can be found by solving the following linear integral equation (for $y>x$):

$$K(x,y;t)+B(x+y;t)+\int_x^{\infty} K(x,z;t)B(z+y;t)dz=0, \tag{3.2.7}$$

from which we derive

$$u(x,t)=2\frac{d}{dx}K(x,x;t). \tag{3.2.8}$$

The asymptotic $t\to\infty$ solution of (3.2.1) when no solitons exist can be described in terms of several different regions connected by matching zones (see Figure 3.1). In the following, we will give the detailed asymptotic analysis.

I. $x\geq O(t)$. For large, positive x, the integral in (3.2.6) can be evaluated asymptotically by the method of steepest descent [3]. The result is

$$B(x,t)=\frac{r(i\kappa/2)e^{-2\kappa^3t}}{4\sqrt{3\pi\kappa t}}[1+O(t^{-1})], \tag{3.2.9}$$

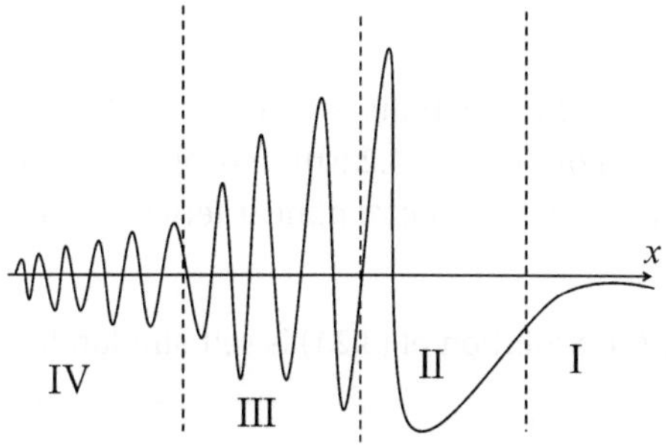

Figure 3.1: Several different regions.

where $\kappa^2 = \frac{x}{6t}$. The integral equation (3.2.7) has the following convergent Neumann series solution:

$$K(x,y;t) = -B(x+y;t) + \int_x^\infty B(x+z;t)B(z+y;t)dz + \cdots.$$

Substituting the representation in (3.2.9) into this series, we show that the series is also asymptotic (in this region only), so that

$$K(x,y;t) \sim -B(x+y;t), \tag{3.2.10}$$

to leading order. From (3.2.8), it then follows that, for $x/t = O(1)$,

$$u(x,t) = \frac{r(\frac{i}{2}\sqrt{\frac{x}{3t}})(\frac{x}{3t})^{\frac{1}{4}}e^{-2(\frac{x}{3t})^{3/2}t}}{2\sqrt{3\pi t}}[1+O(t^{-1})]. \tag{3.2.11}$$

The representations in (3.2.9) and (3.2.11) are not uniformly valid as $x/t \to 0$. To obtain such representations (which one needs in order to match), we write

$$B(x,t) = \frac{1}{4\pi(3t)^{1/3}}\int_{-\infty}^{\infty} r\left(\frac{\kappa}{2(3t)^{1/3}}\right)e^{i(\kappa\eta/2+\kappa^3/3)}d\kappa, \tag{3.2.12}$$

where $\eta = x/(3t)^{1/3}$. After expanding $r(k)$ in a Taylor series near $k = 0$ and using the definition of the Airy function,

$$\mathrm{Ai}(\eta) = \frac{1}{2\pi}\int_{-\infty}^{\infty} e^{i(\kappa\eta+\kappa^3/3)}d\kappa, \tag{3.2.13}$$

we finally obtain a representation corresponding to (3.2.11) which retains its validity as $x/t \to 0$ (but $x/t^{1/3} \gg 1$). We have

$$\begin{aligned} u(x,t) = &-\frac{r(0)}{(3t)^{2/3}}\mathrm{Ai}'(\eta) + \frac{ir'(0)}{2(3t)}\mathrm{Ai}''(\eta) + \frac{r''(0)}{2!2^2(3t)^{4/3}}\mathrm{Ai}'''(\eta) \\ &- \frac{ir'''(0)}{3!2^3(3t)^{5/3}}\mathrm{Ai}^{(4)}(\eta) - \frac{r^{(4)}(0)}{4!2^4(3t)^2}\mathrm{Ai}^{(5)}(\eta) + \cdots, \end{aligned} \tag{3.2.14}$$

where $\eta = x/(3t)^{1/3}$.

In the limit $\eta \to \infty$, the representation in (3.2.14) reproduces that in (3.2.11). This solution still is in the "linear" region (corresponding to (3.2.10)) and provides the boundary conditions for the similarity region, in which the nonlinear terms become important.

II. $|x| \le O(t^{\frac{1}{3}})$. In this region, the asymptotic solution of (3.2.1) is self-similar. It is convenient to define new variables:

$$\eta = \frac{x}{(3t)^{1/3}}, \quad u = (3t)^{-\frac{2}{3}}F(\eta,t),$$

where F satisfies the following partial differential equation

$$F_{\eta\eta\eta} + 6FF_\eta - (2F + \eta F_\eta) + 3tF_t = 0. \tag{3.2.15}$$

As $\eta \to \infty$, the solution of (3.2.15) must match that in (3.2.14). This suggests an expansion of the form

$$F(\eta, t) = f(\eta) + (3t)^{-\frac{1}{3}} f_1(\eta) + (3t)^{-\frac{2}{3}} f_2(\eta) + (3t)^{-1} f_3(\eta) + \cdots. \tag{3.2.16}$$

Substituting into (3.2.15), we obtain the following hierarchy of ordinary differential equations:

$$f''' + 6ff' - (2f + \eta f') = 0, \tag{3.2.17a}$$

$$f_1''' + 6(ff_1)' - (3f_1 + \eta f_1') = 0, \tag{3.2.17b}$$

$$f_2''' + 6(ff_2)' - (4f_2 + \eta f_2') = -3(f_1^2)', \tag{3.2.17c}$$

$$\vdots$$

$$f_m''' + 6(ff_m)' - [(m+2)f_m + \eta f_m'] = -3\sum_{k=1}^{m-1}(f_k f_{m-k})'. \tag{3.2.17d}$$

Comparing with (3.2.14), we obtain the following "initial conditions" for each of these functions as $\eta \to \infty$:

$$f(\eta) \sim -r(0)\mathrm{Ai}'(\eta), \tag{3.2.18a}$$

$$f_1(\eta) \sim \frac{ir'(0)}{2}\mathrm{Ai}''(\eta), \tag{3.2.18b}$$

$$f_2(\eta) \sim \frac{r''(0)}{2!2^2}\mathrm{Ai}'''(\eta), \tag{3.2.18c}$$

$$\vdots$$

$$f_m(\eta) \sim -\frac{r^{(m)}(0)}{m!(2i)^m}\mathrm{Ai}^{(m+1)}(\eta). \tag{3.2.18d}$$

Between (3.2.17) and (3.2.18), each of the functions in (3.2.16) is completely specified. In particular, if

$$r(0) = -1$$

holds, then, as $\eta \to -\infty$,

$$f(\eta) = \frac{1}{2}\eta - \frac{1}{2}(-2\eta)^{-1/2} + \frac{1}{2}(-2\eta)^{-2} - \frac{5}{2}(-2\eta)^{-7/2} + O((-2\eta)^{-5}). \tag{3.2.19}$$

Repeated differentiation of (3.2.17a) yields the following sequence of differential equations, which can be compared with (3.2.17):

$$f''' + 6ff' - [2f + \eta f'] = 0, \tag{3.2.20a}$$

$$(f')''' + 6(ff')' - [3f' + \eta f''] = 0, \tag{3.2.20b}$$

$$(f'')''' + 6(ff'')' - [4f'' + \eta f'''] = -6[(f')^2]'. \tag{3.2.20c}$$

Comparing (3.2.17b) with (3.2.20) and using (3.2.18b), we see that

$$f_1(\eta) = -\frac{ir'(0)}{2r(0)} f'(\eta). \tag{3.2.21}$$

As $\eta \to -\infty$, it follows from (3.2.19) that $f_1(\eta)$ approaches a constant.

Equation (3.2.17c) determines $f_2(\eta)$, which has both a particular solution and a homogeneous solution. The particular solution is apparent from comparison with (3.2.20c), as we have

$$f_{2p}(\eta) = \frac{1}{2}\left(-\frac{ir'(0)}{2r(0)}\right)^2 f''(\eta). \tag{3.2.22}$$

At every order, there is a particular solution in this sequence. Substitution of these particular solutions into (3.2.16) yields the Taylor series expansion for $f(\eta + \eta_0)$,

$$f(\eta) + \eta_0 f'(\eta) + \frac{1}{2}\eta_0^2 f''(\eta) + \frac{1}{3!}\eta_0^3 f'''(\eta) + \cdots, \tag{3.2.23}$$

where

$$\eta_0 = -\frac{ir'(0)}{2r(0)(3t)^{1/3}}.$$

The effect of these terms is to determine an "asymptotically preferred" coordinate system, in which the asymptotic solution is "centered". It is straightforward to show that this coordinate system is obtained from the original one (in which $r(k)$ was calculated) by a translation. We write

$$\tilde{x} = x + x_0, \tag{3.2.24a}$$

$$x_0 = -\frac{ir'(0)}{2r(0)}. \tag{3.2.24b}$$

With respect to this preferred coordinate system, $r'(0)$ vanishes, as do η_0 and all of the terms in (3.2.23) except the first. We note that the reflection coefficient in the translated coordinate system is found to be $r(k)e^{-2ikx_0}$.

There remains a homogeneous solution, $h_2(\eta)$, at the same order as (3.2.22), which satisfies the following problem:

$$h_2''' + 6(fh_2)' - (4h_2 + \eta h_2') = 0, \tag{3.2.25a}$$

$$h_2(\eta) \to M\mathrm{Ai}'''(\eta) \quad \text{as } \eta \to \infty, \tag{3.2.25b}$$

where

$$M = \frac{1}{2!2^2 r(0)}\{r''(0)r(0) - [r'(0)]^2\}. \tag{3.2.25c}$$

The relationship between reflection coefficients in simply translated systems (discussed below (3.2.24)) shows that M is invariant under translations such as (3.2.24). In addition, the fact that $1 - |r(k)|^2$ is nonnegative implies (by the Taylor series about $k = 0$) that M is nonnegative. Using (3.2.19), we can show that, as $\eta \to -\infty$, one solution of (3.2.25a) grows exponentially. Numerical integration confirms that this branch dominates the solution of (3.2.25) as $\eta \to -\infty$. In particular, as $\eta \to -\infty$,

$$h_2(\eta) \sim \kappa M e^{\frac{1}{3}(-2\eta)^{3/2}}(-2\eta)^{-1/4}\left\{1 + \frac{35}{24}(-2\eta)^{-3/2} + O((-2\eta)^{-3})\right\}, \tag{3.2.26}$$

where $\kappa \doteq 0.80$ (κ is obtained by numerical integration). The point here is that, as $\eta \to -\infty$, $f(\eta)$ grows linearly, so $h_2(\eta)$ grows exponentially. Thus, no matter how small $(3t)^{-2/3}$ might be, the third term in (3.2.16) dominates the first for $-\eta$ large enough. We emphasize that this behavior is found only under the condition that $r(0) = -1$. Otherwise, $f(\eta)$ in (3.2.25a) is given by (3.2.17) rather than (3.2.19), $h_2(\eta)$ oscillates as $\eta \to -\infty$, and the expansion does not become disordered.

The next important term in the breakdown of the expansion in (3.2.16) is the particular solution of $f_4(\eta)$ which is forced by $h_2(\eta)$. It satisfies the differential equation

$$f_{4p}''' + 6(ff_{4p})' - (6f_{4p} + \eta f_{4p}') = -6h_2h_2'. \tag{3.2.27}$$

The behavior of f_{4p} as $\eta \to -\infty$ is found by using (3.2.19) and (3.2.26) in (3.2.27). As $\eta \to -\infty$,

$$f_{4p}(\eta) \sim -\frac{(\kappa M)^2}{(-2\eta)^{3/2}} e^{\frac{2}{3}(-2\eta)^{3/2}}, \tag{3.2.28}$$

so this term is exponentially larger than $h_2(\eta)$.

Collecting the dominant terms in the asymptotic expansion in (3.2.16), we find that, under the condition that $r(0) = -1$, as $\eta \to -\infty$,

$$\begin{aligned} u(x,t) = (3t)^{-\frac{2}{3}}(-2\eta)\Big[-\frac{1}{4} - \frac{1}{2}(-2\eta)^{-\frac{3}{2}} + \cdots + \{(3t)^{-\frac{2}{3}}\kappa M(-2\eta)^{-\frac{5}{4}} \\ \cdot e^{\frac{1}{3}(-2\eta)^{3/2}} + \cdots\} - \{(3t)^{-\frac{2}{3}}\kappa M(-2\eta)^{-\frac{5}{4}} e^{\frac{1}{3}(-2\eta)^{3/2}} + \cdots\}^2 + \cdots\Big]. \end{aligned} \tag{3.2.29}$$

III. $-x \geq O(t)$. The central concept in this region is that of a modulated similarity solution. The solution of (3.2.1) tends to a self-similar form, modulated by two functions that depend on the initial data. The more important of these two functions is then determined by the conservation laws. This concept was originally developed in [276] to solve the NLS equation. The procedure here is analogous.

As discussed above, as $\eta \to -\infty$, there is a self-similar solution of (3.2.1) which has the form

$$u = \frac{(-\eta)^{1/4}}{(3t)^{2/3}}\left[\sum_{k=0}^{\infty} A_{2k+1}(-\eta)\cdot(-\eta)^{-3k/2}\cos(2k+1)\theta + \sum_{k=1}^{\infty} B_{2k+1}(-\eta)\cdot(-\eta)^{-3(k+1)/2}\sin(2k+1)\theta\right] + \frac{1}{(3t)^{2/3}(-\eta)^{1/2}}\left[A_0(-\eta) + \sum_{k=1}^{\infty} A_{2k}(-\eta)\cdot(-\eta)^{-\frac{3}{2}(k-1)}\cos 2k\theta + \sum_{k=1}^{\infty} B_{2k}(-\eta)\cdot(-\eta)^{-3k/2}\sin 2k\theta\right],$$

where

$$\theta = \frac{2}{3}(-\eta)^{-3/2} + \kappa\ln(-\eta) + \sum_{j=0}^{\infty}\theta_j(-\eta)^{-3j/2},$$

$$A_k = \sum_{j=0}^{\infty} A_{jk}(-\eta)^{-3j/2}, \quad B_k = \sum_{j=0}^{\infty} B_{jk}(-\eta)^{-3j/2},$$

$\theta_j, A_{jk}, B_{jk}, \kappa$ are constants, $\theta_0, A_{01} = 2d$ are arbitrary, and the others are determined as follows: $A_{0,0} = -2d^2, A_{0,2} = 2d^2, \kappa = -3d^2, \ldots$. Thus,

$$u(x,t) \sim (3t)^{-2/3}\left[2d(-\eta)^{1/4}\cos\theta - 2d^2(-\eta)^{-1/2}(1-\cos 2\theta)\right], \tag{3.2.30}$$

with

$$\theta \sim \frac{2}{3}(-\eta)^{3/2} - 3d^2\ln(-\eta) + \theta_0.$$

IV. $-x = O(t^{\frac{1}{3}}(\ln t)^{\frac{2}{3}})$. This region acts like a collisionless shock wave across which the asymptotic solution changes smoothly from the growing similarity solution in II to oscillations in III. The analysis of this region is rather complicated, so we only list the result:

$$\begin{aligned} u(x,t) &\sim (3t)^{-\frac{2}{3}}(-\eta)^{\frac{1}{4}}\left(\frac{2\ln 3t}{3\pi}\right)^{\frac{1}{2}}\cos\theta, \\ \theta &\sim \frac{2}{3}(-\eta)^{\frac{3}{2}} - \frac{1}{2\pi}(\ln 3t)\ln(-\eta). \end{aligned} \tag{3.2.31}$$

For a detailed derivation, we refer the reader to [5].

3.3 Nonlinear Schrödinger equation

We consider the long-time asymptotic for the Cauchy problem of the defocusing NLS equation

$$\begin{cases} iq_t + q_{xx} - 2|q|^2 q = 0, \\ q(x,0) = q_0(x) \in \mathcal{S}(\mathbb{R}). \end{cases} \tag{3.3.1}$$

We begin by giving a sketch of how to get the RH problem from the inverse scattering transformation of the NLS equation (3.3.1). Equation (3.3.1) admits the following Lax pair representation:

$$\begin{aligned} \partial_x \psi &= (-iz\sigma_3 + Q(x,t))\psi, \\ \partial_t \psi &= (-2iz^2\sigma_3 + \tilde{Q}(x,t,z))\psi, \end{aligned} \tag{3.3.2}$$

where

$$\sigma_3 = \begin{pmatrix} 1 & 0 \\ 0 & -1 \end{pmatrix}, \quad Q(x,t) = \begin{pmatrix} 0 & q(x,t) \\ \overline{q}(x,t) & 0 \end{pmatrix}, \quad \tilde{Q}(x,t,z) = 2zQ - iQ_x\sigma_3 - i|q|^2\sigma_3.$$

Let $\mu(x,t,z) = \psi(x,t,z)e^{i(xz+2tz^2)\sigma_3}$. Then we obtain an equivalent Lax pair

$$\begin{aligned} \mu_x + iz[\sigma_3, \mu] &= Q(x,t)\mu, \\ \mu_t + 2iz^2[\sigma_3, \mu] &= \tilde{Q}(x,t,z)\mu, \end{aligned} \tag{3.3.3}$$

which can be written in full derivative form as follows:

$$d(e^{i(zx+2z^2t)\,\mathrm{ad}\,\sigma_3}\mu(x,t,z)) = W(x,t,z), \tag{3.3.4}$$

where $\mathrm{ad}\,\sigma_3$ denotes the commutator with respect to σ_3. Also, $(\exp \mathrm{ad}\,\sigma_3)A$ can be computed easily as follows:

$$\mathrm{ad}\,\sigma_3 A = [\sigma_3, A], \quad e^{\mathrm{ad}\,\sigma_3}A = e^{\sigma_3}Ae^{-\sigma_3},$$

where A is a 2×2 matrix and the exact 1-form W is defined by

$$W(x,t,z) = e^{i(zx+2z^2t)\,\mathrm{ad}\,\sigma_3}(Q\mu dx + \tilde{Q}\mu dt). \tag{3.3.5}$$

We define two particular solutions of (3.3.3) as the 2×2 matrix-valued solutions of the associated Volterra integral equations. We have

$$\mu_1(x,t,z) = I + \int_{-\infty}^{x} e^{-iz(x-x')\,\mathrm{ad}\,\sigma_3} Q(x',t)\mu_1(x',t,z)dx', \tag{3.3.6}$$

$$\mu_2(x,t,z) = I + \int_{\infty}^{x} e^{-iz(x-x')\,\mathrm{ad}\,\sigma_3} Q(x',t)\mu_2(x',t,z)dx'. \tag{3.3.7}$$

Let the columns of a 2×2 matrix μ be denoted as $([\mu]_1\ [\mu]_2)$. It follows from (3.3.6) and (3.3.7) that, for all (x,t), $[\mu_1]_1$ and $[\mu_2]_2$ are analytic and bounded in $\{z|\operatorname{Im} z > 0\}$, while $[\mu_1]_2$ and $[\mu_2]_1$ are analytic and bounded in $\{z|\operatorname{Im} z < 0\}$.

The fact that Q and $\tilde{Q}$ are traceless implies $\det \mu_j(x,t,z) = 1$ for $j = 1, 2$, so $\det s(z) = 1$. From the symmetry properties of Q and $\tilde{Q}$, it follows that the eigenfunction $\mu(x,t,z)$ satisfies

$$(\mu(x,t,z))_{11} = \overline{(\mu(x,t,\bar{z}))_{22}}, \quad (\mu(x,t,z))_{12} = \overline{(\mu(x,t,\bar{z}))_{21}}.$$

On the other hand, two solutions μ_j of the system of differential equations (3.3.3) must be simply related. We have

$$\mu_1(x,t,z) = \mu_2(x,t,z) e^{-i(zx+2z^2t)\operatorname{ad}\sigma_3} s(z). \tag{3.3.8}$$

From the above symmetry property, we write the spectral matrices $s(z)$ in the following form:

$$s(z) = \begin{pmatrix} a(z) & \overline{b(\bar{z})} \\ b(z) & \overline{a(\bar{z})} \end{pmatrix}. \tag{3.3.9}$$

Define

$$m(x,t,z) = \begin{cases} \left(\frac{[\mu_1]_1}{a(z)} \quad [\mu_2]_2 \right), & \operatorname{Im} z > 0, \\ \left([\mu_2]_1 \quad \frac{[\mu_1]_2}{\overline{a(\bar{z})}} \right), & \operatorname{Im} z < 0. \end{cases} \tag{3.3.10}$$

Then, for each x and t, the 2×2 matrix function $m(x,t,z)$ solves the following RH problem in z:

$$\begin{aligned} &\text{(i)} && m(x,t,z) \quad \text{is analytic in } z \text{ for } \mathbb{C} \setminus \mathbb{R}, \\ &\text{(ii)} && m_+(x,t,z) = m_-(x,t,z) v_{x,t}(z), \quad z \in \mathbb{R}, \\ &\text{(iii)} && \lim_{z \to \infty} m(x,t,z) = I, \end{aligned} \tag{3.3.11}$$

where

$$\begin{aligned} m_\pm(x,t,z) &= \lim_{\varepsilon \to 0} m(x,t,z \pm i\varepsilon), \\ v_{x,t}(z) &= e^{-i(xz+2tz^2)\sigma_3} v(z) e^{i(xz+2tz^2)\sigma_3} \equiv e^{-i(xz+2tz^2)\operatorname{ad}\sigma_3} v(z), \\ v(z) &= \begin{pmatrix} 1 - |r(z)|^2 & -\bar{r}(z) \\ r(z) & 1 \end{pmatrix}, \end{aligned} \tag{3.3.12}$$

where $r(z) = \frac{b(z)}{a(z)}$ lies in a Schwartz space and

$$\sup_{z \in \mathbb{R}} |r(z)| < 1$$

is the reflection coefficient corresponding to the initial data $q_0(x)$. If we expand the limit in (iii), we have

$$m(x,t,z) = I + \frac{m_1(x,t)}{z} + O\left(\frac{1}{z}\right) \tag{3.3.13}$$

and we obtain the following expression for $q(x,t)$:

$$q(x,t) = 2i(m_1(x,t))_{12} = 2i \lim_{z\to\infty}(zm(x,t,z))_{12}. \tag{3.3.14}$$

We now begin the analysis of the long-time asymptotics for the defocusing NLS equation (3.3.1) based on the RH problem (3.3.11). Let $\theta = 2z^2 + \frac{x}{t}z$ with stationary phase point $z_0 = \frac{-x}{4t}$. For simplicity, we restrict ourselves here to the physically interesting region $|z_0| \le M$ for some fixed constant M. The matrix v admits the following triangular factorizations:

$$\begin{aligned} v &= \begin{pmatrix} 1 & -\bar{r} \\ 0 & 1 \end{pmatrix}\begin{pmatrix} 1 & 0 \\ r & 1 \end{pmatrix} \\ &= \begin{pmatrix} 1 & 0 \\ \frac{r}{1-|r|^2} & 1 \end{pmatrix}\begin{pmatrix} 1-|r|^2 & 0 \\ r & \frac{1}{1-|r|^2} \end{pmatrix}\begin{pmatrix} 1 & \frac{-\bar{r}}{1-|r|^2} \\ 0 & 1 \end{pmatrix}. \end{aligned} \tag{3.3.15}$$

We choose $\delta(z)$ analytic and invertible in $\mathbb{C} \setminus \mathbb{R}$ such that

$$\delta_+(z) = \begin{cases} \delta_-(z)(1-|r(z)|^2), & z < z_0, \\ \delta_-(z) \equiv \delta(z), & z > z_0, \\ \delta(z) \to 1, & \text{as } z \to \infty, \end{cases} \tag{3.3.16}$$

where $\pm$ refers to the orientation of $\mathbb{R}$ from $-\infty$ to ∞. The solution to (3.3.16) is given by the formula

$$\delta(z) = \exp\left\{\frac{1}{2\pi i}\int_{-\infty}^{z_0} \frac{\log(1-|r(\xi)|^2)}{\xi - z}d\xi\right\}, \quad z \notin \mathbb{R}. \tag{3.3.17}$$

It is easy to check that $\delta(z)$ and $\delta^{-1}(z)$ are uniformly bounded in z for $|z_0| \le M$.

The function $\tilde{m} = m\delta^{-\sigma_3}$ satisfies an RH problem across $\mathbb{R}$ with the following jump matrix:

$$\begin{aligned} \tilde{v}_{x,t}(z) &= e^{-it\theta \operatorname{ad}\sigma_3}(\delta_-^{\sigma_3} v \delta_+^{\sigma_3}) \\ &= \begin{cases} e^{-it\theta \operatorname{ad}\sigma_3}\begin{pmatrix} 1 & 0 \\ \frac{r\delta_-^{-2}}{1-|r|^2} & 1 \end{pmatrix}\begin{pmatrix} 1 & \frac{-\bar{r}\delta_+^2}{1-|r|^2} \\ 0 & 1 \end{pmatrix}, & z < z_0, \\ e^{-it\theta \operatorname{ad}\sigma_3}\begin{pmatrix} 1 & -\bar{r}\delta^2 \\ 0 & 1 \end{pmatrix}\begin{pmatrix} 1 & 0 \\ r\delta^{-2} & 1 \end{pmatrix}, & z > z_0. \end{cases} \end{aligned} \tag{3.3.18}$$

Note that, as δ^{σ_3} is diagonal, we can replace m by $\tilde{m}$ in (3.3.11).

Having made the above definitions, we now describe the strategy. Suppose that the coefficients

$$\frac{r}{1-|r|^2},\ \frac{\bar{r}}{1-|r|^2},\ r,\ \bar{r} \tag{3.3.19}$$

can be replaced by some rational functions

$$\left[\frac{r}{1-|r|^2}\right],\ \left[\frac{\bar{r}}{1-|r|^2}\right],\ [r],\ [\bar{r}], \tag{3.3.20}$$

respectively. Then, if the poles of these functions are appropriately placed, the RH problem on $\mathbb{R}$ can be deformed to the contour Σ, as shown in Figure 3.2.

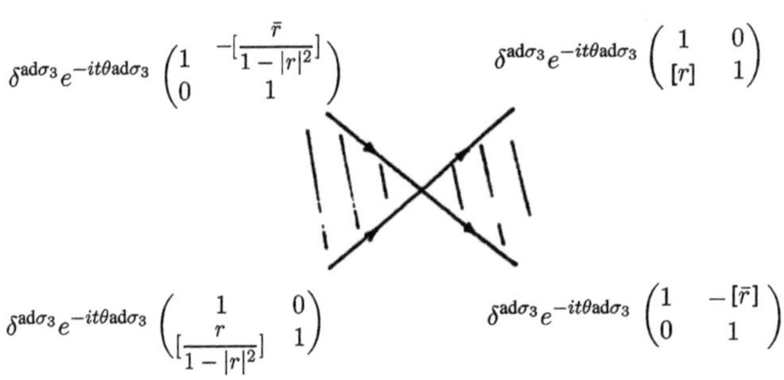

Figure 3.2: The contour Σ.

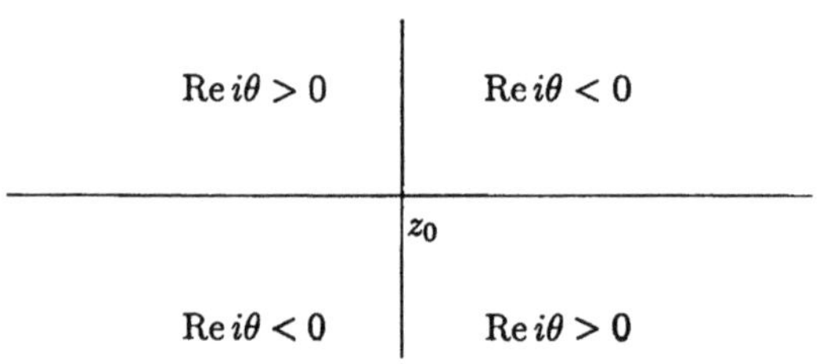

Figure 3.3: The signature table for Re $i\theta$.

Using the signature table for Re $i\theta$ (see Figure 3.3), we now see that the scalar factorization (3.3.16) for δ and the triangular factorizations (3.3.18) for v have been chosen specifically to ensure that the jump matrices for the deformed problem Σ converge rapidly to an identity away from any neighborhood of z_0, as $t \to \infty$.

To verify that the coefficients (3.3.19) can be replaced by the rational functions (3.3.20) with well-controlled errors, we proceed as follows. We expand $(i+z)^{10} r(z)$ in a fifth-order Taylor series around z_0, to obtain

$$(i+z)^{10} r = \mu_0 + \mu_1(z-z_0) + \cdots + \mu_5(z-z_0)^5 + (i+z)^{10} h.$$

We set

$$[r] = \frac{\mu_0 + \mu_1(z - z_0) + \cdots + \mu_5(z - z_0)^5}{(i + z)^{10}} \tag{3.3.21}$$

and

$$\beta = \frac{(z - z_0)^2}{(i + z)^4}. \tag{3.3.22}$$

Observe that

$$\frac{h}{\beta} = \frac{r - [r]}{\beta} = \frac{(z - z_0)^4}{(z + i)^6} g(z; z_0), \quad z \geq z_0, \tag{3.3.23}$$

where

$$|g| + \left|\frac{\partial g}{\partial z}\right| + \left|\frac{\partial^2 g}{\partial z^2}\right| \leq C(M). \tag{3.3.24}$$

Since $\theta(z) = 2(z - z_0)^2 - 2z_0^2$ is one-to-one from (z_0, ∞) to $(-2z_0^2, \infty)$, we consider $\frac{h}{\beta}$ as a function of θ. We have

$$\frac{h}{\beta}(\theta) = \begin{cases} \frac{h}{\beta}(z(\theta)), & \theta > -2z_0^2, \\ 0, & \theta \leq -2z_0^2. \end{cases}$$

It is easy to check that $\frac{h}{\beta} \in H^2(d\theta, -\infty < \theta < \infty)$.

Now, by the Fourier theory with respect to the variable θ, we have

$$\left(\frac{h}{\beta}\right)(\theta) = \frac{1}{\sqrt{2\pi}} \int_{-\infty}^{\infty} e^{-is\theta} \widehat{\left(\frac{h}{\beta}\right)}(s)ds, \tag{3.3.25}$$

where

$$\widehat{\left(\frac{h}{\beta}\right)}(s) = \frac{1}{\sqrt{2\pi}} \int_{-\infty}^{\infty} e^{is\theta} \left(\frac{h}{\beta}\right)(\theta)d\theta. \tag{3.3.26}$$

Thus,

$$\begin{aligned} e^{2it\theta} h &= \frac{1}{\sqrt{2\pi}} \beta \int_t^{\infty} e^{i(2t-s)\theta} \widehat{\left(\frac{h}{\beta}\right)}(s)ds + \frac{1}{\sqrt{2\pi}} \beta e^{it\theta} \int_{-\infty}^{t} e^{i(t-s)\theta} \widehat{\left(\frac{h}{\beta}\right)}(s)ds \\ &\equiv h_I + h_{II}. \end{aligned} \tag{3.3.27}$$

By the Plancherel identity, as $\frac{h}{\beta} \in H^2$,

$$\int_{-\infty}^{\infty} (1 + s^2)^2 \left|\widehat{\left(\frac{h}{\beta}\right)}(s)\right|^2 ds < \infty. \tag{3.3.28}$$

Hence,

$$|h_I(z)| \le \frac{C}{|z+i|^2 t^{3/2}}. \tag{3.3.29}$$

On the other hand, $h_{II}(z)$ has an analytic continuation to the line $z_0 + e^{i\pi/4}\mathbb{R}_+$, where it satisfies the estimate

$$|h_{II}(z)| \le \frac{C}{|z+i|^2 t}, \tag{3.3.30}$$

again by (3.3.28). Thus,

$$\|h_I\|_{L^1\cap L^2\cap L^\infty(z_0\le z<\infty)} = O(t^{-3/2}), \tag{3.3.31}$$

$$\|h_{II}\|_{L^1\cap L^2\cap L^\infty(z_0+e^{i\pi/4}\mathbb{R}_+)} = O(t^{-1}). \tag{3.3.32}$$

An arbitrarily high order of decay in (3.3.31) and (3.3.32) can be obtained by using a higher-order Taylor expansion in (3.3.21) at z_0. This procedure can be found in the following section for the MKdV equation.

Now deform the RH problem for $\tilde{v}$ on $\mathbb{R}$ to the contour, as shown in Figure 3.4.

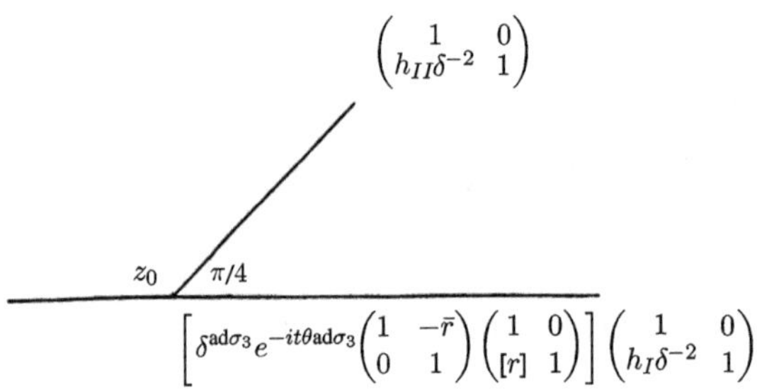

Figure 3.4: Deformation contour.

It turns out that the error estimates (3.3.31) and (3.3.32) are sufficient to ensure that the contributions of $h_{II}\delta^{-2}$ on $z_0+e^{i\pi/4}\mathbb{R}_+$ and $h_I\delta^{-2}$ on (z_0,∞) are negligible for the leading asymptotics as $t\to\infty$. Repeating the above arguments for the remaining functions in (3.3.19), we see that they can all be replaced in the RH problem $\tilde{v}$ by the appropriate rational functions in (3.3.20) with effective error control. Deforming the contour as above, we arrive at the RH problem on Σ, as shown in Figure 3.2.

We define the scaling operator $N : L^2(\Sigma) \to L^2(\Sigma - z_0)$ and

$$f(z) \mapsto Nf(z) = f\left(\frac{z}{\sqrt{8t}} + z_0\right). \tag{3.3.33}$$

Denote the jump matrix in Figure 3.2 by $\delta^{\mathrm{ad}\,\sigma_3} e^{-it\theta\,\mathrm{ad}\,\sigma_3}[\tilde{v}]$. A straightforward computation shows that, as $t\to\infty$,

$$N\delta^{\mathrm{ad}\,\sigma_3} e^{-it\theta\,\mathrm{ad}\,\sigma_3}[\tilde{v}] \to (\delta^0)^{\mathrm{ad}\,\sigma_3} z^{\nu i\,\mathrm{ad}\,\sigma_3} e^{-\frac{iz^2}{4}\,\mathrm{ad}\,\sigma_3}[\tilde{v}](z_0), \tag{3.3.34}$$

where

$$\begin{aligned}\delta^0 &= (8t)^{\frac{-i\nu}{2}} e^{2itz_0^2} e^{\chi(z_0)},\\ \nu = \nu(z_0) &= -\frac{1}{2\pi}\log(1-|r(z_0)|^2) > 0,\\ \chi(z) &= -\frac{1}{2\pi i}\int_{-\infty}^{z_0}\log(z-\xi)\,d\log(1-|r(\xi)|^2).\end{aligned} \tag{3.3.35}$$

$[\tilde{v}](z_0)$ is defined by Figure 3.5.

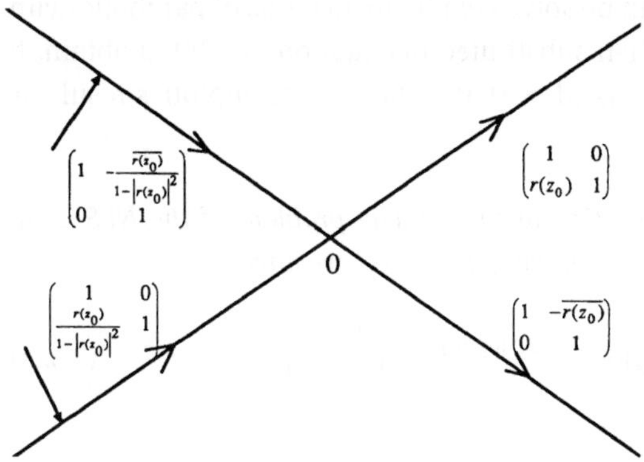

Figure 3.5: $[\tilde{v}](z_0)$.

It follows from the exponential decay of $e^{-\frac{iz^2}{4}\operatorname{ad}\sigma_3}[\tilde{v}](z_0)$ that the asymptotic formula in (3.3.34) has an $L^1\cap L^2\cap L^\infty(\Sigma - z_0)$ error of order $(\log t)/t^{1/2}$. Since δ^0 is independent of z, m^0 is the solution of the RH problem on $\Sigma - z_0$, so

$$\begin{aligned}m_+^0 &= m_-^0 z^{\nu i \operatorname{ad}\sigma_3} e^{-\frac{iz^2}{4}\operatorname{ad}\sigma_3}[\tilde{v}](z_0),\\ m^0 &\to I \quad \text{as } z\to\infty,\end{aligned} \tag{3.3.36}$$

if and only if $(\delta^0)^{\operatorname{ad}\sigma_3} m^0$ is the solution of the RH problem for the jump matrix given by the right-hand side of (3.3.34). Deforming the RH problem (3.3.36) on $\Sigma - z_0$ to the real axis, we obtain precisely the following RH problem:

$$\begin{aligned}\tilde{m}_+^0(z) &= \tilde{m}_-^0(z) e^{-\frac{iz^2}{4}\operatorname{ad}\sigma_3} z_-^{i\nu\sigma_3} v(z_0) z_+^{-i\nu\sigma_3},\\ \tilde{m}^0(z) &\to I \quad \text{as } z\to\infty.\end{aligned} \tag{3.3.37}$$

Reinserting the z-independent factor $(\delta^0)^{\operatorname{ad}\sigma_3}$ and the scaling factor $1/\sqrt{8t}$, we obtain

$$q(x,t) = (2t)^{-1/2} i(\delta^0)^2 (\tilde{m}_1^0)_{12} + O\left(\frac{\log t}{t}\right), \tag{3.3.38}$$

if $\tilde{m}^0 = I + z^{-1}\tilde{m}_1^0 + O(z^{-2})$. We set

$$\Psi(z) = \tilde{m}^0(z) z^{i\nu\sigma_3} e^{-\frac{iz^2}{4}\sigma_3}$$

and we can represent the RH problem (3.3.37) as

$$\begin{aligned} &\Psi_+(z) = \Psi_-(z) v(z_0), \\ &\Psi(z) e^{\frac{iz^2}{4}\sigma_3} z^{-i\nu\sigma_3} \to I \quad \text{as } z \to \infty. \end{aligned} \tag{3.3.39}$$

The factorization problem (3.3.39) can be solved explicitly in terms of parabolic cylinder functions [60]. The solution is then substituted in equation (3.3.38) to obtain, finally, the asymptotics for $q(x,t)$. We fix $M > 0$ and state the asymptotics results as follows.

Theorem 3.3.1. *Let $q(x,t)$ be the solution of the Cauchy problem of the NLS equation* (3.3.1). *If the initial value $q_0(x) \in \mathcal{S}(\mathbb{R})$, then, as $t \to \infty$, we have*

$$q(x,t) = t^{-\frac{1}{2}} \alpha(z_0) e^{\frac{ix^2}{4t} - i\nu(z_0)\log(8t)} + O\left(\frac{\log t}{t}\right), \tag{3.3.40}$$

for $|z_0| = |-x/(4t)| \leq M$, where

$$\begin{aligned} \nu(z_0) &= -\frac{1}{2\pi}\log(1 - |r(z_0)|^2) > 0, \quad |\alpha(z_0)|^2 = \frac{\nu(z_0)}{2}, \\ \arg\alpha(z_0) &= \frac{1}{\pi}\int_{-\infty}^{z_0} \log(z_0 - \xi) d(\log(1 - |r(\xi)|^2)) + \frac{\pi}{4} \\ &\quad + \arg\Gamma(i\nu) - \arg r(z_0). \end{aligned}$$

Here, Γ is the gamma function.

Finally, we will indicate how to use the estimates on the jump matrices obtained above, to obtain error estimates on the asymptotic solutions. In other words, we show how to use the estimates (3.3.31) and (3.3.32) to bound the contribution of h_I and h_{II} to the asymptotic solution. Other error estimates are similar.

We recall the solution procedure for RH problems. For an oriented contour Γ with a factored $n \times n$ jump matrix $v = b_-^{-1} b_+$, the RH problem on Γ,

$$\begin{aligned} &m_+ = m_- v \quad && \text{on } \Gamma, \\ &m \to I && \text{as } z \to \infty, \end{aligned} \tag{3.3.41}$$

is solved as follows [20]. Set $w_\pm = \pm(b_\pm - I)$ and $w = w_+ = w_-$. Let

$$(C_\pm f)(z) = \int_\Gamma \frac{f(\xi)}{\xi - z_\pm}\frac{d\xi}{2\pi i}, \quad z \in \Gamma,\ f \in L^2(\Gamma) \tag{3.3.42}$$

and denote the Cauchy operator on Γ. As is well known, the operators $C_\pm$ are bounded from $L^2(\Gamma)$ to $L^2(\Gamma)$ and $C_+ - C_- = 1$, where 1 denotes the identity operator.

Define

$$C_w f = C_+(fw_-) + C_-(fw_+) \tag{3.3.43}$$

for the 2×2 matrix-valued function f and let μ be the solution of the basic inverse equation

$$\mu = I + C_w \mu. \tag{3.3.44}$$

Then

$$m(z) = I + \int_\Gamma \frac{\mu(\xi) w(\xi)}{\xi - z} \frac{d\xi}{2\pi i}, \quad z \in \mathbb{C} \setminus \Gamma, \tag{3.3.45}$$

is the solution of the RH problem (3.3.41). Substituting equation (3.3.45) into (3.3.14), we learn that

$$\begin{aligned} q(x,t) &= 2i \cdot \left(-\frac{1}{2\pi i}\right) \cdot \left(\int_{\mathbb{R}} \mu(\xi) w(\xi) d\xi\right)_{12} \\ &= -\frac{1}{\pi}\left(\int_{\mathbb{R}} \mu(\xi) w(\xi) d\xi\right)_{12} \\ &= -\frac{1}{\pi}\left(\int_{\mathbb{R}} ((1 - C_w)^{-1} I) w(\xi) d\xi\right)_{12}. \end{aligned} \tag{3.3.46}$$

Here we have, for example (cf. (3.3.15)),

$$\begin{cases} w = w_{x,t} = (w_+)_{x,t} + (w_-)_{x,t}, \\ (w_+)_{x,t} = \begin{pmatrix} 0 & 0 \\ r e^{2it\theta} & 0 \end{pmatrix}, \quad (w_-)_{x,t} = \begin{pmatrix} 0 & -\bar{r} e^{-2it\theta} \\ 0 & 0 \end{pmatrix}. \end{cases} \tag{3.3.47}$$

Let us illustrate how to control the error when we replace r by $[r]$. For the RH problem in Figure 3.4, the error in the jump matrix is controlled by (3.3.31) and (3.3.32). In general, let us assume that two sets of data $w_\pm$ and $w'_\pm$ differ by

$$\|w_\pm - w'_\pm\|_{L^1 \cap L^2 \cap L^\infty} = O(t^{-1})$$

and $\|w_\pm\|_{L^1 \cap L^2 \cap L^\infty} = O(1)$. We also assume that

$$\|(1 - C_w)^{-1}\|_{L^2 \to L^2} = O(1). \tag{3.3.48}$$

The estimate (3.3.48) indeed holds for the NLS equation (see [61]), so the following estimates are easily derived:

$$\|C_w I - C_{w'} I\|_{L^2} = O(\|w - w'\|_{L^2}) = O(t^{-1}), \tag{3.3.49}$$

$$\|C_w - C_{w'}\|_{L^2\to L^2} = O(\|w - w'\|_{L^\infty}) = O(t^{-1}), \tag{3.3.50}$$

$$\|(1 - C_w)^{-1} - (1 - C_{w'})^{-1}\|_{L^2\to L^2} = O(\|C_w - C_{w'}\|) = O(t^{-1}). \tag{3.3.51}$$

We finally obtain the uniform estimate

$$\begin{aligned}\int((1 - C_w)^{-1}I)w &= \int((1 - C_w)^{-1}C_wI)w + \int w\\ &= \int((1 - C'_w)^{-1}C_{w'}I)w' + \int w' + O(t^{-1}).\end{aligned} \tag{3.3.52}$$

If w and w' correspond to r and $[r]$, respectively, this shows that we can replace r by $[r]$ in the asymptotic solution of the inverse problem with a controlled error.

If the initial q_0 lies in the weighted Sobolev space $H^{1,1} = \{f \in L^2(\mathbb{R}) : xf, f' \in L^2(\mathbb{R})\}$, then the following result holds.

Theorem 3.3.2 ([63]). *Let $q(t), t \geq 0$ be the solution of* (3.3.1) *with $q_0 = q(x, t = 0) \in H^{1,1}$ and fix $0 < \kappa < \frac{1}{4}$. Then, as $t \to \infty$,*

$$q(x,t) = t^{-\frac{1}{2}}\alpha(z_0)e^{\frac{ix^2}{4t} - i\nu(z_0)\log(2t)} + O(t^{-(\frac{1}{2}+\kappa)}), \tag{3.3.53}$$

where α and ν are given as above. The error term $O(t^{-\frac{1}{2}+k})$ is uniform for all $x \in \mathbb{R}$.

3.4 MKdV Equation

In this section, we consider the asymptotics of the solution $y(x,t)$ of the following MKdV equation:

$$\begin{aligned}&y_t - 6y^2y_x + y_{xxx} = 0, \quad -\infty < x < \infty,\ t \geq 0,\\ &y(x,0) = y_0(x) \in \mathcal{S}(\mathbb{R}),\end{aligned} \tag{3.4.1}$$

as $t \to \infty$. The MKdV equation (3.4.1) admits the Lax pair formulation

$$\begin{aligned}&\mu_x + iz[\sigma_3, \mu] = Q(x,t)\mu,\\ &\mu_t + 4iz^3[\sigma_3, \mu] = \tilde{Q}(x,t,z)\mu,\end{aligned} \tag{3.4.2}$$

where

$$\sigma_3 = \begin{pmatrix}1 & 0\\ 0 & -1\end{pmatrix}, \quad Q = \begin{pmatrix}0 & y(x,t)\\ y(x,t) & 0\end{pmatrix}, \quad \tilde{Q} = 4z^2Q - 2iz(Q^2 + Q_x)\sigma_3 + 2Q^3 - Q_{xx}.$$

Similar to the method used to construct the RH problem for the defocusing NLS equation (3.3.1), we directly write the RH problem for the MKdV equation as follows:

$$\begin{cases}m_+(z) = m_-(z)v_{x,t}(z), & z \in \mathbb{R},\\ m(z) \to I, & z \to \infty,\end{cases} \tag{3.4.3}$$

where

$$m_\pm(z) = \lim_{\varepsilon\to 0} m(z + i\varepsilon; x, t),$$
$$v_{x,t}(z) = e^{-i(4tz^3+xz)\sigma_3} v(z) e^{i(4tz^3+xz)\sigma_3}, \tag{3.4.4}$$

$$v(z) = \begin{pmatrix} 1 - |r(z)|^2 & -\overline{r(z)} \\ r(z) & 1 \end{pmatrix}, \tag{3.4.5}$$

where $r(z)$ lies in a Schwartz space and satisfies

$$r(z) = -\overline{r(-z)}, \quad \sup_{z\in\mathbb{R}} |r(z)| < 1. \tag{3.4.6}$$

The solution of the inverse problem is given by

$$y(x,t) = 2i \lim_{z\to\infty} (zm(x,t,z))_{12}. \tag{3.4.7}$$

In particular, here $\theta = 4z^3 + \frac{x}{t}z = 4(z^3 - 3z_0^2 z)$ with two stationary phase points

$$\pm z_0 = \pm\sqrt{\frac{-x}{12t}}. \tag{3.4.8}$$

As before, we restrict ourselves to the physically interesting region, here described by $M^{-1} < z_0 < M$, $M > 1$ for any fixed constant $M > 1$. This implies $x < 0$.

In this case, the signature table for $\operatorname{Re} i\theta$ consists of six regions, as shown in Figure 3.6.

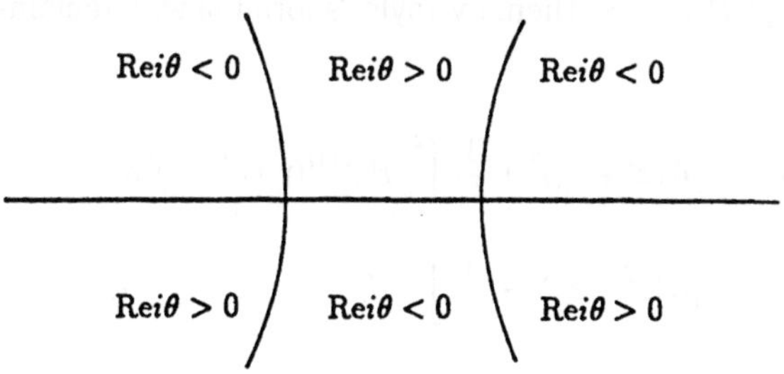

Figure 3.6: The signature table of Re $i\theta$.

Let $\delta(z)$ analytic in $\mathbb{C} \setminus \mathbb{R}$ be the solution of the following scalar RH problem:

$$\begin{cases} \delta_+(z) = \delta_-(z)(1 - |r(z)|^2), & |z| < z_0, \\ \qquad\;\; = \delta_-(z), & |z| > z_0, \\ \delta(z) \to 1, & z \to \infty. \end{cases} \tag{3.4.9}$$

This problem can be solved by the following formula:

$$\delta(z) = \left(\frac{z - z_0}{z + z_0}\right)^{vi} e^{\chi(z)}, \tag{3.4.10}$$

where

$$\nu = -\frac{1}{2\pi}\log(1 - |r(z_0)|^2) > 0, \tag{3.4.11}$$

$$\chi(z) = \frac{1}{2\pi i}\int_{-z_0}^{z_0} \log\left(\frac{1 - |r(\xi)|^2}{1 - |r(z_0)|^2}\right)\frac{d\xi}{\xi - z}. \tag{3.4.12}$$

After conjugation, we have

$$\begin{aligned} \delta_-^{\sigma_3} v_{x,t} \delta_+^{-\sigma_3} &= e^{-it\theta\sigma_3}\begin{pmatrix} 1 & 0 \\ r\delta_-^{-2}(1 - |r|^2)^{-1} & 1 \end{pmatrix}\begin{pmatrix} 1 & -\bar{r}\delta_+^2(1 - |r|^2)^{-1} \\ 0 & 1 \end{pmatrix} e^{it\theta\sigma_3}, \quad |z| < z_0, \\ &= e^{-it\theta\sigma_3}\begin{pmatrix} 1 & -\bar{r}\delta^2 \\ 0 & 1 \end{pmatrix}\begin{pmatrix} 1 & 0 \\ r\delta^{-2} & 1 \end{pmatrix} e^{it\theta\sigma_3}, \quad |z| > z_0. \end{aligned} \tag{3.4.13}$$

Consider $|z| < z_0$ and let $k = 4q + 1$, $q \in \mathbb{Z}_+$ be any positive integer. Splitting

$$\rho(z) \equiv -\bar{r}(z)(1 - |r(z)|^2)^{-1} \tag{3.4.14}$$

into even and odd parts, we obtain

$$\rho(z) = H_e(z^2) + zH_o(z^2)$$

for suitable smooth functions $H_e(\cdot), H_o(\cdot) \in \mathcal{S}$. Then, by Taylor's formula with remainder, we have

$$H_e(z^2) = \mu_0^e + \mu_1^e(z^2 - z_0^2) + \cdots + \mu_k^e(z^2 - z_0^2)^k + \frac{1}{k!}\int_{z_0^2}^{z^2} H_e^{(k+1)}(y)(z^2 - y)^k dy,$$

$$H_o(z^2) = \mu_0^o + \mu_1^o(z^2 - z_0^2) + \cdots + \mu_k^o(z^2 - z_0^2)^k + \frac{1}{k!}\int_{z_0^2}^{z^2} H_o^{(k+1)}(y)(z^2 - y)^k dy.$$

Set

$$R(z) = R_k(z) = \sum_{i=0}^{k} \mu_i^e(z^2 - z_0^2)^i + z\sum_{i=0}^{k} \mu_i^o(z^2 - z_0^2)^i, \quad h(z) = \rho(z) - R(z), \tag{3.4.15}$$

where $R(z)$ is a polynomial in z of order $2k + 1$, such that

$$\left.\frac{d^j\rho(z)}{dz^j}\right|_{z=\pm z_0} = \left.\frac{d^jR(z)}{dz^j}\right|_{z=\pm z_0}, \quad \left.\frac{d^jh(z)}{dz^j}\right|_{z=\pm z_0} = 0, \quad 0 \le j \le k.$$

As $h(z)$ vanishes to arbitrarily high order at $z = \pm z_0$, it is possible to split it further in analogy with (3.3.27). We proceed as follows.

For $0 < z_0 < M$, set $\alpha(z) = (z^2 - z_0^2)^q$. Consider the Fourier transform with respect to θ, so

$$\left(\frac{h}{\alpha}\right)(z) = \frac{1}{\sqrt{2\pi}} \int_{-\infty}^{\infty} e^{is\theta(z)} \widehat{\left(\frac{h}{\alpha}\right)}(s) ds, \quad |z| < z_0, \tag{3.4.16}$$

where

$$\widehat{\left(\frac{h}{\alpha}\right)}(s) = -\frac{1}{\sqrt{2\pi}} \int_{-z_0}^{z_0} e^{-is\theta(z)} \left(\frac{h}{\alpha}\right)(z) d\theta(z), \quad s \in \mathbb{R}. \tag{3.4.17}$$

As $z \mapsto \theta(z)$ is one-to-one in $|z| < z_0$, we define

$$\frac{h}{\alpha}(\theta) = \begin{cases} \frac{h(z(\theta))}{\alpha(z(\theta))}, & -8z_0^3 = \theta(z_0) < \theta < \theta(-z_0) = 8z_0^3, \\ 0, & |\theta| \geq 8z_0^3. \end{cases} \tag{3.4.18}$$

Thus, as $|\theta| \to 8z_0^3$, $|\theta| < 8z_0^3$, we have $\frac{h}{\alpha}(\theta) = O((z^2(\theta) - z_0^2)^{k+1-q})$. As $dz/d\theta = (12(z^2(\theta) - z_0^2))^{-1}$, we see that

$$\frac{h}{\alpha} \in H^j(-\infty < \theta < \infty), \quad 0 \leq j \leq \frac{3q+2}{2}. \tag{3.4.19}$$

Split

$$\begin{aligned} h &= \frac{1}{\sqrt{2\pi}} \alpha(z) \int_t^{\infty} e^{is\theta(z)} \widehat{\left(\frac{h}{\alpha}\right)}(s) ds + \frac{1}{\sqrt{2\pi}} \alpha(z) \int_{-\infty}^{t} e^{is\theta} \widehat{\left(\frac{h}{\alpha}\right)}(s) ds \\ &\equiv h_I(z) + h_{II}(z). \end{aligned} \tag{3.4.20}$$

Then we find

$$|e^{-2it\theta(z)} h_I(z)| \leq \frac{C}{t^{p-1/2}}, \tag{3.4.21}$$

for $|z| \leq z_0 \leq M$ and any $p \leq (3q+2)/2$.

On the other hand, from the signature of $\operatorname{Re} i\theta$, we deduce

$$|e^{-2it\theta(z)} h_{II}(z)| \leq \frac{C}{t^{q/2}}, \tag{3.4.22}$$

for z on the lines

$$z(u) = z_0 + u z_0 e^{3i\pi/4}, \quad 0 \leq u \leq \sqrt{2}, \tag{3.4.23}$$

$$z(u) = -z_0 + u z_0 e^{i\pi/4}, \quad 0 \leq u \leq \sqrt{2}, \tag{3.4.24}$$

where $z_0 < M$. Finally, fix $0 < \epsilon < \sqrt{2}$. Then on the part $\epsilon < u < \sqrt{2}$ of line (3.4.23) and (3.4.24) away from $\pm z_0$, we have

$$|e^{-2it\theta(z)}R(z)| \leq Ce^{-16tz_0^3u^2} \leq Ce^{-16\epsilon^2\tau} \tag{3.4.25}$$

for $0 < z_0 < M$, where

$$\tau = tz_0^3 = (-x/12t^{1/3})^{3/2}. \tag{3.4.26}$$

For $z_0 > M^{-1}$, we set

$$\tilde{\alpha}(z) = \frac{((z/z_0)^2 - 1)^q}{(z+i)^2}, \tag{3.4.27}$$

rescale the phase

$$\tilde{\theta}(z) = \frac{z^3 - 3z_0^2 z}{3z_0^2}, \tag{3.4.28}$$

and, noting that $12tz_0^2 = -x$, obtain $t\theta(z) = |x|\tilde{\theta}(z)$. Thus, in this case, we get, for $z_0 > M^{-1}$,

$$|e^{-2it\theta(z)}h_I(z)| \leq \frac{C}{(1+z^2)|x|^{p-1/2}}, \quad |z| \leq z_0,\ p \leq \frac{3q+2}{2} \tag{3.4.29}$$

and, on the lines (3.4.23) and (3.4.24),

$$|e^{-2it\theta(z)}h_{II}(z)| \leq \frac{C}{|z+i|^2|x|^{q/2}}. \tag{3.4.30}$$

Finally, for $0 < \epsilon < \sqrt{2}$ as above, on the part $\epsilon < u < \sqrt{2}$ of the lines (3.4.23) and (3.4.24) away from $\pm z_0$, we have

$$|e^{-2it\theta(z)}R(z)| \leq C(z_0)e^{-4/3|x|z_0u^2} \leq C(z_0)e^{-(4\epsilon^2/3M)|x|} \tag{3.4.31}$$

for $z_0 > M^{-1}$, where $C(z_0)$ is rapidly decreasing as $z_0 \to \infty$.

For $|z| \geq z_0$, it suffices to consider $z \geq z_0$, as the case where $z \leq -z_0$ is similar. By the method used in Section 3.3 and the above discussion, we can get the estimates for $\rho(z) = \overline{r(z)}$. By taking complex, a similar splitting with similar estimates can be obtained for the factors $r(1-|r|^2)^{-1}$, $r(z)$.

We summarize the results as follows. Let l be an arbitrary positive integer and let $k = 4q+1$ be sufficiently large, so that $(3q+2)/2 - 1/2 > (3q+2)/3 - 1/2 > q/2 > q/3$ are all greater than l. Let L denote the contour

$$L : \{z = z_0 + z_0 u e^{3i\pi/4} : -\infty < u \le \sqrt{2}\}$$
$$\cup \{z = -z_0 + z_0 u e^{i\pi/4} : -\infty < u \le \sqrt{2}\} \qquad (3.4.32)$$

and set

$$L_\epsilon : \{z = z_0 + z_0 u e^{3i\pi/4} : \epsilon < u \le \sqrt{2}\}$$
$$\cup \{z = -z_0 + z_0 u e^{i\pi/4} : \epsilon < u \le \sqrt{2}\}. \qquad (3.4.33)$$

Lemma 3.4.1. *Let*

$$\rho(z) = \begin{cases} -\overline{r(z)}\left(1 - |r(z)|^2\right)^{-1}, & |z| < z_0, \\ \bar{r}(z), & |z| > z_0. \end{cases} \qquad (3.4.34)$$

Then ρ has a decomposition

$$\rho(z) = h_I(z) + h_{II}(z) + R(z), \quad z \in \mathbb{R}, \qquad (3.4.35)$$

where $R(z)$ is piecewise rational and $h_{II}(z)$ has an analytic continuation to L satisfying

$$|e^{-2it\theta(z)} h_I(z)| \le \begin{cases} \frac{C}{(1+|z|^2)t^l}, & z \in \mathbb{R},\ 0 < z_0 < M, \\ \frac{C}{(1+|z|^2)|x|^l}, & z \in \mathbb{R},\ z_0 > M^{-1}, \end{cases} \qquad (3.4.36)$$

$$|e^{-2it\theta(z)} h_{II}(z)| \le \begin{cases} \frac{C}{(1+|z|^2)t^l}, & z \in L,\ 0 < z_0 < M, \\ \frac{C}{(1+|z|^2)|x|^l}, & z \in L,\ z_0 > M^{-1}, \end{cases} \qquad (3.4.37)$$

$$|e^{-2it\theta(z)} R(z)| \le \begin{cases} Ce^{-16\epsilon^2\tau}, & z \in L_\epsilon,\ 0 < z_0 < M, \\ C(z_0)e^{-(4\epsilon^2/3M)|x|}, & z \in L_\epsilon,\ z_0 > M^{-1}. \end{cases} \qquad (3.4.38)$$

Taking conjugates

$$\overline{\rho(z)} = \overline{h_I(z)} + \overline{h_{II}(z)} + \overline{R(z)} \qquad (3.4.39)$$

leads to the same estimates for $e^{2it\theta(z)}\overline{h_I(z)}$, $e^{2it\theta(z)}\overline{h_{II}(z)}$, and $e^{2it\theta(z)}\overline{R(z)}$ on $\mathbb{R} \cup \bar{L}$.

We then get

$$\begin{cases} m_+^\sharp(z) \equiv m_-^\sharp(z) v_{x,t}^\sharp(z), & z \in \Sigma, \\ m_+^\sharp(z) \to I, & z \to \infty, \end{cases} \qquad (3.4.40)$$

on a contour Σ, as shown in Figure 3.7. On $\mathbb{R} \subset \Sigma$, the coefficients of $v_{x,t}^\sharp(z)$ depend on h_I and on $\Sigma \setminus \mathbb{R}$, they depend on both R and h_{II} (in addition to $\delta_\pm$).

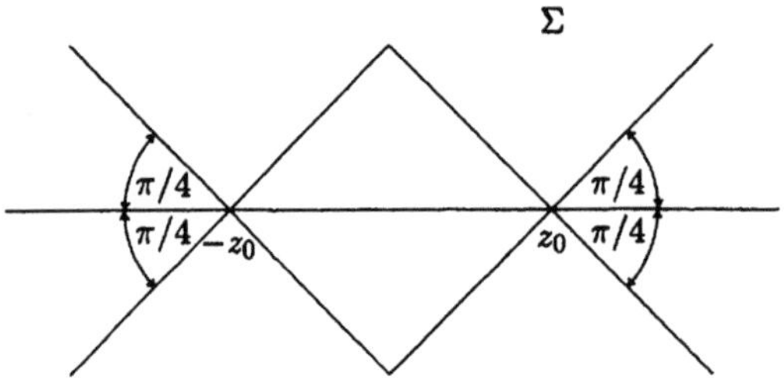

Figure 3.7: The contour Σ.

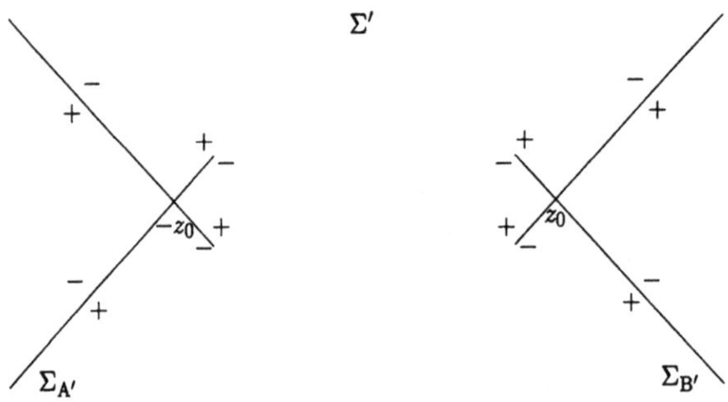

Figure 3.8: The contour Σ′.

As $t \to \infty$, we get $e^{-2it\theta(z)} h_I \to 0$. On $\Sigma \setminus \mathbb{R}$, the term $e^{-2it\theta(z)} h_{II}$ also converges to 0. Similarly, on the finite "triangulaire" part of $\Sigma \setminus \mathbb{R}$ away from $\pm z_0$, the contribution from $R(z)e^{-2it\theta(z)}$ can be neglected. We are left with the RH problem

$$\begin{aligned} &m'_+(z) = m'_-(z) v_{x,t}(z), && z \in \Sigma', \\ &m'(z) \to I, && z \to \infty, \end{aligned} \tag{3.4.41}$$

on Σ', which is a union of two crosses, $\Sigma' = \Sigma_{A'} \cup \Sigma_{B'}$, as depicted in Figure 3.8.

Moreover, as $t \to \infty$, the interaction between the RH problem on $\Sigma_{A'}$ and $\Sigma_{B'}$ tends to zero faster than the leading order of the solution and the contribution of $\Sigma_{A'} \cup \Sigma_{B'}$ to $y(x,t)$ is simply the sum of the separate contributions from the two RH problems on $\Sigma_{A'}$ and $\Sigma_{B'}$. Symmetry implies that we need only to consider the RH problem on $\Sigma_{B'}$.

We first extend $\Sigma_{B'}$ to a full cross by setting the jump matrix on the dotted lines in Figure 3.9 equal to the identity matrix I. Define the scaling operator

$$\begin{aligned} N : L^2(\hat{\Sigma}_{B'}) &\to L^2(\hat{\Sigma}_{B'} - z_0), \\ f(z) &\mapsto Nf(z) = f\left(\frac{z}{\sqrt{48tz_0}} + z_0\right). \end{aligned}$$

Instead of (3.3.34), we have

$$N\delta^{\operatorname{ad}\sigma_3} e^{-it\theta \operatorname{ad}\sigma_3}[\tilde{v}] \to (\delta_B^0)^{\operatorname{ad}\sigma_3} z^{\nu i \operatorname{ad}\sigma_3} e^{-\frac{iz^2}{4}\operatorname{ad}\sigma_3}[\tilde{v}](z_0), \tag{3.4.42}$$

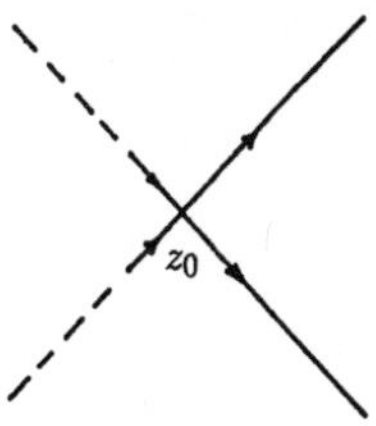

Figure 3.9: The extended contour of $\hat{\Sigma}_{B'}$.

where $[\tilde{v}](z_0)$ appears in Figure 3.5 and

$$\delta_B^0 = (192tz_0^3)^{\frac{-i\nu}{2}} e^{8itz_0^3} e^{\chi(z_0)}. \tag{3.4.43}$$

Thus, the calculation of the long-time behavior of the MKdV equation reduces to the same explicitly solvable isomonodromy problem as in the NLS equation. The result is the following. Let

$$\begin{aligned}\phi(z_0) &= \arg\Gamma(i\nu) + \frac{3\pi}{4} - \arg r(z_0)\\ &\quad - \frac{1}{\pi}\int_{-z_0}^{z_0} \log\left(\frac{1-|r(\xi)|^2}{1-|r(z_0)|^2}\right)\frac{d\xi}{\xi - z_0},\end{aligned} \tag{3.4.44}$$

where Γ is the standard gamma function. Denote

$$y_a = -i\left(\frac{\nu}{3tz_0}\right)^{\frac{1}{2}} \cos(16tz_0^3 - \nu\log(192tz_0^3) + \phi(z_0)). \tag{3.4.45}$$

For all x, let

$$z_0 = \sqrt{\frac{-x}{12t}} \tag{3.4.46}$$

and let M be fixed constant greater than 1.

Theorem 3.4.2. *Suppose $y_0(x)$ lies in a Schwartz space with the reflection coefficient $r(z)$. Then, as $t \to \infty$, the solution $y(x,t)$ of the MKdV equation* (3.4.1) *with initial value $y_0(x)$ has uniform leading asymptotics*

$$y(x,t) = y_a + O\left(\frac{\log t}{t}\right)$$

in the region $M^{-1} \le \sqrt{\frac{-x}{12t}} \le M$.

4 Interaction of solitons and its asymptotic properties

4.1 Interaction of solitons and its asymptotic properties as $t \to \infty$

Solitons can keep their original amplitude and shape after a nonlinear interaction, which was first found by Kruskal and Zabusky using numerical calculations [314]. A few years later, Lax gave a rigorous analytical proof [153]. In addition, Lax analyzed the process of interaction between two solitons in detail and pointed out that:

(i) In the case where the velocity $c_1 \gg c_2$, the first envelop is higher and faster than the second one. If the first one is located on the left side of the second one, then the first one overtakes the other. During the interaction, we find the maximum value (peak value): the bigger one absorbs the smaller and cancels it out (as seen as in Figure 4.1).

(ii) In the case where the velocity $c_1 \approx c_2$, the bigger envelop catches up with the smaller one. With the decreasing of the bigger amplitude and the increasing of the smaller one, there exist two peak values. Next, this process will be exchanged in what follows. In addition, Lax analyzed the behavior at $t \to \infty$. $u(x,t)$ is the solution for the KdV equation

$$u_t + uu_x + u_{xxx} = 0, \tag{4.1.1}$$

where $u \to 0$ as $x = \pm\infty$. There exist n discrete positive numbers $c_1, c_2, \ldots, c_N$ (named u as the eigenvelocity) and the phase $\theta_j^{\pm}$, which satisfy

$$\lim_{t\to\pm\infty} u(x+ct,t) = \begin{cases} s(\xi - \theta_j^{\pm}, c_j), & c = c_j, \\ 0, & c \neq c_j, \end{cases} \tag{4.1.2}$$

where s stands for the soliton solution for equation (4.1.1), $\xi = x - c_j t$.

In the next part, based on the N-soliton solutions obtained through the inverse scattering transform, we will prove Lax's theory via algebraic analysis, which reads as follows. Solutions for the KdV equation at $t \to \infty$ will be made up from the soliton with N eigenvalues, if there exists a soliton with N eigenvalues $k_1, k_2, \ldots, k_N$ at $t \to -\infty$, except for some phase shifts.

From equation (2.2.19) in Chapter 2, we have

$$K(x,y,t) = -\sum_{m=1}^{N} c_m \varphi_m e^{-k_m y},$$

$$u = -2\frac{d}{dx}K(x,x;t) = 2\frac{d}{dx}\sum_{m=1}^{N} c_m \varphi_m e^{-k_m x},$$

https://doi.org/10.1515/9783110549638-004

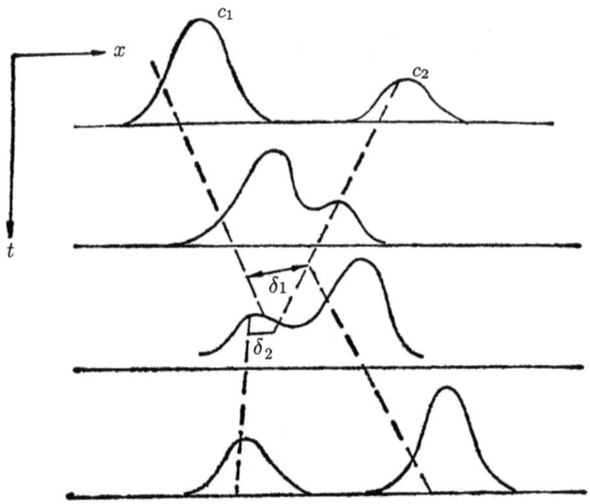

Figure 4.1: The interaction process when $c_1 \gg c_2$.

$$\equiv 2\frac{d}{dx}\sum_{m=1}^{N} f_m(x) \equiv 2\sum_{m=1}^{N} f'_m(x),$$

with

$$f_m(x) = c_m \varphi_m e^{-k_m x}.$$

To compute $\sum_m f'_m$ $(t \to \pm\infty)$, we need to determine f_m and rewrite

$$\varphi_m(x) + \sum_{n=1}^{N} c_m c_n \frac{e^{-(k_m+k_n)x}}{k_m + k_n}\varphi_n = c_m e^{-k_m x} \tag{4.1.3}$$

as

$$c_m^{-2} e^{2k_m x} f_m(x) + \sum_{n=1}^{N} \frac{f_n(x)}{k_m + k_n} = 1 \quad (m = 1, 2, \ldots, N). \tag{4.1.4}$$

Upon derivation with respect to x, we get

$$c_m^{-2} e^{2k_m x} f'_m(x) + \sum_{n=1}^{N} \frac{f'_n(x)}{k_m + k_n} = -2k_m c_m^{-2} e^{2k_m x} f_m. \tag{4.1.5}$$

In order to discuss the asymptotic behavior at $|t| \to \infty$, we choose the motion coordinate system

$$\xi \equiv x - 4k_p^2 t, \quad p = 1, 2, \ldots, N, \tag{4.1.6}$$

where $\lambda_p = -k_p^2$ means the eigenvalue of the pth soliton, $4k_p^2$ is the velocity, and $2k_p^2$ is the amplitude. We have

$$c_m^{-2} e^{2k_m x} = c_m(0)^{-2} \exp\{-8k_m(k_m^2 - k_p^2)t + 2k_m \xi\}$$

$$\equiv c_m(\xi)\exp\{-8k_m(k_m^2 - k_p^2)t\},$$

where $c_m(\xi) = c_m(0)^{-2}e^{2k_m\xi}$. We substitute the above expression into equations (4.1.4) and (4.1.5), to obtain

$$c_m(\xi)e^{-8k_m(k_m^2-k_p^2)t}f_m + \sum_{n=1}^{N}\frac{f_n}{k_m+k_n} = 1, \tag{4.1.7}$$

$$c_m(\xi)e^{-8k_m(k_m^2-k_p^2)t}f_m' + \sum_{n=1}^{N}\frac{f_n'}{k_m+k_n} = -2k_m c_m(\xi)e^{-8k_m(k_m^2-k_p^2)t}f_m. \tag{4.1.8}$$

Now, assume $k_1 > k_2 > \cdots > k_N > 0$.

(1) Asymptotic behavior at $t \to \infty$.
Taking the limit of equation (4.1.7), we have

$$\begin{cases} \sum_{n=1}^{N}\frac{f_n}{k_m+k_n} = 1, \quad m = 1,2,\ldots,p-1, \\ c_pf_p + \sum_{n=1}^{N}\frac{f_n}{k_p+k_n} = 1, \quad m = p, \\ f_m = 0, \quad m = p+1,\ldots,N, \end{cases}$$

which can be simplified as

$$\sum_{n=1}^{p}\frac{f_n}{k_m+k_n} = 1 - c_p\delta_{mp}f_p \quad (m = 1,2,\ldots,p), \tag{4.1.9}$$

$$\sum_{n=1}^{p}\frac{f_n'}{k_m+k_n} = -c_p\delta_{mp}(2k_pf_p + f_p') \quad (m = 1,2,\ldots,p), \tag{4.1.10}$$

$$f_m' = -2k_mf_m = 0, \quad (m = p+1,\ldots,N). \tag{4.1.11}$$

Matrix $K_p = (\frac{1}{k_m+k_n})$ $(m = 1,2,\ldots,p)$ possesses a positive determinant. In fact,

$$\begin{aligned} 0 < \det C &= \det\left(c_mc_n\frac{e^{-(k_m+k_n)x}}{k_m+k_n}\right) \\ &= \det\left(\frac{1}{k_m+k_n}\right)\prod_{m=1}^{N}c_m^2e^{-2\sum_{m=1}^{N}k_mx}, \end{aligned}$$

so $\det = (\frac{1}{k_m+k_n}) > 0$. f_m and f_m' can be solved from equations (4.1.9) and (4.1.10) through Gramer's rule. We have

$$f_m \det K_p = \sum_{n=1}^{p}K_{mn} - c_pK_{pm}f_p \quad (m = 1,2,\ldots,p), \tag{4.1.12}$$

$$f_m' \det K_p = -c_pK_{pm}(2k_pf_p - f_p') \quad (m = 1,2,\ldots,p), \tag{4.1.13}$$

where K_{mn} is the cofactor of the matrix element $(\frac{1}{k_m+k_n})$. Denote L_p as the matrices of K_p in which the elements of the last line are 1. Letting $m = p$, we get

$$f_p = \frac{\det L_p}{\det K_p + c_p \det K_{p-1}},$$
$$f_p' = -\frac{2c_p k_p f_p \det K_{p-1}}{\det K_p + c_p \det K_{p-1}}.$$

Summing up equation (4.1.13) and considering the expressions of f_p, f_p', we obtain

$$\begin{aligned}\sum_{m=1}^{p} f_m' \det K_p &= -\sum_{m=1}^{p} c_p K_{pm}(2k_p f_p - f_p') \\ &= -\sum_{m=1}^{p} c_p K_{pm} 2k_p f_p \left(1 - \frac{c_p \det K_{p-1}}{\det K_p + c_p \det K_{p-1}}\right) \\ &= -\sum_{m=1}^{p} c_p K_{pm} f_p 2k_p \frac{\det K_p}{\det K_p + c_p \det K_{p-1}} \\ &= -c_p \det L_p f_p 2k_p \frac{\det K_p}{\det K_p + c_p \det K_{p-1}} \\ &= -2k_p c_p (\det L_p)^2 \frac{\det K_p}{(\det K_p + c_p \det K_{p-1})^2}.\end{aligned}$$

Therefore,

$$\lim_{t\to\infty,\xi \text{ fixed}} \sum_{m=1}^{p} f_m' = -\frac{2k_p c_p}{[\frac{\det K_p}{\det L_p} + c_p \frac{\det K_{p-1}}{\det L_p}]^2}.$$

Subtracting the last line from the other lines of K_p, we obtain

$$\det K_p = \frac{\prod_{m=1}^{p-1}(k_p - k_m)}{\prod_{m=1}^{p}(k_p + k_m)} \det L_p.$$

In like manner, subtracting the last line from the other lines of L_p,

$$\det L_p = \frac{\prod_{m=1}^{p-1}(k_p - k_m)}{\prod_{m=1}^{p-1}(k_p + k_m)} \det K_{p-1},$$
$$\frac{\det K_p}{\det L_p} = \frac{\prod_{m=1}^{p-1}(k_p - k_m)}{\prod_{m=1}^{p}(k_p + k_m)},$$
$$\frac{\det K_{p-1}}{\det L_p} = \frac{\prod_{m=1}^{p-1}(k_p + k_m)}{\prod_{m=1}^{p-1}(k_p - k_m)}.$$

Noticing that $c_p = c_p(0)^{-2}e^{2k_p\xi}$, we define ξ_p as

$$e^{2k_p\xi_p} \equiv \frac{c_p^2(0)}{2k_p}\prod_{m=1}^{p-1}\left(\frac{k_p-k_m}{k_p+k_m}\right)^2$$

and infer

$$\begin{aligned}
\lim_{t\to\infty,\xi\text{ fixed}} &= \lim_{t\to\infty,\xi\text{ fixed}} 2\sum_{m=1}^{p} f'_m = \frac{4k_pc_p}{\left[\frac{\prod_{m=1}^{p-1}(k_p-k_m)}{\prod_{m=1}^{p}(k_p+k_m)} + c_p\frac{\prod_{m=1}^{p-1}(k_p+k_m)}{\prod_{m=1}^{p-1}(k_p-k_m)}\right]^2} \\
&= -\frac{4k_pc_p}{\prod_{m=1}^{p-1}\left(\frac{k_p-k_m}{k_p+k_m}\right)^2\left[\frac{1}{4k_p^2} + \frac{2c_p}{2k_p}\prod_{m=1}^{p-1}\left(\frac{k_p+k_m}{k_p-k_m}\right)^2 + c_p^2\prod_{m=1}^{p-1}\left(\frac{k_p+k_m}{k_p-k_m}\right)^4\right]} \\
&= -16k_p^3c_p\prod_{m=1}^{p-1}\left(\frac{k_p+k_m}{k_p-k_m}\right)^2\left[1+2k_pc_p\prod_{m=1}^{p-1}\left(\frac{k_p+k_m}{k_p-k_m}\right)^2\right]^{-2} \\
&= -8k_p^2\left[2k_pc_p\prod_{m=1}^{p-1}\left(\frac{k_p+k_m}{k_p-k_m}\right)^2\right]\cdot\left[1+2k_pc_p\prod_{m=1}^{p-1}\left(\frac{k_p+k_m}{k_p-k_m}\right)^2\right]^{-2} \\
&= -8k_p^2e^{2k_p(\xi-\xi_p)}\left[1+e^{2k_p(\xi-\xi_p)}\right]^{-2} \\
&= -2k_p^2\operatorname{sech}^2[k_p(\xi-\xi_p)] \\
&= -2k_p^2\operatorname{sech}^2[k_p(x-4k_p^2t-\xi_p)],
\end{aligned}$$

which means that there exists a soliton at the surroundings of $x = 4k_p^2t$ with amplitude $2k_p^2$ and velocity $4k_p^2$.

(2) Asymptotic behavior at $t \to -\infty$.

From equations (4.1.7) and (4.1.8), we get

$$\begin{aligned}
\sum_{m=p}^{N}\frac{f_n}{k_m+k_n} &= 1 - c_p\delta_{mp}f_p, \quad (m = p, \ldots, N), \\
\sum_{m=p}^{N}\frac{f'_n}{k_m+k_n} &= -c_p\delta_{mp}(2k_pf_p + f'_p), \quad (m = p, \ldots, N), \\
f'_m &= -2k_mf_m = 0, \quad (m = 1, 2, \ldots, p-1).
\end{aligned}$$

The discussion when $t \to \infty$ is similar. Defining ξ_p as

$$\xi_p e^{2k_p\xi_p} \equiv \frac{c_p^2(0)}{2k_p}\prod_{m=p+1}^{N}\left(\frac{k_p-k_m}{k_p+k_m}\right)^2,$$

we obtain

$$\lim_{t\to-\infty} u(x,t) = -2k_p^2\operatorname{sech}^2[k_p(x-4k_p^2t-\bar{\xi}_p)]$$

and phase shifts

$$\xi_p - \bar{\xi}_p = \frac{1}{k_p}\left[\sum_{m=1}^{p-1} \log\left(\frac{k_m - k_p}{k_m + k_p}\right) - \sum_{m=p+1}^{N} \log\left(\frac{k_p - k_m}{k_p + k_m}\right)\right].$$

There is another method to analyze the interaction between finite solitons. The KdV equation reads

$$u_t + \delta u u_x + u_{xxx} = 0. \tag{4.1.14}$$

Via introduction of the transformation $u = p_x$, we find

$$(p_t)_x + \delta\left(\frac{1}{2}p_x^2\right)_x + (p_{xxx})_x = 0.$$

Integrating this with respect to x, we get

$$p_t + \frac{1}{2}\delta p_x^2 + p_{xxx} = 0. \tag{4.1.15}$$

Substituting the transformation $\delta p = 12(\log F)_x$ into the above expression and omitting and meshing some terms, we have

$$F(F_t + F_{xxx})_x - F_x(F_t + F_{xxx}) + 3(F_{xx}^2 - F_x F_{xxx}) = 0. \tag{4.1.16}$$

Take notice of the operator $L = \frac{\partial}{\partial t} + \frac{\partial^3}{\partial x^3}$ in equation (4.1.16), while $F = 1 + e^{-\alpha(x-s)+\alpha^3 t}$ (α, s being real constants) is the special solution for $F_t + F_{xxx} = 0$. Due to the nonlinearity of equation (4.1.16), the linear superposition is not useful. We expand the interaction term as

$$F = 1 + F^{(1)} + F^{(2)} + \cdots$$

and we get the following series of equations by substituting the above expansion into equation (4.1.16):

$$\begin{aligned} &\{F_t^{(1)} + F_{xxx}^{(1)}\}_x = 0, \\ &\{F_t^{(2)} + F_{xxx}^{(2)}\}_x = -3\{F_{xx}^{(1)} + F_x^{(1)} F_{xxx}^{(1)}\}, \\ &\cdots \end{aligned}$$

We terminate the first two terms of $F^{(1)}$ as $F^{(1)} = f_1 + f_2, f_j = e^{-\alpha_j(x-s_j)+\alpha_j^3 t}$ $(j = 1, 2)$, which satisfy the first equation. $F^{(2)}$ can be solved as

$$\begin{aligned} \{F_t^{(2)} + F_{xxx}^{(2)}\}_x &= 3\alpha_1\alpha_2(\alpha_2 - \alpha_1)^2 f_1 \cdot f_2, \\ F^{(2)} &= \frac{(\alpha_2 - \alpha_1)^2}{(\alpha_2 + \alpha_1)^2} f_1 \cdot f_2. \end{aligned}$$

In addition, we obtain $F^{(3)} = F^{(4)} = \cdots = 0$. Consequently, the exact solution for equation (4.1.16) can be expressed as

$$F = 1 + f_1 + f_2 + \frac{(\alpha_2 - \alpha_1)^2}{(\alpha_2 + \alpha_1)^2} f_1 \cdot f_2, \tag{4.1.17}$$

which possesses the terms of f_1, f_2, but not the terms of f_1^2 and f_2^2. Similar results can be generalized to $N - f_j$. Assuming $F^{(1)} = \sum_{j=1}^{N} f_j$, we have

$$F = 1 + \sum_j f_j + \sum_{j \neq k} a_{ij} f_j f_k + \sum_{j \neq k \neq l} a_{jkl} f_j f_k f_l + \cdots + a_{1,2,\ldots,N} f_1 f_2 \cdots f_N.$$

In fact, it can be proved that $F = \det |F_{mn}|$, with $F_{mn} = \delta_{mn} + \frac{2\alpha_m}{\alpha_m + \alpha_n} f_m$, which corresponds to $C = (\delta_{mn} + \frac{e^{-(k_m + k_n)x}}{k_m + k_n})$ from the inverse scattering transform. Taking $N = 2$, $\delta u = \delta p_x = 12(\log F)_{xx}$, and equation (4.1.17), the solution for the KdV equation (4.1.14) reads

$$\begin{aligned} \frac{\delta u}{12} &= \{\alpha_1^2 f_1 + \alpha_2^2 f_2 + 2(\alpha_2 - \alpha_1)^2 f_1 f_2 + [(\alpha_2 - \alpha_1)/(\alpha_2 + \alpha_1)]^2 \\ &\quad \times (\alpha_2^2 f_1^2 f_2 + \alpha_1^2 f_1 f_2^2)\} / \{[1 + f_1 + f_2 + [(\alpha_2 - \alpha_1)/(\alpha_2 + \alpha_1)]^2 f_1 f_2]^2\}, \\ f_j &= \exp[-\alpha_j (x - s_j) + \alpha_j^3 t]. \end{aligned} \tag{4.1.18}$$

The one-soliton solution $\delta u = 3\alpha^2 \operatorname{sech}^2 \frac{\theta - \theta_0}{2}$ can be expressed via $f = e^{-\alpha(x-s)+\alpha^3 t}$, with $\theta = \alpha x - \alpha^3 t$, $\theta_0 = s\alpha$ and

$$\frac{\delta u}{12} = \frac{\alpha^2 f}{(1+f)^2}.$$

When $f = 1$, δu arrives at the maximum amplitude $3\alpha^2$ at $x = s + \alpha^2 t$ with velocity $c = \alpha^2$. For several instances, we discuss the interaction and asymptotic behavior at $t \to \pm\infty$ based on equation (4.1.18).

Case 1. $f_1 \approx 1$, with f_2 being very large or small at area (x, t). We have:

(1) if $f_1 \approx 1$ and $f_2 \ll 1$,

$$\frac{\delta u}{12} \approx \frac{\alpha_1^2 f_1}{(1+f_1)^2} \quad \text{is exactly the } \alpha_1 \text{ wave;}$$

(2) if $f_1 \approx 1$ and $f_2 \gg 1$,

$$\frac{\delta u}{12} \approx \frac{[(\alpha_2 - \alpha_1)/(\alpha_2 + \alpha_1)]^2 \alpha_1^2 f_1 f_2^2}{\{f_2 + [(\alpha_2 - \alpha_1)/(\alpha_2 + \alpha_1)]^2 f_1 f_2\}^2} = \frac{\alpha_1^2 \tilde{f}_1}{(1 + \tilde{f}_1)^2},$$

with

$$\tilde{f}_1 = \left(\frac{\alpha_2 - \alpha_1}{\alpha_2 + \alpha_1}\right)^2 f_1,$$

which is still the following α_1 wave with phase $\tilde{s_1}$:

$$\tilde{s_1} = s_1 - \frac{1}{\alpha_1}\log\left(\frac{\alpha_2+\alpha_1}{\alpha_2-\alpha_1}\right)^2.$$

Case 2. $f_2 \approx 1$, where f_1 is very large or small at area (x,t). We will get the α_2 wave.
Case 3. $\delta u \approx 0$, where both f_1 and f_2 are very large or small.
Case 4. $f_1 \approx 1$ and $f_2 \approx 1$ denote the interaction area.
Taking account of the notion that $\alpha_2 > \alpha_1 > 0$, as $t \to -\infty$, we find an α_1 wave. We have

$$\begin{aligned} &f_1 \approx 1, \quad x = s_1 + \alpha_1^2 t, \\ &f_2 = e^{-\alpha_2(x-s_2)+\alpha_2^3 t} = e^{-\alpha_2(s-s_2)-\alpha_2(\alpha_1^2-\alpha_2^2)t} \ll 1, \end{aligned}$$

which means that there is an α_1 wave at the point of $x = s_1 + \alpha_1^2 t$.
We find an α_2 wave,

$$\begin{aligned} &f_2 \approx 1, \quad x = s_2 - \frac{1}{\alpha_2}\log\left(\frac{\alpha_2+\alpha_1}{\alpha_2-\alpha_1}\right)^2 + \alpha_2^2 t, \\ &f_1 \gg 1, \end{aligned}$$

which means that there is an α_2 wave at the point of $x = s_2 - \frac{1}{\alpha_2}\log(\frac{\alpha_2+\alpha_1}{\alpha_2-\alpha_1})^2 + \alpha_2^2 t$, while $\delta u \approx 0$ at other points.
As $t \to \infty$,

$$\begin{aligned} &\alpha_1 \text{ wave}: \quad x = s_1 - \frac{1}{\alpha_1}\log\left(\frac{\alpha_2+\alpha_1}{\alpha_2-\alpha_1}\right)^2 + \alpha_1^2 t, \quad f_1 \approx 1, \quad f_2 \gg 1, \\ &\alpha_2 \text{ wave}: \quad x = s_2 + \alpha_2^2 t, \quad f_2 \approx 1, \quad f_1 \ll 1, \\ &\delta u \approx 0 \text{ at other points.} \end{aligned}$$

Therefore, we conclude that the parameters α_1 and α_2 keep unchanged after the interaction, apart from some phase shifts, so we have

$$\begin{aligned} &\alpha_2 \text{ moving forward } \frac{1}{\alpha_2}\log\left(\frac{\alpha_2+\alpha_1}{\alpha_2-\alpha_1}\right)^2, \\ &\alpha_1 \text{ moving backward } \frac{1}{\alpha_1}\log\left(\frac{\alpha_2+\alpha_1}{\alpha_2-\alpha_1}\right)^2. \end{aligned}$$

The interaction time and location at $f_1 \approx 1$ and $f_2 \approx 1$ are

$$x = s_1 + \alpha_1^2 t = s_2 + \alpha_2^2 t, \quad t = -\frac{s_2 - s_1}{\alpha_2^2 - \alpha_1^2}, \quad x = \frac{\alpha_2^2 s_1 - \alpha_1^2 s_2}{\alpha_2^2 - \alpha_1^2}.$$

4.2 Solution behavior for KdV equation under weak dispersion action and WKB method

Consider the Burgers equation

$$u_t + uu_x = \epsilon u_{xx} \ (\epsilon > 0),$$

where $u_\epsilon \to u$ as $\epsilon \to 0$ and where u is the generalized solution for

$$u_t + uu_x = 0.$$

As for the KdV equation

$$u_t + uu_x = \epsilon u_{xxx},$$

two problems concern us. Is $u_\epsilon \to u$ established at $\epsilon \to 0$? And is $u(x,t)$ the generalized solution for $u_t + uu_x = 0$? Generally speaking, the answer is negative. That is to say, the solution for the KdV equation cannot tend to any discontinuous solution which contains a shock wave.

The solution u_ϵ for the KdV equation

$$u_t + uu_x = \epsilon u_{xxx} \tag{4.2.1}$$

satisfies the conditions that u_ϵ and its derivatives tend to zero as ($|x| \to \infty$). Also,

$$\int_{-\infty}^{\infty} u_\epsilon(x,t)dx = \int_{-\infty}^{\infty} u(x,0)dx = M_0,$$
$$\int_{-\infty}^{\infty} \frac{1}{2}u_\epsilon^2(x,t)dx = \int_{-\infty}^{\infty} \frac{1}{2}u^2(x,0)dx = E.$$

Integrating $u_t+uu_x = 0$ with respect to $x \in (-\infty, +\infty)$, replacing the order of derivation and integration of t, and mining the jump on the discontinuous line $x = x(t)$, we get

$$\frac{dM}{dt} + D[u] = \frac{1}{2}[u^2],$$

where $D = \frac{dx(t)}{dt}$, $[f] = f_+ - f_- = f(x(t)+0) - f(x(t)-0)$, and $M = \int_{-\infty}^{\infty} u(x,t)dx$. Due to the momentum conservation, we get $\frac{dM}{dt} = 0$ and the shock wave relationship

$$D = \frac{dx}{dt} = \frac{1}{2}\frac{[u^2]}{[u]}.$$

We multiply $u_t + uu_x = 0$ by u and integrate it with respect to x, i.e.,

$$\frac{dE}{dt} + D\left[\frac{1}{2}u^2\right] = \frac{1}{3}[u^3], \quad E = \int_{-\infty}^{\infty} \frac{1}{2}u^2(x,t)dx,$$

from which we derive $\frac{dE}{dt} = \frac{1}{12}[u]^3$. Furthermore, we conclude that $\frac{dE}{dt} < 0$, due to the entropy condition $u_- > u_+$. As to the solution for equation (4.2.1), there exists the identity $\frac{dE}{dt} = 0$ ($E = E_0$). Thus, we declare that, as $\epsilon \to 0$, the solution u_ϵ for the KdV equation (4.2.1) cannot tend to any discontinuous solution which contains a shock wave for equation $u_t + uu_x = 0$. Next, we will examine the asymptotic behavior of the solution for equation (4.2.1). Suppose that the solution for equation (4.2.1) exists in the smooth transition zone with thickness $\triangle\epsilon$, which connects two different states, while $\triangle\epsilon \to 0$ as $\epsilon \to 0$. Introducing the moving coordinate system $\xi = \frac{x-x(t)}{\triangle}$, $t' = t$, $x(t)$ is the unknown shock wave trajectory. As $u(x,t) = u(\xi, t')$, we multiply equation (4.2.1) by $\triangle$, i.e.,

$$\epsilon \triangle^{-2} u_{\xi\xi\xi} - (u - D)u_\xi = \triangle \cdot u_{t'}. \tag{4.2.2}$$

Assume that $u_{t'}$ is bounded and $\triangle \cdot u_{t'} \to 0$ as $\triangle \to 0$. While we set $\triangle = o(\epsilon^{\frac{1}{2}}) = \epsilon^{1/2}$, equation (4.2.2) tends to the following differential equation as $\epsilon \to 0$:

$$u_{\xi\xi\xi} - (u - D)u_\xi = 0,$$

where $u \to u_1$ as $\xi \to +\infty$ and $u \to u_0$ as $\xi \to -\infty$. In this situation, solution $u(\xi, t')$ is composed of the oscillatory solution, as seen in Figure 4.2.

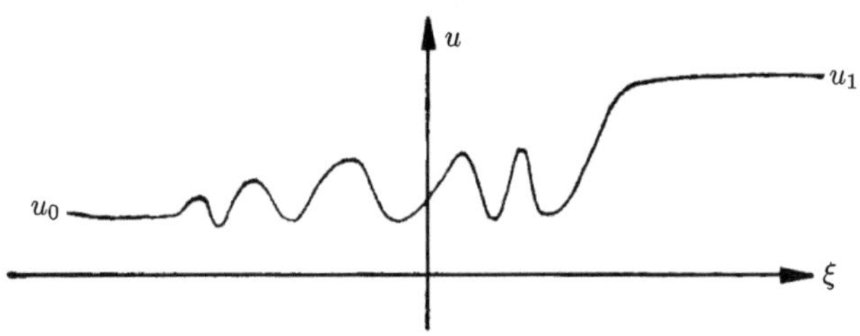

Figure 4.2: The profile of u.

Concerning qualitative theory, we have the following theorem.

Theorem 4.2.1. *The shock wave solution of the quasi-linear hyperbolic equation*

$$u_t + (f(u))_x = 0 \tag{4.2.3}$$

can never be obtained through the limit of the traveling wave solution for the KdV equation

$$u_t + (f(u))_x = \epsilon^2 u_{xxx}. \tag{4.2.4}$$

Proof. Consider the shock wave solution

$$\bar{u}(x,t) = \begin{cases} u_0, & x - Dt < 0, \\ u_1, & x - Dt > 0, \end{cases}$$

where $D(u_1 - u_0) = f(u_1) - f(u_0)$ and $f'(u_1) < D < f'(u_0)$. Take account of the traveling wave solution $u(x,t) = u(\xi)$ for equation (4.2.4), with

$$\xi = \frac{x - Dt}{\epsilon}, \quad u_t = \left(\frac{-D}{\epsilon}\right)\frac{du}{d\xi}, \quad u_x = \frac{1}{\epsilon}\frac{du}{d\xi}.$$

Therefore,

$$u''' = [-Du + f(u)]'.$$

We integrate once and obtain

$$u'' = -Du + f(u) + C,$$

where $u(\xi) \to u_0$ as $\xi \to -\infty$ and $u(\xi) \to u$ as $\xi \to +\infty$. One important question comes to mind. Do the continuous solutions with $f(u_1) - f(u_0) = D(u_1 - u_0)$ for the boundary value problem of the above ordinary differential equation exist?

Under the transformations

$$\begin{cases} u' = v, \\ v' = -Du + f(u), \end{cases}$$

the aforementioned second-order ordinary differential equation can be transformed into the boundary value problem of ordinary differential equations. Making

$$P(u) = \frac{1}{2}Du^2 - F(u), \quad F'(u) = f(u), \quad H(u,v) = \frac{1}{2}v^2 + P(u),$$

they will be converted into the canonical equations

$$\begin{cases} u' = v = Hv, \\ v' = -Du + f(u) = -Hu, \end{cases}$$

where $(u_0, 0)$, $(u_1, 0)$ are two critical points and $v \to 0$ ($|\xi| \to \infty$). If the two critical points $(u_0, 0)$ and $(u_1, 0)$ are connected by a trajectory, they must be on the same energy surface, since $H(u(\xi), v(\xi))$ is a constant along the trajectory. In other words, we can deduce $P(u_0) = P(u_1)$ from $H(u_0, 0) = H(u_1, 0)$. What we can prove is that the equality $P(u_0) = P(u_1)$ cannot be satisfied at least along the weak shock wave curve. In fact, a weak shock wave which is connected to the left side of the state can be represented by a single parameter σ. We have $u = u(\sigma)$, $s(\sigma)(u(\sigma) - u_0) = f(u(\sigma)) - f(u_0)$, and $s(0) = f'(u_0)$. p fulfills

$$p(u(\sigma)) = \frac{1}{2}s(\sigma)u^2(\sigma) - F(u(\sigma)).$$

Differentiating with respect to σ, we find

$$\dot{p}(u(\sigma)) = \frac{1}{2}\dot{s}(\sigma)u^2 + s(\sigma)u\dot{u} - f(u(\sigma))\dot{u}.$$

From the RH condition, $s_u = f$ $(u_0 = f(u_0) = 0)$, the above relationship can be simplified to

$$\dot{p}(u(\sigma)) = \frac{1}{2}\dot{s}(\sigma)u^2 \neq 0.$$

Owing to $f''(u) \neq 0$ and $\dot{s}(\sigma) \neq 0$, $P(u_0) \neq P(u_1)$ can be derived. That is to say, the aforementioned integral curve which connects the two critical points does not exist. □

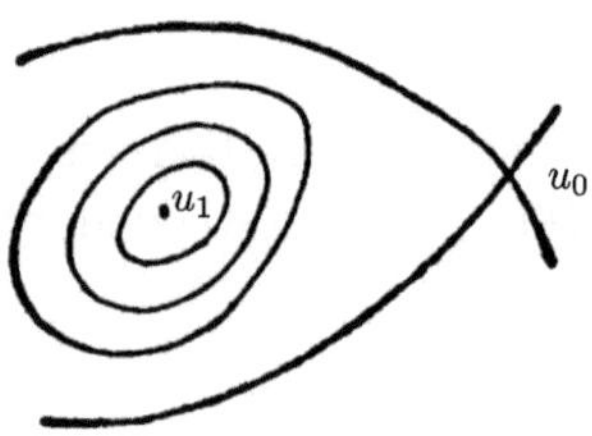

Figure 4.3: The contour of u.

As

$$H(u, v) = \frac{1}{2}v^2 + P(u),$$

we see the equipotential line of H on the (u, v)-plane, as shown in Figure 4.3. There is no trajectory since $(u_1, 0)$ is the center point of the linearized matrix $\begin{pmatrix} H_{uu} & H_{uv} \\ H_{uv} & H_{vv} \end{pmatrix}$.

The one-dimensional isothermal aerodynamics equations

$$\begin{cases} u_t - v_x = 0, \\ v_t + (P(u))_x = 0, \end{cases} \tag{4.2.5}$$

with $P'(u) < 0$ and $P''(u) > 0$ give similar results (see Theorem 4.2.2).

Theorem 4.2.2. *The shock wave solution for equations (4.2.5) cannot be seen as the limit of the traveling wave solution for the corresponding dispersion equation.*

For hyperbolic equations, in order to ensure the convergence of viscosity tends to be zero, the higher-order viscosity term must be of even order, at least for the second and fourth order.

Now, we construct the approximation solution u_ϵ via the Wentzel–Kramers–Brillouin (WKB) method. We have

$$u_t + uu_x = \epsilon^2 u_{xxx}, \tag{4.2.6}$$

where $\epsilon^2 \ll 1$. As we know, the WKB method is an important approximation method in mathematical physics, which has been used to solve the linear ordinary differential equation

$$\epsilon^2 \phi_{xx} + V(x)\phi = 0, \quad \epsilon^2 \ll 1, \tag{4.2.7}$$

where $V(x)$ is the slow-varying function at $o(\delta)$. Equation (4.2.7) has the following form of solution:

$$\phi(x;\epsilon) \sim \phi(\theta, x;\epsilon) \equiv W(x)e^{i\theta}, \quad \theta \equiv \frac{B(x,\epsilon)}{\epsilon}, \tag{4.2.8}$$

which means that the unknown function ϕ is replaced by two unknown functions W and B and the variable x is replaced by two independent variables θ and x. This kind of settings makes the solution to be a cycle of θ. There are no exponentially formed solutions (4.2.8) for nonlinear partial differential equations. We expand $u(x,t;\epsilon)$ as the series

$$u(x,t;\epsilon) \sim U(\theta,x,t;\epsilon) = U^{(0)}(\theta,x,t) + \epsilon U^{(1)}(\theta,x,t) + \cdots, \tag{4.2.9}$$

where $\theta = \theta(x,t;\epsilon)$. Assume that the cycle of θ is 1. Then

$$U(\theta,x,t;\epsilon) = U(\theta+1,x,t;\epsilon).$$

The approximate solution $u(x,t;\epsilon) \simeq U(\theta(x,t;\epsilon),x,t;\epsilon)$ will be obtained if $\theta(x,t;\epsilon)$ is given. In respect of the KdV equation (4.2.6), make

$$\theta = \frac{B(x,t;\epsilon)}{\epsilon},$$

with $B = o(1)$. The variables $L = B_t$, $K = B_x$, and $l = \frac{B_t}{B_x} = \frac{L}{K}$ are independent of θ. We have

$$\frac{\partial}{\partial t} \to \frac{L}{\epsilon}\frac{\partial}{\partial\theta} + \frac{\partial}{\partial t}, \quad \frac{\partial}{\partial x} \to \frac{K}{\epsilon}\frac{\partial}{\partial\theta} + \frac{\partial}{\partial x}.$$

We multiply equation (4.2.6) by ϵ/K, to obtain

$$\begin{aligned} &lU_\theta + UU_\theta + K^2U_{\theta\theta\theta} + \epsilon\left[\frac{1}{K}(U_t + UU_x) + 3(KU)_{x\theta\theta}\right] \\ &+ \epsilon^2\left\{\frac{1}{K}[K_{xx} - U + 3(KU_x)_x]\right\}_\theta + \epsilon^3\frac{1}{K}U_{xxx} = 0. \end{aligned} \tag{4.2.10}$$

Substituting equation (4.2.9) into (4.2.10), we will get a series of differential equations of $U^{(i)}$ $(i = 0, 1, \dots)$. The first one is

$$l^{(0)}U_{\theta}^{(0)} + U^{(0)}U_{\theta}^{(0)} + K^{(0)^2}U_{\theta\theta\theta}^{(0)} = 0, \tag{4.2.11}$$

where $l^{(0)}$ and $K^{(0)}$ are the first terms of the expansion series of l and K, respectively. Integrating twice, we get

$$\frac{1}{2}l^{(0)}[U^{(0)}] + \frac{1}{6}[U^{(0)}]^3 + \frac{1}{2}[K^{(0)}][U_{\theta}^{(0)}]^2 = mU^{(0)} + n, \tag{4.2.12}$$

where m and n are integration constants of x and t. With fixed x and t, the solution for equation (4.2.12) can be expressed as

$$U^{(0)}(\theta, x, t) = -l^{(0)} - (l^{(0)} + 2m)\{a + (b - a){C_n}^2[2\mathscr{K}(k)(\theta - \theta_0); k]\}, \tag{4.2.13}$$

where $k^2 = \frac{b-a}{c-a}$, C_n is a Jacobi elliptic function, $\mathscr{K}$ are Legendre's complete elliptic integrals, and a, b, and c are roots for $U_{\theta}^{(0)} = 0$.

When $i \geq 1$,

$$\mathscr{L}U^{(i)} = N_i(U^{(i-1)}, \dots, l^{(i)}, \dots, K^{(i)}, \dots),$$

with

$$\mathscr{L} \equiv l^{(0)}\frac{\partial}{\partial\theta} + \frac{\partial}{\partial\theta}U^{(0)} + [K^{(0)}]^2\frac{\partial^3}{\partial\theta^3}.$$

In principle, we could go on forever, since the inhomogeneous term N_i only has a lower order. What is important is that, at $\epsilon \to 0$, the limit solution for equation (4.2.6) should be equation (4.2.11), not $u_t + uu_x = 0$.

4.3 The soliton stability problem

In this section, we take account of the linear stability problem of solitons, i.e., the stability problem of the small perturbation of stationary wave solutions for the KdV equation,

$$u_t + uu_x + \mu u_{xxx} = 0, \tag{4.3.1}$$

which possesses the soliton solution

$$u_0(x) = -u_\infty\left(1 - 3\,\mathrm{sech}^2\sqrt{\frac{u_\infty}{4\mu}}x\right), \quad u_\infty > 0. \tag{4.3.2}$$

We add the perturbation term $v(x, t)$ to equation (4.3.2), so that

$$u = u_0(x) + v(x, t), \quad |v| \ll |u_0|. \tag{4.3.3}$$

Substituting the equation (4.3.3) into (4.3.1), we obtain

$$v_t + u_0 v_x + u_{0,x} v + \mu v_{xxx} = 0. \tag{4.3.4}$$

Assuming $v(x,t) = f(x)g(t)$, $g(t) \propto e^{\sigma t}$, and σ = constant, we get the relationship which f satisfies,

$$\frac{d^3 f}{dy^3} - 4(1 - 3\,\mathrm{sech}^2 y)\frac{df}{dy} - (24\mathrm{sech}^2 y \tanh y + \alpha) f = 0, \tag{4.3.5}$$

where $y \equiv \sqrt{\frac{u_\infty}{4\mu}}x$, $\alpha \equiv (-8\sigma/u_\infty)\sqrt{\mu/u_\infty}$. The boundary condition is

$$f \to 0, \quad |y| \to \infty. \tag{4.3.6}$$

Three independent solutions for equation (4.3.5) are given as

$$\begin{aligned} f_k &= \lambda_k(\lambda_k - 2)^2 e^{\lambda_k y} + 4\frac{d^2}{dy^2}[e^{(\lambda_k - 1)y}\mathrm{sech} y] \\ &= e^{\lambda_k y}[\lambda_k(\lambda_k - 2)^2 + 4e^{-y}\mathrm{sech}\{\lambda_k(\lambda_k - 2) \\ &\quad - 2(\lambda_k - 1)\tanh y + 2\tanh^2 y\}], \quad k = 1, 2, 3, \end{aligned} \tag{4.3.7}$$

with λ_k being the roots for the following cubic equation:

$$\lambda^3 - 4\lambda - \alpha = 0. \tag{4.3.8}$$

Specially, when $\alpha = 0$, $\lambda_k = 0, 2, -2$, $f_k \propto \mathrm{sech}^2 y \tanh y$ $(k = 1, 2, 3)$, the three independent solutions are

$$\begin{cases} f_1 = \mathrm{sech}^2 y \tanh y \equiv f_0, \\ f_2 = 3yf_0 + \tanh^2 y - 2\mathrm{sech}^2 y, \\ f_3 = 15yf_0 + 2\sinh^2 y + 7\tanh^2 y - 8\,\mathrm{sech}^2 y. \end{cases} \tag{4.3.9}$$

Notice that equation (4.3.8) has multiple root $\lambda = \pm\frac{2}{\sqrt{13}}$ at $\alpha = \mp\frac{16}{3\sqrt{3}}$, which leads to $f_1 \equiv f_2$ in equation (4.3.9). Therefore, it is essential to choose three new independent solutions. Now we have

$$\begin{cases} f_1 = \lambda_0(\lambda_0 - 2)^2 e^{\lambda_0 y} + 4\dfrac{d^2}{dy^2}[e^{(\lambda_0 - 1)y}\mathrm{sech} y], \\ f_2 = yf_1 + (3\lambda_0^2 - 8\lambda_0 + 4)e^{\lambda_0 y} + 8\dfrac{d}{dy}[e^{(\lambda_0 - 1)y}\mathrm{sech} y], \\ f_3 = \lambda_3(\lambda_3 - 2)^2 e^{\lambda_0 y} + 4\dfrac{d^2}{dy^2}[e^{(\lambda_3 - 1)y}\mathrm{sech} y], \end{cases} \tag{4.3.10}$$

where $(\lambda_0, \lambda_3) = (2/\sqrt{3}, -4/\sqrt{3})$ or $(-2/\sqrt{3}, 4/\sqrt{3})$. The triple root for equation (4.3.8) will not be given. It is easy to deduce that, except for f_0 in the form of equation (4.3.9), the arbitrary combination of solutions (4.3.7), (4.3.9), and (4.3.10) cannot satisfy condition (4.3.6). From the above analysis, we conclude that solutions for the KdV equation are stable under the small perturbation. More details about the nonlinear stabilities of the soliton and the cnoidal wave can be found in [50].

4.4 Water wave and wave equation under weak nonlinear action

The KdV equation was first established in 1895 by Korteweg and de Vries under the assumption of the long wavelength approximation and a small but finite amplitude of the water wave. Next, we give the derivation of the KdV equation, taking account of the incompressible inviscid fluid in constant gravity field with space coordinates system (x_1, x_2, y), velocity $\mathbf{u} = (u_1, u_2, v)$, and acceleration of gravity in the $-y$ direction. We have

$$\nabla \cdot \mathbf{u} = 0, \tag{4.4.1}$$

$$\frac{\partial \mathbf{u}}{\partial t} + (\mathbf{u} \cdot \nabla)\mathbf{u} = -\frac{1}{\rho}\nabla P - g\mathbf{j}. \tag{4.4.2}$$

Considering the irrotational motion, i.e., $\operatorname{rot} \mathbf{u} = 0$, the velocity potential $\mathbf{u} = \nabla\varphi$, and the equality

$$\nabla\left(\frac{1}{2}\mathbf{u}^2\right) = (\mathbf{u} \cdot \nabla)\mathbf{u} - \operatorname{rot} \mathbf{u}_x \cdot \mathbf{u} = (\mathbf{u} \cdot \nabla)\mathbf{u}, \tag{4.4.3}$$

integrating equation (4.4.2), we obtain

$$\frac{p - p_0}{\rho_0} = B(t) - \varphi_t - \frac{1}{2}(\nabla\varphi)^2 - gy,$$

where $B(t)$ is the arbitrary function and p_0 is an arbitrary constant. Making

$$\varphi' = \varphi - \int B(t)dt,$$

we derive

$$\mathbf{u} = \nabla\varphi, \quad \frac{p - p_0}{\rho_0} = -\varphi'_t - \frac{1}{2}(\nabla\varphi')^2 - gy. \tag{4.4.4}$$

In the following steps, we denote $\varphi' = \varphi$. From equation (4.4.1), we find

$$\nabla \cdot \mathbf{u} = 0 \quad \Longrightarrow \quad \nabla^2\varphi = 0. \tag{4.4.5}$$

The surface equation reads

$$f(x_1, x_2, y, t) = 0. \tag{4.4.6}$$

On this surface, the fluid particle cannot pass through the surface, so the velocity of the fluid which is orthogonal to the surface must be equal to the normal velocity of the surface. The normal velocities of equation (4.4.6) and the fluid are written as

$$\frac{-f_t}{\sqrt{f_{x_1}^2 + f_{x_2}^2 + f_y^2}}$$

and

$$\frac{u_1 f_{x_1} + u_2 f_{x_2} + v f_y}{\sqrt{f_{x_1}^2 + f_{x_2}^2 + f_y^2}},$$

respectively. The equal condition between them is

$$f_t + u_1 f_{x_1} + u_2 f_{x_2} + v f_y = 0. \tag{4.4.7}$$

Particularly, equation (4.4.7) transforms into

$$\eta_t + u_1 \eta_{x_1} + u_2 \eta_{x_2} = v, \tag{4.4.8}$$

when $y = \eta(x_1, x_2, t)$ and $f(x_1, x_2, y, t) \equiv \eta(x_1, x_2, t) - y$.

In addition, $p = p_0$ (ignoring the air motion on the free surface) and

$$\begin{cases} \eta_t + \varphi_{x_1} \eta_{x_1} + \varphi_{x_2} \eta_{x_2} = \varphi_y, \\ \varphi_t + \frac{1}{2}(\varphi_{x_1}^2 + \varphi_{x_2}^2 + \varphi_y^2) + g\eta = 0, \end{cases} \tag{4.4.9}$$

where $y = \eta(x_1, x_2, t)$, $u_1 = \varphi_{x_1}$, $u_2 = \varphi_{x_3}$, and $v = \varphi_y$. Under solid boundary conditions, the normal velocity of flow must be zero, i.e., $\mathbf{n} \cdot \nabla\varphi = 0$. Especially $\varphi_y + \varphi_{x_1} h_{0x_1} + \varphi_{x_2} h_{0x_2} = 0$ at the bottom $y = -h_0(x_1, x_2)$, while $\varphi_y = 0$ at the horizontal bottom $y = -h_0$. Thus, we put the problem forward as follows. The velocity potential φ and surface η satisfy

$$\nabla^2 \varphi = 0, \tag{4.4.10}$$

$$\begin{cases} \eta_t + \varphi_{x_1} \eta_{x_1} + \varphi_{x_2} \eta_{x_2} = \varphi_y, \\ \varphi_t + \frac{1}{2}(\varphi_{x_1}^2 + \varphi_{x_2}^2 + \varphi_y^2) - g\eta = 0, \end{cases} \tag{4.4.11}$$

$$\varphi_y = 0, \quad y = -h. \tag{4.4.12}$$

For simplicity, we consider the one-dimensional case, i.e., $\eta = \eta(x, t)$, $\varphi_y = 0$, $y = 0$, and introduce two variables,

$$\alpha = \frac{a}{h_0}, \quad \beta = \frac{h_0^2}{l^2},$$

where a is the amplitude, l is the wavelength, and $y = h_0 + \eta$. Making $x = lx'$, $y = h_0 y'$, $t = \frac{lt'}{c_0}$, $\eta = a\eta'$, $\varphi = \frac{gla\varphi'}{c_0}$, and $c_0^2 = gh_0$ and omitting "′", we obtain the following equations from equations (4.4.5), (4.4.11), and (4.4.12):

$$\beta\varphi_{xx} + \varphi_{yy} = 0, \quad 0 < y < 1 + \alpha\eta, \tag{4.4.13}$$

$$\varphi_y = 0, \quad y = 0, \tag{4.4.14}$$

$$\left.\begin{aligned} &\eta_t + \alpha\varphi_x\eta_x - \frac{1}{\beta}\varphi_y = 0, \\ &\eta + \varphi_t + \frac{1}{2}\alpha\varphi_x^2 + \frac{1}{2}\frac{\alpha}{\beta}\varphi_y^2 = 0, \end{aligned}\right\} y = 1 + \alpha\eta. \tag{4.4.15}$$

The formal solution for equations (4.4.13) and (4.4.14) reads

$$\varphi = \sum_0^\infty (-1)^m \frac{y^{2m}}{(2m)!}\frac{\partial^{2m} f}{\partial x^{2m}}\beta^{2m}, \tag{4.4.16}$$

where $f = f_0(x,t)$. We substitute equation (4.4.16) into the first equation of equation (4.4.15), to obtain

$$\begin{aligned} &\eta_t + \alpha\left[f_x - (1+\alpha\eta)\eta_x f'_{xx}\beta - \frac{(1+\alpha\eta)^2}{2} f_{xxx}\beta + \cdots\right]\eta_x \\ &\quad + (1+\alpha\eta)f'' - \frac{1}{3!}(1+\alpha\eta)^3 f_{xxxx}\beta + o(\beta^2) = 0, \end{aligned}$$

i.e.,

$$\eta_t + \{(1+\alpha\eta)f_x\}_x - \left\{\frac{1}{6}(1+\alpha\eta)^3 f_{xxxx} + \frac{1}{2}\alpha(1+\alpha\eta)^2 f_{xxx}\eta_x\right\}\beta + o(\beta^2) = 0. \tag{4.4.17}$$

Similarly, we substitute equation (4.4.16) into the second equation of equation (4.4.15), to obtain

$$\eta + f_t + \frac{1}{2}\alpha f_x^2 - \frac{1}{2}(1+\alpha\eta)^2\{f_{xxt} + \alpha f_x f_{xxx} - \alpha f_{xx}^2\}\beta + o(\beta^2) = 0. \tag{4.4.18}$$

Ignoring the first order of β in equations (4.4.17) and (4.4.18) and taking the derivative of equation (4.4.18) with respect to x, we find

$$\begin{cases} \eta_t + \{(1+\alpha\eta)w\}_x = 0, \\ w_t + \alpha w w_x + \eta_x = 0, \end{cases} \quad w = f_x. \tag{4.4.19}$$

Otherwise, if we keep the first order of β in equations (4.4.17) and (4.4.18), the following equation will be obtained:

$$\begin{cases} \eta_t + \{(1+\alpha\eta)w\}_x - \frac{1}{6}\beta w_{xxx} + O(\alpha\beta, \beta^2) = 0, \\ w_t + \alpha w w_x + \eta_x - \frac{1}{2}\beta w_{xxt} + O(\alpha\beta, \beta^2) = 0. \end{cases} \tag{4.4.20}$$

It is easy to see that equation (4.4.20) will be transformed into $\eta_t + \eta_x = 0$ under the assumption that $w = \eta$ and omitting the first order of α and β. We expand w with respect to α and β, so we have

$$w = \eta + \alpha A + \beta B + O(\alpha^2 + \beta^2),$$

where A and B are functions of η and its derivatives. Substituting the expansion into equation (4.4.20), we get

$$\begin{cases} \eta_t + \eta_x + \alpha(A_x + 2\eta\eta_x) + \beta\left(B_x - \frac{1}{6}\eta_{xxx}\right) + O(\alpha^2+\beta^2) = 0, \\ \eta_t + \eta_x + \alpha(A_x + \eta\eta_x) + \beta\left(B_t - \frac{1}{2}\eta_{xxt}\right) + O(\alpha^2+\beta^2) = 0. \end{cases}$$

Because $\eta_t = -\eta_x + O(\alpha,\beta)$, the derivative of t in the first order can be changed to the derivative of x. Especially, when we choose $A = -\frac{1}{4}\eta^2$ and $B = \frac{1}{3}\eta_{xx}$, the above two equations will be unified as

$$\eta_t + \eta_x + \frac{3}{2}\alpha\eta\eta_x + \frac{1}{6}\beta\eta_{xxx} + O(\alpha^2+\beta^2) = 0, \tag{4.4.21}$$

with

$$w = \eta - \frac{1}{4}\alpha\eta^2 + \frac{1}{3}\beta\eta_{xx} + O(\alpha^2+\beta^2).$$

Neglecting the second-order term of equation (4.4.21), the classical KdV equation will be derived:

$$\eta_t + \eta_x + \frac{3}{2}\alpha\eta\eta_x + \frac{1}{6}\beta\eta_{xxx} = 0. \tag{4.4.22}$$

Furthermore, we obtain the Benjamin equation,

$$\eta_t + \eta_x + \frac{3}{2}\alpha\eta\eta_x - \frac{1}{6}\beta\eta_{xxt} = 0, \tag{4.4.23}$$

when $\eta_{xxx} = -\eta_{xxt}$.

Here we propose to derive a class of generalized wave equations with weak nonlinear interaction, which we call KdV or Burgers equation [284]. We have

$$n_t + (nu)_x = 0, \tag{4.4.24}$$
$$(nu)_t + (nu^2 + P)_x = 0, \tag{4.4.25}$$
$$P = P(f, n, u, f_i, n_i, u_i, f_{ij}, n_{ij}, u_{ij}, \ldots), \tag{4.4.26}$$
$$F(f, n, u, , f_i, n_i, u_i, f_{ij}, n_{ij}, u_{ij}, \ldots) = 0, \tag{4.4.27}$$

where n and u denote the number density and particle velocity, respectively, and subscripts i, j denote differentiation with respect to the space and time variables x and t. P is the function of the state variables (n, u, f) and their derivatives. The state variable f here serves as a parametric function, which defines P as a functional of n, u and all their derivatives. Equation (4.4.24) is the familiar law of conservation of particles, while equation (4.4.25) represents the law of momentum conservation. To give some idea of possible forms for P and F, we list several examples of physical interest.

(1) Gas dynamics. Here f stands for the thermodynamic pressure. For P and F, we have

$$P = \frac{1}{m}(p - \mu u_x), \quad F = P - A\rho^{\gamma}, \quad mn = \rho, \tag{4.4.28}$$

where ρ is the density and μ is the viscosity coefficient.

(2) Shallow water wave. The number density n now stands for h, the elevation of the water surface above the bottom of a channel. In this case the state is defined by only two functions h, u, as

$$P = \frac{1}{2}gh^2 - \frac{1}{3}h^3(u_{xt} + uu_{xx} + u_x^2). \tag{4.4.29}$$

(3) Hydromagnetic waves in cold plasma. Here f stands for the magnetic field $B(x,t)$ and we have

$$P = \frac{1}{2}B^2, \quad F \equiv B - n - (B_x/n)_x = 0. \tag{4.4.30}$$

(4) Ion-acoustic waves in cold plasma. Here f stands for the electrostatic potential, $\psi(x,t)$ denotes the wave function, and

$$P = e^{\psi} - \frac{1}{2}\psi_x^2, \quad F \equiv n - e^{\psi} + \psi_{xx} = 0. \tag{4.4.31}$$

At equilibrium, all the derivatives in P and F are canceled out and we leave out the dependence of P and F on u to preserve Galilean invariance of the system, i.e.,

$$P = P(f,n), \quad F(f,n) = 0. \tag{4.4.32}$$

Thus, equation (4.4.25) can be rewritten as

$$nu_t + nuu_x + P_x = 0, \quad P_x = \frac{\partial P}{\partial f}\frac{\partial f}{\partial x} + \frac{\partial P}{\partial n}\frac{\partial n}{\partial x},$$

via

$$\frac{\partial F}{\partial f}\frac{\partial f}{\partial x} + \frac{\partial F}{\partial n}\frac{\partial n}{\partial x} = 0.$$

Eliminating $\frac{\partial f}{\partial x}$, we have

$$u_t + uu_x + \frac{a^2}{n}n_x = 0, \quad a^2 \equiv \left[P_n - \frac{F_n}{F_f}P_f\right].$$

If $a^2 > 0$,

$$\begin{cases} n_t + (nu)_x = 0, \\ u_t + uu_x + \dfrac{a^2}{n}n_x = 0 \end{cases} \tag{4.4.33}$$

are hyperbolic equations with two characteristic directions $\frac{dx}{dt} = u \pm a$, where a is defined as the velocity. In the limit of infinitesimal perturbations around a uniform state, we obtain the wave equation with a constant speed of propagation,

$$u_{tt} - a_0^2 u_{xx} = 0,$$

where a_0 is the wave speed of the uniform state. We view the KdV and Burgers equations as designed to describe the slow change of one of these two waves due to both nonlinear and dispersive (or dissipative) effects characterized by the dependence of P and F on the derivatives. Now, we introduce two independent variables,

$$\begin{cases} \xi = \epsilon^{\alpha}(x - a_0 t), \\ \tau = \epsilon^{\alpha+1} t, \end{cases} \tag{4.4.34}$$

where ϵ denotes the amplitude of the initial disturbance and is assumed to be small compared with unity. The exponent $\alpha > 0$ is an undetermined number, while a_0 denotes a certain constant velocity. By virtue of equation (4.4.34), we obtain the following forms of equations (4.4.24) and (4.4.25):

$$\epsilon n_\tau + (u - a_0) n_\xi + n u_\xi = 0, \tag{4.4.35}$$

$$\epsilon u_\tau + (u - a_0) u_\xi + n^{-1} P_\xi = 0. \tag{4.4.36}$$

We now assume that the state variables (n, f, u) can be represented asymptotically as series in powers of ϵ about an equilibrium state $A = (n, f, u) = (n_0, f_0, 0)$, i.e.,

$$\begin{cases} n = n_0 + \epsilon n^{(1)} + \epsilon^2 n^{(2)} + \cdots, \\ f = f_0 + \epsilon f^{(1)} + \epsilon^2 f^{(2)} + \cdots, \\ u = 0 + \epsilon u^{(1)} + \epsilon^2 u^{(2)} + \cdots. \end{cases}$$

We substitute the above transformations and expansions into the Taylor series of P and F around the equilibrium state A_0. In the first order of approximation, all the derivatives of the state variables with respect to x and t are dropped, so we have

$$P = P_0 + P_{f_0}(f - f_0) + P_{n_0}(n - n_0) + P_{u_0}(u - u_0) + O(\epsilon^2),$$
$$F = F_0 + F_{f_0}(f - f_0) + F_{n_0}(n - n_0) + F_{u_0}(u - u_0) + O(\epsilon^2).$$

Due to the Galilean invariance of the system, $P_{u_0} = F_{u_0} = 0$. From $P^{(1)} = P_{f_0} f^{(1)} + P_{n_0} n^{(1)}$ and $F_{f_0} \frac{\partial f^{(1)}}{\partial \xi} + F_{n_0} \frac{\partial n^{(1)}}{\partial \xi} = 0$, we deduce

$$\frac{\partial P^{(1)}}{\partial \xi} = P_{f_0} \frac{\partial f^{(1)}}{\partial \xi} + P_{n_0} \frac{\partial n^{(1)}}{\partial \xi} = \left[P_{n_0} - \frac{F_{n_0}}{F_{f_0}} P_{f_0} \right] \frac{\partial n^{(1)}}{\partial \xi} = a_0^2 \frac{\partial n^{(1)}}{\partial \xi}.$$

Within the second order of the above expansion, we obtain

$$P_\xi^{(2)} \approx a_0^2 n_\xi^{(2)} + A_n^{(1)} n_\xi^{(1)} + \epsilon^{\alpha-1} B n_{\xi\xi}^{(1)} + \epsilon^{2\alpha-1} c n_{\xi\xi\xi}^{(1)},$$

where constants a_0^2, A, B, and C are listed as follows:

	a_0	A	B	C
Gas dynamics	$2KT/m$	0	$-va_0$	0
Water waves	gh_0	0	0	$\frac{1}{3}gh_0^3$
Hydromagnetic	B_0	1	0	1
Ion-acoustic	1	0	0	1

Comparing the first order of ϵ in equations (4.4.35) and (4.4.36), we have

$$a_0 n_\xi^{(1)} = n_0 u_\xi^{(1)}, \quad a_0 u_\xi^{(1)} = \frac{a^2}{n_0} n_\xi^{(1)}.$$

Integrating these above equations and noting the boundary condition for $n^{(1)}$ and $u^{(l)}$ at $\xi \to \pm\infty$, we get

$$a_0 n^{(1)} = n_0 u^{(1)}.$$

In the second-order approximation of equations (4.4.35) and (4.4.36), we obtain

$$n_\tau^{(1)} + u^{(1)} n_\xi^{(1)} + n_0 u_\xi^{(2)} + n^{(1)} u_\xi^{(1)} - a_0 n_\xi^{(2)} = 0,$$

i.e.,

$$n_\tau^{(1)} + 2\frac{a_0}{n_0} n^{(1)} n_\xi^{(1)} - a_0 n_\xi^{(2)} + n_0 u_\xi^{(2)} = 0.$$

Moreover,

$$\frac{a_0}{n_0} n_\tau^{(1)} + \frac{A}{n_0} n^{(1)} n_\xi^{(1)} + \epsilon^{\alpha-1} \frac{B}{n_0} n_{\xi\xi}^{(1)} + \epsilon^{2\alpha-1} \frac{C_0}{n_0} n_{\xi\xi\xi}^{(1)} + \frac{a_0^2}{n} n_\xi^{(2)} + u^{(1)} u_\xi^{(1)} - a_0 u_\xi^{(2)} = 0.$$

We now eliminate $n_\xi^{(2)}$ and $u_\xi^{(2)}$ and obtain the evolution equation for $n^{(1)}$, i.e.,

$$n_\tau^{(1)} + \left(\frac{A}{2a_0} + \frac{3a_0}{2n_0}\right) n^{(1)} n_\xi^{(1)} + \epsilon^{\alpha-1} \frac{B}{2a_0} n_{\xi\xi}^{(1)} + \epsilon^{2\alpha-1} \frac{C}{2a_0} n_{\xi\xi\xi}^{(1)} = 0. \tag{4.4.37}$$

If $B \neq 0$ (for a dissipative system $B < 0$), we set $\alpha = 1$ and $C = 0$. The resulting equation of equation (4.4.37) is the Burgers equation. On the other hand, if $B = 0$ (for a dissipative system) and $\alpha = \frac{1}{2}$, we obtain the KdV equation

$$n_\tau^{(1)} + \left(\frac{A}{2a_0} + \frac{3a_0}{2n_0}\right) n^{(1)} n_\xi^{(1)} + \frac{C}{2a_0} n_{\xi\xi\xi}^{(1)} = 0.$$

5 Hirota method

5.1 Introduction

The Hirota method [127], which obtains the special solution via certain transformations, is an important and direct method to derive the N-soliton solutions for certain nonlinear evolution equations, apart from the inverse scattering transform and structural continuation methods. This method is not only applicable for the KdV, modified KdV (MKdV), sine-Gordon, Toda lattice, and Boussinesq equations, but has also been extended to a multitude of nonlinear evolution equations [118, 303]. The Bäcklund transformation can also be obtained through the Hirota method [46, 302].

First, we take the KdV equation as an example to introduce the fundamental theory of the Hirota method. The KdV equation takes the form of

$$u_t + 6uu_x + u_{xxx} = 0, \tag{5.1.1}$$

which satisfies the boundary condition $u = 0, |x| \to \infty$. We will solve equation (5.1.1) via the perturbation method. Letting $u = w_x$ and integrating equation (5.1.1) with respect to x, we get

$$w_t + 3w_x^2 + w_{xxx} = 0. \tag{5.1.2}$$

Here, the constant of integration is 0. Expanding w as the series of ϵ, we have

$$w = \epsilon w_1 + \epsilon^2 w_2 + \cdots. \tag{5.1.3}$$

Substituting equation (5.1.3) into (5.1.2) and collecting ϵ at the same order, we get the following equations:

$$\left(\frac{\partial}{\partial t} + \frac{\partial^3}{\partial x^3}\right)w_1 = 0, \tag{5.1.4}$$

$$\left(\frac{\partial}{\partial t} + \frac{\partial^3}{\partial x^3}\right)w_2 = -3(w_1)_x^2, \tag{5.1.5}$$

$$\left(\frac{\partial}{\partial t} + \frac{\partial^3}{\partial x^3}\right)w_3 = -6(w_1)_x(w_2)_x, \tag{5.1.6}$$

$$\cdots$$

whose formal solution in the form of perturbation series will be obtained. We consider the similar method of Padé approximation, since these series may converge slowly or even diverge.

Substituting $w = G/F$ into equation (5.1.2), we have

$$\begin{aligned}&(G_tF - GF_t)/F^2 + 3(G_xF - GF_x)^2/F^4\\ &+ (G_{xxx}F - 3G_{xx}F_x - 3G_xF_{xx} - GF_{xxx})/F^2\\ &+ 6(FG_xF_x^2 + FGF_xF_{xx} - GF_x^3)/F^4 = 0.\end{aligned} \tag{5.1.7}$$

https://doi.org/10.1515/9783110549638-005

Introducing a free parameter λ, we notice that the complicated expression (5.1.7) with two unknown functions F and G can be rewritten as

$$[G_tF - GF_t + 3\lambda(G_xF - GF_x) + G_{xxx}F - 3G_{xx}F_x + 3G_xF_{xx} - GF_{xxx}]/F^2 \\ + 3(G_xF - GF_x)[G_xF - GF_x - 2(FF_{xx} - F_x^2) - \lambda F^2]/F^4 = 0. \tag{5.1.8}$$

Hereby, the following two equations are obtained:

$$G_tF - GF_t + 3\lambda(G_xF - GF_x) + G_{xxx}F - 3G_{xx}F_x + 3G_xF_{xx} - GF_{xxx} = 0, \tag{5.1.9}$$

$$2(FF_{xx} - F_x^2) + \lambda F^2 - (G_xF - GF_x) = 0, \tag{5.1.10}$$

which can also be expressed as

$$\left[\frac{\partial}{\partial t} - \frac{\partial}{\partial t'} + 3\lambda\left(\frac{\partial}{\partial x} - \frac{\partial}{\partial x'}\right) + \left(\frac{\partial}{\partial x} - \frac{\partial}{\partial x'}\right)^3\right] \cdot G(x,t)F(x',t')\Big|_{x=x',t=t'} = 0 \tag{5.1.11}$$

and

$$\left[\left(\frac{\partial}{\partial x} - \frac{\partial}{\partial x'}\right)^2 + \lambda\right]F(x,t)F(x',t')\Big|_{x=x',t=t'} \\ - \left(\frac{\partial}{\partial x} - \frac{\partial}{\partial x'}\right)G(x,t)F(x',t')\Big|_{x=x',t=t'} = 0, \tag{5.1.12}$$

where D_x and D_t are the bilinear derivative operators defined by

$$D_t^n D_x^m f \cdot g = \left(\frac{\partial}{\partial t} - \frac{\partial}{\partial t'}\right)^n \left(\frac{\partial}{\partial x} - \frac{\partial}{\partial x'}\right)^m f(x,t)g(x',t')\Big|_{x'=x,t'=t}. \tag{5.1.13}$$

Thus, equations (5.1.11) and (5.1.12) can be transformed into

$$(D_t + 3\lambda D_x + D_x^3)G \cdot F = 0, \tag{5.1.14}$$

$$(D_x^2 + \lambda)F \cdot F - D_xG \cdot F = 0. \tag{5.1.15}$$

Through equation (5.1.15), we get

$$\lambda = (G/F)_x - 2(\log F)_{xx}. \tag{5.1.16}$$

Especially,

$$G = 2F_x \tag{5.1.17}$$

will be obtained when $\lambda = 0$. Therefore, we have

$$u = (G/F)_x = 2(\log F)_{xx}. \tag{5.1.18}$$

Substituting $G = 2F_x$ into equation (5.1.14), the following simplified D-forms can be obtained:

$$(D_t + D_x^3)F_x \cdot F = 0, \tag{5.1.19}$$

or

$$D_x(D_t + D_x^3)F \cdot F = 0. \tag{5.1.20}$$

5.2 Some properties of the *D* operator

Define D_z and the differential operator $\frac{\partial}{\partial z}$ as

$$D_z = \delta D_t + \varepsilon D_x, \quad \frac{\partial}{\partial z} = \delta\frac{\partial}{\partial t} + \varepsilon\frac{\partial}{\partial x}, \tag{5.2.1}$$

where δ and ε are constants. The properties of the D operator can be shown as

$$\begin{aligned}
&\text{(I)} && D_z^m a\cdot 1 = \left(\frac{\partial}{\partial z}\right)^m a,\\
&\text{(II)} && D_z^m a\cdot b = (-1)^m D_z^m b\cdot a,\\
&\text{(II.1)} && D_z^m a\cdot a = 0,\ m \text{ being odd},\\
&\text{(III)} && D_z^m a\cdot b = D_z^{m-1}(a_z\cdot b - a\cdot b_z),\\
&\text{(III.1)} && D_z^m a\cdot a = 2D_z^{m-1}a_z\cdot a,\ m \text{ being even},\\
&\text{(III.2)} && D_xD_t a\cdot a = 2D_x a_t\cdot a = 2D_t a_x\cdot a,\\
&\text{(IV)} && D_x^m \exp(p_1x)\cdot\exp(p_2x) = (p_1-p_2)^m\exp[(p_1+p_2)x].
\end{aligned}$$

Assuming that $F(D_t, D_x)$ is the polynomial of D_t and D_x, we have

$$\begin{aligned}
&\text{(IV.1)} && F(D_t,D_x)\exp(Q_1t+p_1x)\exp(Q_2t+p_2x)\\
& && = F(Q_1-Q_2,p_1-p_2)/F(Q_1+Q_2,p_1+p_2)\\
& && \quad\cdot F(D_t,D_x)\exp[(Q_1+Q_2)t+(p_1+p_2)x],\\
&\text{(V)} && \exp(\epsilon D_x)a(x)\cdot b(x) = a(x+\epsilon)b(x-\epsilon),\\
&\text{(VI)} && \exp(\epsilon D_z)ab\cdot cd = [\exp(\epsilon D_z)a\cdot c]\cdot[\exp(\epsilon D_z)b\cdot d]\\
& && = [\exp(\epsilon D_z)a\cdot d]\cdot[\exp(\epsilon D_z)b\cdot c],\\
&\text{(VI.1)} && D_z ab\cdot c = \left(\frac{\partial a}{\partial z}\right)bc + a(D_z b\cdot c),\\
&\text{(VI.2)} && D_z^2 ab\cdot c = \left(\frac{\partial^2 a}{\partial z^2}\right)bc + 2\left(\frac{\partial a}{\partial z}\right)D_z b\cdot c + a(D_z^2 b\cdot c),\\
&\text{(VI.3)} && D_z^3 ac\cdot bc = (D_z^3 a\cdot b)c^2 + 3(D_z a\cdot b)D_z^2 b\cdot c,\\
&\text{(VI.4)} && D_x^m \exp(px)a\cdot\exp(px)b = \exp(2px)D_x^m a\cdot b,\\
&\text{(VII)} && \exp(\delta D_t)[\exp(\epsilon D_x)a\cdot b]\cdot[\exp(\epsilon D_x)c\cdot d]\\
& && = \exp(\epsilon D_x)[\exp(\delta D_t)a\cdot c]\cdot[\exp(\delta D_t)b\cdot d]\\
& && = [\exp(\delta D_t+\epsilon D_x)a\cdot d]\cdot[\exp(-\delta D_t+\epsilon D_x)c\cdot b].
\end{aligned}$$

The following expressions will be useful during the transformation from nonlinear differential equations to bilinear forms:

$$\text{(VIII)}\quad \exp\left(\epsilon\frac{\partial}{\partial z}\right)(a/b) = [\exp(\epsilon D_z)a\cdot b]/[\cosh(\epsilon D_z)b\cdot b],$$

$$\text{(VIII.1)}\quad \frac{\partial}{\partial z}(a/b) = \frac{D_z a\cdot b}{b^2},$$

$$\text{(VIII.2)}\quad \frac{\partial^2}{\partial z^2}(a/b) = \frac{D_z^2 a\cdot b}{b^2} - \left(\frac{a}{b}\right)\frac{D_z^2 b\cdot b}{b^2},$$

$$\text{(VIII.3)}\quad \frac{\partial^3}{\partial z^3}(a/b) = \frac{D_z^3 a\cdot b}{b^2} - 3\frac{D_z a\cdot b}{b^2}\frac{D_z^2 b\cdot b}{b^2},$$

$$\text{(IX)}\quad 2\cosh\left(\epsilon\frac{\partial}{\partial z}\right)\log f = \log[\cosh(\epsilon D_z)f\cdot f],$$

$$\text{(IX.1)}\quad \frac{\partial^2}{\partial z^2}\log f = \frac{D_z^2 f\cdot f}{2f^2},$$

$$\text{(IX.2)}\quad \frac{\partial^4}{\partial z^4}\log f = \frac{D_z^4 f\cdot f}{2f^2} - 6\left(\frac{D_z^2 f\cdot f}{2f^2}\right)^2.$$

Using the following expression, we transform the bilinear forms to the original nonlinear equations:

$$\text{(X)}\quad \exp(\epsilon D_x)a\cdot b = \left\{\exp\left[2\cosh\left(\epsilon\frac{\partial}{\partial x}\right)\log b\right]\right\}\left[\exp\left(\epsilon\frac{\partial}{\partial x}\right)(a/b)\right].$$

Making $\psi = a/b$ and $u = 2(\log b)_{xx}$, we have

$$\text{(X.1)}\quad (D_x a\cdot b)/b^2 = \psi_x,$$

$$\text{(X.2)}\quad (D_x^2 a\cdot b)/b^2 = \psi_{xx} + u\psi,$$

$$\text{(X.3)}\quad (D_x^3 a\cdot b)/b^2 = \psi_{xxx} + 3u\psi_x,$$

$$\text{(X.4)}\quad (D_x^4 a\cdot b)/b^2 = \psi_{xxxx} + 6u\psi_{xx} + (u_{xx} + 3u^2)\psi,$$

$$\text{(XI)}\quad \exp(\epsilon D_x)a\cdot b = \exp\left[\sinh\left(\epsilon\frac{\partial}{\partial x}\right)\log(a/b) + \cosh\left(\epsilon\frac{\partial}{\partial x}\right)\cdot\log(ab)\right].$$

Letting $\varphi = \log(a/b)$ and $\rho = \log(ab)$, we have

$$\text{(XI.1)}\quad (D_x a\cdot b)/ab = \varphi_x,$$

$$\text{(XI.2)}\quad (D_x^2 a\cdot b)/ab = \rho_{xx} + \varphi_x^2,$$

$$\text{(XI.3)}\quad (D_x^3 a\cdot b)/ab = \varphi_{xxx} + 3\varphi_x\rho_{xx} + \varphi_x^3,$$

$$\text{(XI.4)}\quad (D_x^4 a\cdot b)/ab = \rho_{xxxx} + 4\varphi_x\varphi_{xxx} + 3(\rho_{xx})^2 + 6\varphi_x^2\rho_{xx} + \varphi_x^4.$$

We take the verification of expression (X) as an example, since all the above properties can easily be verified. We have

$$2\cosh\left(\epsilon\frac{\partial}{\partial x}\right)\log b = \log b(x+\epsilon) + \log b(x-\epsilon),$$

$$\exp\left(\epsilon\frac{\partial}{\partial x}\right)(a/b) = a(x+\epsilon)/b(x+\epsilon).$$

From expression (V), we deduce

$$\exp(\epsilon D_x)a \cdot b = a(x+\epsilon)b(x-\epsilon).$$

Therefore,

$$\exp(\epsilon D_x)a \cdot b = \exp\left[2\cosh\left(\epsilon\frac{\partial}{\partial x}\right)\log b\right] \cdot \left[\exp\left(\epsilon\frac{\partial}{\partial x}\right)(a/b)\right],$$

which is exactly expression (X). Expanding expression (X) as the power series ϵ and truncating at the same power, expressions (X.1)–(X.4) are obtained.

5.3 Solution of the bilinear differential equation

In order to solve equation (5.1.20), we expand F as the power series of ϵ, to obtain

$$F = 1 + \epsilon f_1 + \epsilon^2 f_2 + \cdots. \tag{5.3.1}$$

Substituting the above expression into equation (5.1.20) and collecting the same order of ϵ, we get

$$2\frac{\partial}{\partial x}\left(\frac{\partial}{\partial t} + \frac{\partial^3}{\partial x^3}\right)f_1 = 0, \tag{5.3.2}$$

$$2\frac{\partial}{\partial x}\left(\frac{\partial}{\partial t} + \frac{\partial^3}{\partial x^3}\right)f_2 = -D_x(D_t + D_x^3)f_1 \cdot f_1, \tag{5.3.3}$$

$$2\frac{\partial}{\partial x}\left(\frac{\partial}{\partial t} + \frac{\partial^3}{\partial x^3}\right)f_3 = -D_x(D_t + D_x^3)(f_2 \cdot f_1 + f_1 \cdot f_2), \tag{5.3.4}$$

$$\cdots.$$

Next, we focus on two kinds of solutions: (I) polynomial solutions and (II) exponential solutions.

As regards case (I), we find the following solution for expression (5.3.2):

$$f_1 = a_0 + a_1 x + a_2 x^2 + a_3 x^3 + a_4 x^4 + bt - 24a_4 tx. \tag{5.3.5}$$

We choose $f_2 = 0$ when $a_4 = 0$, $3a_1a_3 = a_2^2$, and $b = 12a_3$. Therefore, the explicit solution for equation (5.1.20) writes

$$F = 1 + \epsilon\left[a_0 + a_1 x + (3a_1a_3)^{1/2}x^2 + a_3(x^3 + 12t)\right]. \tag{5.3.6}$$

Under the boundary condition of $u|_{x=0} = 0$, $a_1 = 0$. Without loss of generality, choosing $\epsilon = 1$, we obtain

$$F = a_3\left[x^3 + 12(t + \text{constant})\right] \tag{5.3.7}$$

and

$$u = 2(\log F)_{xx} = -6x(x^3 - 24t)/(x^3 + 12t)^2. \tag{5.3.8}$$

As regards case (II), we solve expression (5.3.2) as

$$f_1 = \sum_{j=1}^{N} a_j \exp(\Omega_j t + p_j x). \tag{5.3.9}$$

Here, $\Omega_j + p_j^3 = 0$ where p_j and a_j are constants. Substituting equation (5.3.9) into (5.3.3) and by virtue of (IV) and (IV.1), we obtain

$$f_2 = \sum_{i>j}^{N} \exp(A_{ij} + \eta_i + \eta_j), \tag{5.3.10}$$

where $\exp(\eta_j) = a_j \exp(\Omega_j t + p_j x)$ and

$$\begin{aligned}\exp(A_{ij}) &= -\frac{(p_i - p_j)[\Omega_i - \Omega_j + (p_i - p_j)^3]}{(p_i + p_j)[\Omega_i + \Omega_j + (p_i + p_j)^3]} \\ &= (p_i - p_j)^2/(p_i + p_j)^2.\end{aligned} \tag{5.3.11}$$

Substituting equation (5.3.10) into (5.3.4), we get f_3 via (VI.4) and equation (5.3.2), so we have

$$f_3 = \sum_{i>j>k}^{N} \exp(A_{ijk} + \eta_i + \eta_j + \eta_k), \tag{5.3.12}$$

with

$$\exp(A_{ijk}) = \exp(A_{ij} + A_{ik} + A_{jk}). \tag{5.3.13}$$

Following similar procedures, we get f_N and the explicit solution of F as follows:

$$F = \sum_{\mu=0,1} \exp\left(\sum_{i>j}^{N} A_{ij}\mu_i\mu_j + \sum_{j} \mu_j\eta_j \right), \tag{5.3.14}$$

where $\sum_{\mu=0,1}$ denotes the summation over all possible combinations of $\mu_1 = 0, 1,\ \mu_2 = 0, 1, \ldots, \mu_N = 0, 1$. $\sum_{i>j}^{N}$ means the summation over all possible combinations taken from the N-elements. The parameter ϵ has been included in a_i. Equation (5.3.14) and $u = 2(\log F)_{xx}$ give the N-soliton solutions for the KdV equation.

5.4 Application to the sine-Gordon equation, MKdV Equation

Firstly, we take the sine-Gordon equation

$$\varphi_{xx} - \varphi_{tt} = \sin\varphi, \tag{5.4.1}$$

with the condition $\frac{\partial\varphi}{\partial x} \to 0$ ($|x| \to \infty$), into account. We set

$$\varphi(x,t) = 4\tan^{-1}[g(x,t)/f(x,t)], \tag{5.4.2}$$

where

$$f(x,t) = \sum_{n=0}^{[N/2]} \sum_{N^c n} a(i_1, i_2, \ldots, i_{2n}) \cdot \exp(\eta_{i_1} + \eta_{i_2} + \cdots + \eta_{i_{2n}}), \tag{5.4.3}$$

$$g(x,t) = \sum_{m=0}^{[(N-1)/2]} \sum_{N^c 2m+1} a(j_1, j_2, \ldots, j_{2m+1}) \cdot \exp(\eta_{j_1} + \eta_{j_2} + \cdots + \eta_{j_{2m+1}}), \tag{5.4.4}$$

$$a(i_1, i_2, \ldots, i_{2n}) : \begin{cases} \prod_{k<l}^{(n)} a(i_k, i_l), & n \geq 2, \\ 1, & n = 0, 1, \end{cases}$$

$$\begin{aligned} a(i_k, i_l) &= \frac{(p_{ik} - p_{il})^2 - (\Omega_{ik} - \Omega_{il})^2}{(p_{ik} + p_{il})^2 - (\Omega_{ik} + \Omega_{il})^2} \\ &= -\frac{(p_{ik} - p_{il} + \Omega_{ik} - \Omega_{il})^2}{(p_{ik} - p_{il} + \Omega_{ik} + \Omega_{il})^2}, \\ \eta_i &= p_i x - \Omega_i t - \eta_i^0, \quad p_i^2 - \Omega_i^2 = 1, \end{aligned}$$

where p_i and η_i^0 are both finite arbitrary real constants, which determine the amplitude and phase of the ith soliton, respectively. For example, in the case where $N = 3$, the solution is written as

$$\begin{aligned} &f(x,t) = 1 + a(1,2)\exp(\eta_1 + \eta_2) + a(1,3)\exp(\eta_1 + \eta_3) + a(2,3)\exp(\eta_2 + \eta_3), \\ &g(x,t) = \exp(\eta_1) + \exp(\eta_2) + \exp(\eta_3) + a(1,2,3)\exp(\eta_1 + \eta_2 + \eta_3), \\ &a(1,2,3) = a(1,2)a(1,3)a(2,3), \\ &\eta_i = p_i x - \Omega_i t, \quad p_i^2 - \Omega_i^2 = 1, \end{aligned}$$

where η_1 remains finite as $t \to \infty$. When $\Omega_3/p_3 > \Omega_2/p_2 > \Omega_1/p_1 > 0$, $p_i > 0$, and $g(x,t)/f(x,t) = \exp(\eta_1)$, the soliton solution will be obtained as follows:

$$i(x,t) = -\frac{\partial\varphi}{\partial x} = -2p_1 \operatorname{sech}(\eta_1).$$

In the following part, we will show a brief proof that solutions (5.4.2), (5.4.3), and (5.4.4) solve equation (5.4.1). Substituting solution (5.4.2) into equation (5.4.1), we have

$$fg_{xx} - 2f_x g_x + f_{xx} g - (fg_{tt} - 2f_t g_t + f_{tt} g) = fg, \tag{5.4.5}$$

$$\begin{aligned} &f_{xx} f - 2f_x^2 + ff_{xx} - (f_{tt} f - 2f_t^2 + ff_{tt}) \\ &\quad = g_{xx} g - 2g_x^2 - (g_{tt} g - 2g_t^2 + gg_{tt}). \end{aligned} \tag{5.4.6}$$

As the next step, consider solutions (5.4.3) and (5.4.4). Then the above expressions can be rewritten as

$$\sum_{l=0}^{n}\sum_{n^c l} a(i_1,i_2,\dots,i_l)a(i_{l+1},i_{l+2},\dots,i_n)$$
$$\cdot h_1(i_1,i_2,\dots,i_l;i_{l+1},i_{l+2},\dots,i_n)=0,\quad \text{for } n=1,3,5,\dots,\le N, \tag{5.4.7}$$
$$\sum_{l=0}^{n}\sum_{n^c l}(-1)^l a(i_1,i_2,\dots,i_l)a(i_{l+1},i_{l+2},\dots,i_n)$$
$$\cdot h_2(i_1,i_2,\dots,i_l;i_{l+1},i_{l+2},\dots,i_n)=0,\quad \text{for } n=2,4,6,\dots,\le N, \tag{5.4.8}$$

with

$$\begin{aligned}h(i_1,i_2,\dots,i_l;i_{l+1},i_{l+2},\dots,i_n)\\ &=(p_{i_1}+p_{i_2}+\dots+p_{i_l}-p_{i_{l+1}}-p_{i_{l+2}}-\dots-p_{i_n})^2\\ &\quad-(\Omega_{i_1}+\Omega_{i_2}+\dots+\Omega_{i_l}-\Omega_{i_{l+1}}-\Omega_{i_{l+2}}-\dots-\Omega_{i_n})^2.\end{aligned}$$

For certain n, the following identities are formed from expressions (5.4.7) and (5.4.8):

$$\sum_{\sigma_1,\sigma_2,\dots,\sigma_n=\pm1}\left(\prod_{i=1}^{n}\sigma_i\right)\hat{b}(\sigma_1x_1,\sigma_2x_2,\dots,\sigma_nx_n)$$
$$\cdot\hat{h}_1(\sigma_1x_1,\sigma_2x_2,\dots,\sigma_nx_n)=0,\quad \text{if } n \text{ is odd}, \tag{5.4.9}$$
$$\sum_{\sigma_1,\sigma_2,\dots,\sigma_n=\pm1}\left(\prod_{i=1}^{n}\sigma_i\right)\hat{b}(\sigma_1x_1,\sigma_2x_2,\dots,\sigma_nx_n)$$
$$\cdot\hat{h}_2(\sigma_1x_1,\sigma_2x_2,\dots,\sigma_nx_n)=0,\quad \text{if } n \text{ is even}, \tag{5.4.10}$$

with

$$\begin{aligned}\hat{b}(\sigma_1x_1,\sigma_2x_2,\dots,\sigma_nx_n)&=\prod_{k<l}^{(n)}(\sigma_kx_k-\sigma_lx_l)^2,\\ \hat{h}_1(\sigma_1x_1,\sigma_2x_2,\dots,\sigma_nx_n)&=\left(\prod_{i=1}^{n}\sigma_ix_i\right)\left(\sum_{i=1}^{n}\prod_{l=1,l\neq i}^{n}{}'\sigma_lx_l\right)-\prod_{i=1}^{n}\sigma_ix_i,\\ \hat{h}_2(\sigma_1x_1,\sigma_2x_2,\dots,\sigma_nx_n)&=\left(\sum_{i=1}^{n}\sigma_ix_i\right)\left(\sum_{i=1}^{n}\prod_{l=1,l\neq i}^{n}{}'\sigma_lx_l\right),\\ x_i&=p_i+\Omega_i.\end{aligned}$$

In fact, denoting the left side of equation (5.4.9) as $D_1(x_1,x_2,\dots,x_n)$, we see it has the following two properties:

(i) D_1 is a symmetrical homogeneous polynomial,

(ii)

$$D_1(x_1,\dots,x_n)_{x_1=\pm x_2}=8x_1^4\prod_{i=3}^{n}(x_1^2-x_i^2)^2D(x_3,x_4,\dots,x_n),$$

at $x_1=\pm x_2$.

It is easy to see that equation (5.4.9) is satisfied at $n = 1$. Assuming that equation (5.4.9) is satisfied at $n-2$, via (i) and (ii), we find that D_1 is the $2n(n-1)$th-order symmetrical homogeneous polynomial

$$\prod_{k<l}^{(n)} (x_k^2 - x_l^2)^2.$$

On the other hand, we conclude that $D_1 = 0$ for certain n, since it is straightforward to find that D_1 is the polynomial of the n^2th order. Similarly, we find $D_2 = 0$.

Secondly, we consider the following MKdV equation:

$$v_t + 24v^2 v_x + v_{xxx} = 0, \tag{5.4.11}$$

$$v(x,t) = \frac{\partial \varphi}{\partial x}, \tag{5.4.12}$$

$$\tan \varphi(x,t) = g(x,t)/f(x,t), \tag{5.4.13}$$

with

$$f(x,t) = \sum_{n=0}^{[N/2]} \sum_{N^c n} a(i_1, i_2, i_3, \ldots, i_{2n}) \exp(\xi_{i_1} + \xi_{i_2} + \cdots + \xi_{i_{2n}}), \tag{5.4.14}$$

$$g(x,t) = \sum_{n=0}^{[(N-1)/2]} \sum_{N^c 2m+1} a(i_1, i_2, \ldots, i_{2m+1}) \exp(\xi_{i_1} + \xi_{i_2} + \cdots + \xi_{i_{2m+1}}), \tag{5.4.15}$$

$$a(i_1, i_2, \ldots, i_n) = \begin{cases} \prod_{k<l}^{(n)} a(i_k, i_l), & n \geq 2, \\ 1, & n = 0, 1, \end{cases}$$

$$a(i_k, i_l) = -\frac{(p_{ik} - p_{il})^2}{(p_{ik} + p_{il})^2}, \quad \xi_i = p_i x - \Omega_i t - \xi_i^0, \quad \Omega_i = p_i^3.$$

When $N = 3$,

$$\begin{aligned} f(x,t) &= 1 + a(1,2)\exp(\xi_1 + \xi_2) + a(1,3)\exp(\xi_1 + \xi_3) + a(2,3)\exp(\xi_2 + \xi_3), \\ g(x,t) &= \exp(\xi_1) + \exp(\xi_2) + \exp(\xi_3) + a(1,2,3)\exp(\xi_1 + \xi_2 + \xi_3), \\ a(1,2,3) &= a(1,2)a(1,3)a(2,3), \quad \xi_i = p_i x - p_i^3 t. \end{aligned}$$

When $t \to \infty$, the soliton solution reads

$$v(x,t) = p_1/2 \operatorname{sech} \xi_1,$$

with

$$g/f = \exp(\xi_1), \quad p_3 > p_2 > p_1 > 0,$$

$$g_t f - g f_t + g_{xxx} f - 3g_{xx} f_x + 3g_x f_{xx} - g f_{xxxx} = 0, \tag{5.4.16}$$

$$f f_{xx} - 2f_x^2 + f_{xx} f + g g_{xx} - 2g_x^2 + g_{xx} g = 0. \tag{5.4.17}$$

The above two expressions can be written in the following forms:

$$\left[\left(\frac{\partial}{\partial t}-\frac{\partial}{\partial t'}\right)-\left(\frac{\partial}{\partial x}-\frac{\partial}{\partial x'}\right)^3\right]g(x,t)f(x',t')\bigg|_{t'=t,x'=x}=0, \tag{5.4.18}$$

$$\left(\frac{\partial}{\partial t}-\frac{\partial}{\partial x'}\right)^2[f(x,t)f(x',t')+g(x,t)g(x',t')]\bigg|_{t'=t,x'=x}=0. \tag{5.4.19}$$

Reconsidering equations (5.4.14) and (5.4.15), we rewrite equations (5.4.16) and (5.4.17) as

$$\begin{aligned}&\sum_{l=0}^{n}\sum_{n^c l}\hat{a}(i_1,i_2,\dots,i_l)\hat{a}(i_{l+1},i_{l+2},\dots,i_n)\\&\quad\cdot h_1(i_1,i_2,\dots,i_l;i_{l+1},\dots,i_n)=0,\quad n=1,3,5,\ldots\le N,\end{aligned} \tag{5.4.20}$$

$$\begin{aligned}&\sum_{l=0}^{n}\sum_{n^c l}(-1)^l\hat{a}(i_1,i_2,\dots,i_l)\hat{a}(i_{l+1},i_{l+2},\dots,i_n)\\&\quad\cdot h_2(i_1,i_2,\dots,i_l;i_{l+1},\dots,i_n)=0,\quad n=2,4,6,\ldots\le N,\end{aligned} \tag{5.4.21}$$

$$\hat{a}(i_1,i_2,\dots,i_n)=\begin{cases}\prod_{k<l}^{(n)}\hat{a}(i_k,i_l), & n\ge 2,\\ 1, & n=0,1,\end{cases}$$

$$\begin{aligned}&\hat{a}(i_k,i_l)=\frac{(p_{ik}-p_{il})^2}{(p_{ik}+p_{il})^2},\\&h_1(i_1,i_2,\dots,i_l;i_{l+1},\dots,i_n)\\&\quad=-(p_{i_1}^3+p_{i_2}^3+\cdots+p_{i_l}^3-p_{i_{l+1}}^3-\cdots-p_{i_n}^3)\\&\qquad+(p_{i_1}+p_{i_2}+\cdots+p_{i_l}-p_{i_{l+1}}-\cdots-p_{i_n})^3,\\&h_2(i_1,i_2,\dots,i_l;i_{l+1},\dots,i_n)\\&\quad=(p_{i_1}+p_{i_2}+\cdots+p_{i_l}-p_{i_{l+1}}-\cdots-p_{i_n})^2.\end{aligned}$$

Furthermore, we get

$$\begin{aligned}&\sum_{\sigma_1,\sigma_2,\dots,\sigma_n=\pm1}\hat{b}(\sigma_1p_1,\sigma_2p_2,\dots,\sigma_np_n)\\&\quad\cdot h_1(\sigma_1p_1,\sigma_2p_2,\dots,\sigma_np_n)=0,\quad\text{if } n \text{ is odd},\end{aligned} \tag{5.4.22}$$

$$\begin{aligned}&\sum_{\sigma_1,\sigma_2,\dots,\sigma_n=\pm1}\left(\prod_{i=1}^{n}\sigma_i\right)\hat{b}(\sigma_1p_1,\sigma_2p_2,\dots,\sigma_np_n)\\&\quad\cdot h_2(\sigma_1p_1,\sigma_2p_2,\dots,\sigma_np_n)=0,\quad\text{if } n \text{ is even},\end{aligned} \tag{5.4.23}$$

$$\begin{aligned}&\hat{b}(\sigma_1p_1,\sigma_2p_2,\dots,\sigma_np_n)=\prod_{k<l}^{(n)}(\sigma_kp_k-\sigma_lp_l)^2,\\&h_1(\sigma_1p_1,\dots,\sigma_np_n)=-(\sigma_1p_1^3+\sigma_n^3p_n^3)(\sigma_1p_1+\sigma_2p_2+\cdots+\sigma_np_n)^3,\\&h_2(\sigma_1p_1,\dots,\sigma_np_n)=(\sigma_1p_1+\sigma_2p_2+\cdots+\sigma_np_n)^2.\end{aligned}$$

Denoting the left side of equation (5.4.22) as $D_1(p_1, p_2, \ldots, p_n)$, we see it has the following three properties:

(i) D_1 is a symmetrical homogeneous polynomial,

(ii) D_1 is the even function of $p_1, p_2, \ldots, p_n$,

(iii)

$$D_1(p_1, \ldots, p_n) = 2(2p_1)^2 \prod_{m=2}^{n} (p_1^2 - p_m^2)^2 D(p_3, p_4, \ldots, p_n),$$

if $p_1 = p_2$.

It is easy to see that equation (5.4.22) is satisfied at $n = 1$. Assuming that equation (5.4.22) is satisfied at $n - 2$, we find that D_1 is the $2n(n-1)$th-order symmetrical homogeneous polynomial. On the other hand, it can be seen that D_1 is the polynomial of the $n(n-1) + 3$th order. Thus, we conclude that $D_1 = 0$ for certain n. In a similar way, we can prove equation (5.4.23).

Thirdly, we consider the nonlinear lattice equation

$$m\frac{d^2 r_n}{dt^2} = a\left[e^{-br_n} - e^{-br_{n+1}}\right], \quad n = 1, 2, \tag{5.4.24}$$

where $r_n = y_n - y_{n-1}$ and a and b are constants. Via the transformation

$$\frac{ab}{m}(e^{-br_n} - 1) = (\log f_n)_t, \tag{5.4.25}$$

Hirota obtained the following N-soliton solution for equation (5.4.24):

$$f_n(t) = \sum_{\mu=0,1} \exp\left[\sum_{i<j}^{N} B_{ij}\mu_i\mu_j + \sum_{i=1}^{N} \mu_i x_i\right], \tag{5.4.26}$$

where

$$x_i = \beta_i t - k_i n + \gamma_i, \quad k_i, \gamma_i \text{ being constants,}$$

$$\beta_i = \pm\left(\frac{ab}{m}\right)^{1/2} 2\sin\frac{k_i}{2},$$

$$e^{B_{ij}} = \frac{\frac{m}{ab}(\beta_i - \beta_j)^2 - 4\sinh^2\frac{k_i+k_j}{2}}{\frac{m}{ab}(\beta_i + \beta_j)^2 - 4\sinh^2\frac{k_i+k_j}{2}},$$

where $\sum_{\mu=0,1}$ denotes the summation over all possible combinations of $\mu_1 = 0, 1$, $\mu_2 = 0, 1, \ldots, \mu_N = 0, 1$.

Fourthly, we take the nonlinear electric filter equations

$$\frac{d^2}{dt^2}\log(1 + V_n(t)) = V_{n+1}(t) - 2V_n(t) + V_{n-1}(t) \tag{5.4.27}$$

and

$$\begin{aligned}\frac{dV_n}{dt} &= (1+V_n^2)(I_n - I_{n-1}),\\ \frac{dI_n}{dt} &= (1+I_n^2)(V_{n-1} - V_n)\end{aligned} \tag{5.4.28}$$

into account. N-soliton solutions for equation (5.4.27) have been obtained under the transformation

$$V_n = [\tan^{-1} g_n/f_n]_t \tag{5.4.29}$$

and f_n, g_n are in the forms of

$$\begin{aligned}f_n(t) &= \sum_{\mu=0,1}^{(l)} \exp\left[\sum_{i<j}^{N} B_{ij}\mu_i\mu_j + \sum_{i=1}^{N} \mu_i x_i\right],\\ g_n(t) &= \sum_{\mu=0,1}^{(e)} \exp\left[\sum_{i<j}^{N} B_{ij}\mu_i\mu_j + \sum_{i=1}^{N} \mu_i x_i\right],\end{aligned}$$

where

$$\begin{aligned}\alpha_i &= \beta_i t - k_i n + y_i,\\ \beta_i &= \pm 2\sin\frac{k_i}{2},\\ e^{B_{ij}} &= -\frac{(\beta_i-\beta_j)^2 - 4\sinh^2\frac{k_i-k_j}{2}}{(\beta_i+\beta_j)^2 - 4\sinh^2\frac{k_i+k_j}{2}},\end{aligned}$$

where $\sum_{\mu=0,1}^{(l)}$ and $\sum_{\mu=0,1}^{(e)}$ denote the summation over all possible combinations of $\mu_1 = 0,1$, $\mu_2 = 0,1,\ldots,\mu_N = 0,1$. Particularly, we require that

$$\sum_{i=1}^{N(l)} \mu_i = \text{even integer}, \quad \sum_{i=1}^{N(e)} \mu_i = \text{odd integer}.$$

Fifthly, as for the Hirota equation

$$i\varphi_t + 3i\alpha|\varphi|^2\varphi_x + \rho\varphi_{xx} + i\sigma\varphi_{xxx} + \delta|\varphi|^2\varphi = 0, \tag{5.4.30}$$

the N-envelop soliton solutions can be expressed as

$$\begin{aligned}\varphi &= g/f,\\ f(x,t) &= \sum_{\mu=0,1}{}' \exp\left[\sum_{i<j}^{2N} B_{ij}\mu_i\mu_j + \sum_{i=1}^{2N} \mu_i x_i\right],\\ g(x,t) &= \sum_{\mu=0,1}{}'' \exp\left[\sum_{i<j}^{2N} B_{ij}\mu_i\mu_j + \sum_{i=1}^{2N} \mu_i x_i\right],\end{aligned}$$

$$g^*(x,t) = \sum_{\mu=0,1}{}''' \exp\left[\sum_{i<j}^{2N} B_{ij}\mu_i\mu_j + \sum_{i=1}^{2N}\mu_i x_i\right],$$
$$x_j = k_j x - \beta_j t + \gamma_j, \quad k_j, \gamma_j \text{ all being constants,}$$
$$\beta_j = -i\rho k_j^2 + \sigma k_j^3, \quad j = 1,2,\ldots,2N,\ i = \sqrt{-1},$$
$$k_{j+N} = k_j^*, \quad \beta_{j+N} = \beta_j^*, \quad j = 1,2,\ldots,2N.$$

Here,

$$B_{ij} = \log\left[\frac{\alpha}{2\sigma}(k_i + k_j)^2\right], \quad \text{for } i = 1,2,\ldots,N;\ j = N+1, N+2,\ldots,2N$$
$$\text{or } i = N+1, N+2,\ldots,2N;\ j = 1,2,\ldots,N,$$
$$B_{ij} = -\log\left[\frac{\alpha}{2\sigma}(k_i - k_j)^2\right], \quad \text{for } i = 1,2,\ldots,N;\ j = 1,2,\ldots,N$$
$$\text{or } i = N+1, N+2,\ldots,2N;\ j = N+1, N+2,\ldots,2N,$$

where $\sum_{\mu=0,1}$ denotes the summation over all possible combinations of $\mu_1 = 0,1$, $\mu_2 = 0,1,\ldots,\mu_{2N} = 0,1$. The sums $\sum'_{\mu=0,1}$, $\sum''_{\mu=0,1}$, and $\sum'''_{\mu=0,1}$ satisfy

$$\sum_{\mu=0,1}{}'\mu_i = \sum_{\mu=0,1}{}'\mu_{i+N}, \quad \sum_{\mu=0,1}{}''\mu_i = 1 + \sum_{\mu=0,1}{}''\mu_{i+N}, \quad 1 + \sum_{\mu=0,1}{}'''\mu_i = \sum_{\mu=0,1}{}'''\mu_{i+N}.$$

5.5 Bilinear Bäcklund transformation

As we know, we can obtain the bilinear forms for some nonlinear evolution equations through the Hirota method. We consider the bilinear differential equation

$$F(D_t, D_x)f \cdot f = 0 \tag{5.5.1}$$

and construct a new differential equation

$$[F(D_t, D_x)f' \cdot f']ff - f'f'[F(D_t, D_x)f \cdot f] = 0. \tag{5.5.2}$$

Obviously, if f satisfies equation (5.5.1), we deduce that f' is another solution for equation (5.5.1), corresponding to equation (5.5.2). Therefore, equation (5.5.2) gives the Bäcklund transformation of f' and f for equation (5.5.1). The following exchange formulas should be mentioned and used:

(1) $$\exp(D_1)[\exp(D_2)a \cdot b] \cdot [\exp(D_3)c \cdot d]$$
$$= \exp\frac{D_2 - D_3}{2}\left[\exp\left(\frac{D_2 + D_3}{2} + D_1\right)a \cdot d\right] \cdot \left[\exp\left(\frac{D_2 + D_3}{2} - D_1\right)c \cdot b\right],$$

where $D_i = \epsilon_i D_x + \delta_i D_t$, ϵ_i and δ_i are constants, and $i = 1,2,3$,

(2) $(D_x^2 a \cdot b)cd - ab(D_x^2 c \cdot d) = D_x[(D_x a \cdot d)cb + ad(D_x c \cdot b)]$,

(3) $(D_xD_tf'\cdot f')ff - f'f'(D_x^2f\cdot f) = 2D_x(D_tf'\cdot f)\cdot ff'$,

(4) $(D_x^2f'\cdot f')ff - f'f'(D_x^2f\cdot f) = 2D_x(D_xf'\cdot f)\cdot ff'$,

(5) $(D_x^4f'\cdot f')ff - f'f'(D_x^4f\cdot f) = 2D_x(D_x^3f'\cdot f)\cdot ff' + 6D_x(D_x^2f'\cdot f)\cdot D_x(f\cdot f')$.

We take the KdV equation as an example, whose bilinear form is given by

$$D_x(D_t + c_0D_x + D_x^3)f\cdot f = 0, \tag{5.5.3}$$

where c_0 is a constant. Assuming that f is a solution for equation (5.5.3), f' is another one. f and f' satisfy

$$[D_x(D_t + c_0D_x + D_x^3)f'\cdot f]ff - f'f'[D_x(D_t + c_0D_x + D_x^3)f\cdot f] = 0. \tag{5.5.4}$$

By virtue of exchange formulas (2)–(5), equation (5.5.4) becomes

$$\begin{aligned} &2D_x\{[D_t + (c_0 + 3\lambda)D_x + D_x^3]f'\cdot f\}\cdot ff' \\ &\quad + 6D_x[(D_x^2 - \mu D_x - \lambda)f'\cdot f](D_xf\cdot f') = 0, \end{aligned} \tag{5.5.5}$$

where λ and μ are arbitrary constants. Supposing that f is a solution for equation (5.5.3), f' is another solution for equation (5.5.3) in the case where

$$[D_t + (c_0 + 3\lambda)D_x + D_x^3]f'\cdot f = 0, \tag{5.5.6}$$

$$(D_x^2 - \mu D_x - \lambda)f'\cdot f = 0, \tag{5.5.7}$$

which are exactly the Bäcklund transformations for equation (5.5.3). Similarly, we can obtain the Bäcklund transformation for the following nonlinear equations:

(I) Boussinesq equation. We have

$$(D_t^2 - D_x^2 - D_x^4)f\cdot f = 0, \tag{5.5.8}$$

whose BTs read as

$$\begin{cases} (D_t + aD_x^2)f'\cdot f = 0, \\ (aD_tD_x + D_x + D_x^3)f'\cdot f = 0, \end{cases} \tag{5.5.9}$$

where $a^2 = -3$.

(II) Kadomtsev–Petviashvili equation. We have

$$(D_tD_x + D_y^2 + D_x^4)f\cdot f = 0, \tag{5.5.10}$$

whose BTs read as

$$\begin{cases} (D_y + aD_x^2)f'\cdot f = 0, \\ (-aD_yD_x + D_t + D_x^3)f'\cdot f = 0, \end{cases} \tag{5.5.11}$$

where $a^2 = 3$.

(III) Higher-order KdV equation. We have

$$D_x(D_t + D_x^5)f \cdot f = 0, \tag{5.5.12}$$

whose BTs read as

$$\begin{cases} D_x^3 f' \cdot f = \lambda f' \cdot f, \\ \left[D_t - \frac{15}{2}\lambda D_x^2 - \frac{3}{2}D_x^5\right] f' \cdot f = 0. \end{cases} \tag{5.5.13}$$

(IV) Shallow water wave equation. We have

$$D_x(D_t - D_t D_x^2 + D_x)f \cdot f = 0, \tag{5.5.14}$$

whose BTs read as

$$\begin{cases} (D_x^3 - D_x)f' \cdot f = \lambda f' \cdot f, \\ (3D_x D_t - 1)f' \cdot f = \mu D_x f' \cdot f. \end{cases} \tag{5.5.15}$$

6 Bäcklund transformations and the infinitely many conservation laws

Bäcklund transformations were first discovered for the famous sine-Gordon equation at the end of the 19th century. They are usually treated as nonlinear superpositions to create a new solution from a known one.

6.1 Sine-Gordon equation and Bäcklund transformation

We consider the nonlinear Klein–Gordon equation

$$\varphi_{tt} - \varphi_{xx} + F'(\varphi) = 0. \tag{6.1.1}$$

If $F'(\varphi) = \varphi$, it is called Klein-Gordon equation; if $F'(\varphi) = \sin\varphi$, it will be the sine-Gordon equation

$$\varphi_{tt} - \varphi_{xx} + \sin\varphi = 0. \tag{6.1.2}$$

In the case where $\sin\varphi \sim \varphi$, the above two equations are equivalent. When $\sin\varphi \sim \varphi - \frac{1}{3!}\varphi^3$, i.e., $F(\varphi) = \frac{1}{2}\varphi^2 - \frac{1}{24}\varphi^4$, we have the following φ^4-field equation:

$$\varphi_{tt} - \varphi_{xx} + \varphi - \frac{1}{3!}\varphi^3 = 0. \tag{6.1.3}$$

When we choose $F'(\varphi) = \sin\varphi + \lambda \sin 2\varphi$, equation (6.1.1) becomes

$$\varphi_{tt} - \varphi_{xx} + \sin\varphi + \lambda \sin 2\varphi = 0, \tag{6.1.4}$$

which is called a double sine-Gordon equation.

Passing to the light cone coordinate,

$$\xi = \frac{x - t}{2}, \quad \eta = \frac{x + t}{2},$$

equation (6.1.2) transforms into

$$\varphi_{\xi\eta} = \sin\varphi. \tag{6.1.5}$$

The sine-Gordon equation was first derived in the course of investigation of surface geometry with constant Gaussian curvature $K = -1$. Many physical problems can be reduced to equations of this type, such as the propagation of vortices in Josephson junctions. In studies of general superconducting junctions, Josephson found that the current flow through a superconducting junction satisfies

$$J = J_0 \sin\varphi, \quad \frac{d\varphi}{dt} = \frac{2e}{\hbar}v,$$

where v denotes the voltage and $\varphi = \varphi_1 - \varphi_2$ is the phase offset between two superconducting wave functions. φ satisfies the next sine-Gordon equation

https://doi.org/10.1515/9783110549638-006

$$\varphi_{xx} + \varphi_{yy} - LC\varphi_{tt} = \frac{2eLJ_0}{\hbar} \sin\varphi,$$

where $L, C, e, J_0, \hbar$ are all the physical constants. Other related physical problems include crystal dislocations, wave propagation produced in the direction of magnetization in ferromagnetic material, etc.

It is easy to find the traveling wave solutions for equation (6.1.2). Setting $\varphi = \Phi(\xi)$, $\xi = x - Dt$, and $D = \text{constant} > 0$, equation (6.1.2) converts to

$$(D^2 - 1)\Phi_{\xi\xi} + \sin\Phi = 0. \tag{6.1.6}$$

Multiplying Φ_ξ and integrating, we obtain

$$\frac{1}{2}(D^2 - 1)\Phi_\xi^2 + 2\sin^2\frac{1}{2}\Phi = A, \tag{6.1.7}$$

where A is the integral constant. The soliton and periodic wave solutions for equation (6.1.2) are obtained from equation (6.1.7). If we take $A = 0$, $D^2 - 1 < 0$, we have

$$\text{tg}\Phi/4 = \pm\exp\{\pm(1 - D^2)^{-\frac{1}{2}}(\xi - \xi_0)\},$$

or

$$\Phi = 4\text{tg}^{-1} \pm \{\pm(1 - D^2)^{-\frac{1}{2}}(x - Dt)\}, \quad \text{for } \xi_0 = 0, \tag{6.1.8}$$

which is the soliton solution for equation (6.1.2). Periodic wave solutions for equation (6.1.2) can be expressed in detail as follows:

(1) The periodic wave solution will be obtained when $0 < A < 2, D^2-1 > 0$, Φ oscillates about $\Phi = 0$ in the interval $-\Phi_0 < \Phi < \Phi_0$, and $\Phi_0 = 2\sin^{-1}(\frac{A}{2})^{\frac{1}{2}}$.

(2) The periodic wave solution will also be obtained when $0 < A < 2$, $D^2 - 1 < 0$, and Φ oscillates about $\Phi = \pi$ in the interval $\pi - \Phi_0 < \Phi < \pi + \Phi_0$.

(3) The helicon wave solution will be obtained when $A < 0$ and $D^2 - 1 < 0$, as

$$\Phi_\xi = \pm\left\{\frac{2}{1 - D^2}\left(|A| + 2\sin^2\frac{\Phi}{2}\right)\right\}^{\frac{1}{2}}.$$

(4) The helicon wave solution when $A > 2$ and $D^2 - 1 > 0$ will be expressed as

$$\Phi_\xi = \pm\left\{\frac{2}{D^2 - 1}\left(A - 2\sin^2\frac{\Phi}{2}\right)\right\}^{\frac{1}{2}}.$$

(5) The kink solution will be obtained when $A = 2$ and $D^2 - 1 > 0$, as

$$\tan\left(\frac{\Phi + \pi}{4}\right) = \exp\{\pm(D^2 - 1)^{-\frac{1}{2}}(\xi - \xi_0)\},$$

where $-\pi < \Phi < \pi$.

Lamb solved the soliton solutions for the sine-Gordon equation via Bäcklund transformation. As for equation (6.1.5), we introduce the following Bäcklund transformation:

$$\begin{cases} \dfrac{\partial \varphi'}{\partial \xi} = \dfrac{\partial \varphi}{\partial \xi} + 2\lambda \sin\left(\dfrac{\varphi + \varphi'}{2}\right), \\ \dfrac{\partial \varphi'}{\partial \eta} = -\dfrac{\partial \varphi}{\partial \eta} + \dfrac{2}{\lambda} \sin\left(\dfrac{\varphi' - \varphi}{2}\right), \end{cases} \tag{6.1.9}$$

where λ is an arbitrary parameter. From equation (6.1.9), considering the derivative of the first equation with respect to η, the derivative of the second equation with respect to ξ, and the original equation $\varphi_{\xi\eta} = \sin\varphi$, we deduce the new equation $\varphi'_{\xi\eta} = \sin\varphi'$, which has the exact same form as the original. Choosing $\varphi = 0$ as the seed solution, another solution φ_1 can be expressed as

$$\varphi_1 = 4\mathrm{tg}^{-1}\left[\exp\left(\pm\frac{x - Dt}{\sqrt{1 - D^2}}\right)\right], \quad D = \frac{1 - \lambda^2}{1 + \lambda^2}. \tag{6.1.10}$$

φ_1 is a soliton solution for the sine-Gordon equation.

Generally speaking, for the sine-Gordon equation, Bäcklund transformation is a method structuring a new solution from a known one. In other words, Bäcklund transformation is a transformation between N-soliton and $N + 1$-soliton solutions. In addition, with the so-called "commutation principle", we can get a new solution from several existing solutions based on algebraic manipulation, while the integration of equation (6.1.9) is not essential. That is what the principle says: "Based on the initial solution φ_0 for equation (6.1.5), we will get the same φ_2, no matter the orders of equation (6.1.9) with respect to λ_1 and λ_2." From the above analysis, we get the following nonlinear superposition formulas:

$$\mathrm{tg}\left(\frac{\varphi_3 - \varphi_0}{4}\right) = \frac{D_1 + D_2}{D_1 - D_2}\mathrm{tg}\left(\frac{\varphi_1 - \varphi_2}{4}\right). \tag{6.1.11}$$

If $\varphi_0 = 0$, φ_1 and φ_2 take the forms of equation (6.1.10). Inserting them into equation (6.1.11), the following Perring–Skyrme solution for equation (6.1.5) will be obtained:

$$\mathrm{tg}\varphi/4 = \frac{\mathrm{sh}(x/\sqrt{1 - D^2})}{\mathrm{ch}(Dt/\sqrt{1 - D^2})}, \tag{6.1.12}$$

which can be seen as the following superposition of two kink solutions:

$$\mathrm{tg}\varphi/4 = \frac{\mathrm{sh}(Dt/\sqrt{1 - D^2})}{D\mathrm{ch}(x/\sqrt{1 - D^2})}, \tag{6.1.13}$$

which can be seen as the interaction between kink and anti-kink solutions. If we take $D = ib$ in equation (6.1.13), another soliton solution for equation (6.1.5) will be shown as

$$\mathrm{tg}\varphi/4 = \frac{\sin(bt/\sqrt{1 + b^2})}{\mathrm{ch}(x/\sqrt{1 + b^2})}, \tag{6.1.14}$$

whose evolution state displays the periodic attraction and repulsion between kink and anti-kink solutions, named "breather".

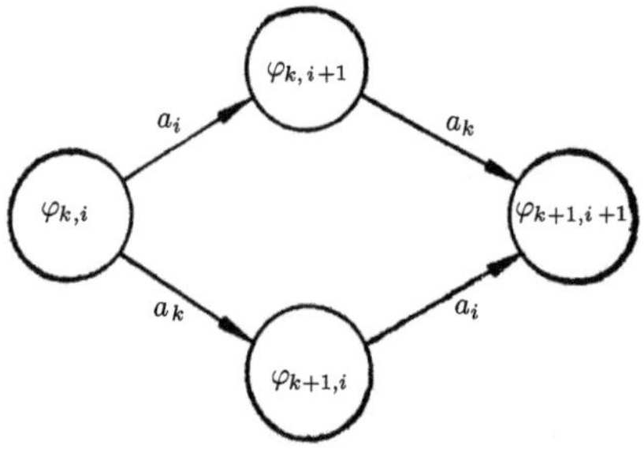

Figure 6.1: The relationships between solutions $\varphi_{k,j}$ and $\varphi_{k+1,j+1}$.

Now we generalize the procedure of finding soliton solutions for the sine-Gordon equation via Bäcklund transformation. We assume that the special solutions $\varphi_1, \varphi_2, \dots, \varphi_k$ for equation (6.1.5) will be obtained from the following Bäcklund transformations:

$$\begin{cases} \frac{1}{2}(\varphi_{j,x} - \varphi_{j-1,x}) = a_j \sin\frac{1}{2}(\varphi_j + \varphi_{j-1}), \\ \frac{1}{2}(\varphi_{j,t} - \varphi_{j-1,t}) = -\frac{1}{a_j}\sin\frac{1}{2}(\varphi_j - \varphi_{j-1}), \end{cases} \tag{6.1.15}$$

where $\varphi_0 = 0$. Under suitable initial conditions, solutions for equation (6.1.5) can be related by two parameters a_k and a_j as follows:

$$\begin{cases} \varphi_{k,j+1} = B_{a_j}\varphi_{k_j}, \\ \varphi_{k+1,j} = B_{a_k}\varphi_{k_j}, \\ \varphi_{k+1,j+1} = B_{a_j}B_{a_k}\varphi_{k_j} = B_{a_k}\cdot B_{a_j}\varphi_{k_j}, \end{cases} \tag{6.1.16}$$

where B_{a_j} is the Bäcklund transformation for a_j. The relationship between the solutions can be found in Figure 6.1.

N-soliton solutions for the sine-Gordon equation can be expressed as follows:

(i) if $j > 0$, then $\varphi_{j+1,j} = 0$ and

$$\varphi_{j,j} = 4\mathrm{tg}^{-1}[e^{-k_j x + \frac{1}{k_j}t + \gamma_{jj}}], \quad \gamma_{jj} \text{ is constant}, \ (-1)^j/k_j < 0; \tag{6.1.17}$$

(ii) if $j > l$, then

$$\varphi_{l,j} = \varphi_{l+1,j-1} + 4\mathrm{tg}^{-1}\left[\frac{\frac{1}{k_j} - \frac{1}{k_l}}{\frac{1}{k_j} + \frac{1}{k_l}}\right]\mathrm{tg}\left(\frac{\varphi_{l,j-1} - \varphi_{l+1,j}}{4}\right), \tag{6.1.18}$$

$$(-1)^l\frac{1}{k_l} < (-1)^j\frac{1}{k_j}.$$

We note that, in order to derive the N-soliton solutions through Bäcklund transformation, it is essential to know all the lower-order soliton solutions, which can be found in Figure 6.2.

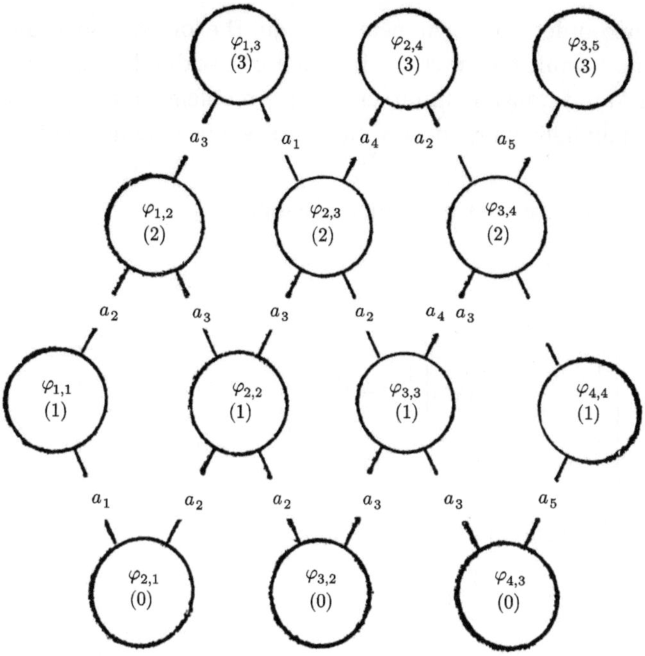

Figure 6.2: Lamb profile for *N*-solitons. The numbers in brackets stand for the numbers of solitons.

6.2 Bäcklund transformations for a class of nonlinear evolution equations

We presented the Bäcklund transformation for the sine-Gordon equation in the last section. In fact, a lot of nonlinear evolution equations have corresponding Bäcklund transformations. The transformation $z(x,y) \to z'(x',y')$ satisfies

$$\begin{cases} P = f(x',y',z',p',q'), \\ q = g(x',y',z',p',q'), \end{cases} \tag{6.2.1}$$

where $p = \frac{\partial z}{\partial x}$, $q = \frac{\partial z}{\partial y}$, $p' = \frac{\partial z'}{\partial x'}$, and $q' = \frac{\partial z'}{\partial y'}$. In a similar way, we denote $r = \frac{\partial^2 z}{\partial x^2}$, $s = \frac{\partial^2 z}{\partial x \partial y}$, $t = \frac{\partial^2 z}{\partial y^2}$, $r' = \frac{\partial^2 z'}{\partial x'^2}$, $s' = \frac{\partial^2 z'}{\partial x' \partial y'}$, and $t' = \frac{\partial^2 z'}{\partial y'^2}$. Setting $x = x'$ and $y = y'$, considering the integrable condition $\frac{dp}{dy} = \frac{dq}{dx}$ of z, we have

$$\Omega = f_{y'} - q_{x'} + f_{z'}q' - q_{z'}p' + (f_{p'} - g_{q'}) + f_{q'}t' + g_{p'}r' = 0. \tag{6.2.2}$$

For equation (6.2.2), two cases exist: (1) the equation is identical to zero, i.e.,

$$\begin{aligned} & f_{p'} - g_{q'} = f_{q'} = g_{p'} = 0, \\ & f_{y'} - q_{x'} + f_{z'}q' - q_{z'}p' = 0, \end{aligned}$$

or (2) $\Omega = 0$ gives the second-order Monge–Ampère equation. The former case results from contact transformation, while the latter one results from Bäcklund transformation. By virtue of the Bäcklund transformation, eigenvalue problems of the inverse scattering transform and infinitely many conservation laws will be obtained. For example:

(1) Bäcklund transformation for the sine-Gordon equation

$$S = \sin z \tag{6.2.3}$$

can be expressed as

$$\begin{cases} \frac{1}{2}(p - p') = a \sin\left[\frac{1}{2}(z + z')\right], \\ \frac{1}{2}(q + q') = a^{-1} \sin\left[\frac{1}{2}(z - z')\right]. \end{cases} \tag{6.2.4}$$

Making $\Gamma = \tan[(z + z')/4]$, from equation (6.2.4), we have

$$\Gamma_x + a\Gamma - \frac{1}{2}p(1 + \Gamma^2) = 0. \tag{6.2.5}$$

We know the Riccati equation

$$\Gamma_x + 2p\Gamma + Q\Gamma^2 + R = 0, \tag{6.2.6}$$

which is equivalent to

$$\begin{cases} w_{1,x} + pw_1 = -Rw_2, \\ w_{2,x} - pw_2 = Qw_1, \end{cases} \tag{6.2.7}$$

with $\Gamma = w_1/w_2$. Therefore, equation (6.2.5) is equal to

$$\begin{cases} w_{1,x} + \frac{1}{2}aw_1 = \frac{1}{2}pw_2, \\ w_{2,x} - \frac{1}{2}aw_2 = -\frac{1}{2}pw_1, \end{cases} \tag{6.2.8}$$

which are exactly the eigenvalue problems of the inverse scattering transform corresponding to equation (6.2.3).

If a is very small in equation (6.2.4), infinitely many conservation laws for equation (6.2.3) are obtained. In fact, taking $z'(x, y, a)$ in the form of

$$z'(x, y, a) \approx \sum_{j=0}^{\infty} z'_j(x, y)a^j, \quad a \to 0, \tag{6.2.9}$$

substitution of equation (6.2.9) into (6.2.4) gives

$$\sum_{j=0}^{\infty} z'_{jy}a^j = z_y + \frac{2}{a} \sin\left[\frac{1}{2}\left(\sum_{j=0}^{\infty} z'_j a^j - z\right)\right],$$

where $z_0' = z$ and $z_1' = 2z_y$ as $a \to 0$. Because the coefficients of higher-order terms are equal, we get

$$\begin{cases} z_2' = 2z_{yy}, \\ z_3' = 2z_{yyy} + \frac{1}{3}(z_y)^3, \\ z_4' = 2z_{yyyy} + 2(z_y)^2 z_{yy}, \\ z_5' = 2z_{yyyyy} + 3(z_y)^2 z_{yyy} + 5z_y(z_{yy})^2 + \frac{3}{20}(z_y)^5. \end{cases} \tag{6.2.10}$$

The energy-conserved form of equation (6.2.3) is written as

$$\frac{1}{2}(z_x'^2)_y + (\cos z' - 1)_x = 0. \tag{6.2.11}$$

Reconsidering equation (6.2.10) and collecting the same order of a, we get infinitely many conservation laws. Next, we give the following terms of conserved density:

$$\begin{aligned} T_0 &= \frac{1}{2}z_x^2, \\ T_1 &= 2z_{yyyx}z_x + 4z_{yyx}z_{yx} + z_y^2 z_{yx} z_x, \\ T_2 z7 &= 2z_{yyyyyx}z_x + 4z_{yyyyx}z_{yx} + 4z_{yyx}z_{yyyx} + 6z_{yyy}z_y z_x z_{yx} \\ &\quad + 3z_{xyyy}z_y^2 z_x + 10z_{yyx}z_{yy}z_y z_x + 5z_{yy}^2 z_{yx} z_x + 8z_{yy}z_{yx}^3 z_y \\ &\quad + 8z_y^2 z_{yx} z_{yyx} + \frac{3}{4}z_y^4 z_{yx} z_x + \cdots. \end{aligned}$$

(2) Recall the KdV equation

$$u_y + 6uu_x + u_{xxx} = 0. \tag{6.2.12}$$

Making $z = \int_{-\infty}^{x} u(x', y)dx'$, we get

$$q + 3p^2 + \alpha = 0, \quad \alpha = z_{xxx}. \tag{6.2.13}$$

Assuming the Bäcklund transformation for equation (6.2.12) in the following form:

$$\begin{cases} p = f(z, z', p'), \\ q = \varphi(z, z', q', r, r', p, p'), \end{cases} \tag{6.2.14}$$

the second equation of equation (6.2.14) can be simplified to

$$q = \varphi(z, z', q', p', r'), \tag{6.2.15}$$

since

$$r = f_z f + f_{z'} p' + f_p r'. \tag{6.2.16}$$

The second-order mixed derivative of z can be written as

$$\frac{dp}{dy} = f_z q + f_{z'} q' + f_{p'} s'$$

or

$$\frac{dq}{dx} = \varphi_z p + \varphi_{z'} p' + \varphi_{q'} s' + \varphi_{p'} r' + \varphi_{r'} \alpha'.$$

Due to the equality of the mixed derivatives and z' satisfying equation (6.2.13), we define function $\Omega(z, z', p, p', q, q', r, r')$ as

$$\Omega = (f_{p'} - \varphi_{q'})s' + f_z q + f_{z'} q' - \varphi_z p - \varphi_{z'} p' - \varphi_{p'} r' + \varphi_{r'}(q' + 3p'^2) = 0. \tag{6.2.17}$$

Selecting $\Omega_{s'} = f_{p'} - \varphi_{q'} = 0$, with f independent of q' and r', we have

$$\begin{aligned} &\varphi_{q'q'} = \varphi_{q'r'} = 0, \\ &\Omega_{q'} = f_{p'} f_z + f_{z'} - f f_{p'z} - p' f_{p'z'} - r' f_{pp'} + \varphi_{r'} = 0. \end{aligned} \tag{6.2.18}$$

Because $\Omega_{q'r'} = -f_{p'p'} + \varphi_{r'r'} = 0$, we deduce

$$f_{p'p'} = \varphi_{r'r'} = a(z, z', r'), \tag{6.2.19}$$

where $a(z, z', r')$ is undetermined. Considering

$$\Omega_{r'r'r'} = -3 f_{p'p'p'} = 0 \tag{6.2.20}$$

and equation (6.2.18), equation(6.2.19) can be written as

$$\begin{aligned} f(z, z', p') &= b(z, z')p' + c(z, z'), \\ \varphi(z, z', q', p', r') &= b(z, z')q' + \lambda(z, z', p')r' + v(z, z', p'). \end{aligned} \tag{6.2.21}$$

In the following steps, we will determine λ, c, and v. From equation (6.2.17), we have

$$\Omega_{r'r'} = -2\varphi_{r'p'} = 0,$$

from which we deduce that λ is independent of p'. In addition,

$$v(z, z', p') = v_2(z, z')p'^2 + v_1(z, z')p' + v_0(z, z'), \tag{6.2.22}$$

since $\Omega_{p'p'p'} = 0$. We deem $b(z, z')$ as a constant, so

$$\begin{cases} p = bp' + c, \\ q = bq' + \lambda r' + v_2 p'^2 + v_1 p' + v_0, \end{cases} \tag{6.2.23}$$

where b is constant and c, λ, and v_i $(i = 0, 1, 2)$ are the undetermined functions of z and z'. Substituting equation (6.2.23) into Q, we get the following seven equations to solve five unknown functions and a constant:

$$2v_2 = -(b\lambda_z + \lambda_{z'}), \tag{6.2.24}$$

$$\lambda = -(bc_z + c_{z'}), \tag{6.2.25}$$

$$v_1 = \lambda c_z - c\lambda_z, \tag{6.2.26}$$

$$v_2 c_z - cv_{2z} + 3\lambda - bv_{1z} - v_{1z'} = 0, \tag{6.2.27}$$

$$v_1 c_z - cv_{1z} - v_{0z'} - bv_{0z} = 0, \tag{6.2.28}$$

$$bv_{2z} + v_{2z'} = 0, \tag{6.2.29}$$

$$v_0 c_z - cv_{0z} = 0. \tag{6.2.30}$$

For z belonging to equation (6.2.13), we find the third derivatives of equation (6.2.23) as follows:

$$\begin{aligned} \alpha = b\alpha' - \lambda r' + 2v_2 p'^2 + p'[2bcc_{zz} + 2cc_{zz} + c_z(bc_z + c_{z'})] \\ + c^2 c_{zz} + cc_z^2 = 0, \end{aligned} \tag{6.2.31}$$

$$v_2 - b + b^2 = 0, \tag{6.2.32}$$

$$bc_{zz} + 2b + c_{zz'} = 0, \tag{6.2.33}$$

$$c^2 c_{zz} + cc_z^2 + v_0 + 3c^2 = 0. \tag{6.2.34}$$

From equation (6.2.30), we derive $v_0 = \psi(z')c(z, z')$, where $\psi(z')$ is undetermined. Integrating equation (6.2.34) once, we have

$$c_z^2 + 2c + \psi + Kc^{-2} = 0. \tag{6.2.35}$$

Choosing $K = 0$, we obtain $c_{zz} = -1$ from equation (6.2.35) and $c_{zz'} = -b$ from equation (6.2.33). Combining equations (6.2.32), (6.2.24), and (6.2.25), we get

$$c_{z'z'} = 2b + b^2. \tag{6.2.36}$$

Integrating equation (6.2.36), we have

$$c(z, z') = m - \frac{1}{2}[z^2 + 2bzz' - b(2 + b)z'^2] + kz + lz',$$

where k, l, m are integral constants. Supposing $m \neq 0$, $k = l = 0$, the following relationships are obtained from equations (6.2.24)–(6.2.30):

$$\lambda = 2b(z - z'),$$
$$v_1 = -2bm - b(z^2 - 2zz' + b^2 z'^2),$$
$$v_2 = b - b^2,$$

$$\psi = -2m - 2b(1+b)z'^2.$$

Taking $b = -1$, Bäcklund transformation for the KdV equation (6.2.12) can be expressed as

$$\begin{cases} p + p' = m - \frac{1}{2}(z - z')^2, \\ q + q' = (z - z')(r - r') - 2(p^2 + pp' + p'^2). \end{cases} \tag{6.2.37}$$

If $z' = 0$ for the KdV equation, equation (6.2.37) can be simplified to

$$\begin{cases} p = m - \frac{1}{2}z^2, \\ q = zr - 2p^2 = -2mp, \end{cases}$$

whose solutions can be solved as

$$z = (2m)^{\frac{1}{2}} \tanh\left[\left(\frac{m}{2}\right)^{\frac{1}{2}}(x - 2my)\right]$$

and

$$u = p = m\mathrm{sech}^2\left[\left(\frac{m}{2}\right)^{\frac{1}{2}}(x - 2my)\right], \tag{6.2.38}$$

where u is a solution for the KdV equation (6.2.12). If we set $\Gamma = z - z'$, from equation (6.2.37), we have

$$\Gamma_x - \frac{1}{2}\Gamma^2 + m - 2p = 0,$$

which is equal to

$$\begin{cases} v_{1x} = (2p - m)v_2, \\ v_{2x} = -\frac{1}{2}v_1. \end{cases} \tag{6.2.39}$$

(3) Recall the modified KdV (MKdV) equation

$$u_y + 6u^2u_x + u_{xxx} = 0. \tag{6.2.40}$$

In a procedure similar to the KdV equation, the integrated form of equation (6.2.40) becomes

$$q + 3p^2 + \alpha = 0.$$

The Bäcklund transformation writes

$$\begin{cases} p = bp' + a\sin v, \\ q = bq' - 2a\left[br'\cos v + p'^2\sin v + \frac{1}{2}a(p + bp')\right], \end{cases} \tag{6.2.41}$$

where $b = \pm 1$. Setting $\Gamma = \tan[\frac{1}{2}(z + bz')]$, we have

$$\Gamma_z + a\Gamma - p(1 + \Gamma^2) = 0,$$

which is equivalent to the following eigenvalue equations of the scattering problem for equation (6.2.40):

$$\begin{cases} w_{1x} + \dfrac{1}{2}aw_1 = pw_2, \\ w_{2x} - \dfrac{1}{2}aw_2 = pw_1. \end{cases} \tag{6.2.42}$$

(4) The complex conjugation form of the nonlinear Schrödinger equation reads

$$\begin{cases} iq + r + z^2\bar{z} = 0, \\ -i\bar{q} + \bar{r} + \bar{z}^2 z = 0, \end{cases} \tag{6.2.43}$$

where "–" denotes the complex conjugation. The Bäcklund transformation for equation (6.2.43) is

$$\begin{cases} p = p' - \dfrac{1}{2}iw\tau + ikv, \\ q = q' + \dfrac{1}{2}\tau(p + p') - k_n + \dfrac{1}{4}iv(|w|^2 + |v|^2), \end{cases} \tag{6.2.44}$$

where $w = z + z'$, $v = z - z'$, $\tau = \pm i(b - 2|v|^2)^{1/2}$, and n, b, k are real constants. Making $\Gamma = (b - 2|v|^2)^{1/2}/\sqrt{2}y$, from equation (6.2.44), we get

$$z[\Gamma_x + ik\Gamma + \tau^{-\frac{1}{2}}(z\Gamma^2 + \bar{z})] = z'[\Gamma_x + ik\Gamma + \tau^{-\frac{1}{2}}(z'\Gamma^2 + z^{-1})],$$

which is equivalent to the following equations:

$$\begin{cases} w_{1x} + \dfrac{1}{2}ikw_1 = -\tau^{-\frac{1}{2}}\bar{z}w_2, \\ w_{2x} - \dfrac{1}{2}ikw_1 = \tau^{-\frac{1}{2}}zw_1. \end{cases} \tag{6.2.45}$$

6.3 The commutativity of Bäcklund transformation for the KdV equation

The KdV equation

$$u_y + 6uu_x + u_{xxx} = 0, \tag{6.3.1}$$

remains invariant under the boundary terms

$$B_{\beta u'}: \begin{cases} u_x = \beta - u'_x - \dfrac{1}{2}(u - u')^2, \\ u_y = -u'_y + (u - u')(u_{xx} - u'_{xx}) - 2[u_x^2 + u_x u'_x + u'^2_x], \end{cases} \tag{6.3.2}$$

where β is the arbitrary Bäcklund transformation parameter. The integrated form of equation (6.3.1) reads

$$u_y + 3u_x^2 + u_{xxx} = 0. \tag{6.3.3}$$

Theorem 6.3.1. *If $u_{\beta_i} = B_{\beta_i} u_0$ $(i = 1, 2)$ is the solution for equation (6.3.3), which is obtained by Bäcklund transformation with seed solution u_0 and parameter β_i, we get another new solution φ for equation (6.3.3),*

$$\varphi = u_0 + 2(\beta_1 - \beta_2)/(u_{\beta_1} - u_{\beta_2}), \tag{6.3.4}$$

where $\varphi = B_{\beta_1} B_{\beta_2} u_0 = B_{\beta_2} B_{\beta_1} u_0$.

Proof. Obviously, we have

$$u_{0x} + u_{\beta_1,x} = \beta_1 - \frac{1}{2}(u_0 - u_{\beta_1})^2, \tag{6.3.5}$$

$$u_{0x} + u_{\beta_2,x} = \beta_2 - \frac{1}{2}(u_0 - u_{\beta_2})^2, \tag{6.3.6}$$

$$u_{\beta_1,x} + u_{\beta_1\beta_2,x} = \beta_2 - \frac{1}{2}(u_{\beta_1} - u_{\beta_1\beta_2})^2, \tag{6.3.7}$$

$$u_{\beta_2,x} + u_{\beta_2\beta_1,x} = \beta_1 - \frac{1}{2}(u_{\beta_2} - u_{\beta_2\beta_1})^2, \tag{6.3.8}$$

where $u_{\beta_1\beta_2} = B_{\beta_2} B_{\beta_1} u_0$ and $u_{\beta_2\beta_1} = B_{\beta_1} B_{\beta_2} u_0$. If we set $\varphi = u_{\beta_1\beta_2} = u_{\beta_2\beta_1}$, from equations (6.3.5)–(6.3.8), we obtain

$$u_{\beta_1,x} - u_{\beta_2,x} = \beta_1 - \beta_2 + \frac{1}{2}(u_{\beta_1} - u_{\beta_2})(2u_0 - u_{\beta_1} - u_{\beta_2}), \tag{6.3.9}$$

$$u_{\beta_1,x} - u_{\beta_2,x} = \beta_2 - \beta_1 + \frac{1}{2}(u_{\beta_2} - u_{\beta_1})(u_{\beta_1} + u_{\beta_2} - 2\varphi). \tag{6.3.10}$$

Equation (6.3.9) minus equation (6.3.10) is

$$\varphi = u_0 + 2(\beta_1 - \beta_2)/(u_{\beta_1} - u_{\beta_2}).$$

It is easy to verify that equation (6.3.4) is a solution for equation (6.3.3).

Similarly, Bäcklund transformation for the MKdV equation

$$v_y + 6v^2 v_x + v_{xxx} = 0 \tag{6.3.11}$$

can be expressed as

$$B_{\beta u'}: \begin{cases} u_x = \alpha u'_x + \beta \sin(u + \alpha u'), \\ u_y = \alpha u'_y - \beta[2\alpha u'_{xx} \cos(u + \alpha u') + 2u_x'^2 \sin(u + \alpha u') \\ \qquad + \beta(u_x + \alpha u'_x)], \quad \alpha = \pm 1, \end{cases} \tag{6.3.12}$$

where β is the arbitrary Bäcklund transformation parameter. □

Theorem 6.3.2. *If u_{β_i} $(i = 1, 2)$ is the solution for equation (6.3.11), which is obtained by Bäcklund transformation with seed solution u_0 and parameter $\beta = \beta_i$ $(i = 1, 2)$, we get another new solution for equation (6.3.11),*

$$\tan\left(\frac{\varphi - u_0}{2}\right) = \alpha\left(\frac{\beta_1 + \beta_2}{\beta_1 - \beta_2}\right)\tan\left(\frac{u_{\beta_1} - u_{\beta_2}}{2}\right), \tag{6.3.13}$$

where $\varphi = B_{\beta_1} B_{\beta_2} u_0 = B_{\beta_2} B_{\beta_1} u_0$.

Example. Taking $u_0 = 0$, we get from equation (6.3.12)

$$u_{\beta_i} = 2\tan^{-1} e^{\mu_i}, \tag{6.3.14}$$

where

$$\mu_i = \beta_i x - \beta_i^3 y + \gamma_i \tag{6.3.15}$$

and γ_i $(i = 1, 2)$ is an integral constant. The following solutions is obtained from equation (6.3.13):

$$\varphi = \pm 2\tan^{-1}\left[\left(\frac{\beta_1 + \beta_2}{\beta_1 - \beta_2}\right)\frac{\sinh\{\frac{1}{2}(\mu_1 - \mu_2)\}}{\cosh\{\frac{1}{2}(\mu_1 + \mu_2)\}}\right]. \tag{6.3.16}$$

6.4 Bäcklund transformations for the higher-order KdV equation and multi-dimensional sine-Gordon equation

In [270], Sawada and Kotera pointed out the following higher-order KdV equation:

$$u_t + 180u^2 u_x + 30(uu_{xxx} + u_x u_{xx}) + u_{xxxxx} = 0, \tag{6.4.1}$$

which possesses the bilinear form

$$D_x(D_t + D_x^5) f \cdot f = 0. \tag{6.4.2}$$

Based on equation (6.4.2), Sawada and Kaup [269] constructed the Bäcklund transformations

$$\left(D_t - \frac{15}{2}\beta D_x^2 - \frac{3}{2}D_x^5\right) f' \cdot f = 0, \tag{6.4.3}$$

$$(D_x^3 - \beta) f' \cdot f = 0, \tag{6.4.4}$$

where β is the Bäcklund transformation parameter. To prove the Bäcklund transformations, we need to verify that

$$P \equiv f' \cdot f' D_x(D_t + D_x^5) f \cdot f - ff D_x(D_t + D_x^5) f' \cdot f' = 0. \tag{6.4.5}$$

By virtue of the exchange formulas, equation (6.4.5) can be developed to

$$
\begin{aligned}
P &= D_x\Big[2(f'\cdot f)\cdot(D_tf'\cdot f)+\frac{3}{4}(f'\cdot f)\cdot(D_x^5f'\cdot f)\\
&\quad -\frac{15}{4}(D_xf'\cdot f)\cdot(D_x^4f'\cdot f)+\frac{15}{2}(D_x^2f'\cdot f)\cdot(D_x^3f'\cdot f)\Big]\\
&\quad +\frac{5}{4}D_x^3[(f'\cdot f)\cdot(D_x^3f'\cdot f)-3(D_xf'\cdot f)\cdot(D_x^2f'\cdot f)]\\
&= D_x[2(f'\cdot f)\cdot(D_tf'\cdot f)-3(f'\cdot f)\cdot(D_x^5f'\cdot f)\\
&\quad +15(D_x^2f'\cdot f)\cdot(D_x^3f'\cdot f)]+5D_x^3(f'\cdot f)\cdot(D_x^3f'\cdot f) && (6.4.6)\\
&= D_x[15\beta(f'\cdot f)\cdot(D_x^2f'\cdot f)+15\beta(D_x^2f'\cdot f)\cdot(f'\cdot f)]=0. && (6.4.7)
\end{aligned}
$$

Bäcklund transformation for the three-dimensional sine-Gordon equation

$$
\left(\sum_{i=1}^{3}\frac{\partial^2}{\partial x_i^2}-\frac{\partial^2}{\partial t^2}\right)u=\sin u \tag{6.4.8}
$$

has been carried out as follows:

$$
\begin{aligned}
&\left\{I\frac{\partial}{\partial x}+i\sigma_1\frac{\partial}{\partial x^2}+i\sigma_3\frac{\partial}{\partial x^3}+\sigma_2\frac{\partial}{\partial t}\right\}\left\{\frac{\alpha-i\beta}{2}\right\}\\
&\quad=\exp\{i\theta\sigma_1\exp[-i\varphi\sigma_2\exp(-\tau\sigma_1)]\}\sin\left\{\frac{\alpha+i\beta}{2}\right\},
\end{aligned} \tag{6.4.9}
$$

where $\sigma_1,\sigma_2,\sigma_3$ are Pauli matrices, I is a 2×2 identity matrix, $0\le\theta\le2\pi$, $0\le\varphi\le2\pi$, and $-\infty<\tau<\infty$. α and β satisfy

$$
\left(\sum_{i=1}^{3}\frac{\partial^2}{\partial x^{i2}}-\frac{\partial^2}{\partial t^2}\right)\alpha(x^1,x^2,x^3,t)=\sin\alpha(x^1,x^2,x^3,t), \tag{6.4.10}
$$

$$
\left(\sum_{i=1}^{3}\frac{\partial^2}{\partial x^{i2}}-\frac{\partial^2}{\partial t^2}\right)\beta(x^1,x^2,x^3,t)=\sinh\beta(x^1,x^2,x^3,t), \tag{6.4.11}
$$

respectively. We rewrite equation (6.4.9) in the new form

$$
\left\{I\frac{\partial}{\partial x}+iP\right\}\left\{\frac{\alpha-i\beta}{2}\right\}=[A_1+iA_2]\sin\left\{\frac{\alpha+i\beta}{2}\right\}, \tag{6.4.12}
$$

where

$$
\begin{cases}
P=\sigma_1\dfrac{\partial}{\partial x^2}+\sigma_3\dfrac{\partial}{\partial x^3}-i\sigma_2\dfrac{\partial}{\partial t},\\
\sigma_1=\begin{pmatrix}0&1\\1&0\end{pmatrix},\quad \sigma_2=\begin{pmatrix}0&-i\\i&0\end{pmatrix},\quad \sigma_3=\begin{pmatrix}1&0\\0&-1\end{pmatrix}
\end{cases} \tag{6.4.13}
$$

and

$$\begin{cases} A_1 = I\cos\theta, \\ A_2 = \begin{pmatrix} \sin\theta\sin\varphi\cosh\tau & (\cos\varphi - \sin\varphi\sinh\tau)\sin\theta \\ (\cos\varphi + \sin\varphi\sinh\tau)\sin\theta & -\sin\theta\sin\varphi\cosh\tau \end{pmatrix}. \end{cases} \quad (6.4.14)$$

Dividing equation (6.4.12) into the real and imaginative parts, we get

$$I\frac{\partial}{\partial x}\left\{\frac{\alpha}{2}\right\} + P\left\{\frac{\beta}{2}\right\} = A_1\sin\left(\frac{\alpha}{2}\right)\cosh\left(\frac{\beta}{2}\right) - A_2\cos\left(\frac{\alpha}{2}\right)\sinh\left(\frac{\beta}{2}\right), \quad (6.4.15)$$

$$P\left\{\frac{\alpha}{2}\right\} - I\frac{\partial}{\partial x}\left\{\frac{\beta}{2}\right\} = A_1\cos\left(\frac{\alpha}{2}\right)\sinh\left(\frac{\beta}{2}\right) + A_2\sin\left(\frac{\alpha}{2}\right)\cosh\left(\frac{\beta}{2}\right). \quad (6.4.16)$$

We construct the most simple nontrivial solutions for equations (6.4.10) and (6.4.11). Making $\beta = \beta_0 = 0$ and $\alpha = \alpha_0 = 0$ in equations (6.4.15) and (6.4.16), we derive

$$\alpha_1(x^1, x^2, x^3; \theta, \varphi, \tau) = 4\tan^{-1}\{a_0 \exp R\}, \quad (6.4.17)$$

$$\beta_1(x^1, x^2, x^3; \theta, \varphi, \tau) = \begin{cases} 4\tan^{-1}\{a_1 \exp R\}, & R \le 0, \\ 4\cosh^{-1}\{a_1 \exp R\}, & R > 0, \end{cases} \quad (6.4.18)$$

where

$$R = x\cos\theta + x^2\sin\theta\cos\varphi + \sin\theta\sin\varphi(x^3\cosh\tau + t\sinh\tau) \quad (6.4.19)$$

and a_i $(i = 0, 1)$ are integral constants. It can be verified that equation (6.4.17) is the soliton solution for equation (6.4.10), while β is not amongst the soliton candidates.

6.5 Bäcklund transformation for the Benjamin–Ono equation

The Benjamin–Ono (BO) equation, which describes one-dimensional internal waves in deep water, is expressed as

$$u_t + 2uu_x + H[u_{xx}] = 0, \quad (6.5.1)$$

where H is the Hilbert operator, defined by the Cauchy principal value integral

$$Hf(x) = \frac{1}{\pi}P\int_{-\infty}^{\infty}\frac{f(z)}{z - x}dz.$$

To get the bilinear form for equation (6.5.1), we introduce the following transformation:

$$u(x, t) = i\frac{\partial}{\partial x}(\log[f'/f]), \quad (6.5.2)$$

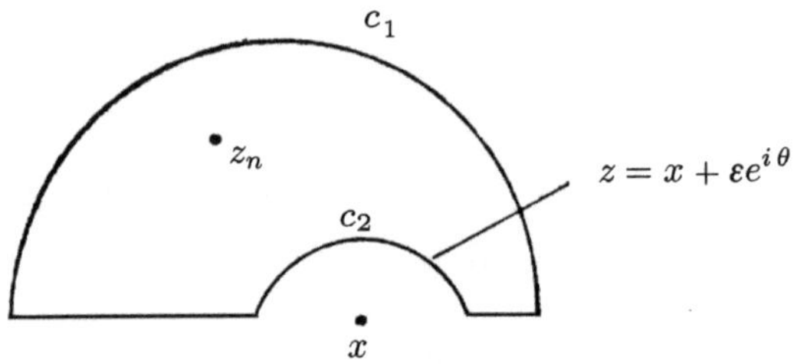

Figure 6.3: Profile of contour C.

where

$$f \propto \prod_{n=1}^{N}(x - z_n(t)), \quad f' \propto \prod_{n=1}^{N}(x - z_n'(t)), \tag{6.5.3}$$

z_n, z_n' are complex functions, and $\operatorname{Im} z_n > 0$, $\operatorname{Im} z_n' < 0$, $N \in Z^+$. Thus,

$$u = i\left(\frac{f_x'}{f'} - \frac{f_x}{f}\right) = i\sum_{n=1}^{N}\left\{\frac{1}{x - z_n'} - \frac{1}{x - z_n}\right\} \tag{6.5.4}$$

and

$$Hu = \frac{i}{\pi}P\int_{-\infty}^{\infty}\frac{1}{z - x}\sum_{n=1}^{N}\left\{\frac{1}{z - z_n'} - \frac{1}{z - z_n}\right\}dz. \tag{6.5.5}$$

To compute equation (6.5.5), taking the contour C (as shown in Figure 6.3) and by virtue of the residue theorem, we have

$$\frac{1}{2\pi i}\oint_C \frac{1}{z - x}\left[\frac{1}{z - z_n'} - \frac{1}{z - z_n}\right]dz = \operatorname{res}(z = z_n).$$

Therefore,

$$\begin{aligned}
&\lim_{\epsilon\to 0}\frac{1}{2\pi i}\int_{-\infty}^{x-\epsilon}\frac{1}{z - x}\left[\frac{1}{z - z_n'} - \frac{1}{z - z_n}\right]dz \\
&\quad + \lim_{\epsilon\to 0}\frac{1}{2\pi i}\int_{C_2}\frac{1}{z - x}\left[\frac{1}{z - z_n'} - \frac{1}{z - z_n}\right]dz \\
&\quad + \lim_{\epsilon\to 0}\frac{1}{2\pi i}\int_{x+\epsilon}^{\infty}\frac{1}{z - x}\left[\frac{1}{z - z_n'} - \frac{1}{z - z_n}\right]dz \\
&\quad = \frac{1}{x - z_n},
\end{aligned}$$

$$\begin{aligned}
&\frac{1}{2\pi i}P\int_{-\infty}^{\infty}\frac{1}{z - x}\left[\frac{1}{z - z_n'} - \frac{1}{z - z_n}\right]dz \\
&\quad = \frac{1}{x - z_n} - \lim_{\epsilon\to 0}\frac{1}{2\pi i}\int_{C_2}\frac{1}{z - x}\left[\frac{1}{z - z_n'} - \frac{1}{z - z_n}\right]dz
\end{aligned}$$

$$
\begin{aligned}
&= \frac{1}{x-z_n} - \lim_{\epsilon\to 0}\left\{\frac{1}{2\pi i}\int_\pi^0 \epsilon^{-1}e^{-i\theta}\left[\frac{1}{x+\epsilon e^{i\theta}-z_n'} - \frac{1}{x+\epsilon e^{i\theta}-z_n}\right]\epsilon i e^{i\theta}\,d\theta\right\}\\
&= \frac{1}{x-z_n} + \frac{1}{2}\left[\frac{1}{x-z_n'} - \frac{1}{x-z_n}\right]\\
&= \frac{1}{2}\left[\frac{1}{x-z_n'} + \frac{1}{x-z_n}\right].
\end{aligned} \tag{6.5.6}
$$

As a result, we derive

$$
\begin{aligned}
Hu &= \frac{i}{\pi}\sum_{n=1}^{N}\pi i\left(\frac{1}{x-z_n'} + \frac{1}{x-z_n}\right)\\
&= -\sum_{n=1}^{N}\left(\frac{1}{x-z_n'} + \frac{1}{x-z_n}\right)\\
&= -\left[\frac{f_x'}{f'} + \frac{f_x}{f}\right] = -\frac{\partial}{\partial x}(\log[f'f]).
\end{aligned} \tag{6.5.7}
$$

Substituting equations (6.5.2) and (6.5.7) into equation (6.5.1), we have

$$
\frac{\partial}{\partial x}\left[i\frac{\partial}{\partial t}(\log[f'/f]) - \left[\frac{\partial}{\partial x}(\log[f'/f])\right]^2 - \frac{\partial^2}{\partial x^2}(\log[f'f])\right] = 0.
$$

Integrating the above expression with respect to x and taking the integration constant as 0, we have

$$
i\frac{\partial}{\partial t}(\log[f'/f]) - \left[\frac{\partial}{\partial x}(\log[f'/f])\right]^2 - \frac{\partial^2}{\partial x^2}(\log[f'f] = 0,
$$

i.e.,

$$
i(ff_t' - f'f_t) - ff_{xx}' + 2f_x'f_x - f'f_{xx} = 0, \tag{6.5.8}
$$

which can be rewritten as

$$
[i(f_t'f - f'f_{t'}) - (f_{xx}'f - 2f_x'f_{x'} + f'f_{x'x'})]_{x'=x,t'=t} = 0, \tag{6.5.9}
$$

or

$$
\left[i(f_t'f - f'f_{t'}) - \left(\frac{\partial}{\partial x} - \frac{\partial}{\partial x'}\right)(f_x'f - f'f_{x'})\right]_{x'=x,t'=t} = 0.
$$

Thus, the following bilinear form for equation (6.5.1) is obtained:

$$
(iD_t - D_x^2)f'\cdot f = 0. \tag{6.5.10}
$$

Assuming that (f, f') are a pair of solutions for equation (6.5.10), while (g, g') are defined by the following Bäcklund transformations:

$$
(iD_t - 2i\lambda D_x - D_x^2 - \mu)f\cdot g = 0, \tag{6.5.11}
$$

$$(iD_t - 2i\lambda D_x - D_x^2 - \mu)f' \cdot g' = 0, \tag{6.5.12}$$
$$(D_x + i\lambda)f \cdot g' = ivf'g, \tag{6.5.13}$$

where λ, μ, v are undetermined parameters, we can prove that g and g' satisfy

$$(iD_t - D_x^2)g' \cdot g = 0. \tag{6.5.14}$$

In fact, equation (6.5.14) can be satisfied logically, if we derive

$$P = g'g(iD_t - D_x^2)f' \cdot f - f'f(iD_t - D_x^2)g' \cdot g = 0. \tag{6.5.15}$$

Taking account of

$$g'g(D_tf' \cdot f) - f'f(D_tg' \cdot g) = fg(D_tf' \cdot g') - f'g'(D_tf \cdot g),$$

equation (6.5.15) can be rewritten as

$$\begin{aligned} P &= fg(iD_tf' \cdot g') - f'g'(D_tf \cdot g) - gg'(D_x^2f' \cdot f) + f'f(D_x^2g' \cdot g) \\ &= 2i\lambda[fgD_xf' \cdot g' - f'g'D_xf \cdot g] + [fgD_x^2f' \cdot g' - f'g'D_x^2f \cdot g] \\ &\quad - g'gD_x^2f' \cdot f + f'fD_x^2g' \cdot g. \end{aligned}$$

By virtue of the exchange formulas

$$fgD_xf' \cdot g' - f'g'D_xf \cdot g = D_xf'g \cdot fg', \tag{6.5.16}$$
$$fgD_x^2f' \cdot g' - f'g'D_x^2f \cdot g = D_x[(D_xf' \cdot g) \cdot fg' + f'g \cdot (D_xf \cdot g')], \tag{6.5.17}$$
$$f'fD_x^2g' \cdot g - g'gD_x^2f' \cdot f = D_x[(D_xg' \cdot f) \cdot f'g + fg' \cdot (D_xf' \cdot g)] \tag{6.5.18}$$

and equations (6.5.17) and (6.5.18), we have

$$\begin{aligned} &fgD_x^2f' \cdot g' - f'g'D_x^2f \cdot g + f'fD_x^2g' \cdot g - g'gD_x^2f' \cdot f \\ &\quad = 2D_x[f'g \cdot (D_xf \cdot g')]. \end{aligned} \tag{6.5.19}$$

In addition,

$$cd(D_xa \cdot b) - ab(D_xc \cdot d) = D_xad \cdot bc, \tag{6.5.20}$$

so we derive

$$\begin{aligned} P &= 2i\lambda D_xf'g \cdot fg' + 2D_x[f'g \cdot (D_xf \cdot g')] \\ &= 2D_x[f'g \cdot (D_x + i\lambda)f \cdot g']. \end{aligned} \tag{6.5.21}$$

Considering equation (6.5.13), we get

$$P = 2D_x[f'g \cdot (ivf'g)] = 2ivD_x(f'g \cdot f'g) = 0,$$

which is exactly what we need to prove. As

$$v = i\frac{\partial}{\partial x}(\log[g'/g]), \tag{6.5.22}$$

a new solution for the BO equation (6.5.1) is obtained.

Consider the modified BO equation

$$u_t - 2\lambda u_x + 2ve^u u_x + Hu_{xx} + u_x Hu_x = 0, \tag{6.5.23}$$

where H is the Hilbert operator and λ, v are constants. In a similar way, via the following transformation:

$$u(x,t) = u_0 + \log\left[\frac{f'g}{fg'}\right], \tag{6.5.24}$$

we can prove that $u(x,t)$ is another solution for equation (6.5.23) under the following Bäcklund transformations:

$$(iD_t - 2i\lambda D_x - D_x^2 - \mu)f \cdot g = 0, \tag{6.5.25}$$

$$(iD_t - 2i\lambda D_x - D_x^2 - \mu)f' \cdot g' = 0, \tag{6.5.26}$$

$$(D_x + i\lambda)f \cdot g' = iv'f'g, \quad v' = ve^{u_0}. \tag{6.5.27}$$

Consider the wave propagation equation of streamline flow in finite depth,

$$u_t + 2uu_x + G[u_{xx}] = 0, \tag{6.5.28}$$

where G is the integral operator and

$$G[u(x,t)] = \frac{1}{2}\lambda \int_{-\infty}^{\infty} \left[\coth\frac{\pi}{2}\lambda(x'-x) - \operatorname{sgn}(x'-x)\right] u(x',t)dx', \tag{6.5.29}$$

where λ^{-1} is the parameter related to the depth of the flow. For the shallow water wave $\lambda \to \infty$, it becomes the KdV equation. For the deep water wave, $\lambda = 0$, it reduces to the BO equation. We make

$$u(x,t) = i\frac{\partial}{\partial x}(\log[\bar{f}/f]), \tag{6.5.30}$$

where

$$f(x,t) = \prod_{n=1}^{N}[1 + \exp\{\lambda[\lambda(\mathrm{Im}z_n)(x - \lambda t) - \bar{z}_n]\}], \tag{6.5.31}$$

where $z_n (n = 1, 2, \ldots, N)$ are complex, $0 < \lambda \mathrm{Im} z_n < \pi$, and $\bar{f}$ is the complex conjugation of f. Bäcklund transformations for equation (6.5.28) read

$$(iD_t + i(\lambda - 2\lambda')D_x - D_x^2 - \mu')f \cdot g = 0, \tag{6.5.32}$$

$$(iD_t + i(\lambda - 2\lambda')D_x - D_x^2 - \mu')\bar{f} \cdot g = 0, \tag{6.5.33}$$

$$(D_x + i\lambda')f \cdot \bar{g} = iv'\bar{f}g, \tag{6.5.34}$$

where λ', μ', v' are undetermined parameters.

6.6 The infinitely many conservation laws for the KdV equation

As is well known, mass, momentum, and energy conservation laws are the three important conservation laws in physics. In mathematics, if a physical problem can be described by a differential equation in the form of

$$u_t = K(u), \tag{6.6.1}$$

the corresponding conservation law can be written in the divergence form

$$\frac{\partial T}{\partial t} + \frac{\partial X}{\partial x} = 0, \tag{6.6.2}$$

where T and X are related to the unknown function $u(x,t)$. T denotes the density conservation and X is named flow conservation. When X is zero at the area boundary, it is sure that the invariant $I = \int T dx$ is independent of time.

Infinitely many conservation laws are closely related to the existence of soliton solution. More and more instances indicate that the nonlinear equations which have soliton solutions have a high possibility of possessing infinitely many conservation laws. On the other hand, the conservation integral is an important mathematical tool, based on which we can make a priori estimate on the solutions for differential equations. The a priori estimate is the core and key of the existence and uniqueness theorems of solutions for differential equations. As pointed out by Lax, infinitely many conservation laws are an important characteristic to distinguish the KdV equation from other nonlinear evolution equations.

For the KdV equation

$$u_t - 6uu_x + u_{xxx} = 0, \tag{6.6.3}$$

the first corresponding conservation is

$$u_t - (3u^2 + u_{xx})_x = 0, \tag{6.6.4}$$

from which we obtain the momentum conservation

$$\int_{-\infty}^{\infty} u(x,t)dx = \int_{-\infty}^{\infty} u(x,0)dx = M_0, \tag{6.6.5}$$

where u and its derivatives are zero as $|x| \to \infty$.

Multiplying the KdV equation by u, we get the second conservation form

$$\left(\frac{1}{2}u^2\right)_t + \left(-2u^3 + uu_x - \frac{1}{2}u_x^2\right)_x = 0 \tag{6.6.6}$$

and the energy conservation

$$E = \int_{-\infty}^{\infty} \frac{1}{2}u^2(x,t)dx = \int_{-\infty}^{\infty} \frac{1}{2}u^2(x,0)dx = E_0. \tag{6.6.7}$$

The third conservation law can be expressed as

$$\left(u^3+\frac{1}{2}u_x^2\right)_t+\left(-\frac{9}{2}u^4+3u^2u_{xx}-6uu_x^2+u_xu_{xxx}-\frac{1}{2}u_{xx}^2\right)_x=0. \tag{6.6.8}$$

Furthermore,

$$\int_{-\infty}^{\infty}\left(u^3+\frac{1}{2}u_x^2\right)dx=\int_{-\infty}^{\infty}\left(u^3(x,0)+\frac{1}{2}u_x^2(x,0)\right)dx. \tag{6.6.9}$$

These conservation laws were first derived by Whitham [306] and Miura [180], who gave the conservation forms explicitly. In addition, the infinitely many conservation laws have been obtained via a certain function transformation by Miura.

The MKdV equation

$$Qv\equiv v_t-6vv_x+v_{xxx}=0 \tag{6.6.10}$$

is related to the KdV equation by the following relationship.

Theorem 6.6.1. *If v solves equation (6.6.10), we conclude that $u=v^2+v_x$ satisfies equation (6.6.3), i.e.,*

$$Pu\equiv u_t-6uu_x+u_{xxx}=0.$$

Proof. It is obvious that $Pu=(2u+\frac{\partial}{\partial x})Qv$, so

$$Qv=0\quad\Rightarrow\quad Pu=0.$$

Equation (6.6.3) remains invariant under the following scale transformations:

$$t\to t',\quad x\to x'-6ct',\quad u\to u'+c,$$

where c is a constant. Here, setting

$$t'=t,\quad x'=x+\frac{3}{2\epsilon^2}t,\quad u(x,t)=u(x',t')+\frac{1}{4\epsilon^2},\quad \epsilon>0,$$

and $v(x,t)=\epsilon w(x',t')+\frac{1}{2\epsilon}$, we rewrite the transformation $u=v^2+v_x$ as

$$u(x',t')=w(x',t')+\epsilon w_{x'}(x',t')+\epsilon^2w^2(x',t').$$

Omitting $'$, we have

$$\begin{aligned}0=Pu&=u_t-6uu_x+u_{xxx}\\&=\left(1+\epsilon\frac{\partial}{\partial x}+2\epsilon^2w\right)[w_t-6(w+\epsilon^2w^2)w_x+w_{xxx}]\equiv LRw,\end{aligned}$$

where

$$Rw \equiv w_t - 6(w + \epsilon^2 w^2)w_x + w_{xxx}.$$

We claim that u is independent of ϵ, since ϵ is not present in the KdV equation. Therefore, from $u = w + \epsilon w_x + \epsilon^2 w^2$ ($\epsilon \ll 1$), we derive that w is a function of u and ϵ. Expanding w with respect to ϵ, we have

$$\begin{aligned} w &= w_0 + \epsilon w_1 + \epsilon^2 w_2 + \cdots \\ &= u - \epsilon u_x - \epsilon^2(u^2 - u_{xx}) + \cdots, \end{aligned}$$

where w_i ($i = 0, 1, 2, \ldots$) are polynomials of $u, u_x, u_{xx}, \ldots$. Substituting the above expansion into

$$Rw = w_t + (-3w^2 - 2\epsilon^2 w^3 + w_{xx})_x = 0$$

and collecting the coefficients of higher-order terms that are zero, infinitely many conservation laws for the KdV equation are obtained. □

6.7 Infinitely many conservation quantities for the AKNS equation

For the generalized Ablowitz–Kaup–Newell–Suger equation

$$\begin{cases} v_{1,x} = -i\zeta v_1 + q v_2, \\ v_{2,x} = i\zeta v_2 + r v_1, \end{cases} \tag{6.7.1}$$

the boundary conditions of the eigenfunctions $\varphi, \bar{\varphi}, \psi, \bar{\psi}$ are

$$\begin{aligned} &\varphi \sim \begin{pmatrix} 1 \\ 0 \end{pmatrix} e^{-i\xi x}, \quad \bar{\varphi} \sim \begin{pmatrix} 0 \\ -1 \end{pmatrix} e^{i\xi x}, \quad x \to -\infty, \\ &\psi \sim \begin{pmatrix} 0 \\ 1 \end{pmatrix} e^{i\xi x}, \quad \bar{\psi} \sim \begin{pmatrix} 1 \\ 0 \end{pmatrix} e^{-i\xi x}, \quad x \to +\infty, \end{aligned} \tag{6.7.2}$$

where $q, r \to 0$ as $|x| \to \infty$ and $\zeta = \xi + i\eta$ is the eigenvalue. Applying the Wentzel–Kramers–Brillouin method, we find

$$\begin{aligned} \psi e^{-i\zeta x} &\sim \begin{pmatrix} 0 \\ 1 \end{pmatrix} + \frac{1}{2i\zeta} \begin{pmatrix} q \\ -\int_x^{\infty} qr dx' \end{pmatrix} + \cdots, \\ \bar{\psi} e^{i\zeta x} &\sim \begin{pmatrix} 1 \\ 0 \end{pmatrix} + \frac{1}{3i\zeta} \begin{pmatrix} \int_x^{\infty} qr dx' \\ -r \end{pmatrix} + \cdots, \\ \varphi e^{i\zeta x} &\sim \begin{pmatrix} 1 \\ 0 \end{pmatrix} - \frac{1}{2i\zeta} \begin{pmatrix} \int_{-\infty}^{x} qr dx' \\ r \end{pmatrix} + \cdots, \end{aligned} \tag{6.7.3}$$

$$\begin{aligned}
\bar{\varphi}e^{-i\zeta x} &\sim \begin{pmatrix} 0 \\ -1 \end{pmatrix} - \frac{1}{2i\zeta}\begin{pmatrix} q \\ \int_{-\infty}^{x} qrdx' \end{pmatrix} + \cdots, \\
a(\zeta) &= w(\varphi, \psi) \sim 1 - \frac{1}{2i\zeta}\int_{-\infty}^{\infty} qrdx' + \cdots, \\
\bar{a}(\zeta) &= w(\bar{\varphi}, \bar{\psi}) \sim 1 + \frac{1}{2i\zeta}\int_{-\infty}^{\infty} qrdx' + \cdots,
\end{aligned} \tag{6.7.4}$$

where $w(u, v) = u_1v_2 - v_1u_2$. At $x \to \infty$, we have

$$\psi \sim \begin{pmatrix} 0 \\ 1 \end{pmatrix} e^{i\zeta x}, \tag{6.7.5}$$

which leads to

$$a(\zeta) = \lim_{x\to\infty} (\varphi_1 e^{i\zeta x}). \tag{6.7.6}$$

Making $e^{\hat{\varphi}} = \varphi_1 e^{i\zeta x}$, from equation (6.7.1), we get

$$\begin{aligned}
(\varphi_1 e^{i\zeta x})_x &= q\varphi_2 e^{i\zeta x}, \\
(\varphi_2 e^{-i\zeta x})_x &= r\varphi_1 e^{-i\zeta x}.
\end{aligned} \tag{6.7.7}$$

Eliminating φ_2 and using the definition of $\hat{\varphi}$, equation (6.7.7) develops to

$$\left(\frac{1}{qe^{2i\zeta x}}(e^{\hat{\varphi}})_x\right)_x = re^{-2i\zeta x}e^{\hat{\varphi}}, \tag{6.7.8}$$

or

$$\hat{\varphi}_x = \frac{1}{2i\zeta}\left[-qr + \hat{\varphi}_x^2 + q\left(\frac{\hat{\varphi}_x}{q}\right)_x\right]. \tag{6.7.9}$$

Making $\zeta \to \infty$ in equation (6.7.3), we have

$$\varphi e^{i\zeta x} \sim \begin{pmatrix} 1 \\ 0 \end{pmatrix} + O(1/\zeta). \tag{6.7.10}$$

We expand $\hat{\varphi}$ with respect to ζ, to obtain

$$\hat{\varphi} = \frac{\hat{\varphi}_1}{2i\zeta} + \frac{\hat{\varphi}_2}{(2i\zeta)^2} + \frac{\hat{\varphi}_3}{(2i\zeta)^3} + \cdots = \sum \frac{\hat{\varphi}_n}{(2i\zeta)^n}. \tag{6.7.11}$$

Meanwhile, the first examples of a series of solvable equations produced by equation (6.7.9) read

$$\hat{\varphi}_{1,x} = -qr \quad \Rightarrow \quad \hat{\varphi}_1 = -\int_{-\infty}^{x} qrdy,$$

$$\hat{\varphi}_{2,x} = q\left(\frac{\hat{\varphi}_{0,x}}{q}\right)_x = qr_x \quad \Rightarrow \quad \hat{\varphi}_2 = -\int_{-\infty}^{x} qr_y dy, \tag{6.7.12}$$
$$\hat{\varphi}_3 = -\int_{-\infty}^{x} (qr_{yy} - q^2r^2)dy.$$

The general recursion formula for $\hat{\varphi}_n$ is

$$\hat{\varphi}_{n+1} = q\left(\frac{\hat{\varphi}_n}{q}\right)_x + \sum_{k=1}^{n-1} \hat{\varphi}_k \varphi_{n-k}, \quad (n \geq 1), \tag{6.7.13}$$

where $\hat{\varphi}_0 = 0$ and $\hat{\varphi}_1 = -qr$. Noticing that $a(\zeta)$ is independent of t, the following conserved quantity will be obtained:

$$\ln a(\zeta) = \lim_{x\to\infty} \ln(\varphi_1 e^{i\zeta x}) = \lim_{x\to\infty} \hat{\varphi} = \sum_1^{\infty} \lim_{x\to\infty} \frac{\hat{\varphi}_n}{(2i\zeta)^n}. \tag{6.7.14}$$

For simplicity, we write $C_n = \lim_{x\to\infty} \hat{\varphi}_n$, $n = 1, 2, \ldots$, and

$$\begin{cases} C_1 = \int_{-\infty}^{\infty} qrdy, \\ C_2 = \int_{-\infty}^{\infty} qr_y dy, \\ C_3 = \int_{-\infty}^{\infty} (qr_{yy} - q^2r^2)dy. \end{cases} \tag{6.7.15}$$

For a kind of higher-order KdV equation,

$$u_t + u^q u_x + u_{x^p} = 0, \tag{6.7.16}$$

where p, q are nonnegative integers, $p \geq 2$. Kruskal and Miura [148] have predicted the number of conservation laws for equation (6.7.16) in 1970, as shown in Table 6.1.

This prediction has been proved perfectly through symmetric function methods. For the more generalized KdV equation

$$u_t + (f(u))_x = \beta u_{xxx}, \tag{6.7.17}$$

p \ q		0	1	2	⩾3
偶		1	1	1	1
奇	3	∞	∞	∞	∞
	⩾5	∞	3	3	3

Table 6.1: Numbers of conservation laws for equation (6.7.16).

there exist three conservation laws for $f(u)$ being the polynomials of u:

$$\begin{cases} T_1 = u, \quad X_1 = f(u) - \beta u_{xx}, \\ T_2 = \frac{1}{2}u^2, \quad X_2 = \int_0^u f'(u)udu - \beta uu_{xx} + \frac{1}{2}\beta u_x^2, \\ T_3 = \frac{\beta}{2}u_x^2 + \int_0^u f(u)du, \\ X_3 = \beta f'(u)u_x^2 + \beta^2 u_x u_{xxx} - \frac{\beta^2}{2}u_{xx}^2 + \frac{1}{2}f^2(u) - \beta f(u)u_{xx}. \end{cases} \tag{6.7.18}$$

By virtue of infinitely small transformation and symmetry, it has been proved that there exist at least three conservation laws for the following nonlinear evolution equation:

$$u_t = H(u, u_1, \ldots, u_n), \quad u_i = \mathscr{D}^i u, \quad \mathscr{D} = \frac{d}{dx}, \tag{6.7.19}$$

when $H = \mathscr{D}g$, where g is the grad polynomial and $H(u, u_1, \ldots, u_n)$ is the constant coefficient polynomial of u_i.

Infinitely many conservation laws for the Boussinesq, the nonlinear Schrödinger, and the derivative of the nonlinear Schrödinger equations have also been obtained through Bäcklund transformation [175].

6.8 Darboux transformations

Darboux transformations were originally developed by Gaston Darboux while studying the linear Sturm–Liouville problem [175]. Since the analysis of the inverse scattering transform is more difficult, Darboux transformations provided a convenient way to study solitons and their interactions for linear and nonlinear partial differential equations, including the nonlinear Schrödinger, KdV, Kadomtsev–Petviashvili, Toda lattice, and sine-Gordon equations, amongst others.

In 1882, Darboux [175] studied the eigenvalue of the Sturm–Liouville equation

$$-\Phi_{xx} + u(x)\Phi = \lambda\Phi, \tag{6.8.1}$$

which is usually referred to as the one-dimensional stationary Schrödinger equation in quantum mechanics, where $u(x)$ is the given potential function and λ is the spectral parameter. The following transformation was applied:

$$\begin{aligned} u[1] &= u - 2(\ln \Phi_1)_{xx}, \\ \Phi[1] &= \left(\partial_x - \frac{\Phi_{1,x}}{\Phi_1}\right)\Phi = \frac{\mathrm{Wr}(\Phi_1, \Phi)}{\Phi_1}, \end{aligned} \tag{6.8.2}$$

where Wr is the Wronskian determinant and Φ_1 is the special solution for equation (6.8.1) with $\lambda = \lambda_1$. We easily verify that

$$-\Phi_{xx}[1] + u[1]\Phi[1] = \lambda\Phi[1]. \tag{6.8.3}$$

The above transformation (6.8.2) is the classical Darboux transformation.

One way to understand the above transformation is the operator factorization method. We consider the second-order differential operator

$$L = -D^2 + u, \quad D = \frac{\mathrm{d}}{\mathrm{d}x},$$

introducing a factorization $L - \lambda = AA^*$, where

$$A = D - v, \quad A^* = -D - v$$

are the first-order operators, which are formally adjoint to each other. The following construction is entirely algebraic and all operators are considered as formal differential operators without boundary condition.

Definition 6.8.1. For any differential operator P of arbitrary finite order, which has possibly complex-valued coefficients, we define P^* as

$$(P^*f)g - f(Pg) = \frac{\mathrm{d}}{\mathrm{d}x}Q(f,g)$$

for arbitrary C^∞ functions, where Q is the polynomial of f, g.

For example, when $P = A$, we have

$$(A^*f)g - f(Ag) = -\frac{\mathrm{d}}{\mathrm{d}x}(fg).$$

To construct the general operator factorization, we take $\Phi \neq 0$ as a solution for the eigenvalue problem

$$(L - \lambda)\Phi = 0 \tag{6.8.4}$$

and make

$$A = \Phi D \Phi^{-1}, \quad A^* = -\Phi^{-1} D \Phi. \tag{6.8.5}$$

Next, we need to verify

$$A^* A = L - \lambda. \tag{6.8.6}$$

In fact, A^*A is a formally self-adjoint second-order differential operator, whose leading term is $-D^2$. Hence, A^*A is of the form $-D^2 + q$. From the form of A, we find $A^*A\Phi = 0$, i.e., $(-D^2 + q)\Phi = 0$. Therefore, through (6.8.4) and (6.8.5),

$$q = \frac{\Phi_{xx}}{\Phi} = u - \lambda.$$

The proof of (6.8.6) is complete.

Thus, every solution Φ for (6.8.4) may result such a factorization, but it is not the Darboux transformation we need. Similarly, we conclude that AA^* has the same leading term with A^*A, namely,

$$L[1] = -D^2 + u[1].$$

It is easy to find that

$$u[1] = u - 2(\ln \Phi)_{xx}$$

and the eigenfunction for $L[1]$ is

$$(L[1] - \lambda)\Phi^{-1} = 0.$$

Obviously, the above invertible transformation is the Darboux transformation (6.8.2).

However, one problem of the classical Darboux transformation is that the higher iteration cannot be carried out with the same spectral parameter. Recently, generalized Darboux transformation has been improved by using the limit technique to construct the rogue-wave solutions, especially the higher-order rogue-wave solutions. More details as regards generalized Darboux transformation are listed in [121, 120, 115].

7 Multi-dimensional solitons and their stability

7.1 Introduction

After having encountered a large number of problems as regards one-dimensional solitons, we now consider the following question. Do multi-dimensional solitons exist? If so, how do they behave? This is a widespread concern and a very important question. A lot of work has been done on the multi-dimensional solitons problem and some meaningful results have been obtained which are, but a lot more work will need to be carried out. Of course, the problem of multi-dimensional solitons is a complex and difficult problem, which involves a series of problems that must be solved. In the current situation, there are at least the following questions to be dealt with. (1) Do solitary waves and standing wave solutions exist? From a mathematical point of view, this problem is related to the existence of nonzero solutions to some boundary value problems for nonlinear elliptic equations. (2) Are these solitary waves and standing wave solutions stable? And can they collapse in a finite time? This question is currently a big issue in physics. (3). Are these solutions soliton solutions, i.e., do their wave forms and amplitudes remain unchanged (or changed slightly)? Some of these specific questions have been answered in part. In [65, 128], the authors pointed out that there does not exist a multi-dimensional fully stable solution to the nonlinear wave equations for a class of real (uncharged) scalar fields. In other words, if there exists a fully stable solution, it can be stable only in the case of plane geometry. Some sufficient conditions for the existence of solitary wave and standing wave solutions of the nonlinear Klein–Gordon equation were studied in [24, 283]. The existence conditions of multi-dimensional nonlinear Langmuir solitary wave and periodic wave solutions were discussed in [89, 301]. The conditions for the existence of solitons formed by a three-dimensional scalar field were studied in detail in [82] and a general theorem for its stability was given. Numerical results were given for some special problems. The existence and stability of the three-dimensional ion acoustic solitons in low-voltage magnetized plasmas was shown by Zakharov. Three expressions of soliton solutions to the two-dimensional sine-Gordon equation were given in [126]. The solitary wave problem for the multi-dimensional nonlinear Schrödinger equation was discussed in [2] and [171]. The cylindrical solitons in water waves were considered in [176] and the numerical results were also given. Guo et al. have derived the two-dimensional Boussinesq equation and the KdV equation and also discussed the problem of their solitary wave solutions. The stability of the soliton of the nonlinear Klein–Gordon equation in nonlinear field theory was considered in [13]. It was shown that the soliton was unstable with the nonlinear cubic term, but it was stable when the nonlinear term is quintic. The problem of the existence and collapse of the multi-dimensional plasma solitons was studied and discussed in [315] and [58]. From the point of view of current research, considering the existence of multi-dimensional solitons, a large number of

https://doi.org/10.1515/9783110549638-007

papers are devoted to the study of fully symmetric stable solutions, such that the problem is reduced to one dimension in mathematics, i.e., to study the spherical symmetric or column-symmetric model. As for the dynamics of multi-dimensional solitons and the investigation of the formation of the interaction process, the use of computer numerical calculation is widespread. From this point of view, it promotes the development of computational methods and computational mathematics for a new class of evolution equations. In this chapter, we introduce the existence results of the multi-dimensional solitons for several important nonlinear evolution equations and their stability and collapses are briefly introduced and commented on.

7.2 Existence problem for multi-dimensional solitons

We define the solitary wave as a solution of the wave equations when its maximal amplitude $\sup_x |\varphi(x,t)|$ is not vanishing as $t \to \infty$, but, for any t, it vanishes as $|x| \to \infty$. From the point of view of physics, some physical quantities, such as electric charge, energy, etc., are concentrated in a limited area of space in any time (i.e., nondispersive). Solitary waves generally have two special forms: (1) traveling waves, $\varphi = u(x-ct)$, where c is a constant vector, and (2) standing waves, $\varphi = \exp(i\omega t)u(x)$, where ω is a real number and $i = \sqrt{-1}$. Solitary waves are usually referred to as traveling waves, but in recent years, standing waves with oscillation factor are also known as solitary waves. For example, the traveling wave solution with oscillation factor for the nonlinear Schrödinger equation has been called envelope solitary wave. In some literature, solitary waves are confused with solitons, but solitons should be understood as solitary waves with "some certain safety factor", that is, the amplitude and shape of the solitary waves are not changed or only changed slightly by an interaction. In the following, we discuss the existence of solitary waves and solitons for several important nonlinear wave equations.

(I) The real nonlinear Klein–Gordon (NLKG) equation.
We have

$$\varphi_{tt} - \Delta\varphi + m^2\varphi + f(\varphi) = 0, \tag{7.2.1}$$

where $x = (x_1, x_2, \dots, x_n) \in R^n$, Δ is the Laplace operator, and $m > 0$. We assume that $f(0) = 0, f(re^{i\theta}) = f(r)e^{i\theta}$. If φ posses the standing wave from solution (2), then equation (7.2.1) becomes

$$-\Delta u + (m^2 - \omega^2)u + f(u) = 0. \tag{7.2.2}$$

We will show that there exists a nontrivial solution to (7.2.2), and the solution vanishes exponentially as $|x| \to \infty$ when $f(u)$ satisfies some growth conditions and $|\omega| < m$.

If (7.2.1) has the traveling wave solution (1), then we have

$$-\sum_{ij} a_{ij}\frac{\partial^2 u}{\partial x_i \partial x_j} + m^2 u + f(u) = 0, \tag{7.2.3}$$

where $a_{ij} = \delta_{ij} + c_i c_j$. If $|c| < 1$, then (a_{ij}) is positive definite. In fact, we have $\sum_{ij} a_{ij}\xi_i\xi_j = |\xi|^2 - (c\cdot\xi)^2 \geq (1-(c)^2)|\xi|^2$ for all $\xi = (\xi_1, \xi_2, \ldots, \xi_n) \in R^n$. Through some rotation transformations, (7.2.2) and (7.2.3) can be reduced to the following equation:

$$-\Delta u + F(u) = 0, \quad x \in R^n, \tag{7.2.4}$$

where $F(u) = f(u) + (\text{constant})u$. We always assume $F(0) = 0$, which means that equation (7.2.4) always has the trivial solution $u = 0$. We also assume F is a real continuous function. Let $G' = F$, $G(0) = 0$. Then we easily obtain some necessary conditions for the solution of equation (7.2.4).

Theorem 7.2.1. *If $u(x,t)$ is the solution of (7.2.4) and vanishes as $|x| \to \infty$, then we have*

$$\begin{aligned}(n-2)\int |\nabla u|^2 dx &= -(n-2)\int uf(u)dx \\ &= -2n\int G(u)dx. \end{aligned}\tag{7.2.5}$$

Therefore, if $sF(s)$ or $G(s)$ $(n \neq 1)$ or $H(s) = (n-2)sF(s) - 2nG(s)$, or $-H(s)$ is positive $(s \neq 0)$, then (7.2.4) only has a trivial solution. For any nontrivial solutions, the energy is positive, as we have

$$\begin{aligned}E(t) &= \int\left[\frac{1}{2}|\nabla u|^2 - G(u)\right]dx \\ &= \frac{1}{n}\int |\nabla u|^2 dx.\end{aligned}$$

Proof. We prove (7.2.5). Assuming $\bar{u}$ represents the complex conjugation of u, we have

$$-(\Delta u)\bar{u} = \nabla(\nabla u\cdot\bar{u}) + |\nabla u|^2.$$

Multiplying (7.2.4) by $\bar{u}$ and integrating with respect to x and assuming u and its derivative vanish as $|x| \to \infty$, we get

$$\int [|\nabla u|^2 + \operatorname{Re}\bar{u}F(u)]dx = 0.$$

On the other hand, $r\frac{\partial \bar{u}}{\partial r} = \sum x_i\bar{u}_i$ and the identities

$$\begin{aligned}&-\operatorname{Re} u_{ij}\bar{u}_i = -\operatorname{Re}(u_j x_i \bar{u}_i)_j + \left(\frac{1}{2}x_i|u_j|^2\right)_i + \left(1-\frac{n}{2}\right)|u_j|^2, \\ &\operatorname{Re} F(u)x_i\bar{u}_i = (x_i G(u))_i - nG(u), \\ &\int [(n-2)|\nabla u|^2 + 2nG(u)]dx = 0.\end{aligned}$$

This immediately yields (7.2.5). □

Let $L \equiv -\sum a_{ij}\frac{\partial^2}{\partial x_i \partial x_j} + a_0$, where the constant matrix a_{ij} is positive definite and a_0 is a positive constant. Assume $F_1(s)$, $F_2(s)$ are real continuous functions, $s \in [0, \infty)$, and $G_1(s)$, $G_2(s)$ are the indefinite integrals of F_1, F_2, respectively. We assume they satisfy the following conditions:

$$F_1(s) \geq 0, \quad F_2(s) > 0, \quad s > 0, \tag{7.2.6}$$

$$F_1(s) = O(s), \quad F_2(s) = o(s), \quad \text{when } s \to 0, \tag{7.2.7}$$

$$F_2(s) = o(s^l + F_1(s)), \quad s \to \infty, \tag{7.2.8}$$

$$F_2(s) = o(s^l + G_1(s)/s), \quad s \to \infty, \tag{7.2.9}$$

where $l = \frac{n+2}{n-2}$ and $n > 3$.

Theorem 7.2.2. *Assume that the conditions* (7.2.6), (7.2.7), (7.2.8), *and* (7.2.9) *hold. Then there exist* $\lambda > 0$ *and the solution* $u \in H^1$ *of*

$$Lu + F_1(u) = \lambda F_2(u), \tag{7.2.10}$$

where u is nonnegative; u exponentially decays to zero as $|x| \to \infty$ *and* $\int G_1(u(x))dx < \infty$.

Remark. Theorem 7.2.2 is still true when $n = 1$ or $n = 2$ and it can be established in weaker conditions at this time. In what follows, we give a few examples of how to use the results of Theorems 7.2.1 and 7.2.2.

Example 7.2.3. Given $-\Delta u + u - |u|^{q-1}u = 0$, $x \in R^n$, $n \geq 3$, $q > 1$.

Applying Theorem 7.2.1, we have

$$F(s) = s - |s|^{q-1}s, \quad G(s) = \frac{s^2}{2} - \frac{|s|^{q+1}}{q+1}.$$

Let $\alpha^{-1} = 2^{-1} - (q+1)^{-1}$, so while the coefficients of

$$\left(\frac{n-2}{2}\right)sF(s) - nG(s) = -s^2 + (1 - \alpha^{-1}n)|s|^{q+1}$$

have the same symbol, i.e., $\alpha \leq n$ or $q \geq \frac{n+2}{n-2}$, there does not exist a nontrivial solution. Thus, we assume $1 < q < \frac{n+2}{n-2}$. Any solutions must satisfy the identity (7.2.5), that is,

$$\begin{aligned} \alpha(n-2)\int |\nabla u|^2 dx &= \frac{n\alpha}{\alpha - n}\int |u|^2 dx \\ &= n\int |u|^{q+1} dx. \end{aligned}$$

If we set $F_1(s) = 0$, $F_2(s) = |s|^{q-1}s$, $L = -\Delta + I$, $\lambda = 1$, then, from Theorem 7.2.2, we know that the nonnegative solution exists.

Example 7.2.4. Given $-\Delta u+(m^2-\omega^2)u+|u|^{p-1}u-\lambda|u|^{q-1}u=0$, where $x\in R^n$, $m^2-\omega^2>0$, and p, q are two numbers which are different from each other and bigger than 1. We discuss four situations.

Case A: $1<q<\max(p,\frac{n+2}{n-2})$. Theorem 7.2.2 asserts that there exist nontrivial solutions for some $\lambda>0$. Note that

$$G(s)=\frac{1}{2}(m^2-\omega^2)s^2+\frac{1}{p+1}|s|^{p+1}-\frac{\lambda}{q+1}|s|^{q+1}$$

is bounded from below, so there exists λ_* such that $G(s)$ is nonnegative while $\lambda\ge\lambda_*$. By Theorem 7.2.1, there only exists the trivial solution. From Theorem 7.2.1, we know that, if the nontrivial solution exists, the energy integral must be positive. At this time, we have the energy density

$$\begin{aligned}&\frac{1}{2}|\varphi_t|^2+\frac{1}{2}|\nabla\varphi|^2+G(\varphi)\\&\quad=\frac{1}{2}|\nabla u|^2+\omega^2u^2+G(u).\end{aligned}$$

If $\omega>0$ and λ is slightly larger than λ_*, then it easy to prove the above is positive. These solutions have been calculated in [315] when $n=3$, $p=5$, and $q=3$. Interesting results were obtained. When the energy density is positive, the perturbation of the positive solution with respect to the initial conditions is stable. Under the above selection of p, q, n and choosing $m^2-\omega^2=1$, we have

$$\begin{aligned}\lambda s^4&=(2s)\left(\frac{1}{2}\lambda s^3\right)\le\frac{1}{2}(2s)^2+\frac{1}{2}\left(\frac{1}{2}\lambda s^3\right)^2\\&=2s^2+\frac{1}{8}\lambda^2s^6,\\G(s)&=\frac{1}{2}s^2+\frac{1}{6}s^6-\frac{\lambda}{4}s^4\ge\left(\frac{1}{6}-\frac{1}{32}\lambda^2\right)s^6.\end{aligned}$$

Then we get $\lambda_*=(4.3)^{-\frac{1}{2}}$.

Case B: $p<q<\frac{n+2}{n-2}$. Applying Theorem 7.2.2, we can show that there exists an infinite sequence of nontrivial solutions for each $\lambda>0$.

Case C: $p\le\frac{n+2}{n-2}\le q$. Let $\alpha^{-1}=2^{-1}+(q+1)^{-1}$ and $\beta^{-1}=2^{-1}+(p+1)^{-1}$. Then $\alpha\le n\le\beta$ and

$$\begin{aligned}&\frac{n-2}{2}sF(s)-nG(s)\\&\quad=-s^2-\left(1-\frac{n}{\beta}\right)|s|^{p+1}+\left(1-\frac{n}{\alpha}\right)\lambda|s|^{\alpha+1}.\end{aligned}$$

By Theorem 7.2.1, there exist nontrivial solutions.

Case D: $\frac{n+2}{n-2}<p<q$. We do not know whether there exist nontrivial solutions.

Remark. For the existence of nontrivial solutions to the equation $-\Delta u + F(u) = 0$, we substantially required $F'(0) \geq 0$. In fact, assume $-\alpha = F'(0) < 0$, let $f(s) = F(s)/s + \alpha$, assume $u(x)$ to be a nontrivial solution and small at infinity, and let $q(x) = f(u(x))$. Then the equation can be written as $-\Delta u + qu = \alpha u$. Assume $u(x)$ to be sufficiently small at infinity, such that $q(x) = O(|x|^{-1})$. Then it is easy to see that the operator $-\Delta + q$ has no positive eigenvalues, which is a contradiction. Therefore,

$$F'(0) \geq 0.$$

We now consider the solution of the axisymmetric problem. The axial solution $u(r)$ is continuous for $r = |x| \neq 0$ and satisfies the following equation:

$$u_{rr} + \frac{n-1}{r}u_r - F(u) = 0, \quad 0 < r < \infty,$$

where $F(u) = u + F_1(u) - \lambda F_2(u)$. Then

$$r^{1-n}(r^{n-1}u_r)_r = u_{rr} + \frac{n-1}{r}u_r$$

is continuous, so $u \in C^2$, $r \neq 0$. Let

$$q(r) = \frac{F(u(r))}{u(r)} = 1 + F_1'(0) + p(r).$$

From (7.2.6) and (7.2.7), we have $p(r) \to 0$ and $r \to \infty$. Thus, we have $q(r) \geq \frac{1}{2}$ for sufficiently large r. Set $v = r^{(n-1)/2}u$. Then v satisfies

$$v_{rr} - \left[q(r) + \frac{(n-1)(n-3)}{4r^2}\right]v = 0,$$
$$\left(\frac{1}{2}v^2\right)_{rr} = v_r^2 + \left[q(r) + \frac{(n-1)(n-3)}{4r^2}\right]v^2.$$

Thus $\omega = v^2$ satisfies $\omega_{rr} \geq \omega$ for sufficiently large r. From this, we can derive the exponential decay of ω and u.

In fact, for the large r, we deduce that $Q = e^{-r}(\omega_r + \omega)$ is nondecreasing. If Q remains nonpositive for large r, then we have $(e^r\omega)_r = e^{2r}Q \leq 0$, so we can derive $\omega = O(e^{-r})(r \to \infty)$. If $Q \geq 2\delta > 0$, then $\omega_r + \omega$ must not be integrable near infinity, but since $u \in H^1$, functions v^2, v_r^2, ω, and ω_r are all integrable in the interval $k < r < \infty$. This is a contradiction and, therefore, we have proved the exponential decay of the solutions.

The relation of the axial solution and a class of definite solutions is as follows.

Theorem 7.2.5. *Assume $L = I - \Delta$, $F_1 = 0$, and F_2 is a continuous real function such that*

(i) $\quad sF_2(s) > 0, \quad s \neq 0,$

(ii) $F_2(s) = O(|s|^p)$, $|s| \to \infty$, *and*

$$p < (n+2)/(n-2) = l,$$

(iii) F_2 *is an odd function and* $F_2(0) = 0$.

Then, for any $\gamma > 0$, *there exist infinite axial solutions* $(\lambda_k, \pm u_k)$, $k = 0, 1, 2, \ldots$, *to* $Lu = \lambda F_2(u)$, *while*

$$(Lu_k, u_k) = \gamma.$$

Theorem 7.2.6. *Assume F is a real continuous function and satisfies*

(i) $F(s)/s \to -\infty$, $s \to \infty$,

(ii) $sF(s) \geq \alpha G(s)$, $\alpha > 2$,

(iii) $F(s) = o(s)$, $s \to 0$,

(iv) $F_2(s) = O(|s|^p)$, $|s| \to \infty$, $p < (n+2)/(n-2) = l$.

Then there exists at least one nontrivial solution for the equation $Lu + F(u) = 0$. *If F is an odd function, then there exist infinitely different solutions* $\pm u_k$ $(k = 0, 1, 2, \ldots)$.

Example 7.2.7. Given $-\Delta u + u - |u|^{q-1}u = 0$, $1 < q < \frac{n+2}{n-2}$. By Theorems 7.2.5 and 7.2.6, we can derive that there exist many axial solutions $u_0, u_1, u_2, \ldots$.

Example 7.2.8. Given $-\Delta u + u + |u|^{p-1}u - \lambda|u|^{q-1}u = 0$, $1 < p < q < \frac{n+2}{n-2}$. For any $\lambda > 0$, we can derive that there exists at least one nontrivial solution by Theorem 7.2.6. For large p, q, this problem remains unresolved.

(II) Consider the multi-dimensional nonlinear Langmuir wave

$$\begin{cases} i\dfrac{\partial \vec{E}}{\partial t} = -\nabla^2 \vec{E} + n\vec{E}, \\ \dfrac{\partial^2 n}{\partial t^2} = \nabla^2 [n + g(|\vec{E}|^2)], \end{cases} \tag{7.2.11}$$

where $i = \sqrt{-1}$, $\vec{E} = (E_1, E_2, \ldots, E_N)$ is the complex amplitude of the high-frequency electric field, n is the low-frequency disturbance of the ion density with respect to its constant equilibrium state, and g is the given function of $|\vec{E}|^2$. When $g(|\vec{E}|^2) = |\vec{E}|^2$, (7.2.11) represents the Zakharov equations. When $g(|\vec{E}|^2) = \chi(1 - \exp(-|\vec{E}|^2))$, where χ is a positive constant, the equation corresponds to the saturated state of the ion density. Assume $\vec{k}$ to be the unit vector of R^n and v to be the traveling wave velocity. We look for the following traveling wave solutions of (7.2.11):

$$\vec{E}(\vec{x}, t) = \vec{h}(\vec{k} \cdot \vec{x} - vt), \quad n(x, t) = s(\vec{k} \cdot \vec{x} - vt), \tag{7.2.12}$$

where $\vec{h}$ and s are the vector function and the function which is to be demanded, respectively. Assume $|\vec{h}(\xi)|$, $|s(\xi)|$ to be uniformly bounded, where $\xi = \vec{k}\cdot\vec{x} - vt$ and assume

$$|\vec{h}(\xi)|,\ |s(\xi)| \to 0,\ (|\xi| \to \infty).$$

Inserting (7.2.12) into (7.2.11), we find $\vec{h} = (h_1, \dots, h_N)$ and s satisfy the equations

$$-iv\frac{d\vec{h}}{d\xi} + \frac{d^2\vec{h}}{d\xi^2} = s(\xi)\vec{h}(\xi), \tag{7.2.13}$$

$$(v^2-1)\frac{d^2 s}{d\xi^2} = \frac{d^2}{d\xi^2}g(|\vec{h}|^2). \tag{7.2.14}$$

Integrating (7.2.14), we have

$$(v^2-1)s(\xi) = g(|\vec{h}|^2) + \hat{c}\xi + c, \tag{7.2.15}$$

where $\hat{c}$ and c are integration constants. Assuming $v^2 \neq 1$, from the boundedness requirement of $s(\xi)$, we set $\hat{c} = 0$. We solve $s(\xi)$ from (7.2.15) and insert the result in (7.2.13), to find $\vec{h}$ satisfies the complex equation

$$\frac{d^2\vec{h}}{d\xi^2} - iv\frac{d\vec{h}}{d\xi} = (v^2-1)^{-1}\left[g\left(\left|\vec{h}(\xi)\right|^2\right) + c\right]\vec{h}.. \tag{7.2.15'}$$

For the sake of convenience, we write (7.2.15') in polar form. Let

$$h_j(\xi) = A_j(\xi)\exp\left[i\theta(\xi)\right],\quad j = 1, 2, \dots, N.$$

Then we have

$$\frac{d^2A_j}{d\xi^2} + A_j\theta_j'[v - \theta_j'(\xi)]$$
$$= (v^2-1)^{-1}A_j[g(\|A\|^2) + c] \tag{7.2.16}$$

$$\frac{d^2\theta}{d\xi^2} = [v - 2\theta_j'(\xi)]\frac{d}{d\xi}\ln A_j,\quad j = 1, 2, \dots, N, \tag{7.2.17}$$

where $A = (A_1, A_2, \dots, A_N)$, $\|A\| = |\vec{h}|$, $\theta_j' = \frac{d\theta_j}{d\xi}$. Integrating (7.2.17), we get

$$\begin{aligned} \theta_j'(\xi) &= (v - \mu_j A_j^{-2}(\xi))/2, \\ \mu_j &= A_j^2(0)(v - 2\theta_j'(0)). \end{aligned} \tag{7.2.18}$$

Inserting (7.2.18) into (7.2.16), we find that A_j satisfies the differential equations

$$\frac{d^2A_j}{d\xi^2} = f(\mu_j, c, \vec{A})A_j,\quad j = 1, 2, \dots, N, \tag{7.2.19}$$

where

$$f(\mu_j, c, \vec{A}) = (\mu_j^2 A_j^{-4} - v^2)/4 + (v^2 - 1)^{-1}\left[g(\|A\|^2) + c\right]. \tag{7.2.20}$$

Equation (7.2.19) can be written in the following form:

$$\frac{d^2 A_j}{d\xi^2} = \frac{\partial U}{\partial A_j}, \quad j = 1, 2, \ldots, N, \tag{7.2.21}$$

where

$$U(\vec{A}, \vec{\mu}, c) = U_1(\|\vec{A}\|^2, c) - \sum_{j=1}^{N} \frac{\mu_j^2}{8A_j^2}, \tag{7.2.22}$$

$$2U_1(\|\vec{A}\|^2, c) = \int_0^{\|A\|^2} [Kg(\eta) + \gamma]\, d\eta, \tag{7.2.23}$$

$$K = (v^2 - 1)^{-1}, \quad \gamma = (v^2 - 1)^{-1} c - \frac{v^2}{4}, \tag{7.2.24}$$

where $\vec{\mu} = (\mu_1, \mu_2, \ldots, \mu_N)$. The first integral of (7.2.21) is

$$I(\vec{A}(\xi), \vec{A}'(\xi)) = \|\vec{A}'(\xi)\|^2 - 2U_1(\|\vec{A}(\xi)\|^2, c) - \sum_{j=1}^{N} \mu_j^2 A_j^{-2}(\xi)/4 = c_1, \tag{7.2.25}$$

where $\|\vec{A}'(\xi)\|^2 = \sum_{j=1}^{N} [\frac{dA_j(\xi)}{d\xi}]^2$ and

$$c_1 = \|\vec{A}'(0)\|^2 - 2U_1(\|\vec{A}(0)\|^2, c) - \sum_{j=1}^{N} [v - 2\theta_j'(0)]. \tag{7.2.26}$$

Obviously, if $\mu_j \neq 0$ for some j, since c_1 is limited, then, when

$$\|\vec{A}'(\xi)\| \to 0, \quad \|\vec{A}(\xi)\| \to 0,$$

where $I(\vec{A}(\xi), \vec{A}'(\xi)) \to -\infty$, the solitary wave solution of (7.2.21) or (7.2.16) does not exist, such that $\|\vec{A}(\xi)\| \to 0$, $\|\vec{A}'(\xi)\| \to 0$ ($|\xi| \to \infty$). When $\mu = 0$, we have

$$\theta_j(\xi) = \theta_j(0) + \frac{1}{2} v\xi \quad j = 1, 2, \ldots, N. \tag{7.2.27}$$

At this time, (7.2.21) becomes

$$\frac{d^2 A_j}{d\xi^2} = \frac{\partial U_1}{\partial A_j}, \quad j = 1, 2, \ldots, N. \tag{7.2.28}$$

The equilibrium point of (7.2.28) is $(A_e, 0) \in R^{2N}$ such that A_e is the stationary point of U_1 or satisfies the equation $f(0, c, \vec{A})\vec{A} = 0$. A_e obviously includes the cases of $A = 0$

and A satisfying $g(\|\vec{A}\|^2) = v^2(v^2 - 1)/4c$. For equation (7.2.28), letting $u(\xi) = \|\vec{A}(\xi)\|^2$, a direct calculation yields

$$\begin{aligned} \frac{d^2u}{d\xi^2} &= 2\|\vec{A}'(\xi)\|^2 + 2\vec{A}(\xi) \cdot \frac{d^2\vec{A}}{d\xi^2} \\ &= 2\|\vec{A}'(\xi)\|^2 + 2u\tilde{f}(c,u), \end{aligned} \tag{7.2.29}$$

where $\tilde{f}(c, \|\vec{A}\|^2) = f(0, c, A)$. For a fixed c_1 and $\mu = 0$, (7.2.29) is

$$\frac{d^2u}{d\xi^2} = 2[u\tilde{f}(c,u) + c_1 + 2U_1(u,c)] = p(u,c,c_1). \tag{7.2.30}$$

The initial conditions are

$$u(0) = \|\vec{A}(0)\|^2, \quad u'(0) = 2\vec{A}(0) \cdot \vec{A}'(0), \tag{7.2.31}$$

$$\|\vec{A}'(0)\|^2 = c_1 + 2U_1(\|\vec{A}(0)\|^2, c) \geq 0. \tag{7.2.32}$$

The first integral of (7.2.30) is

$$\begin{aligned} [u'(\xi)]^2 &= [u'(0)]^2 + \int_{u(0)}^{u(\xi)} p(\eta, c, c_1) d\eta \\ &= Q(u, c, c_1, u'(0)), \end{aligned} \tag{7.2.33}$$

where $u' = \frac{du}{d\xi}$. Equation (7.2.33) is true only if its right-hand side term is nonnegative. The implicit representation of $\|\vec{A}\|^2$ can be obtained from the integral of (7.2.33). We have

$$\int_{\|A(0)\|^2}^{\|A(\xi)\|^2} Q(\eta, c, c_1, u'(0))^{-\frac{1}{2}} d\eta = \pm\xi. \tag{7.2.34}$$

We now discuss the existence of solutions with $\|A(\xi)\| \to 0$ ($|\xi| \to \infty$) to equation (7.2.28).

Theorem 7.2.9. *If*

$$Kg(u) + \gamma \geq 0, \quad \forall u \geq 0, \tag{7.2.35}$$

then there does not exist a solution of (7.2.28) *such that*

$$\|\vec{A}(0)\| > 0, \quad \|\vec{A}(\xi)\| \to 0, \ (|\xi| \to \infty).$$

Proof. The condition (7.2.35) is equivalent to $\tilde{f}(c,u) \geq 0$ for $u \geq 0$. By (7.2.29), we have $\frac{d^2u}{d\xi^2} \geq 0$. We deduce that any solution $\|\vec{A}(\xi)\|^2$ corresponding to (7.2.28) is a convex function of ξ. Therefore, it is impossible to have the solution such that $\|\vec{A}(0)\| > 0$ and $\|\vec{A}(\xi)\| \to 0$ ($|\xi| \to \infty$). □

Theorem 7.2.10. *Assume the following conditions hold:*
(i) $v^2(v^2 - 1) > 4c, v^2 < 1$;
(ii) *$g(s)$ is a strictly monotone increasing function, $g(0) = 0$, and there exists a positive number $u_1 < \infty$ such that*

$$\int_0^{u_1} g(\eta)d\eta = [v^2(v^2 - 1)/4 - c]u_1, \tag{7.2.36}$$

$$\int_0^{u} g(\eta)d\eta > [v^2(v^2 - 1)/4 - c]u_1, \quad \forall u > u_1. \tag{7.2.37}$$

Then (7.2.28) has a solution $\vec{A}(\xi) \geq 0, \forall \xi \in R$ with $\|\vec{A}(0)\| > 0, \|\vec{A}'(0)\| = 0$, so we obtain $\|\vec{A}(\xi)\| \to 0, \|\vec{A}'(\xi)\| \to 0\ (|\xi| \to \infty)$.

For equation (7.2.28), in order to look for the module of the solution $\vec{A}(\xi)$ being the periodic function of ξ, we have the following theorem.

Theorem 7.2.11. *Suppose the conditions of Theorem 7.2.10 hold and there exists a real number $\gamma_e > 0$ satisfying*

$$Kg(\gamma_e^2) + \gamma = 0, \quad or \quad g(\gamma_e^2) = v^2(v^2 - 1)/4 - c. \tag{7.2.38}$$

Then there exists a solution $\vec{A}(\xi)$ of (7.2.28) and $\|\vec{A}(\xi)\|$ is the periodic function of ξ.

Theorem 7.2.12. *Assume the following conditions hold:*
(i) *$g(u)$ is a real-valued increasing function and $g(0) = 0$;*
(ii) $v^2(v^2 - 1) < 4c, v^2 > 1$;
(iii) *the initial conditions $\vec{A}(0), \vec{A}'(0)$ satisfy $\|\vec{A}(0)\| > 0$ and*

$$\tilde{c}_1 = \|\vec{A}'(0)\|^2 - 2U_1(\|\vec{A}(0)\|^2, c) \geq 0.$$

Then there exists a solution $\vec{A}(\xi)$ of (7.2.28) and $\|\vec{A}(\xi)\|$ must not be the periodic function of ξ.

(III) Consider the three-dimensional Friedberg–Lee–Sirlin (FDS) [82] nonlinear wave equations

$$\Box\varphi + \alpha^2\chi^2\varphi = 0, \tag{7.2.39}$$

$$\Box\chi + \alpha^2\chi|\varphi|^2 + \frac{1}{2}\chi(\chi^2 - 1) = 0, \tag{7.2.40}$$

where $\Box \equiv \frac{\partial^2}{\partial t^2} - \Delta$ and $\Delta \equiv \frac{\partial^2}{\partial x_1^2} + \frac{\partial^2}{\partial x_2^2} + \frac{\partial^2}{\partial x_3^2}$. For the complex field, we consider the following traveling wave solutions with oscillating factors:

$$\varphi(r,t) = \frac{1}{\sqrt{2}}\psi(r)e^{-i\omega t}. \tag{7.2.41}$$

From (7.2.39) and (7.2.40), we get

$$\nabla^2\psi - \alpha^2\chi^2\psi + \omega^2\psi = 0, \tag{7.2.42}$$

$$\nabla^2\chi - \alpha^2\psi^2\chi - \frac{1}{2}(\chi^2 - 1)\chi = 0. \tag{7.2.43}$$

The electric charge is $Q = \omega \int \psi^2 d^3x$ and the energy of the system is $E = \int \varepsilon d^3\chi$, where

$$\varepsilon = \frac{1}{2}(\nabla\chi)^2 + (\nabla\psi)^2 + \frac{1}{2}(\omega^2 + \alpha^2\chi^2)\psi^2 + \frac{1}{4}(\chi^2 - 1)^2.$$

Let

$$\begin{aligned} &\xi = (\alpha^2 - \omega^2)^{\frac{1}{2}}, \quad \chi = 1 - \frac{1}{2}(\xi/\alpha)^2 x, \\ &\psi = 2^{-\frac{1}{2}}\frac{\xi}{\alpha}y. \end{aligned} \tag{7.2.44}$$

Considering the spherically symmetric solution x, y as the function of r, inserting (7.2.44) into (7.2.43), and comparing the lowest-order terms of ξ, we get

$$x = y^2.$$

Furthermore, from (7.2.42) we obtain

$$\frac{1}{r^2}\frac{d}{dr}\left(r^2\frac{dy}{dr}\right) - y + y^3 = 0 \tag{7.2.45}$$

and the boundary conditions are

$$\begin{cases} \dfrac{dy}{dr} = 0, & r = \infty, \\ y = 0, & r \to \infty. \end{cases} \tag{7.2.46}$$

It is easy to see that (7.2.45) and (7.2.46) have infinitely many solutions, whereas the solution with the lowest energy and no intersection in the radial direction corresponding to problem (7.2.42) and (7.2.43) is the soliton solution, which is stable.

(IV) Consider the multi-dimensional nonlinear Schrödinger equation

$$iu_t + \nabla^2 u + q(|u|^2)u = 0. \tag{7.2.47}$$

Let

$$\begin{aligned} &u = f(x_1', x_2', \dots, x_n')e^{i\theta}, \quad x_j' = x_j - c_j t, \\ &\theta = \sum_{j=1}^{N} k_j x_j - \omega t. \end{aligned} \tag{7.2.48}$$

Then

$$u_t = \left[f(-i\omega) - \sum_{j-1}^{N} f_{x'_j} c_j\right] e^{i\theta},$$

$$u_{x_i x_j} = [-k_i k_j f + i(k_i f_{x'_j} + k_j f_{x'_i}) + f_{x'_i x'_j}] e^{i\theta}$$
$$(j = 1, 2, \ldots, N).$$

Inserting this into (7.2.47), we have

$$\omega f - i \sum_{j=1}^{N} f_{x'_j} c_j + 2i \sum_{j=1}^{N} k_j f_{x'_j} - \sum_{j=1}^{N} k_j^2 f + \nabla'^2 f + q(f^2) f = 0.$$

Choosing $c_j = 2k_j$ and eliminating the imaginary part, we get

$$\nabla'^2 f + \left(\omega - \sum_{j=1}^{N} k_j^2\right) f + q(f^2) f = 0. \tag{7.2.49}$$

For the spherical symmetry case, we have

$$\frac{1}{\rho^{n-1}} \frac{\partial}{\partial \rho}\left(\rho^{n-1} \frac{\partial f}{\partial \rho}\right) + \left(\omega - \sum_{j=1}^{N} k_j^2\right) f + q(f^2) f = 0. \tag{7.2.50}$$

When $n = 1$, $\omega = k_i^2 - \eta^2$, we get the solitary solution of the one-dimensional nonlinear Schrödinger equation, which is stable. For the solitary solutions of $n > 1$, they are unstable.

(V) In the low-voltage magnetized plasma, the three-dimensional ion acoustic wave equations are

$$\begin{cases} \dfrac{\partial n}{\partial t} + \operatorname{div} n\vec{V} = 0, \\ \dfrac{\partial \vec{V}}{\partial t} + (\vec{V} \cdot \nabla)\vec{V} = -e\nabla\varphi/M + [\vec{V}, W_{H_i}], \\ \Delta\varphi = -4\pi e(n - n_0)\exp(e\varphi/T_e). \end{cases} \tag{7.2.51}$$

We reduce these to the dimensionless form, so we have

$$\frac{\partial u}{\partial \tau} + \frac{\partial}{\partial \xi_z}(\Delta_{\xi\xi} + u)u = 0, \tag{7.2.52}$$

where $\tau = \frac{1}{2}\omega_{p_i} t$ and $u = \frac{V_z}{2C_s}$. Letting $u = u(\xi_z - \lambda\tau)$, we have

$$\Delta_{\xi\xi} u - (\lambda - u)u = 0. \tag{7.2.53}$$

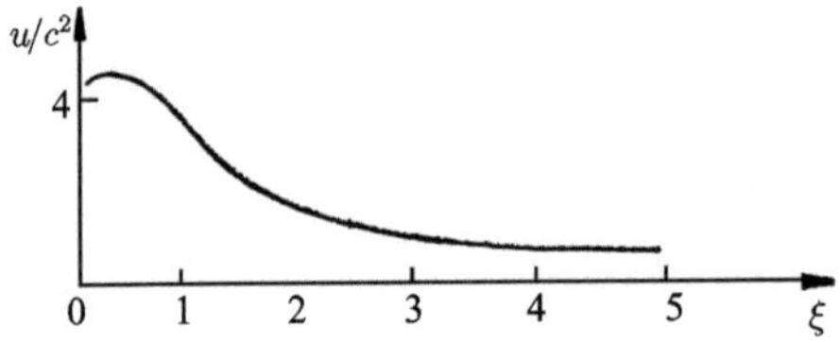

Figure 7.1: The graph of the solution of (7.2.54).

When $\lambda = c^2 > 0$, it has the solution of exponential decay as $|\xi| \to \infty$. The simplest is the spherical symmetry case:

$$\frac{1}{\xi^2}\frac{d}{d\xi}\left(\xi^2\frac{du}{d\xi}\right) - (c^2 - u)u = 0. \tag{7.2.54}$$

The solution of (7.2.54) exists, as it is the three-dimensional soliton solution and stable. Through numerical calculations, we get the graphics as shown in Figure 7.1.

(VI) Consider the two-dimensional sine-Gordon equation

$$\varphi_{xx} - \varphi_{yy} - \varphi_{tt} = \sin\varphi. \tag{7.2.55}$$

Its three soliton solutions were considered in [126]. The formal solution of (7.2.55) is

$$\varphi(x,y,t) = 4\tan^{-1}[g(x,y,t)/f(x,y,t)],$$

where

$$\begin{aligned} f &= 1 + a(1,2)e^{\eta_1+\eta_2} + a(1,3)e^{\eta_1+\eta_3} + a(2,3)e^{\eta_1+\eta_3}, \\ g &= e^{\eta_1} + e^{\eta_2} + e^{\eta_3} + a(1,2)a(1,3)a(2,3)e^{\eta_1+\eta_2+\eta_3}, \\ a(i,j) &= \frac{(p_i - p_j)^2 + (q_i - q_j)^2 - (\Omega_i - \Omega_j)^2}{(p_i + p_j)^2 + (q_i + q_j)^2 - (\Omega_i + \Omega_j)^2}, \\ \eta_i &= p_i x + q_i y - \Omega_i t - \eta_i^0 \ (\eta_i^0 \text{ is a constant}), \\ p_i^2 + q_i^2 - \Omega_i^2 &= 1, \quad i = 1,2,3, \end{aligned}$$

and we have

$$\begin{vmatrix} p_1 & q_1 & \Omega_1 \\ p_2 & q_2 & \Omega_2 \\ p_3 & q_3 & \Omega_3 \end{vmatrix} = 0.$$

(VII) Consider the complex nonlinear field equation

$$\nabla^2\psi - c^{-2}\frac{\partial^2\psi}{\partial t^2} = k^2\psi - \mu^2|\psi|^2\psi. \tag{7.2.56}$$

Assume $\psi = \varphi(r)e^{i\omega t}$, where $\varphi(r)$ is real and spherically symmetric. Then (7.2.56) is

$$\frac{d^2\varphi}{dr^2} + \frac{2}{r}\frac{d\varphi}{dr} = \left(k^2 - \frac{\omega^2}{c^2}\right)\varphi - \mu^2\varphi^3, \tag{7.2.57}$$

$$\left.\frac{d\varphi}{dr}\right|_{r=0} = 0, \quad \varphi \to 0,\ r \to \infty. \tag{7.2.58}$$

Supposing $k^2 - \omega^2/c^2 > 0$ and making the transformation

$$r' = r(k^2 - \omega^2/c^2)^{\frac{1}{2}}, \quad \varphi' = \mu\varphi(k^2 - \omega^2/c^2)^{-\frac{1}{2}},$$

we find that φ' satisfies the equation

$$\frac{d^2\varphi'}{dr'^2} + \frac{2}{r'}\frac{d\varphi'}{dr'} = \varphi' - \varphi'^3.$$

It can be shown that the solution $\psi = \varphi(r)e^{i\omega t}$ of (7.2.56) is unstable for small perturbations.

7.3 The stability and collapse of multi-dimensional solitons

One of the most important and natural requirements for solitons in plasma physics and various field models is that it must be stable, that is, from the "process" point of view, the solitons must have a sufficiently long lifetime. In other words, the soliton lifetime must be much longer than the interaction characteristic of solitons. The stability of this aspect has longitudinal stability and lateral stability by the direction of disturbance. From the analysis and processing of the stability, we know there are linear stability and nonlinear stability. Nonlinear stability generally refers to the stability according to some functional. This stability, which is usually considered physically, means that the energy of the system is minimized. In the multi-dimensional case, many solitary waves are unstable.

1. Let us start with the simplest case, a real (uncharged) scalar field, as described by the following nonlinear wave equation:

$$\Box\varphi + F'(\varphi) = 0 \quad \left(\Box \equiv \frac{\partial^2}{\partial t^2} - \nabla^2,\ F'(\varphi) = \frac{dF}{d\varphi}\right). \tag{7.3.1}$$

The Hamiltonian of the stationary field is

$$E + \int\left[\frac{1}{2}(\nabla\varphi)^2 + F(\varphi)\right]dx = K + V. \tag{7.3.2}$$

After the scaling transformation, $\varphi_\alpha = \varphi(\alpha x)$, we have

$$E[\varphi_\alpha] = \alpha^{2-n}K + \alpha^{-n}V, \tag{7.3.3}$$

$$\left.\frac{dE}{d\alpha}\right|_{\alpha=1} = 0, \quad V = \frac{2-n}{n}K, \tag{7.3.4}$$

$$\left.\frac{d^2E}{d\alpha^2}\right|_{\alpha=1} = -2(n-2)K. \tag{7.3.5}$$

From (7.3.4) and (7.3.5), we see that the minimal value of E is given at $n = 1$, which is stable; the maximal value of $E[\varphi]$ is given at $n > 2$, which is unstable; an inflection point is given at $n = 2$.

2. If the charged field is present, then the above situation will emerge qualitative change. For example, for the three-dimensional FDS nonlinear wave equation (7.2.39), (7.2.40), while

$$Q > Q_s = \frac{1}{2}\left(\frac{4}{3}\frac{\pi}{\alpha}\right)^4,$$

we can prove that $E_{\min} < Q_m$ (corresponding to the free meson solution). Its solitary solution is absolutely stable.

3. Consider the φ^5 nonlinear wave equation

$$\nabla^2\psi - \frac{1}{c^2}\frac{\partial^2\psi}{\partial t^2} = k^2\psi - \mu^2|\psi|^2\psi + \lambda|\psi|^4\psi, \tag{7.3.6}$$

where λ is real and usually positive. Considering the spherically symmetric solution, (7.3.6) becomes

$$\frac{d^2\varphi'}{dr'^2} + \frac{2}{r'}\frac{d\varphi'}{dr'} = \varphi' - \varphi'^3 + \beta\varphi'^5, \tag{7.3.7}$$

where

$$\begin{aligned} &\varphi' = \mu\varphi(1-\omega'^2)^{-\frac{1}{2}}, \quad r' = kr(1-\omega'^2)^{\frac{1}{2}}, \\ &\beta = \lambda k^2(1-\omega'^2)/\mu^4. \end{aligned} \tag{7.3.8}$$

Both the first-order perturbation theory and the direct perturbation method can be used to prove the existence of the stable solution to (7.3.6).

4. Consider the three-dimensional ion acoustic equation (7.2.53) in low-pressure magnetized plasma. Its energy is

$$\mathcal{H} = \int\left[\frac{1}{2}(D_\xi u)^2 - \frac{1}{3}u^3\right]d\xi.$$

Using the Hölder inequality

$$\int u^3 d\xi \le \left(\int u^2 d\xi\right)^{\frac{1}{2}}\left(\int u^4 d\xi\right)^{\frac{1}{2}}$$

and the interpolation inequality of $\int u^4 d\xi$,

$$\int u^4 d\xi \le 2\left(\int u^2 d\xi\right)^{\frac{1}{2}}\left(\int |\nabla u|^2 d\xi\right)^{\frac{3}{2}}, \tag{7.3.9}$$

we get

$$\begin{aligned}\mathcal{H} &\ge \int \frac{(\nabla u)^2}{2} d\xi - \frac{2}{3}\left(\int u^2 d\xi\right)^{\frac{3}{4}}\left(\int (\nabla u)^2 d\xi\right)^{\frac{3}{4}} \\ &\ge -\frac{1}{6}\left(\int u^2 d\xi\right)^3.\end{aligned}$$

Then we deduce that the functional $\mathcal{H}$ has a lower bound, so the three-dimensional spherical soliton solution reaches an absolute minimum. Hence, it is stable.

5. Here, we turn to the collapse of the Langmuir wave. Recalling the mechanics of the spherical shock wave concentrating effect, a similar phenomenon occurs in the dissipation mechanism of the “Langmuir condensation” – the turbulence energy condenses on the long wave region of the frequency spectrum. This coalescence indicates the instability of the multi-dimensional Langmuir solitons.

Example 7.3.1. In the φ^3 approximation, the collapse of the Langmuir wave is described by the equation

$$\nabla^2(i\psi_t + \nabla^2\psi) - \operatorname{div}(|\nabla\psi|^2\nabla\psi) = 0, \tag{7.3.10}$$

where ψ is the high-frequency potential envelope. Considering the spherically symmetric case, (7.3.10) is

$$i\varphi_t + \nabla_{rr}^2\varphi - \frac{n-1}{r^2}\varphi + |\varphi|^2\varphi = 0, \tag{7.3.11}$$

where $\varphi = -\nabla\psi$ and $\varphi(0) = 0$. Equations (7.3.10) and (7.3.11) have the conserved quantities

$$s = \int |\nabla\psi|^2 d^3r, \quad s_2 = \int\left[|\nabla^2\psi|^2 - \frac{1}{2}|\nabla\psi|^4\right] d^3r, \tag{7.3.12}$$

$$\begin{aligned} s &= \int_0^\infty |\varphi|^2 r^2 dr, \\ s_2 &= \int_0^\infty \left[|(r\varphi)_r|^2 + 2|\varphi|^2 - \frac{1}{2}r^2|\varphi|^4\right] dr,\end{aligned} \tag{7.3.13}$$

respectively. Consider the acceleration motion of the quasi-plane soliton of (7.3.11) to the origin (see Figure 7.2).

Let $D = \langle r^2\rangle\varphi = \int_0^\infty |\varphi|^2 r^4 dr$. Then, by (7.3.11), we have

$$\frac{d^2D}{dt^2} = 6s_2 - 2\int_0^\infty |(r\varphi)_r|^2 dr - 4\int_0^\infty |\varphi|^4 r^2 dr < 6s_2.$$

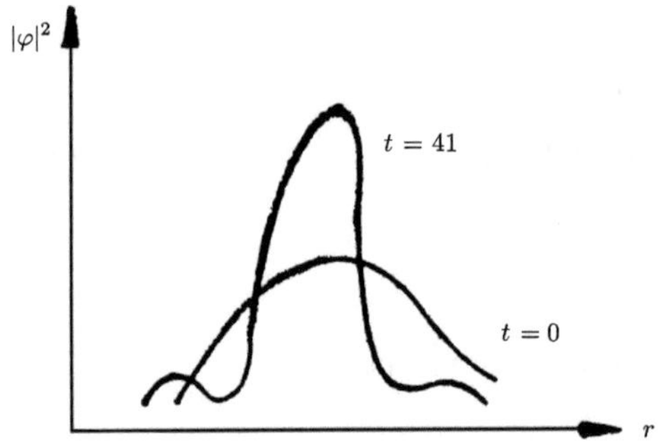

Figure 7.2: The acceleration motion to the origin of the quasi-plane soliton of (7.3.11).

Supposing $s_2 \le 0$ and integrating the above with respect to t, we have

$$D \le 3s_2 t^2 + c_1 t + c_2.$$

When $t \to t_0 = \frac{c_1+(c_1^2+12c_2|s_2|)^{\frac{1}{2}}}{6|s_2|}$, the local solution of the initial value problem causes singularity. If $c_1 > 0$, the wave packet dispersion occurs for small t. On the contrary, when $c_1 < 0$, it leads to contraction.

Example 7.3.2. Consider the system of equations

$$\operatorname{div}(-2i\nabla\psi_t - \nabla\nabla^2\psi + \Phi\nabla\psi) = 0, \tag{7.3.14}$$

$$\left(\frac{\partial^2}{\partial t^2} - \nabla^2\right)\Phi = \nabla^2(|\nabla\psi|^2). \tag{7.3.15}$$

Introducing the low-frequency potential u,

$$u_t = \Phi + |\varphi|^2 = \Phi + |\nabla\psi|^2, \tag{7.3.16}$$

$$\nabla^2 u = \Phi_t, \tag{7.3.17}$$

$$\nabla^2(i\psi_t + \nabla^2\psi) = \operatorname{div}(\Phi\nabla\psi), \tag{7.3.18}$$

we easily get

$$s_2 = \int\left[\left|\nabla^2\psi\right|^2 + \Phi|\nabla\psi|^2 + \frac{1}{2}\Phi^2 + \frac{1}{2}(\nabla u)^2\right]d^3r. \tag{7.3.19}$$

Assuming $s_2 \le 0$, for equations (7.3.16), (7.3.17), and (7.3.18), we can obtain the self-type transformation in two limit cases. Under the quasi-static limit, they reduce to (7.3.10). In the ultrasound limit, the right-hand side of (7.3.16) may ignore Φ. In the first case, the self-type transformation of (7.3.10) is

$$\psi = \exp\{-i\mu^2 \ln(t_0 - t)\}\chi(\vec{\xi}), \quad \vec{\xi} = \frac{\vec{x}}{\sqrt{t_0 - t}}, \tag{7.3.20}$$

where $\chi(\vec{\xi})$ satisfies the equation

$$\nabla^2\left(-\mu^2\chi + \frac{1}{2}i\vec{\xi}\cdot\nabla\chi + \nabla^2\chi\right) + \operatorname{div}(|\nabla\chi|^2\nabla\chi) = 0. \tag{7.3.21}$$

In [58], Zakharov obtained a spherically symmetric solution in the region of $|\vec{\xi}| \geq \frac{1}{\mu}$, where χ satisfies the equation

$$i\vec{\xi}\cdot\nabla\chi = 2\mu^2\chi.$$

Thus, we have

$$\chi \approx |\vec{\xi}|^{-2i\mu^2}\chi_0.$$

From (7.3.17), we get

$$s_2(t) = s_2(0)/\sqrt{t_0 - t}.$$

In the ultrasound limit, making the transformation $i\psi_t \to -\mu^2(t)\psi$, from (7.3.16), (7.3.17), and (7.3.18), we have

$$\nabla^2(-\mu^2(t)\chi + \nabla^2\chi) - \operatorname{div}(\Phi\nabla\chi) = 0, \tag{7.3.22}$$

$$\Phi_{tt} = \nabla^2|\nabla\chi|^2. \tag{7.3.23}$$

This system of equations allows for the following transformations:

$$\begin{aligned} &\mu^2(t) = \frac{\mu_0^2}{t_0 - t}, \quad \chi = \frac{\eta(\xi)}{(t_0 - t)^{1-\frac{2}{n}}}, \\ &\Phi = \frac{D(\xi)}{(t_0 - t)}, \\ &\xi = r(t_0 - t)^{-\frac{2}{n}}, \end{aligned} \tag{7.3.24}$$

where n is the dimension of the space. The solution of (7.3.24) has the following properties:

(i) (7.3.24) leads to $s_2 = 0$;

(ii) $\frac{\nabla^2\Phi}{\Phi_{tt}} \approx \frac{(t_0-t)^2}{r^2} \approx \frac{(t_0-t)^{2-\frac{4}{n}}}{|\xi|^2}$;

(iii) $|\varphi(0,t)|^2 = |\nabla\chi(0,t)|^2 = f(t) = \frac{f_0}{(t_0-t)^2}$;

(iv) $|\Phi(0,t)| = \varphi_0(t_0 - t)^{\frac{4}{3}}$.

For the plane soliton, we deduce from (7.3.24) that $\frac{\nabla^2\Phi}{\Phi_{tt}} \to \infty$ as $t \to t_0$. For the two-dimensional collapse $\nabla^2\Phi/\Phi_{tt} \to$ constant and in the three-dimensional case $\nabla^2\Phi/\Phi_{tt} \to 0$.

We denote the initial selection in the three-dimensional space (in r, z coordinates)

$$\begin{aligned} \rho = \nabla^2\psi &= \begin{cases} \rho_0\sqrt{\omega}\sin\frac{\pi z}{2}, & \omega > 0, \\ 0, & \omega \le 0, \end{cases} \\ \omega &= 1 - \frac{1}{4}(r^2 + z^2). \end{aligned} \tag{7.3.25}$$

For the system of equations (7.3.16), (7.3.17), and (7.3.18), we also give

$$\Phi(r, z, 0) = -|\nabla\psi|^2, \quad \Phi_t(r, z, 0) = 0.$$

Equation (7.3.25) describes a dipole-type charge directed along the z-axis at $t = 0$.

The results show the collapse behavior at some time. In fact, it is a good description of the self-type transformation (7.3.24) (see Figure 7.3).

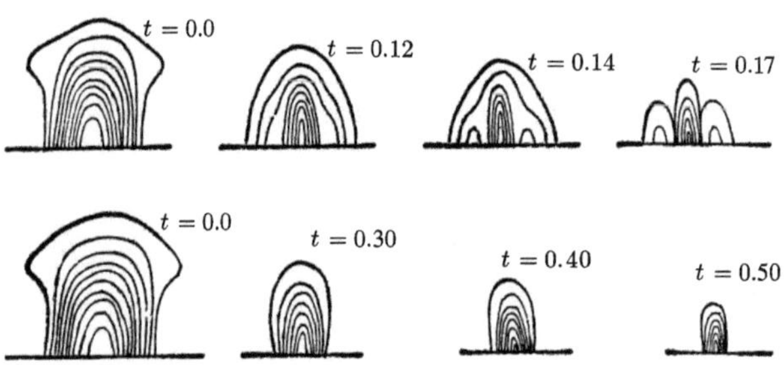

Figure 7.3: The collapse behavior of the plane soliton.

8 Numerical computation methods for some nonlinear evolution equations

8.1 Introduction

With the development of the soliton problem, the numerical computation methods for a large number of nonlinear evolution equations with soliton solutions (which have dispersion properties in general) are also currently emerging and have been developing vigorously. In fact, the numerical computation results of nonlinear equations have played an important role when the soliton problem began to give compelling results. For example, for the KdV equation, although the analytical processing for solitary waves was done in 1895 by Korteweg and de Vries, the rich content of the nonlinear phenomenon was unknown. It was not until 1965 that Zabusky and Kruskal obtained the Korteweg–de Vries (KdV) equation by the harmonic lattice model and after the discovery of the maximum stability of wave forms remaining unchanged after soliton interaction, people started to become much more interested in the soliton. Other calculations, such as the Fermi–Pasta–Ulam problem and the calculation of the two solitary solutions (kink) of the sine-Gordon equation by Perring and Skryme, provide an important basis for analyzing the existence of solitons in physics. With the deepening and complexity of the study of soliton problems, especially the interaction of multiple solitons and quasi-solitons, as well as the qualitative and quantitative research for the problems of the existence and interaction of multi-dimensional solitons, numerical calculations have already played an increasingly important role. It is no exaggeration to say that, for the soliton problems in laser and plasma physics, numerical computation has become the main tool to investigate stability.

For the numerical computation for nonlinear evolution equations with soliton solutions, it is generally required that the calculation is stable and can adapt to the large gradient change of soliton solutions. Also, the computational schemes must satisfy the characteristics of conservation laws to an acceptable extent. There are two commonly used numerical methods. One is the finite difference method, the other is the function approximation method, that is, the finite element method and the collocation method.

Now we consider the general evolution equation

$$u_t = L(u), \tag{8.1.1}$$

where $L(u)$ is the general nonlinear differential operator. For the finite difference method, we use the difference operator $L_h(u_m^n)$ to approximate $L(u)$, where $u_m^n = u(x_m, t_n)$, $x_m = mh$, and $t_n = nk$. We usually use the following equations to discretize the time derivative:

$$u_m^{n+1} - u_m^n = kL_h(u_m^n), \tag{8.1.2}$$

https://doi.org/10.1515/9783110549638-008

$$u_m^n - u_m^{n-1} = kL_h(u_m^n), \tag{8.1.3}$$

$$u_m^{n+1} - u_m^{n-1} = 2kL_h(u_m^n). \tag{8.1.4}$$

We know that (8.1.2) is a simple display format, (8.1.3) is a simple implicit format, and (8.1.4) is the leapfrog format. There are two more complex and important formats. One of these is the Crank–Nicolson format, which is the sum of (8.1.2) and (8.1.3), while the other is a jump point format (Hopscotch format); when $n+m$ is odd, it is calculated in the format of (8.1.2) and when $n+m$ is an even number, it is calculated in the format of (8.1.3), so that the result of the simultaneous computation becomes the display, as shown in Figure 8.1. In order to make (8.1.3) explicit, the nonlinear part of $L_h(u_m^n)$ must be averaged by the space

$$L_h\left(\frac{1}{2}(u_{m+1}^n + u_{m-1}^n)\right).$$

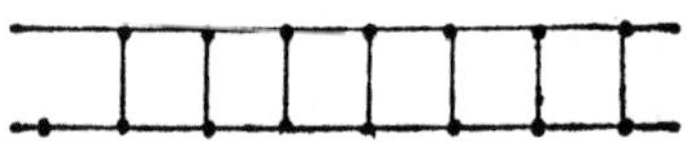

Figure 8.1: The jump point format.

All of these methods must satisfy the stability conditions, because otherwise the calculation cannot proceed. For linear equations, the Crank–Nicolson scheme may be the most efficient, but for nonlinear equations, it is troublesome to solve a large number of nonlinear simultaneous equations at each step. It is well known that the leapfrog scheme (8.1.4) is not stable for the linear heat conduction equations, but it is very suitable for the second-order hyperbolic equations. The Hopscotch scheme is simple, fast, and stable, but for the parabolic equations, the time step size must be limited, $k \approx h^2$, to ensure a reasonable accuracy.

The function approximation method, as the name suggests, uses the approximation solution defined in the finite-dimensional subspace to approximate the exact solution $u(x,t)$ as follows:

$$u(x,t) \approx \tilde{u}(x,t) = \sum_{i=1}^{N} c_i(t)\varphi_i(x), \tag{8.1.5}$$

where $\varphi_i(x)$ are the base functions of the approximation space. They are usually selected as trigonometric functions, which lead to the finite F-transform or pseudo-spectral method. If we use the fragment polynomials as the local bases, we get the finite element method. Supposing $\varphi_i(x)$ satisfy the boundary condition, let

$$r(x,t) = \tilde{u}_t - L(\tilde{u}) = \sum_{i=1}^{N} \dot{c}_i(t)\varphi_i(x) - L(\tilde{u}). \tag{8.1.6}$$

The residual $r(x, t)$ is required to be small in a certain sense, if one requires

$$\int_0^1 r(x,t)\varphi_j(x)dx = 0, \quad j = 1, 2, \ldots, N, \tag{8.1.7}$$

at this point, that is, the Galerkin method, (8.1.7) leads to a series of ordinary differential equations.

If, in a given set of points (such as Gauss points), the following condition is strictly satisfied:

$$r(x_j, t) = 0, \quad j = 1, 2, \ldots, N, \tag{8.1.8}$$

we get the collocation method.

In the following, we discuss the numerical methods and the calculated results for some special nonlinear evolution equations.

8.2 Finite difference method and finite element method for the KdV equation

In [314], the numerical calculation for the definite solution problem of the KdV equation in the following form:

$$u_t + uu_x + \delta^2 u_{xxx} = 0, \tag{8.2.1}$$

$$u|_{t=0} = \cos \pi x, \tag{8.2.2}$$

$$u(x+2, t) = u(x, t) \tag{8.2.3}$$

is proceeded, using the following difference scheme:

$$\begin{aligned} u_m^{n+1} &= u_m^{n-1} - \frac{1}{3}\frac{k}{h}(u_{m+1}^n + u_m^n + u_{m-1}^n)(u_{m+1}^n - u_{m-1}^n) \\ &\quad - \left(\frac{\delta^2 k}{h^3}\right)(u_{m+2}^n - 2u_{m+1}^n + 2u_{m-1}^n - u_{m-2}^n) \\ &\quad (m = 0, 1, 2, \ldots, 2N-1), \end{aligned} \tag{8.2.4}$$

$$u_m^0 = \cos \pi x_m, \tag{8.2.5}$$

$$u_m^n = u_{m+2N}^n, \tag{8.2.6}$$

where k is the time step size, $h = \frac{1}{N}$ is the space step size, and $u_m^n = u(mh, nk)$. The momentum $\sum_{m=0}^{2N-1} u_m^n$ of this difference scheme is conserved and the energy $\sum_{m=0}^{2N-1} \frac{1}{2}(u_m^n)^2$ is almost conserved. Choosing $\delta = 0.022$, in this case, the initial dispersion is small with respect to the nonlinear term, because

$$\{\max|\delta^2 u_{xxx}| / \max|uu_x|\}_{t=0} = 0.004.$$

The results are divided into three time periods.

(i) At first, the first and second terms of equation (8.2.1) play a dominant role, which leads to the usual catch-up phenomenon. The solution is essentially determined by the hyperbolic equation $u_t + uu_x = 0$ at the moment when $u \approx \cos \pi(x - ut)$.

(ii) When u is sufficiently steep, the third term becomes important, which destroys the formation of the discontinuous solution. At this time, the small wavelength of vibration on the left is developed, the amplitude of the dispersion vibration is increased, and finally a series of single solitons is formed.

(iii) Each soliton moves at a uniform velocity, which is proportional to the amplitude, and two or more solitons overlap in space due to the periodicity, producing nonlinear interactions. After a short period of interaction, they show no influence on their size and shape. In Figure 8.2, curve A represents the initial value (8.2.2) ($t = 0$), curve B shows the image of the solution (8.2.1) when $u = (\cos \pi x - ut)$ generates multiple values in $x = \frac{1}{2}, t = t_B = \frac{1}{\pi}$, and curve C shows the image of the dispersion structure being fully developed into a series of solitons at $t = 3.6t_B$. At this point, the maximum positions of the solitons form a straight line.

For the Gaussian initial function, the KdV equation was calculated in [23], considering the initial value problem

$$v_t + vv_x + \frac{1}{\sigma^2} v_{xxx} = 0, \quad (-\infty < x < \infty,\ t > 0), \tag{8.2.7}$$

$$v|_{t=0} = \varphi(x) = e^{-x^2}, \quad (-\infty < x < \infty). \tag{8.2.8}$$

It was found that two solitons were formed when $4 < \sigma < 7$, there were three solitons when $7 < \sigma < 11$, four solitons when $\sigma \approx 11$, and six solitons when $\sigma \approx 16$. However, there do not exist solitons and there only exist dispersion vibration waves when $\sigma \ll \sigma_3 = \sqrt{12}$. For some intermediate values σ, there are both solitons and dispersive oscillations, as shown in Figure 8.3. Here,

$$\sigma_c = 6\sigma_s^2 \int_{-\infty}^{\infty} [\varphi(\xi)]^2 d\xi \Big/ \left(\int_{-\infty}^{\infty} \varphi(\xi) d\xi \right)^3, \quad \varphi(\xi) = e^{-\xi^2},$$

$$\sigma_s = \sqrt{12}.$$

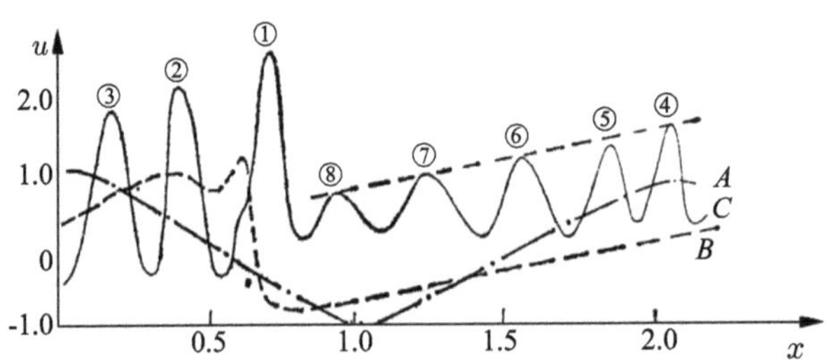

Figure 8.2: The curves A, B and C as stated above.

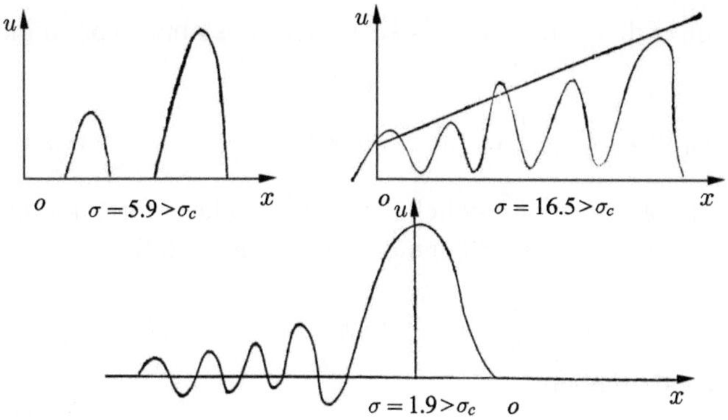

Figure 8.3: The figures of the solutions for the initial value problem (8.2.7)–(8.2.8) with different values of σ.

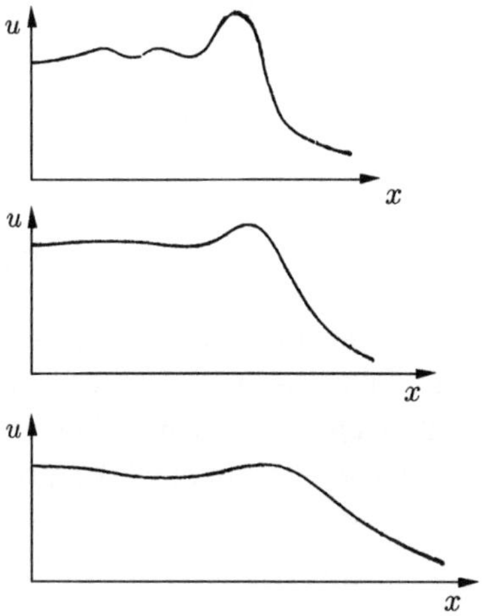

Figure 8.4: The numerical results of initial value problem (8.2.9)–(8.2.10) using the format (8.2.4).

For the initial value problem of the KdV equation

$$u_t + \varepsilon u u_x + \mu u_{xxx} = 0, \tag{8.2.9}$$

$$u|_{t=0} = \frac{1}{2}[1 - \tanh(x-25)/25], \tag{8.2.10}$$

using the format (8.2.4), the calculation results are shown in Figure 8.4.

Equation (8.2.4) is a three-tier format. The first step can be taken as (noncenter format)

$$\begin{aligned} u_m^1 = u_m^0 &- \frac{1}{6}\varepsilon\frac{k}{h}(u_{m+1}^0 + u_m^0 + u_{m-1}^0)(u_{m+1}^0 - u_{m-1}^0) \\ &- \frac{1}{2}\mu\frac{k}{h^3}[u_{m+2}^0 - 2u_{m+1}^0 + 2u_{m-1}^0 - u_{m-2}^0]. \end{aligned}$$

The truncation error equation of this format is $O[k^3 + kh^2]$. Its linear stability condition is

$$\frac{k}{h}(\varepsilon|u_0| + (4r/h^2)) \leq 1, \quad |u_0| \leq \max|u|. \tag{8.2.11}$$

In order to reduce the amount of storage, we switch to use the two-layer format – the Hopscotch method. For equation (8.2.9), the difference schemes are as follows:

$$\begin{aligned} v_m^{n+1} &= v_m^n - \frac{1}{2}\frac{k}{h}\varepsilon(f_{m+1}^n - f_{m-1}^n) - \frac{k\mu}{2h^3}(v_{m+2}^n - 2v_{m+1}^n + 2v_{m-1}^n - v_{m-2}^n), \\ &\quad m+n \text{ is an odd number}, \\ v_m^{n+1} &= v_m^n - \frac{1}{2}\frac{k}{h}\varepsilon(f_{m+1}^{n+1} - f_{m-1}^{n+1}) - \frac{k\mu}{2h^3}(v_{m+2}^{n+1} - 2v_{m+1}^{n+1} + 2v_{m-1}^{n+1} - v_{m-2}^{n+1}), \\ &\quad m+n \text{ is an even number}. \end{aligned} \tag{8.2.12}$$

It is not difficult to verify that the truncation error of this and equation (8.2.9) is

$$kO\left(k^2 + \left(\frac{k}{h}\right)^2 + h^2\right).$$

Its linear stability condition is

$$\frac{k}{h}|\varepsilon|u_0| - (2\mu/h^2)| \leq 1. \tag{8.2.13}$$

It can be seen that, although the Hopscotch method has an advantage of less storage, the time step that is required is smaller than that for the Zabusky–Kruskal scheme.

Another numerical method that is used for the KdV equation is the function approximation method. We consider the periodic initial value problem for the KdV equation as follows:

$$\begin{cases} u_t + uu_x + u_{xxx} = 0, & 0 < t \leq T,\ x \in R, \\ u(x,0) = u_0(x), & x \in R, \\ u(x+1,t) = u(x,t), & \forall x,t. \end{cases} \tag{8.2.14}$$

Suppose that $u_0(x)$ is a function with period 1 which is sufficiently smooth and the solution of (8.2.14) exists and is sufficiently smooth. We take the finite-dimensional subspace S^μ as

$$\begin{aligned} S^\mu = \{&\chi(x), \chi \in [0,1]; \chi \text{ can be periodic extended to } C^k(R), \\ &\chi(x) \text{ is a polynomial of order less than } \mu - 1 \text{ in the interval} \\ &[ih, (i+1)h](i = 0,1,2,\ldots,h^{-1})\} \end{aligned}$$

and suppose μ, k are integers, $\mu-1 > k \geq 0$, and $k \geq 2$. Now the Galerkin approximation of problem (8.2.14) is defined as

$$\begin{aligned} &(U_t + U_{xxx} + UU_x, \chi + h^3\chi_{xxx}) = 0, \quad \chi \in S^\mu, \\ &0 \leq t \leq T, \quad U(0) \in S^\mu. \end{aligned} \tag{8.2.15}$$

We have the following result.

Theorem 8.2.1. *Assume $k \geq 2$ and the initial value $U(0)$ satisfies*

$$\|U(0) - u_0\|_{L^2[0,1]} \leq C_1 h^\mu.$$

Then there exist constants C, h_0 depending on T, u_0, and C_1 such that the Galerkin approximation solution of problem (8.2.15) *exists for $0 \leq t \leq T$, $0 \leq h \leq h_0$. We have the estimate*

$$\|U(t) - u(\cdot,t)\|_{L^2[0,1]} \leq Ch^\mu. \tag{8.2.16}$$

For the numerical calculation and the study of the method for the KdV equation, we refer the reader to [23, 72, 73, 99, 300, 314] and the articles of [1, 11, 16, 114].

8.3 Finite difference method for the nonlinear Schrödinger equation

We consider the following definite solution problem of a class of nonlinear Schrödinger equations:

$$iu_t - [a(x)u_x]_x + \beta|u|^2u + f(x)u = 0, \quad 0 < x < 1,\ t > 0, \tag{8.3.1}$$

$$u|_{x=0} = u|_{x=1} = 0, \quad t \geq 0, \tag{8.3.2}$$

$$u|_{t=0} = u_0(x), \quad 0 \leq x \leq 1, \tag{8.3.3}$$

where $i = \sqrt{-1}$, $\beta > 0$, $a(x)$, $f(x)$ are known functions, $a(x) \geq \alpha > 0$, $u_0(x)$ is a given complex function, and $u(x,t)$ is the unknown complex function. Assuming $Q = [0,1] \times [0,1]$ is a rectangular region, we use the straight lines $t = mk$, $x = ph$ to divide the region into many small grids, where m is an integer, $m \in [0, [T/h]]$, p is an integer, and $p \in [0, [h^{-1}]]$, as shown in Figure 8.5. Supposing that all the interior point grids

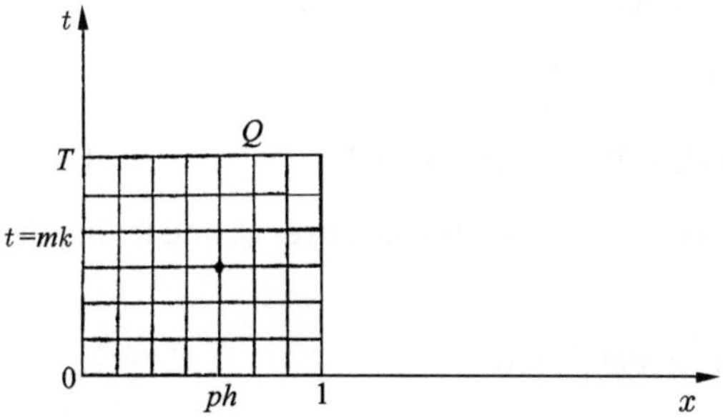

Figure 8.5: Regional division diagram.

are Q_h, the remaining grids including the boundary points are s_h and $\Omega(t)$ denotes $(t = \text{constant}) \cap Q_h$.

Let

$$\varphi_x(x,t) = \frac{1}{h}[\varphi(x+h,t) - \varphi(x,t)] = D_+\varphi,$$
$$\varphi_{\bar{x}}(x,t) = \frac{1}{h}[\varphi(x,t) - \varphi(x-h,t)],$$
$$\varphi_{\hat{x}}(x,t) = \frac{1}{2h}[\varphi(x+h,t) - \varphi(x-h,t)].$$

We similarly define φ_t and $\varphi_{\bar{t}}$. We define the discrete norm as follows:

$$\|\varphi\|^2_{\Omega(t)} = h\sum_{\Omega} |\varphi(x,t)|^2,$$
$$\|\varphi\|^2_{Q_h} = kh\sum_{Q_h} |\varphi(x,t)|^2,$$
$$\|\varphi\|^2_{l,\Omega} = \|\varphi\|^2_{\Omega(t)} + \sum_{|s|\le l} \|D_t^s\varphi\|^2_{\Omega(t)},$$
$$\|\varphi\|_{L_\infty(\Omega)} = \sup_{x_i\in\Omega} |\varphi(x_i)|, \quad \|D_t^l\varphi\|_{L_\infty} = \sup_{x_i\in\Omega} |D_t^l\varphi|.$$

We consider the following definite solution problem to the four-point implicit difference equation

$$i\varphi_{\bar{t}} - [b(x)\varphi_x]_{\bar{x}} + \beta|\varphi|^2\varphi + f(x)\varphi = 0, \tag{8.3.4}$$
$$\varphi|_{s_h} = 0, \quad \varphi|_{t=0} = u_0(x). \tag{8.3.5}$$

For the solution of (8.3.4), (8.3.5), we have the following estimates.

Lemma 8.3.1. *If the following conditions hold:*
(i) $\beta > 0, f(x), b(x)$ *are real functions;*
(ii) $u_0(x) \in C^0$;

then we have

$$\|\varphi\|^2_{\Omega(T)} \le 2\|u_0(x)\|^2_{L_2} = E_0. \tag{8.3.6}$$

Proof. Multiplying (8.3.4) by $\bar{\varphi}$, we get

$$i\varphi_{\bar{t}}\bar{\varphi} - \bar{\varphi}[b(x)\varphi_x]_{\bar{x}} + \beta|\varphi|^4 + f(x)|\varphi|^2 = 0. \tag{8.3.7}$$

Using the partial sum and the boundary conditions, summing over Q, and taking the imaginary part by (8.3.7), we get

$$\sum_{Q}(|\varphi|^2_{\bar{t}} + k|\varphi_{\bar{t}}|^2) = 0.$$

Then we immediately obtain (8.3.6). □

Lemma 8.3.2. *If the conditions of Lemma 8.3.1 are satisfied and $u_0(x) \in H^1$, $0 \le \alpha \le b(x) \le M$, $f|u_0|^2 \in C^0$, and $f(x) \ge 0$, then we have the following estimate:*

$$\frac{\alpha}{2}\|\varphi_x\|^2_{\Omega(T)} + \frac{\beta}{4}\|\varphi^2\|^2_{\Omega(T)} + \frac{1}{2}\|f^{\frac{1}{2}}\varphi\|^2_{\Omega(T)} \le M\|u_0(x)\|^2_{L^2} + \frac{\beta}{2}\|u_0^2\|^2_{L^2} + \|f(x)u_0^2\|_{L^1}. \tag{8.3.8}$$

Proof. Multiplying (8.3.4) by $\bar{\varphi}_t$, we have

$$i|\varphi_{\bar{t}}|^2 - \bar{\varphi}_t[b(x)\varphi_x]_{\bar{x}} + \beta|\varphi|^2\varphi\bar{\varphi}_t + f(x)\varphi\bar{\varphi}_t = 0. \tag{8.3.9}$$

By (8.3.9), taking the imaginary part, multiplying by kh, and summing over Q, we get

$$\frac{\alpha h}{2}\sum_{\Omega(T)}|\varphi_x(T)|^2 + \frac{\beta h}{4}\sum_{\Omega(T)}|\varphi(T)|^4 + \frac{h}{2}\sum_{\Omega(T)} f(x)|\varphi(T)|^2 \le \frac{Mh}{2}\sum_{\Omega(0)}|\varphi_x(0)|^2 + \frac{\beta h}{4}\sum_{\Omega(0)}|\varphi(0)|^4 + \frac{h}{2}\sum_{\Omega(0)} f(x)|\varphi(0)|^2.$$

When $h \le h_0$, we immediately get (8.3.8). □

Now we consider the following differential definite solution problem:

$$i\varphi_{\bar{t}} - [b(x)\varphi_x]_{\bar{x}} + C(x,t,\varphi) + g(x,t) = 0, \tag{8.3.10}$$

$$\varphi|_{S_h} = 0, \quad \varphi|_{t=0} = u_0(x). \tag{8.3.11}$$

For the solution φ of (8.3.10), (8.3.11), we have the following estimate.

Lemma 8.3.3. *If the following conditions are satisfied:*
(i) *β is a real number, $b(x)$ is a real function;*
(ii) *$|C(x,t,\varphi)\bar{\varphi}| \le M|\varphi|^2$, M is a positive constant;*
(iii) *$u_0(x) \in C^0$, $g(x,t) \in C^0(Q)$;*

then we have

$$\|\varphi\|^2_{\Omega(T)} \le 2(\|u_0\|^2_{L^2} + \|g\|^2_{L^2(Q)})e^{(2M+1)T} = C_2. \tag{8.3.12}$$

Proof. The proof is similar to Lemma 8.3.1. □

We assume that the smooth solution of the definite solution problem (8.3.1), (8.3.2), (8.3.3) exists in the region $Q = [0,1] \times [0,T]$. Now we consider the definite solution problem of the difference equation corresponding to the definite solution problem (8.3.1)–(8.3.3). We have

$$i\varphi_{\bar{t}} - [b(x)\varphi_x]_{\bar{x}} + \beta|\varphi|^2\varphi + f(x)\varphi = 0, \tag{8.3.13}$$

$$\varphi|_{S_h} = 0, \quad \varphi|_{t=0} = u_0(x), \tag{8.3.14}$$

where $b(x) = a(x + \frac{h}{2})$. We have the following convergence theorem.

Theorem 8.3.4. *If the conditions of Lemma* 8.3.2 *are satisfied, assuming* $u(x,t)$, $\varphi(x,t)$ *are the solutions of problem* (8.3.1)–(8.3.3) *and problem* (8.3.13), (8.3.14), *respectively, we have*

$$\|u-\varphi\|_{\Omega(T)} = O(k+h^2). \tag{8.3.15}$$

Proof. Because $u(x,t)$ is the smooth solution of the definite solution problem (8.3.1)–(8.3.3), we have, according to the Taylor expansion,

$$iu_t - [b(x)u_x]_{\bar{x}} + \beta|u|^2u + fu = O(k+h^2),$$

where $b(x) = a(x+\frac{h}{2})$. Let $\varepsilon(x,t) = u(x,t) - \varphi(x,t)$. Then we have

$$i\varepsilon_{\bar{t}} - [b(x)\varepsilon_x]_{\bar{x}} + \beta(|u|^2u - |\varphi|^2\varphi) + f\varepsilon = O(k+h^2), \tag{8.3.16}$$

$$\varepsilon|_{S_h} = 0, \quad \varepsilon|_{t=0}, \tag{8.3.17}$$

$$\begin{aligned}\beta(|u|^2u - |\varphi|^2\varphi) &= \beta|u|^2(u-\varphi) + \beta\varphi(|u|^2 - |\varphi|^2)\\ &= \beta|u|^2\varepsilon + \beta\varphi(|u| + |\varphi|)(|u| - |\varphi|).\end{aligned}$$

Since it is assumed that the solutions of (8.3.1)–(8.3.3) are smooth and bounded, from Lemma 8.3.2 we can get the uniformly bounded estimate of the differential solution φ, so we have

$$\left|\bar{\varepsilon}\beta(|u|^2u - |\varphi|^2\varphi)\right| \le M|\varepsilon|^2,$$

where $M = |\beta|[\|u\|^2_{L_\infty} + \|\varphi\|_{L_\infty}(\|u\|_{L_\infty} + \|\varphi\|_{L_\infty})]$. Note that, when f is a real function, the estimate is irrelevant to f. Using Lemma 8.3.3, we get

$$\|\varepsilon\|^2_{\Omega(T)} \le 2\|O(k+h^2)\|^2_{\Omega}e^{(2M+1)T}.$$

The proof is complete. □

Theorem 8.3.5. *The solution* φ *of the difference equations* (8.3.13), (8.3.14) *is stable with respect to the norm* $\|\cdot\|_\Omega$ *in accordance with the initial value.*

Proof. The proof is similar to Theorem 8.3.4. □

For the six-point symmetric (Crank–Nicolson) format, we have

$$\begin{aligned} &i\varphi_{\bar{t}} - \frac{1}{2}[(b\varphi_x)_{\bar{x}} + (b\varphi_x(t-k))_{\bar{x}}] + \frac{\beta}{2}[|\varphi(t)|^2\varphi(t) + |\varphi(t-k)|^2\varphi(t-k)]\\ &\quad + f(x)\frac{1}{2}[\varphi(t) + \varphi(t-k)] = 0, \end{aligned} \tag{8.3.18}$$

$$\varphi|_{S_h} = 0, \quad \varphi|_{t=0} = u_0(x), \tag{8.3.19}$$

where $b(x) = a(x+\frac{h}{2})$. We have the following results.

Theorem 8.3.6. *Assuming $u(x,t)$, $\varphi(x,t)$ are the solutions of problem* (8.3.1)–(8.3.3) *and problem* (8.3.18), (8.3.19), *respectively, we have*

$$\|u-\varphi\|_{\Omega(T)} = O(k^2+h^2). \tag{8.3.20}$$

Proof. The proof is similar to Theorem 8.3.4. □

Theorem 8.3.7. *The difference equations* (8.3.18), (8.3.19) *are stable with respect to the norm $\|\cdot\|_\Omega$ in accordance with the initial value.*

For the solution of the nonlinear algebraic equations (8.3.13), (8.3.14), we can generally use the chase-after iterative method.

Numerical calculations show that the following conservation scheme of problem (8.3.1)–(8.3.3):

$$\begin{aligned} i\varphi_{\bar t} - \frac{1}{2}[(b\varphi_x)_{\bar x} + (b\varphi_x(t-k))_{\bar x}] + \frac{\beta}{4}[|\varphi(t)|^2 + |\varphi(t-k)|^2]\cdot(\varphi(t)+\varphi(t-k)) \\ + f(x)\frac{1}{2}[\varphi(t)+\varphi(t-k)] = 0, \end{aligned} \tag{8.3.21}$$

$$\varphi|_{S_h} = 0, \quad \varphi|_{t=0} = u_0(x), \tag{8.3.22}$$

has better conserved properties than the six-point symmetric scheme. Therefore, the calculation results are good. For example, in [45], the definite solution problem

$$\begin{cases} iu_t + u_{xx} + 2|u|^2u = 0, \\ u|_{t=0} = \operatorname{sech}(x+10)\exp[2i(x+10)], \\ u|_{x=\pm 15} = 0 \end{cases}$$

was calculated using the format of (8.3.21) and it has better accuracy than the exact solution

$$u(x,t) = \operatorname{sech}(x+10-4t)\exp[2i(x+10)-3it].$$

The numerical methods for solving the nonlinear Schrödinger equation and their system (including multi-dimensional) can be found in [102, 113, 119].

8.4 Numerical study of the RLW equation

For the regularized long wave (RLW) equation

$$u_t + u_x + uu_x - u_{xxt} = 0, \tag{8.4.1}$$

the existence and uniqueness of its solution have been proved by Benjamin, Bona, and Mahony in [22]. We use the following three-layer difference scheme to approximate it:

$$\begin{aligned}\omega_{m-1}^{n+1} - (2+h^2)\omega_m^{n+1} + \omega_{m+1}^{n+1} = {}& \omega_{m-1}^{n-1} - (2+h^2)\omega_m^{n-1} \\ & + \omega_{m+1}^{n-1} - kh(1+\omega_m^n)\cdot(\omega_{m+1}^n - \omega_{m-1}^n).\end{aligned} \tag{8.4.2}$$

Obviously, the truncation error of the format of (8.4.2) and equation (8.4.1) is

$$\frac{h^2}{6}u_{xxx}(1+u) + (k^2/6)u_{ttt}.$$

In the actual calculation, if $u \ll 1$ and $u_t \sim u_x$, then, when $h = k$, the two terms may be canceled. Using this scheme, we calculated the interaction of two solitons and three solitons, respectively, as shown in Figure 8.6.

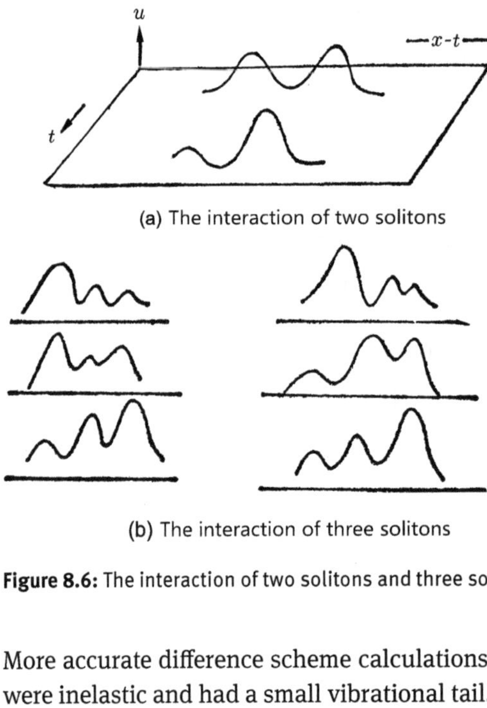

Figure 8.6: The interaction of two solitons and three solitons.

More accurate difference scheme calculations showed that the stronger two solitons were inelastic and had a small vibrational tail.

Other differential formats of (8.4.1), such as

$$u_{\bar{t}} - u_{\hat{x}} + \frac{1}{2}(u^2)_{\hat{x}} - u_{x\bar{x}\bar{t}} = 0 \tag{8.4.3}$$

and

$$u_{\bar{t}} - u_{\hat{x}} + \frac{1}{3}((u^2)_{\hat{x}} + uu_{\hat{x}}) - u_{x\bar{x}\bar{t}} = 0, \tag{8.4.4}$$

can be used.

8.5 Numerical study of the nonlinear Klein–Gorden equation

For the nonlinear wave equation

$$u_{tt} - u_{xx} + F'(u) = 0, \tag{8.5.1}$$

the different forms of $F'(u)$ play an important role in the study of solitons, such as

$$F'(u,\lambda) = \sin u + \lambda \sin 2u \quad \text{(double sine-Gordon)}, \tag{8.5.2}$$

$$F'(u,\lambda) = \sin u \quad \text{(sine-Gordon)}, \tag{8.5.3}$$

$$F'(u) = -u - u^3 \quad (\varphi_-^4), \tag{8.5.4}$$

$$F'(u) = u - u^3 \quad (\varphi_+^4), \tag{8.5.5}$$

$$F'(u) = \begin{cases} \frac{\pi}{4}, & 2n\pi < u \le (2n+1)\pi, \\ 0, & u = n\pi, \\ -\frac{\pi}{4}, & (2n+1)\pi < u < (2n+2)\pi, \\ & \quad n = 0, \pm 1, \pm 2, \dots . \end{cases} \tag{8.5.6}$$

For the sine-Gordon equation, two simple calculation methods have been considered, one of which is the simple leapfrog format, written

$$u_m^{n+1} = -u_m^{n-1} + \frac{k^2}{h^2}[u_{m+1}^n + u_{m-1}^n] + 2\left[1 - \frac{k^2}{h^2}\right]u_m^n - k^2 \sin u_m^n. \tag{8.5.7}$$

The linear stability analysis shows that this form is unstable when $k = h$ and can be overcome when $k = 0.95h$. It makes a numerical calculation for the two-kink case. Another format is to convert the original equation into a system of first-order equations, i.e.,

$$\begin{cases} u_x + u_t = v, \\ v_x - v_t = \sin u. \end{cases} \tag{8.5.8}$$

Introducing $\xi = t - x$ and $\eta = t + x$, (8.5.8) can be reduced to

$$u_\eta = \frac{1}{2}v, \quad v_\xi = -\frac{1}{2}\sin u. \tag{8.5.9}$$

This is called the characteristic form. The characteristic line is a straight line, using the pre-correction format to solve the ordinary differential equations. Although this format is more accurate, the iterative is time-consuming.

In [7], Ablowitz et al. proposed a new scheme. For equation (8.5.1), denote $u_m^n = u(mh, nh)$, $v_m^n = u((m+\frac{1}{2})h, (n+\frac{1}{2})h)$, $\omega_m = u_t(mh, 0)$, and

$$v_m^0 = \frac{1}{2}(u_m^0 + u_{m+1}^0) + \frac{h}{4}(\omega_m + \omega_{m+1}) - \frac{h^2}{8}F'\left(\frac{u_m^0 + u_{m+1}^0}{2}\right) + O(h^3), \tag{8.5.10}$$

$$u_m^{n+1} = -u_m^n + v_m^n + v_{m-1}^n - \frac{h^2}{4}F'\left(\frac{v_m^n + v_{m-1}^n}{2}\right) + O(h^4), \tag{8.5.11}$$

$$v_m^{n+1} = -v_m^n + u_{m+1}^{n+1} + u_m^{n+1} - \frac{h^2}{4}F'\left(\frac{u_{m+1}^{n+1} + u_m^{n+1}}{2}\right) + O(h^4) \tag{8.5.12}$$

and choose the periodic condition $u_{2p+h}^n = u_h^n$, where $2p$ is the period.

The collision of the two solitons of the double sine-Gordon equation and the φ^4-equation are calculated by the formats of (8.5.10), (8.5.11), and (8.5.12). The results are shown in Figure 8.7.

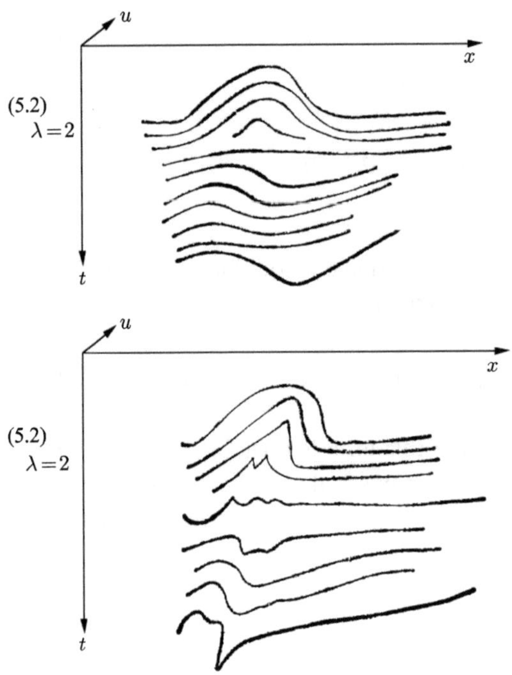

Figure 8.7: The collision of the two solitons of the double sine-Gordon equation and the φ^4-equation.

8.6 Numerical study of the stability problem for a class of nonlinear waves

The plasma dynamics equations are a set of complex equations. Min Yu has given the high-low-frequency and two-fluid plasma dynamics equations with no external magnetic field and uniform initial density. In [86], under a certain assumption, the author gives the one-dimensional plane form of this system, i.e.,

$$\frac{\partial n_i}{\partial t} + \frac{\partial(n_i v_i)}{\partial x} = 0, \tag{8.6.1}$$

$$\frac{\partial v_i}{\partial t} + v_i \frac{\partial v_i}{\partial x} + \frac{\partial \varphi}{\partial x} = 0, \tag{8.6.2}$$

$$\frac{\partial^2 \varphi}{\partial x^2} - e^{\varphi - \psi_1^2 - \psi_2^2} + n_i = 0, \tag{8.6.3}$$

$$\mu \frac{\partial \psi_1}{\partial t} + \frac{\partial^2 \psi_2}{\partial x^2} - (e^{\varphi - \psi_1^2 - \psi_2^2} - 1)\psi_2 = 0, \tag{8.6.4}$$

$$\mu \frac{\partial \psi_2}{\partial t} - \frac{\partial^2 \psi_1}{\partial x^2} + (e^{\varphi - \psi_1^2 - \psi_2^2} - 1)\psi_1 = 0, \tag{8.6.5}$$

where n_i is the ion density, v_i is the ion velocity, $n_e = e^{\varphi - \psi_1^2 - \psi_2^2}$ is the low frequency electron number density, φ is the potential function, ψ_1, ψ_2 are the amounts describing the high-frequency field amplitude, and μ is a constant.

If we let $n_i(x,t)$, $v_i(x,t)$, $\varphi(x,t)$, and

$$\psi(x,t) = \sqrt{\psi_1^2(x,t) + \psi_2^2(x,t)}$$

be the functions of $\xi = x - ct$, then we find that the equations for the solitary wave solution of this system must be satisfied. We have

$$\frac{d^2\varphi}{d\xi^2} = e^{\varphi - \psi^2} - \left(1 - \frac{2}{c^2}\right)\varphi^{-\frac{1}{2}}, \tag{8.6.6}$$

$$\frac{d^2\psi}{d\xi^2} = (e^{\varphi - \psi^2} - 1 + a^2)\psi, \tag{8.6.7}$$

$$n_i = \left(1 - \frac{2}{c^2}\right)\varphi^{-\frac{1}{2}}, \tag{8.6.8}$$

$$v_i = c\left[1 - \left(1 - \frac{2}{c^2}\varphi\right)^{-\frac{1}{2}}\right], \tag{8.6.9}$$

where a, c are the parameters, c represents the propagation velocity of the solitary wave, and a represents the deviation of the frequency of the high-frequency electric field with respect to some fixed frequency.

Equations (8.6.6) and (8.6.7) form a closed system, whose definite solution conditions are

$$\varphi = 0, \quad x \to \pm\infty, \tag{8.6.10}$$

$$\psi = 0, \quad x \to \pm\infty. \tag{8.6.11}$$

The solution of (8.6.6), (8.6.7), (8.6.10), and (8.6.11) and its properties are discussed in [23], in which the solitary wave solutions of various parameters a and c are given by numerical calculation. Are these solitary waves stable? This is a matter of concern and the question must be answered. Because the equations (8.6.1)–(8.6.5) are quite complex, it is difficult to get the answer by the analytical qualitative analysis. Shen et al. have considered the stability of such waves through actual numerical calculation.

The initial conditions of (8.6.1)–(8.6.5) are

$$n_i(x,0) = n_i^0(x), \quad v_i(x,0) = v_i^0(x), \quad \psi_1|_{t=0} = \psi_1^0(x), \quad \psi_2|_{t=0} = \psi_2^0(x). \tag{8.6.12}$$

Assume, when $x \to \pm\infty$,

$$n_i \to 1, \quad v_i \to 0, \quad \varphi \to 0, \quad \psi_1 \to 0, \quad \psi_2 \to 0. \tag{8.6.13}$$

We see that the Cauchy problem of problem (8.6.1)–(8.6.5) is an infinite-interval problem. For finite-difference calculations, the finite-interval approximate calculation always brings some errors. Therefore, the transformation of the spatial variable can be implemented by

$$\xi = \operatorname{th} \lambda x. \tag{8.6.14}$$

This transformation changes the interval $(-\infty, \infty)$ of x to the interval $(-1, +1)$ of ξ. Then the equations correspondingly becomes

$$\frac{\partial n_i}{\partial t} + \lambda(1-\xi^2)\frac{\partial n_i v_i}{\partial \xi} = 0, \tag{8.6.15}$$

$$\frac{\partial v_i}{\partial t} + \lambda(1-\xi^2)v_i\frac{\partial v_i}{\partial \xi} + \lambda(1-\xi^2)\frac{\partial \varphi}{\partial \xi} = 0, \tag{8.6.16}$$

$$\lambda^2(1-\xi^2)\frac{\partial}{\partial \xi}(1-\xi^2)\frac{\partial \varphi}{\partial \xi} - e^{\varphi-\psi_1^2-\psi_2^2} + n_i = 0, \tag{8.6.17}$$

$$\mu\frac{\partial \psi_1}{\partial t} + \lambda^2(1-\xi^2)\frac{\partial \psi_2}{\partial \xi} - (e^{\varphi-\psi_1^2-\psi_2^2} - 1)\psi_2 = 0, \tag{8.6.18}$$

$$\mu\frac{\partial \psi_2}{\partial t} - \lambda^2(1-\xi^2)\frac{\partial}{\partial \xi}(1-\xi^2)\frac{\partial \psi_1}{\partial \xi} + (e^{\varphi-\psi_1^2-\psi_2^2} - 1)\psi_1 = 0. \tag{8.6.19}$$

Obviously, $\xi \to \pm 1$, $n_i \to 1$, $v_i \to 0$, $\varphi \to 0$, $\psi_1 \to 0$, and $\psi_2 \to 0$.

Equations (8.6.15) and (8.6.16) are the hydrodynamic equations. We use the method of Richtmyer as follows:

$$\begin{aligned} &\frac{v_{j+\frac12}^{k+1} - v_{j+\frac12}^k}{\Delta t} + \lambda(1-\xi_{j+\frac12}^2)v_{j+\frac12}^k \frac{v_{j+\frac12}^k - v_{j-\frac12}^k}{\Delta \xi_j} \\ &\quad + \lambda(1-\xi_{j+\frac12}^2)\frac{\varphi_{j+1}^k - \varphi_j^k}{\Delta \xi_{j+\frac12}} = 0, \quad \text{when } v_{j+\frac12}^k \geq 0, \end{aligned} \tag{8.6.20}$$

$$\begin{aligned} &\frac{v_{j+\frac12}^{k+1} - v_{j+\frac12}^k}{\Delta t} + \lambda(1-\xi_{j+\frac12}^2)v_{j+\frac12}^k \frac{v_{j+\frac32}^k - v_{j+\frac12}^k}{\Delta \xi_{j+1}} \\ &\quad + \lambda(1-\xi_{j+\frac12}^2)\frac{\varphi_{j+1}^k - \varphi_j^k}{\Delta \xi_{j+\frac12}} = 0, \quad \text{when } v_{j+\frac12}^k < 0, \end{aligned} \tag{8.6.21}$$

$$\frac{n_j^{k+1}-n_j^k}{\Delta t}+\lambda(1-\xi_j^2)v_{j-\frac{1}{2}}^{k+1}\frac{n_j^k-n_{j-1}^k}{\Delta\xi_{j-\frac{1}{2}}}$$

$$+\lambda(1-\xi_j^2)n_j^k\frac{v_{j+\frac{1}{2}}^{k+1}-v_{j-\frac{1}{2}}^{k+1}}{\Delta\xi_j}=0,\quad \text{when } v_{j-\frac{1}{2}}^{k+1}\geq 0, \tag{8.6.22}$$

$$\frac{n_j^{k+1}-n_j^k}{\Delta t}+\lambda(1-\xi_j^2)v_{j-\frac{1}{2}}^{k+1}\frac{n_{j+1}^k-n_j^k}{\Delta\xi_{j+\frac{1}{2}}}$$

$$+\lambda(1-\xi_j^2)n_j^k\frac{v_{j+\frac{1}{2}}^{k+1}-v_{j-\frac{1}{2}}^{k+1}}{\Delta\xi_j}=0,\quad \text{when } v_{j-\frac{1}{2}}^{k+1}<0. \tag{8.6.23}$$

We know that, for (8.6.17), (8.6.18), the display format is absolutely unstable and the implicit format is absolutely stable, but in order to facilitate the calculation, we can use the following semi-implicit format:

$$\mu\frac{\psi_{1j}^{k+1}-\psi_{1j}^k}{\Delta t}=-\lambda^2(1-\xi_j^2)\frac{1}{\Delta\xi_j}\left[\frac{1-\xi_{j+\frac{1}{2}}^2}{\Delta\xi_{j+\frac{1}{2}}}\psi_{2j+1}^k-\left(\frac{1-\xi_{j+\frac{1}{2}}^2}{\Delta\xi_{j+\frac{1}{2}}}+\frac{1-\xi_{j-\frac{1}{2}}^2}{\Delta\xi_{j-\frac{1}{2}}}\right)\psi_{2j}^k\right.$$

$$\left.+\frac{1-\xi_{j-\frac{1}{2}}^2}{\Delta\xi_{j-\frac{1}{2}}}\psi_{2j-1}^k\right]+\left(e^{\varphi_j^k-(\psi_{1j}^k)^2-(\psi_{2j}^k)^2}-1\right)\psi_{2j}^k, \tag{8.6.24}$$

$$\mu\frac{\psi_{2j}^{k+1}-\psi_{2j}^k}{\Delta t}=\lambda^2(1-\xi_j^2)\frac{1}{\Delta\xi_j}\left[\frac{1-\xi_{j+\frac{1}{2}}^2}{\Delta\xi_{j+\frac{1}{2}}}\psi_{1j+1}^{k+1}-\left(\frac{1-\xi_{j+\frac{1}{2}}^2}{\Delta\xi_{j+\frac{1}{2}}}+\frac{1-\xi_{j-\frac{1}{2}}^2}{\Delta\xi_{j-\frac{1}{2}}}\right)\psi_{1j}^{k+1}\right.$$

$$\left.+\frac{1-\xi_{j-\frac{1}{2}}^2}{\Delta\xi_{j-\frac{1}{2}}}\psi_{1j+1}^{k+1}\right]-\left(e^{\varphi_j^k-(\psi_{1j}^k)^2-(\psi_{2j}^k)^2}-1\right)\psi_{1j}^k. \tag{8.6.25}$$

In this way, the two equations can be solved directly without iteration. It is easy to derive that such a format needs to satisfy the stability requirements, so we have

$$\Delta t\leq\frac{\mu}{2}\Delta x^2.$$

After obtaining n_i, ψ_1, ψ_2, we use the iterative method to solve (8.6.17) as follows:

$$\lambda^2(1-\xi_j^2)\frac{1}{\Delta\xi_j}\left[\frac{1-\xi_{j+\frac{1}{2}}^2}{\Delta\xi_{j+\frac{1}{2}}}\varphi_{j+1}^{k+1,s+1}-\left(\frac{1-\xi_{j+\frac{1}{2}}^2}{\Delta\xi_{j+\frac{1}{2}}}+\frac{1-\xi_{j-\frac{1}{2}}^2}{\Delta\xi_{j-\frac{1}{2}}}\right)\varphi_j^{k+1,s+1}\right.$$

$$\left.+\frac{1-\xi_{j-\frac{1}{2}}^2}{\Delta\xi_{j-\frac{1}{2}}}\psi_{j-1}^{k+1,s+1}\right]-e^{\varphi_j^{k+1,s}-(\psi_{1j}^{k+1})^2-(\psi_{2j}^{k+1})^2}+n_j^{k+1}=0. \tag{8.6.26}$$

Shen et al. made numerical calculations by using the difference scheme (8.6.20)–(8.6.26) for problem (8.6.15)–(8.6.18) of a class of so-called nonlinear initial conditions with a single peak. Numerical results show that this type of wave form has no change, which shows the better stability of the solitary wave of problem (8.6.1)–(8.6.5), as shown in Figure 8.8.

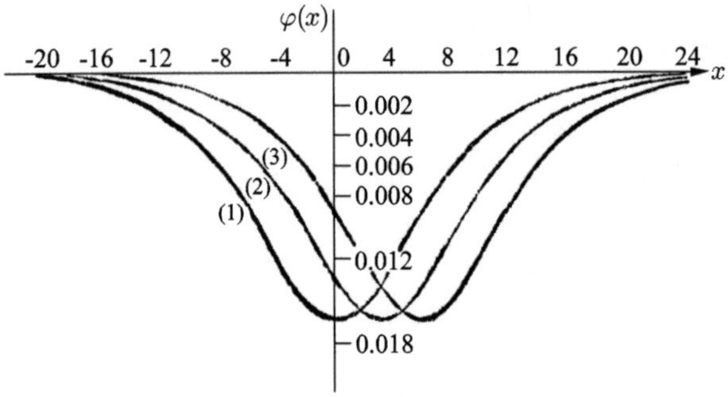

Figure 8.8: The waveforms of φ under a class of weakly nonlinear initial value conditions.

In addition, we obtain the following conserved quantities for equations (8.6.1)–(8.6.5):

$$\int_{-\infty}^{\infty}(n_i - 1)dx = \text{constant},$$
$$\int_{-\infty}^{\infty} v_i dx = \text{constant},$$
$$\int_{-\infty}^{\infty}(e^{\varphi-\psi_1^2-\psi_2^2} - 1)dx = \text{constant},$$
$$\int_{-\infty}^{\infty} |\psi|^2 dx = \text{constant},$$
$$\int_{-\infty}^{\infty}\left(\psi_{1x}^2 - \psi_{2x}^2 + \frac{1}{2}\varphi(n + e^{\varphi-\psi_1^2-\psi_2^2}) + \frac{1}{2}nv^2\right)dx = \text{constant}.$$

We can use the above conserved quantities to test the conservation of the difference scheme.

9 The geometric theory of solitons

9.1 Bäcklund transformation and the surface with total curvature $K = -1$

As we have seen earlier, the Bäcklund transform can be used to find another soliton solution from one soliton solution of the sine-Gordon equation. By the nonlinear superposition principle, it is easy to get the special solution of nonlinear equations by algebraic operation; this is a very clever method. In the following, we present the geometric approach taken by Chern and Terng. The relationship between the sine-Gordon equation and the $K = -1$ surface and the study of the geometric properties of its solution show that the problem of solving the sine-Gordon equation can be reduced to the problem of finding another $K = -1$ surface from a $K = -1$ surface.

Let $M(u, v)$ be a planar region, R^3 be a three-dimensional Euclidean space, and $x : M \to R^3$ be a surface. At each point on the surface, we take a unit right-handed orthogonal frame $[x; e_1, e_2, e_3]$, $(e_\alpha, e_\beta) = \delta_{\alpha\beta}$, $(e_1, e_2, e_3) = 1$, $1 \le \alpha, \beta \le 3$. Suppose e_3 is the normal vector. Then we have the motion equations

$$dx = \sum_\alpha w_\alpha e_\alpha, \quad w_3 = 0, \tag{9.1.1}$$

$$de_\alpha = \sum_\beta w_{\alpha\beta} e_\beta, \quad w_{\alpha\beta} + w_{\beta\alpha} = 0, \tag{9.1.2}$$

where w_α, $w_{\alpha\beta}$ are two differential one-forms.

We call

$$\mathrm{I} = w_1^2 + w_2^2 (= dx \cdot dx),$$
$$\mathrm{II} = w_1 w_{13} + w_2 w_{23} (= -de_3 dx)$$

the first and second fundamental forms of the surface, respectively.

We take the exterior differential for (9.1.1) and (9.1.2) and obtain the following structural equation:

$$\begin{cases} dw_1 = w_{12} \wedge w_2, \\ dw_2 = w_1 \wedge w_{12}, \\ \mathrm{II} = -(dx, de_3) = aw_1^2 + 2bw_1w_2 + cw_2^2, \end{cases} \tag{9.1.3}$$

so

$$\begin{cases} w_{13} = aw_1 + bw_2, \\ w_{23} = bw_1 + cw_2, \\ dw_{12} = -Kw_1 \wedge w_2 \quad \text{(Gauss equation)}, \end{cases} \tag{9.1.4}$$

https://doi.org/10.1515/9783110549638-009

where $K = ac - b^2$ is the Gauss curvature. We have

$$\begin{cases} dw_{13} = w_{12} \wedge w_{23}, \\ dw_{23} = w_{12} \wedge w_{13}. \end{cases} \tag{9.1.5}$$

Therefore, w_{12} is only related to the first fundamental form, which is called the contact form of the surface. Equation (9.1.5) is called Codazzi's equation. Now we consider the surface of $K = -1$. Taking the coordinates of the curvature line

$$\begin{cases} w_1 = \sin\psi du, \quad w_2 = \cos\psi dv, \\ w_{13} = \cos\psi du, \quad w_{23} = -\sin\psi dv, \\ w_{12} = -\psi_v du - \psi_u dv \end{cases} \tag{9.1.6}$$

and inserting these into the Gauss equation, we get

$$\psi_{uu} - \psi_{vv} = -\cos\psi\sin\psi. \tag{9.1.7}$$

We know from (9.1.7) that

$$\begin{aligned} \mathrm{I} &= \sin^2\psi du^2 + \cos^2\psi dv^2, \\ \mathrm{II} &= \sin\psi\cos\psi(dv^2 - du^2), \end{aligned} \tag{9.1.8}$$

so 2ψ is the angle of the asymptote $\frac{d\psi}{du} = \pm 1$. Moreover, it can be seen from equation (9.1.8), being a function of (u, v), 2ψ is the solution of the sine-Gordon equation. On the contrary, from the basic theorem of surface theory, any solution of the sine-Gordon equation can be regarded as the angle between the asymptotes of a $K = -1$ surface. Thus, the problem of finding the solution to the sine-Gordon equation comes down to the problem of finding the $K = -1$ surface. We know that the $K = -1$ surface is a pseudo-spherical surface, that is, a singular horn-shaped surface. We can give the correspondence between $K = -1$ surfaces through a match-fixing clue.

The clue refers to the family of straight lines depending on two parameters (u, v) of the Euclidean space. We have

$$y = \chi(u, v) + \lambda n(u, v), \tag{9.1.9}$$

where $n^2 = 1$ and $\chi(u, v)$ is generally expressed as a surface. If we fix (u, v), then (9.1.9) denotes a straight line which passes through point $\chi(u, v)$ and has the direction $n(u, v)$. If $(u(t), v(t))$ is a curve on the surface, then

$$y(t, \lambda) = \chi(u(t), v(t)) + \lambda n(u(t), v(t))$$

is a ruled surface. The sufficient and necessary condition of the surface is

$$|n, dx, dn| = 0, \tag{9.1.10}$$

where $|n, dx, dn|$ represents the mixed product determinant of n, dx, and dn and (9.1.10) is a binary homogeneous equation with respect to du, dv. If its ratio has two different real solutions, then there should be two family curves corresponding to the surface. The straight lines passing through the clue of each of the curves in the family form a developable surface. Then two families of developable surfaces are obtained and each developable surface has a ridge line. The whole of the ridge lines of each family of developable surfaces constitutes a surface, known as the focal surface. Thus, we obtain two focal surfaces, denoted s and s', where each of the lines in clue (9.1.9) is the common tangent of s and s'. Thus, by means of these common tangents, clue (9.1.9) gives a transformation between the focal surface s and the focal surface s', i.e.,

$$l : s \to s',$$

that is, if $p' = l(p)$, then $p(\in s)$ and $p'(\in s')$ have the common tangent l, which belongs to the clue (9.1.9).

Definition 9.1.1. The clue is called match-fixing if (1) $|pp'| = \gamma$ (constant), that is, the distance between the corresponding points is a fixed value and (2) $\langle e_3(p), e_3'(p)\rangle = \tau$ (constant), that is, the angle between the normal direction of the corresponding points is a fixed value.

Theorem 9.1.2 (Bäcklund). *The two focal surfaces of the match-fixing clue have corresponding constant Gauss curvatures*

$$K = -\frac{\sin^2 \tau}{r^2}.$$

In particular, if $r = \sin \tau$, then $K = -1$, so the problem of constructing another $K = -1$ surface from a given $K = -1$ surface comes down to the problem of constructing a match-fixing clue from a given $K = -1$ surface. In this case, one only needs to determine the direction of each line in the desired clues. At this time, the problem is reduced to solving a complete integrable differential equation,

$$d\alpha + \sin \alpha w = \cos \tau w_{13}, \tag{9.1.11}$$

where α represents the angle between the straight line and the main direction of the tangent point (i.e., the direction of the u curve) in the desired clue. We get the following completely integrable first-order partial differential equations from (9.1.6), (9.1.11):

$$\begin{cases} \sin \tau(\alpha_u - \psi_v) = \cos \tau \cos \alpha \cos \psi + \sin \alpha \sin \psi, \\ \sin \tau(\alpha_v - \psi_u) = -\cos \tau \sin \alpha \sin \psi - \cos \alpha \cos \psi. \end{cases} \tag{9.1.12}$$

At this time, the first and second fundamental forms of the desired surface are

$$\mathrm{I}' = \cos^2 \alpha du^2 + \sin^2 \alpha dv^2,$$

$$\mathrm{II}' = \cos\alpha \sin\alpha(du^2 - dv^2),$$

respectively. It can be seen that 2α is a solution of the sine-Gordon equation. By (9.1.12), if a solution 2ψ of the sine-Gordon equation is given, then another solution 2α of the sine-Gordon equation can be solved. Since (9.1.12) is completely integrable, it is only needed to solve an ordinary differential equation.

Tenenblat and Terng further discuss the Bäcklund theorem of n-dimensional submanifolds in a $2n-1$-dimensional Euclidean space and the high-dimensional generalization of the sine-Gordon equation [291, 292].

9.2 Lie group and the nonlinear evolution equations

In Chapter 2, we have pointed out in detail the inverse scattering method of solving the nonlinear evolution equations established by Gardner, Greene, Kruskal, and Miura and Ablowitz, Kaup, Newell, and Suger (AKNS). Chern and Peng [47] pointed out that the algebraic basis of these equations is formed by Lie groups and their structural equations. They started from the structure of the 2×2 real unimodular Lie group SL(2) and naturally and concretely gave the high-order Korteweg–de Vries (KdV) equation and the modifiedi KdV (MKdV) equation, so that the geometric significance of these equations is clear. The use of group operations is also more convenient; Sasaki [268] further established the relationship between the AKNS equations and the negative constant curvature surfaces.

Assume

$$SL(2;R) = \{X = \begin{pmatrix} a & b \\ c & d \end{pmatrix} | ad - bc = 1\} \tag{9.2.1}$$

is the group of all 2×2 real unimodular matrices. Its right invariant Maurer–Cartan form is

$$w = dxX^{-1} = \begin{pmatrix} w_1^1, & w_1^2 \\ w_2^1 & w_2^2 \end{pmatrix}, \tag{9.2.2}$$

where

$$w_1^1 + w_2^1 = 0.$$

The structural equation of $SL(2;R)$ or the Maurer–Cartan equation is

$$dw = w \wedge w \tag{9.2.3}$$

or, in more detail,

$$\begin{cases} dw_1^1 = w_1^2 \wedge w_2^1, \\ dw_1^2 = 2w_1^1 \wedge w_1^2, \\ dw_2^1 = 2w_2^1 \wedge w_1^1. \end{cases} \tag{9.2.4}$$

Suppose v is a neighborhood in the (x,t)-plane and consider the smooth map

$$f : v \to SL(2; R). \tag{9.2.5}$$

Then these equations become the functions of (x,t) after mapping in Lie groups. We have

$$\begin{cases} w_1^1 = \eta dx + Adt, \\ w_1^2 = qdx + Bdt, \\ w_2^1 = rdx + Cdt, \end{cases} \tag{9.2.6}$$

where the coefficients are all functions of (x,t). Because

$$\begin{aligned} dw_1 &= w_1^2 \wedge w_2^1 = (qdx + Bdt) \wedge (rdx + Cdt) \\ &= qrdx \wedge dx + (qC - Br)dx \wedge dt + BCdt \wedge dt, \end{aligned}$$

on the other hand,

$$\begin{aligned} dw_1^1 &= A_x dx \wedge dt + A_t dt \wedge dt + \eta d^2x + Ad^2t \\ &\quad + \eta_t dt \wedge dx + \eta_x dx \wedge dx, \\ dw_1^2 &= 2w_1^1 \wedge w_1^2 = 2(\eta dx + Adt) \wedge (qdx + Bdt) \\ &= 2\eta qdx \wedge dx + 2(\eta B - Aq)dx \wedge dt \\ &\quad + 2ABdt \wedge dt. \end{aligned}$$

In addition,

$$\begin{aligned} dw_1^2 &= q_x dx \wedge dx + q_t dt \wedge dx + qd^2x + B_x dxdt \\ &\quad + B_t dt \wedge dt + Bd^2t. \end{aligned}$$

Of course, dw_2^1 can be calculated similarly. We obtain

$$\begin{cases} -\eta_t + A_x - qC + rB = 0, \\ -q_t + B_x - 2\eta B + 2qA = 0, \\ -r_t + C_x - 2rA + 2\eta A = 0. \end{cases} \tag{9.2.7}$$

Assume η = constant, that is, η is a parameter independent of x, t. Now we consider some special cases:

(1) $r = +1$, η is a constant, $q = u(x,t)$. At this time, solving A from the third formula of (9.2.7) and B from the first formula of (9.2.7), we have

$$\begin{cases} A = \eta C + \frac{1}{2}C_x, \\ B = -\frac{1}{2}C_{xx} - \eta C_x + uC. \end{cases} \tag{9.2.8}$$

Inserting (9.2.8) into the second equation of (9.2.7), we get

$$u_t = K(u), \tag{9.2.9}$$

where

$$K(u) = u_x C + 2uC_x + 2\eta^2 C_x - \frac{1}{2} C_{xxx}. \tag{9.2.10}$$

As an example, we choose

$$C = \eta^2 - \frac{1}{2}u. \tag{9.2.11}$$

Then we obtain, from (9.2.9),

$$u_t = \frac{1}{4}u_{xxx} - \frac{3}{2}uu_x. \tag{9.2.12}$$

This is the well-known KdV equation.

We can naturally choose C to be the arbitrary polynomial of η. Because (9.2.10) only contains η^2, we can assume that C is the polynomial of η^2. Letting

$$C = \sum_{0\le j\le n} C_j(x,t)\eta^{2(n-j)}, \tag{9.2.13}$$

where $C_j(x,t)$ are the functions of x, t, inserting (9.2.13) into (9.2.10), and letting the coefficient of η^2 be zero, we get

$$C_0 = \text{constant}, \tag{9.2.14}$$

$$C_{j+1,x} = -\frac{1}{2}u_x C_j - uC_{j,x} + \frac{1}{4}C_{j,xxx}. \tag{9.2.15}$$

We note that the latter is just a cyclic formula for the conservation density of the KdV equation. The right-hand side of (9.2.9) can be written as

$$K_n(x) = u_x C_n + 2uC_{n,x} - \frac{1}{2}C_{n,xxx} = -2C_{n+1,x}, \tag{9.2.16}$$

where the last equality is introduced as a definition. Furthermore, an infinite sequence of C_j can be introduced. Assume that (9.2.14) holds for all j, $0 \le j < \infty$, and the equation

$$u_t = K_n(u) \tag{9.2.17}$$

is called the nth-order KdV equation. It can be shown that C is a polynomial of u and the derivative of x. For example, we obtain

$$2C_1 = -u,$$

$$
\begin{aligned}
2C_{j+1} = &-\sum_{1\le k\le j} C_k C_{j+1-k} - u \sum_{0\le k\le j} C_k C_{j-k} \\
&+ \frac{1}{2} \sum_{0\le k\le j-1} C_k C_{j-k,xx} \\
&- \frac{1}{4} \sum_{1\le k\le j} C_{k,x} C_{j-k,x}, \quad j = 1, 2, \dots .
\end{aligned} \tag{9.2.18}
$$

In particular,

$$
\begin{aligned}
2C_2 &= \frac{3}{4}u^2 - \frac{1}{4}u_{xx}, \\
2C_3 &= -\frac{5}{8}u^3 + \frac{5}{16}u_x^2 + \frac{5}{8}uu_{xx} - \frac{1}{16}u_{xxxx}.
\end{aligned} \tag{9.2.19}
$$

(2) $q = r = V(x,t)$ and η is a parameter independent of x, t. Then (9.2.7) becomes

$$
\begin{cases}
A_x = V(C - B), \\
V_t = B_x - 2\eta B + 2VA, \\
V_t = C_x + 2\eta C - 2VA.
\end{cases} \tag{9.2.20}
$$

The last two equations in (9.2.20) can be written as

$$
\begin{cases}
(C - B)_x = 4VA - 2\eta(B + C), \\
V_t = \frac{1}{2}(B + C)_x + \eta(C - B).
\end{cases} \tag{9.2.21}
$$

Letting

$$
C - B = \eta P, \quad C + B = Q, \quad A = \eta R, \tag{9.2.22}
$$

the above equations become

$$
\begin{cases}
R_x = VP, \\
P_x = 4VR - 2Q, \\
V_t = \tfrac{1}{2}Q_x + \eta^2 P.
\end{cases} \tag{9.2.23}
$$

Eliminating P, Q, we get

$$
V_t = M(V), \tag{9.2.24}
$$

where

$$
M(V) = \eta^2 \frac{R_x}{V} + (VR)_x - \frac{1}{4}\left(\frac{R_x}{V}\right)_{xx}.
$$

Choosing $R = \eta^2 - \frac{1}{2}V^2$, equation (9.2.24) becomes

$$
V_t = \frac{1}{4}V_{xxx} - \frac{3}{2}V^2 V_x. \tag{9.2.25}
$$

This is the well-known MKdV equation.

9.3 Prolongation structure of the nonlinear equations

The differential manifold M and the n-form ideal I, if the so-called exterior extension refers to the $n-1$-form P on M and the coefficients take on the differentiable function on M, also satisfy

$$dP \subset F^*(M) \wedge P + I, \tag{9.3.1}$$

where $F^*(M)$ is the 1-form on M. The concept of exterior extension was first proposed by Wohlquist and Estabrook in 1975 and applied to the KdV equation. They presented the KdV equation as a set of equivalent external differential forms of closed ideal, extended this closed ideal, successfully found the inverse scattering problem and Bäcklund transform for the KdV equation, and made a similar discussion for the nonlinear Schrödinger equation [76]. Morris, Corones, and Gibbon et al. discussed the exterior extension structure of the shallow water wave equation with gravitation, the Hirota equation, the nonlinear Schrödinger, the high-order KdV equation, and the self-dual Yang–Mills equation [184, 52, 53, 48, 187, 69, 68, 186, 185, 188]. It can be seen that the prolongation structure method is not only suitable for a large number of nonlinear evolution equations, but can also be naturally extended to the high-dimensional space, so it has more advantages than the inverse scattering method in this respect. This differential geometry method may become the theoretical basis of the inverse scattering method.

Now we consider the prolongation structure method for the KdV equation. Assume we have the following KdV equation:

$$u_t + u_{xxx} + 12uu_x = 0. \tag{9.3.2}$$

Letting $z = u_x$ and $p = z_x = u_{xx}$, (9.3.2) can be written as the following first-order equation:

$$u_t + p_x + 12uz = 0. \tag{9.3.3}$$

For the five-dimensional manifold $M\{x, t, u, z, p\}$, the basis of the dual space $T^*(M)$ of the tangent space is $\{dx, dt, du, dz, dp\}$. We introduce the two-forms in the two-dimensional submanifold $\{x, t, u(x, t), z(x, t), p(x, t)\}$ of M to obtain

$$\begin{cases} \alpha_1 = du \wedge dt - zdx \wedge dt, \\ \alpha_2 = dz \wedge dt - pdx \wedge dt, \\ \alpha_3 = -du \wedge dx + dp \wedge dt + 12uzdx \wedge dt, \end{cases} \tag{9.3.4}$$

where d represents the exterior derivative and $\wedge$ denotes the exterior product. The first two items of (9.3.4) correspond to the entries of the new variables and the latter one corresponds to the item of the original equation. By direct calculation, we obtain

$$\begin{cases} d\alpha_1 = dx \wedge \alpha_2, \\ d\alpha_2 = dx \wedge \alpha_3, \\ d\alpha_3 = -12dx \wedge (z\alpha_1 + u\alpha_2). \end{cases} \tag{9.3.5}$$

Therefore, $\{\alpha_1, \alpha_2, \alpha_3\}$ forms a closed ideal on manifold M and the two-forms equation (9.3.4) is zero when restricted to the manifold $S_2 = \{u(x,t), z(x,t), p(x,t)\}$. Then the KdV equation is derived from the external form. For the given five-dimensional differentiable manifold M and the closed ideal generated by α_i and $d\alpha_i$, there exist additional extension variables $y^i (i = 1, 2, \ldots, m)$, while $\{y^i\}$ spans an m-dimensional manifold at each point of the original manifold $M(x, t, u, z, p)$, which extends an $m+5$-dimensional fiber bundle. Then the enlarged ideal I' can be generated in the fiber bundle. The generators of I' include not only α_i, but also include the introduced m one-forms w_i due to the extension variables y^i. These w_i are called the exterior extension forms. For exterior extension variables y^i, we have the following Pfaff form w_K:

$$\begin{aligned} w_K = dy^K &+ F^K(x, t, u, z, p, y^i)dx \\ &+ G^K(x, t, u, z, p, y^i)dt. \end{aligned} \tag{9.3.6}$$

This must satisfy the closed ideal condition

$$dw_K = \sum_{i=1}^{3} f_k^i \alpha_i + \sum_{i=1}^{m} \eta_k^i \wedge w_i, \tag{9.3.7}$$

where η_k^i are one-forms. From (9.3.5) and (9.3.6), we can obtain the first-order partial differential equations of F^K and G^K. These equations are generally nonlinear, because they contain the commutator terms

$$\sum_i \left(G^i \frac{\partial F^k}{\partial y^i} - F^i \frac{\partial G^k}{\partial y^i} \right) dx \wedge dt. \tag{9.3.8}$$

If F^k and G^k only depend on y^k, this y^k determines a general conservation law and y^i is called potential. If F^K and G^K depend on extension variables $y^i (i \neq k)$, then y^k is called pseudo-potential. The existence of the pseudo-potential is the key that leads to the Bäcklund transform.

We define the following commutator:

$$[F \cdot G]_k \equiv F^i G^k_{,yi} - G^i F^k_{,yi}. \tag{9.3.9}$$

By (9.3.7) and eliminating f_k^i, we find that $F^k(u, z, p, y^i)$ and $G^k(u, z, p, y^i)$ satisfy the following partial differential equations:

$$F^k_{,z} = 0, \quad F^k_{,p} = 0, \quad F^k_{,u} + G^k_{,p} = 0, \tag{9.3.10}$$

$$zG^k_{,u} + pG^k_{,z} - 12uzG^k_{,p} + G^iF^k_{,yi} - F^iG^k_{,yi} = 0. \tag{9.3.11}$$

From the integrable conditions of equations (9.3.10) and (9.3.11), it is easy to find the following expressions of F^k and G^k:

$$\begin{cases} F^k = 2X^k + 2uX_2^k + 3u^2X_3^k, \\ G^k = -2(p+6u^2)X_2^k + 3(z^2 - 8u^3 - 2up)X_3^k \\ \qquad +8X_4^k + 8uX_5^k + 4u^2X_6^k + 4zX_7^k. \end{cases} \tag{9.3.12}$$

Inserting the forms of F^k, G^k given by (9.3.12) into (9.3.11), we get the following series of commutator relations:

$$\begin{cases} [X_1, X_3] = [X_2, X_3] = [X_1, X_4] = [X_2, X_6] = 0, \\ [X_1, X_2] = -X_7, \quad [X_1, X_7] = X_5, \quad [X_2, X_7] = X_6, \\ [X_1, X_5] + [X_2, X_4] = 0, \quad [X_3, X_4] + [X_1, X_6] + X_7 = 0. \end{cases} \tag{9.3.13}$$

Forcing this open algebraic structure to be close to a finite-dimensional Lie algebra and using the Jacobi identity, we obtain a further relation. We introduce the new generators X_8, X_9. We have

$$[X_3, X_4] = -X_8, \quad [X_1, X_5] = X_8$$

and demand

$$X_9 = \sum_{m=1}^{8} c_m X_m, \tag{9.3.14}$$

where c_m are constants. We also demand the generators 1 through 8 in (9.3.13) to be linearly independent. Using the Jacobi identity, we obtain

$$c_m = 0 (m \neq 7, 8), \quad c_7 = -c_8 \equiv \lambda,$$

where λ is an arbitrary constant. Finally, we get the closed Lie algebra constituted by $\{X_1, \ldots, X_8\}$. We have

$$\begin{cases} [X_1, X_2] = -X_7, \quad [X_2, X_5] = -X_9/\lambda, \\ [X_4, X_7] = -\lambda X_5, \\ [X_1, X_5] = X_9, \quad [X_2, X_3] = X_6, \quad [X_5, X_6] = X_9/\lambda, \\ [X_1, X_6] = -X_9/\lambda, \quad [X_3, X_4] = -X_8, \\ [X_5, X_7] = -X_5 - \lambda X_6, \\ [X_1, X_7] = X_5, \quad [X_4, X_5] = -\lambda X_9, \quad [X_6, X_7] = X_6, \\ [X_2, X_4] = -X_9, \quad [X_4, X_6] = X_9, \quad X_9 \equiv \lambda(X_7 - X_8). \end{cases} \tag{9.3.15}$$

It is not difficult to get the eight-dimensional relationship of this algebra. Choose the basis vectors to be

$$b_k = \frac{\partial}{\partial y^k}(k = 1, \ldots, 8), \tag{9.3.16}$$

where y^k are the coordinate sets of extension variables and the nondegenerate representations of generators are the following:

$$\begin{cases} X_1 = \frac{1}{2}[b_1 + \exp(2y_3)^{-b_2} + y_8 b_3 + y_7 b_5 + (y_8^2 - \lambda)b_8], \\ X_2 = \frac{1}{2}[b_7 + 2b_8], \\ X_3 = \frac{1}{3}b_6, \\ X_4 = -\frac{1}{2}\lambda\Big[b_1 + \exp(2y_3)b_2 + y_8 b_3 - b_4 \\ \qquad + \left(\frac{3}{2\lambda}\right)y_6 b_5 + (y_8^2 - \lambda)b_8\Big], \\ X_5 = -\frac{1}{2}[\exp(2y_3)b_2 + y_8 b_3 + (y_8^2 + \lambda)b_8], \\ X_6 = b_8, \\ X_7 = \frac{1}{2}\left[b_3 + \frac{1}{2}b_5 + 2y_8 b_8\right], \\ X_8 = \frac{1}{4}b_5. \end{cases} \tag{9.3.17}$$

By (9.3.12), writing the eight display expressions in the Pfaff form, we have

$$\begin{cases} w_k = dy^k + F^k dx + G^k dt, \\ w_1 = dy_1 + dx - 4\lambda dt, \\ w_2 = dy_2 + \exp(2y_3)dx - 4\exp(2y_3)(u + \lambda)dt, \\ w_3 = dy_3 + y_8 dx + [2z - 4y_8(u + \lambda)]dt, \\ w_4 = dy_4 + 4\lambda dt, \\ w_5 = dy_5 + y_7 dx + (z - 6y_6)dt, \\ w_6 = dy_6 + u^2 dx + (z^2 - 8u^3 - 2up)dt, \\ w_7 = dy_7 + udx - (p + 6u^2)dt, \\ w_8 = dy_8 + (2u + y_8^2 - \lambda)dx - 4 \\ \qquad \cdot \left[(u + \lambda)(2u + y_8^2 - \lambda) - \frac{1}{2}p - zy_8\right]dt. \end{cases} \tag{9.3.18}$$

Using (9.3.18), we obtain the soliton solution, Bäcklund transformation of the KdV equation, and the corresponding inverse scattering problem.

In fact, from $w_8 = 0$, letting $y_8 = y$, we have

$$\begin{cases} y_x = -(2u + y^2 - \lambda), \\ y_t = -4\left[(u+\lambda)(2u + y^2 - \lambda) + \frac{1}{2}p - zy\right]. \end{cases} \tag{9.3.19}$$

The first equation of (9.3.19) is Riccati's equation. Letting

$$y = \psi_x/\psi, \tag{9.3.20}$$

we get

$$\psi_{xx} + (2u - x)\psi = 0. \tag{9.3.21}$$

This corresponds to the one-dimensional Schrödinger equation of the KdV equation. From the Pfaff form w_3, we have

$$y = -y_{3,x}.$$

By (9.3.20) and w_2, we have

$$y_3 = -\ln\psi.$$

Letting $\varphi = \psi_x$, we deduce from (9.3.20) that

$$y = \varphi/\psi.$$

Letting

$$w_9 \equiv \psi w_8 - \varphi w_3,$$
$$w_{10} \equiv -\psi w_3,$$

we have

$$\begin{cases} w_9 = d\varphi - (2u - \lambda)\psi dx + \{2z\varphi - [4(u+\lambda) \\ \qquad \times (2u - \lambda) + 2p]\psi\}dt, \\ w_{10} = d\varphi - \varphi dx - [2z\psi - 4(u+\lambda)\varphi]dt. \end{cases} \tag{9.3.22}$$

From (9.3.22), we see that $\psi_x = \varphi$ and $\varphi_x + (2u - \lambda)\varphi = 0$ are the first-order scattering equations of the KdV equation.

On the other hand, assume the KdV equation has another solution $u' = u'(u, z, p, y^i)$ and satisfies

$$\begin{cases} \alpha_1' = du' \wedge dt - z'dx \wedge dt, \\ \alpha_2' = dz' \wedge dt - p'dx \wedge dt, \\ \alpha_3' = -du' \wedge dx + dp' \wedge dt + 12u'z'dx \wedge dt. \end{cases} \tag{9.3.23}$$

By direct calculation, we get

$$u' = -u - y^2 + \lambda. \tag{9.3.24}$$

Since $u = 0$ is a solution of the KdV equation, $u_0' = -y^2 + \lambda$ must be the solution. From (9.3.19), we have

$$\begin{cases} y_x = -(y^2 - \lambda), \\ y_t = 4\lambda(y^2 - \lambda) = -4\lambda y_x. \end{cases} \tag{9.3.25}$$

Its analytic integral is $y = \lambda^{1/2}\tanh[\lambda^{1/2}(x - x_0 - 4\lambda t)]$ and u_0' is the analytic soliton solution.

From the Pfaff form w_7, we obtain

$$u = -y_{7,x} = -w_x.$$

Equation (9.3.24) can be written as

$$-w_x' = w_x - y^2 + \lambda = w_x + y_x - 2w_x. \tag{9.3.26}$$

After integrating and incorporating the integral constant in the potential, we have

$$y = w - w', \tag{9.3.27}$$

so (9.3.24) can finally be written as

$$-w_x' - w_x = u' + u = \lambda - (w' - w)^2. \tag{9.3.28}$$

Then, letting $\lambda = k^2$, by (9.3.25) as well as (9.3.27), we write the second equation of (9.3.19) as

$$w_t' + w_t = 4(u'^2 u^2 + u'u + u^2) + 2(w' - w)(z' - z). \tag{9.3.29}$$

Combining (9.3.28) and (9.3.29), we immediately obtain the Bäcklund transform of the KdV equation.

10 Global existence and blow up for the nonlinear evolution equations

10.1 Nonlinear evolution equations and the integral estimation method

Recently, with the development of the soliton problem and theory, a multitude of nonlinear evolution equations with soliton solution have attracted more and more attention, such as the Korteweg–de Vries (KdV) equation, nonlinear Schrödinger equation, regularized long wave (RLW) equation, and nonlinear Klein–Gordon equation. Apart from the important feature of having solitons, these equations have other obvious physical properties, such as the unity of dispersion and nonlinearity, some degree of volatility but also a certain degree of smoothness of their solutions, and the decay and dispersion of the solution as $t \to \infty$ (or $x \to \infty$). Because of the intimate connection between these equations and the physical problems, the solving methods and the theoretical research of their characteristics have already gone beyond the traditional research methods. For instance, the occurrence of the inverse scattering method, which is completely new, accurate, and very important, has opened up a new way for the theoretical research of differential equations, the Bäcklund transform method, and the extension structure method established by exterior differential forms in differential geometry and Lie groups. At the same time, we cannot imitate some traditional methods with regard to the theoretical research itself of such kinds of nonlinear partial differential equations. For example, for the KdV equation and the nonlinear Schrödinger equation, although their solutions have a good smoothness, we can only make estimates with the energy integration because of the nonexistence of the maximum value principle. However, an unusual aspect of this kind of integral estimate is that we must fully use the various conservation laws. As Lax put it: "The main feature of the KdV equation is the infinite number of conservation laws." In view of proving the existence and uniqueness of the global solution of this kind of nonlinear equations, we have the following methods. (1) Make a good integral a priori estimate and establish the local solution in $[0, t_1]$ by the use of various approximate methods, where t_1 depends on the initial data. Next, establish the local solution in $[t_1, t_2]$, where $t_2 - t_1$ depends on the norm $\|u(t_1)\|$; since we have the a priori estimate $\|u(t)\| \le$ constant, the solution can be extended from $t_1, t_2, \dots$ to any finite interval $[0, T]$. (2) The method of vanishing viscosity (or parabolic regularization method) goes as follows. Look for the global solution u_ε of the viscous approximate equation and use the uniform boundedness of u_ε and its derivatives with regard to the small parameter ε and let $\varepsilon \to 0$. Then we obtain the desired solution. (3) The functional analysis method consists in transforming the original equation to the standard differential operator form and using some known theorem of the differential operator to obtain the existence of the global solution, where it needs

https://doi.org/10.1515/9783110549638-010

to be validated specifically whether the existence conditions of the differential operator are satisfied. (4) The Galerkin approximate method makes use of the notion that we can use the uniform estimation of the Galerkin approximate solution to obtain the global solution. In these various methods of proving the existence of the global solution, whatever it is, the integral a priori estimate plays a decisive role. Indeed, we cannot obtain the global solution of all nonlinear evolution equations by use of the a priori estimate. For instance, the "blow up" phenomenon exists in some multi-dimensional nonlinear wave equations and the nonlinear Schrödinger equation, as the L_2 norm of the solution or its first-order derivative tends to infinity as $(t \to t_1)$ (t_1 is finite). However, we can obtain its global solution when the L_2 norm of the initial data is suitably small. All these problems have attracted the attention and interest of many people. At present, for the problems of definite solutions of this kind of nonlinear equation, we mainly focus on the periodic initial value problem or the initial value problem and less emphasis is put on the initial boundary value problem. Generally speaking, there exist many difficulties concerning the formulation of the boundary value problem and the existence study of the solution for this kind of equations (for example, the KdV equation). Therefore, the results on this subject are relatively scarce. For the initial value problem, we usually assume its solution tends to zero as $|x| \to \infty$, which is a reasonable requirement. For instance, for the KdV equation, we have proved that, as long as the initial data tend to zero with certain decay rate as $|x| \to \infty$, its solution decays with the corresponding rate. Of course, this requirement is not necessary. If the initial condition is approximated by the periodic boundary condition, the requirement could be substituted by another initial condition. We still make this assumption for the initial value problem. In addition, the following results, obtained for the definite solution problems, also hold for the periodic initial value problem or the initial value problem.

10.2 Periodic initial value problem and initial value problem for the KdV equation

The existence and uniqueness of the solution to the KdV equation were first obtained by Sjöberg [280]. He considered the following definite solution problem:

$$\begin{cases} u_t = uu_x + \delta u_{xxx}, & \delta \neq 0, \\ u(x,0) = f(x), & \forall x \in R, \\ u(x,t) = u(x+1,t), & \forall x, t, \end{cases} \tag{10.2.1}$$

and obtained the following result.

Theorem 10.2.1. *Assume $f(x)$ is a function with period 1 and its derivatives until the third derivative belong to L^2. Then, if $\delta \neq 0$, there exists a unique solution to problem* (10.2.1).

We use the following differential difference scheme corresponding to (10.2.1):

$$\begin{cases} \frac{\partial}{\partial t}u_N(x,t) = [u_N(x_r,t)D_0u_N(x_r,t) + D_0u_N^2(x_r,t)]/3 \\ \qquad\qquad + \delta D_+D_-^2u_N(x_r,t) \quad (r = 1,2,\dots,N), \\ u_N(x_r,0) = f(x_r) \qquad\qquad\qquad (r = 1,2,\dots,N), \\ u_N(x_r,t) = u_N(x_{r+N},t), \qquad\qquad \forall x,t, \end{cases} \tag{10.2.2}$$

where $h = \frac{1}{N}$, $x_r = rh$, and D_+, D_-, and D_0 represent the difference operators. We define

$$hD_+g(x_r) = g(x_{r+1}) - g(x_r),$$
$$hD_-g(x_r) = g(x_r) - g(x_{r-1}),$$
$$2hD_0g(x_r) = g(x_{r+1}) - g(x_{r-1}).$$

Then we can prove the local existence of the solution to problem (10.2.1). Using the three conservation laws of (10.2.1),

$$\int_0^1 u^2(x,t)dx = \int_0^1 f^2(x)dx = \alpha_1 = \text{constant}, \tag{10.2.3}$$

$$\int_0^1 \left(\frac{u^2}{3} - \delta u_x^2\right)dx = \int_0^1 \left(\frac{f^3}{3} - \delta f'^2(x)\right)dx = \alpha_2 = \text{constant}, \tag{10.2.4}$$

$$\int_0^1 (u^4 - 12\delta uu_x^2 + 36\delta^2 u_{xx}/5)dx$$
$$= \int_0^1 (f^4 - 12\delta ff'^2 + 36\delta^2 f''^2/5)dx = \alpha_3 = \text{constant} \tag{10.2.5}$$

and making a priori estimates of integration, we can prove the existence of the global solution for problem (10.2.1) and the uniqueness can easily be obtained by the energy inequality. Lax first proved the uniqueness of solution to the Cauchy problem of the KdV equation

$$u_t + uu_x + u_{xxx} = 0, \quad (-\infty < x < +\infty,\ t > 0), \tag{10.2.6}$$

$$u|_{t=0} = u_0(x), \quad (-\infty < x < +\infty), \tag{10.2.7}$$

over $(-\infty, +\infty)$ [153], where the solution refers to $u(\cdot,t) \in C^\infty\ (-\infty < x < +\infty)$ and where u and its derivatives to x all tend to zero ($|x| \to \infty$). Assume v is other solution of problem (10.2.6), (10.2.7). We have

$$\begin{cases} v_t + vv_x + v_{xxx} = 0, \\ v|_{t=0} = u_0(x) \end{cases}$$

and we let $w = u - v$, so we obtain the following linear equation of w:

$$w_t + uw_x + wv_x + w_{xxx} = 0.$$

We multiply the above equation by w and integrate with respect to $x \in (-\infty, +\infty)$. Then, after integration by parts, we get the following relation:

$$\frac{d}{dt}\frac{1}{2}\int_{-\infty}^{+\infty} w^2 dx + \int_{-\infty}^{+\infty}\left(v_x - \frac{1}{2}u_x\right)w^2 dx = 0. \tag{10.2.8}$$

Let $E(t) = \frac{1}{2}\int_{-\infty}^{+\infty} w^2 dx$, $\max|2v_x - u_x| = m$. From (10.2.8), we get

$$\frac{d}{dt}E(t) \le mE(t),$$

and then

$$E(t) \le E(0)e^{mt}.$$

Since $E(0) = 0$, we find $E(t) = 0$ $(t > 0)$. Thus, $w \equiv 0$. As described in [290], for the following periodic initial value problem of the KdV equation:

$$\begin{cases} u_t + uu_x + \mu u_{xxx} = 0, & (0 \le t \le T, 0 < x < 1), \\ u(x,0) = u_0(x), & (0 \le x \le 1), \\ u(0,t) = u(1,t), & (0 \le t \le T), \\ u_x(0,t) = u_x(1,t), & (0 \le t \le T), \\ u_{xx}(0,t) = u_{xx}(1,t), & (0 \le t \le T), \end{cases} \tag{10.2.9}$$

using the fourth-order small parameter method, that is, considering the solution $u_\varepsilon(x,t)$ of the following definite problem corresponding to problem (10.2.9):

$$u_{\varepsilon t} + uu_{\varepsilon x} + \mu u_{\varepsilon xxx} + \varepsilon u_{\varepsilon xxxx} = 0, \quad (\varepsilon > 0), \ (0 \le t \le T, \ 0 < x < 1), \tag{10.2.10}$$

$$u_\varepsilon(x,0) = u_{0\varepsilon}(x), \quad (0 \le x \le 1), \tag{10.2.11}$$

$$\frac{\partial^j u_\varepsilon}{\partial x^j}(0,t) = \frac{\partial^j u_\varepsilon}{\partial x^j}(1,t), \quad (0 \le t \le T), \ (j = 0,1,2,3), \tag{10.2.12}$$

it tends to the solution of (10.2.9) as $\varepsilon \to 0$. We assume $u_{0\varepsilon} \in C^\infty([0,1])$, such that

$$\frac{d^j u_{0\varepsilon}(0)}{dx^j} = \frac{d^j u_{0\varepsilon}(1)}{dx^j}, \quad \forall j \ge 0,$$

and we suppose $u_{0\varepsilon}$ converges weakly to u_0 in $H^1(\Omega)(\varepsilon \to 0)$. Here we use Ω to denote the interval $(0,1)$, while $H^s(\Omega)(s \ge 0$ and is integer) is the Sobolev space

$$\{v \mid v(x) \in L^2(\Omega), D^j v(x) \in L^2(\Omega), 0 \le j \le s\},$$

$$\|v\|_{H^s(\Omega)} = \left\{\sum_{j=0}^{s}\left\|\frac{\partial^j v}{\partial x^j}\right\|_{L^2(\Omega)}^2\right\}^{1/2}.$$

We know from [163] that the solution u_ε of problem (10.2.10)–(10.2.12) exists and satisfies

$$u_\varepsilon \in L^\infty(0,T;L^2(\Omega)) \cap L^2(0,T;H^2(\Omega)), \tag{10.2.13}$$

where $L^\infty(0,T;H^s)$ represents the function space defined over $[0,T]$ and takes values in H^s. We define $u(x,t)$ as the function of x belonging to H^s. For $t \in [0,T]$,

$$\sup_{0\le t\le T} \|u(\cdot,t)\|_s < \infty.$$

The term $L^2(0,T;H^s)$ represents the function space in which $u(x,t)$ as a function of x belongs to H^s for each $t \in [0,T]$ and $\int_0^T \|u(x,t)\|_s^2 dt < \infty$. From (10.2.13), we have

$$\begin{cases} \dfrac{\partial u_\varepsilon}{\partial x} \in L^2(0,T;H^1(\Omega)) \subset L^2(0,T;L^\infty(\Omega)), \\ u_\varepsilon \dfrac{\partial u_\varepsilon}{\partial x} \in L^2(0,T;L^2(\Omega)). \end{cases} \tag{10.2.14}$$

We deduce from (10.2.10) that

$$\frac{\partial u_\varepsilon}{\partial t} + \mu \frac{\partial^3 u_\varepsilon}{\partial x^3} + \varepsilon \frac{\partial^4 u_\varepsilon}{\partial x^4} = -u_\varepsilon \frac{\partial u_\varepsilon}{\partial x} \in L^2. \tag{10.2.15}$$

From this, the boundary conditions, and the smoothness theorem of linear equation, we deduce

$$\frac{\partial u_\varepsilon}{\partial t} \in L^2(Q), \quad u_\varepsilon \in L^2(0,T;H^4(\Omega)),$$
$$\text{where } Q = \Omega \times [0,T]. \tag{10.2.16}$$

We now make a priori estimates for the solution of problem (10.2.10)–(10.2.12).

Lemma 10.2.2. *If $u_0 \in L^2(\Omega)$, then we have*

$$\|u_\varepsilon\|_{L^\infty(0,T;L^2(\Omega))} \le c, \tag{10.2.17}$$

$$\sqrt{\varepsilon}\left\|\frac{\partial^2 u_\varepsilon}{\partial x^2}\right\|_{L^2(Q)} \le c, \tag{10.2.18}$$

where the constant c is independent of ε.

Proof. Multiplying (10.2.10) by u_ε and integrating with respect to x, we have, under the periodic conditions (10.2.12),

$$\frac{1}{2}\frac{d}{dt}\|u_\varepsilon(t)\|_{L^2}^2 + \varepsilon \int_\Omega \left(\frac{\partial^2 u_\varepsilon}{\partial x^2}\right)^2 dx = 0.$$

From this, we immediately get (10.2.17) and (10.2.18). □

Lemma 10.2.3. *For all functions $v(x) \in H^3(\Omega)$, we have*

$$\|v\|_{L^4(\Omega)} \le c\|v\|_{L^2(\Omega)}^{11/12}\left(\|v\|_{L^2(\Omega)} + \left\|\frac{d^3v}{dx^3}\right\|_{L^2(\Omega)}\right)^{\frac{1}{12}}, \tag{10.2.19}$$

$$\left\|\frac{dv}{dx}\right\|_{L^4(\Omega)} \le c\|v\|_{L^2(\Omega)}^{7/12}\left(\|v\|_{L^2(\Omega)} + \left\|\frac{d^3v}{dx^3}\right\|_{L^2(\Omega)}\right)^{\frac{5}{12}}. \tag{10.2.20}$$

Proof. By the interpolation representation in Chapter 1 of [164],

$$[H^3(\Omega), H^0(\Omega)]_{\frac{11}{12}} = H^{\frac{1}{4}}(\Omega), H^0(\Omega) = L^2(\Omega),$$

and $H^{\frac{1}{4}}(\Omega) \subset L^4(\Omega)$, we deduce

$$\|v\|_{L^4(\Omega)} \le c\|v\|_{H^{\frac{1}{4}}(\Omega)} \le c\|v\|_{H^3(\Omega)}^{\frac{1}{12}}\|v\|_{H^0(\Omega)}^{\frac{11}{12}},$$

that is, we get (10.2.19). Similarly, $[H^3(\Omega), H^0(\Omega)]_{\frac{7}{12}} = H^{\frac{5}{4}}(\Omega)$,

$$\left\|\frac{dv}{dx}\right\|_{L^4(\Omega)} \le c\|v\|_{H^{\frac{5}{4}}(\Omega)},$$

so we obtain (10.2.20). □

Lemma 10.2.4. *If $u_0(x) \in H^1(\Omega)$, then we have*

$$\left\|\frac{\partial u_\varepsilon}{\partial x}\right\|_{L^\infty(0,T;L^2(\Omega))} \le c, \tag{10.2.21}$$

$$\sqrt{\varepsilon}\left\|\frac{\partial^3 u_\varepsilon}{\partial x^3}\right\|_{L^2(Q)} \le c, \tag{10.2.22}$$

where the constant c is independent of ε.

Proof. Multiplying (10.2.10) by

$$\psi_1(u_\varepsilon) = u_\varepsilon^2 + 2\mu\frac{\partial^2 u_\varepsilon}{\partial x^2},$$

integrating with respect to x, and using the periodic boundary conditions, we get

$$\frac{d}{dt}\int_\Omega\left[\frac{1}{3}u_\varepsilon^3 - \mu\left(\frac{\partial u_\varepsilon}{\partial x}\right)^2\right]dx + \varepsilon\int_\Omega \frac{\partial^4 u_\varepsilon}{\partial x^4}\left(u_\varepsilon^2 + 2\mu\frac{\partial^2 u_\varepsilon}{\partial x^2}\right)dx = 0,$$

or

$$\frac{d}{dt}\int_\Omega\left[\frac{1}{3}u_\varepsilon^3 - \mu\left(\frac{\partial u_\varepsilon}{\partial x}\right)^2\right]dx - 2\varepsilon\int_\Omega \frac{\partial^3 u_\varepsilon}{\partial x^3}u_\varepsilon\frac{\partial u_\varepsilon}{\partial x}dx - 2\mu\varepsilon\int_\Omega\left(\frac{\partial^3 u_\varepsilon}{\partial x^3}\right)^2 dx = 0. \tag{10.2.23}$$

Then we get

$$\mu\frac{d}{dt}\left\|\frac{\partial u_\varepsilon}{\partial x}\right\|^2_{L^2(\Omega)} + 2\mu\varepsilon\left\|\frac{\partial^3 u_\varepsilon}{\partial x^3}\right\|^2_{L^2(\Omega)} = \frac{1}{3}\frac{d}{dt}\int_\Omega u_\varepsilon^3 dx - 2\varepsilon\int_\Omega u_\varepsilon\frac{\partial u_\varepsilon}{\partial x}\frac{\partial^3 u_\varepsilon}{\partial x^3}dx. \tag{10.2.24}$$

Integrating (10.2.24) with respect to t and dividing by μ yields

$$\begin{aligned}\left\|\frac{\partial u_\varepsilon}{\partial x}\right\|^2_{L^2(\Omega)} &+ 2\varepsilon\int_0^t\left\|\frac{\partial^3 u_\varepsilon}{\partial x^3}(\sigma)\right\|^2_{L^2} d\sigma \\ &= \left\|\frac{du_0}{dx}\right\|^2_{L^2} + \frac{1}{3\mu}\int_\Omega u_\varepsilon^3(x,t)dx - \frac{1}{3\mu}\int_\Omega u_0^3(x)dx \\ &\quad - \frac{2\varepsilon}{\mu}\int_0^t\int_\Omega u_\varepsilon\frac{\partial u_\varepsilon}{\partial x}\frac{\partial^3 u_\varepsilon}{\partial x^3}dxd\sigma.\end{aligned} \tag{10.2.25}$$

Since

$$\begin{aligned}\left|\int_\Omega u_\varepsilon^3(t)dt\right| &\le \|u_\varepsilon(t)\|_{L^\infty(\Omega)}\|u_\varepsilon(t)\|^2_{L^2(\Omega)} \\ &\le c_1\|u_\varepsilon(t)\|_{L^\infty(\Omega)} \quad \text{(due to (10.2.17))} \\ &\le c_2\|u_\varepsilon(t)\|^{\frac{1}{2}}_{L^2(\Omega)}\left(\|u_\varepsilon(t)\|_{L^2(\Omega)} + \left\|\frac{\partial u_\varepsilon(t)}{\partial x}\right\|_{L^2(\Omega)}\right)^{1/2} \\ &\le c_3\left(1 + \left\|\frac{\partial u_\varepsilon(t)}{\partial x}\right\|_{L^2(\Omega)}\right)^{1/2} \quad \text{(due to (10.2.17))} \\ &\le c_4 + \frac{3|\mu|}{2}\left\|\frac{\partial u_\varepsilon(t)}{\partial x}\right\|^2_{L^2(\Omega)},\end{aligned}$$

$$\begin{aligned}\left|\int_\Omega u_\varepsilon\frac{\partial u_\varepsilon}{\partial x}\cdot\frac{\partial^3 u_\varepsilon}{\partial x^3}dx\right| &\le \|u_\varepsilon(t)\|_{L^4(\Omega)}\left\|\frac{\partial u_\varepsilon}{\partial x}\right\|_{L^4(\Omega)}\left\|\frac{\partial^3 u_\varepsilon}{\partial x^3}\right\|_{L^2(\Omega)} \\ &\le c_5\|u_\varepsilon(t)\|^{3/2}_{L^2}\left(\|u_\varepsilon(t)\|_{L^2} + \left\|\frac{\partial^3 u_\varepsilon(t)}{\partial x^3}\right\|_{L^2}\right)^{1/2} \\ &\quad\cdot\left\|\frac{\partial^3 u_\varepsilon(t)}{\partial x^3}\right\|_{L^2(\Omega)} \quad \text{(by (10.2.19), (10.2.20))} \\ &\le c_6\left(1 + \left\|\frac{\partial^3 u_\varepsilon(t)}{\partial x^3}\right\|_{L^2}\right)^{1/2}\left\|\frac{\partial^3 u_\varepsilon(t)}{\partial x^3}\right\|_{L^2(\Omega)} \\ &\le c_7 + \frac{|\mu|}{2}\left\|\frac{\partial^3 u_\varepsilon(t)}{\partial x^3}\right\|^2_{L^2(\Omega)} \quad \text{(by (10.2.17))}.\end{aligned}$$

(Note that, for $v \in H^1(\Omega)$, we have $\|v\|_{L^\infty(\Omega)} \le c\|v\|^{1/2}_{L^2}(\|v\|_{L^2} + \|\frac{\partial u}{\partial x}\|_{L^2})^{1/2}$.) Considering the last inequality, from (10.2.25) we deduce

$$\left\|\frac{\partial u_\varepsilon(t)}{\partial x}\right\|^2_{L^2(\Omega)} + 2\varepsilon\int_0^t\left\|\frac{\partial^3 u_\varepsilon(\sigma)}{\partial x^3}\right\|^2_{L^2(\Omega)} d\sigma$$

$$\leq \left\| \frac{du_0}{dx} \right\|_{L^2(\Omega)}^2 + \frac{1}{3|\mu|} \int_\Omega |u_0|^3 dx$$
$$+ \frac{1}{2} \left\| \frac{\partial u_\varepsilon(t)}{\partial x} \right\|_{L^2(\Omega)}^2 + \varepsilon \int_0^t \left\| \frac{\partial^3 u_\varepsilon(\sigma)}{\partial x^3} \right\|_{L^2(\Omega)}^2 d\sigma + c. \qquad (10.2.26)$$

Thus, we obtain

$$\left\| \frac{\partial^3 u_\varepsilon(t)}{\partial x^3} \right\|_{L^2(\Omega)}^2 + \varepsilon \int_0^t \left\| \frac{\partial^3 u_\varepsilon(\sigma)}{\partial x^3} \right\|_{L^2(\Omega)}^2 d\sigma \leq c.$$

This immediately yields (10.2.21) and (10.2.22). □

Using equation (10.2.10), we have

$$\frac{\partial u_\varepsilon}{\partial t} = -u_\varepsilon \frac{\partial u_\varepsilon}{\partial x} - \mu \frac{\partial^3 u_\varepsilon}{\partial x^3} - \varepsilon \frac{\partial^4 u_\varepsilon}{\partial x^4}.$$

By (10.2.21) and (10.2.22), we have

$$\frac{\partial u_\varepsilon}{\partial t} \quad \text{uniformly bounded in } L^2(0, T; H^{-2}(\Omega)). \qquad (10.2.27)$$

Thus, by (10.2.17), (10.2.18), (10.2.21), (10.2.22), and (10.2.17), we choose a subsequence of u_ε, still denoted by u_ε, such that

$$u_\varepsilon \to u \quad \text{weakly in } L^\infty(0, T; L^2(\Omega)),$$
$$\frac{\partial u_\varepsilon}{\partial x} \to \frac{\partial u}{\partial x} \quad \text{weakly in } L^\infty(0, T; L^2(\Omega)),$$
$$\frac{\partial u_\varepsilon}{\partial t} \to \frac{\partial u}{\partial t} \quad \text{weakly in } L^\infty(0, T; H^{-2}(\Omega)).$$

From the first and second results, we infer that $u_\varepsilon \to u$ weakly in $L^\infty(0, T; H^1(\Omega))$. From the second result, we deduce that $u_\varepsilon \to u$ strongly in $L^\infty(0, T; L^2(\Omega))$. Using (10.2.13), taking the limit in (10.2.10), (10.2.11), and (10.2.12), and applying

$$u_\varepsilon \frac{\partial u_\varepsilon}{\partial x} \to u \frac{\partial u}{\partial x} \qquad (10.2.28)$$

weakly in $L^\infty(0, T; L^1(\Omega))$, it is not difficult to obtain equation (10.2.10), $\to u_t + uu_x + \mu u_{xxx} = 0$, so we get the desired solution. We have the following theorem.

Theorem 10.2.5. *Suppose that $\mu \in R$, $\mu \neq 0$, $u_0(x) \in H^1(\Omega)$, and $u_0(0) = u_0(1)$. Then there exists a function $u(x, t)$, $u \in L^\infty(0, T; H^1(\Omega))$, that satisfies*

$$u, u_x \in L^\infty(0, T; L^2(\Omega)), \qquad (10.2.29)$$
$$u_t + uu_x + \mu u_{xxx} = 0, \qquad (10.2.30)$$
$$u(x, 0) = u_0(x), \qquad (10.2.31)$$
$$u(0, t) = u(1, t). \qquad (10.2.32)$$

Theorem 10.2.6. *Assume that* $\mu \in R, \mu \neq 0, u_0(x) \in H^2(\Omega)$, *and* $\frac{d^j u(0)}{dx^j} = \frac{d^j u(1)}{dx^j}$ $(j = 0, 1)$. *Then the function of the definite solution problem* (10.2.9) *is unique.*

Remark. If $u_0 \in L^\infty(\bar{\Omega})$ and

$$\frac{d^j u_0(0)}{dx^j} = \frac{d^j u_0(1)}{dx^j}, \quad \forall j \geq 0,$$

then the solution of (10.2.9) satisfies $u \in L^\infty(\bar{\Omega})$.

As described in [31], for the initial value problem

$$\begin{cases} u_t + uu_x + u_{xxx} - \varepsilon u_{xxt} = 0, & (t > 0, -\infty < x < +\infty), \\ u(x,0) = g(x), & (-\infty < x < +\infty), \end{cases} \tag{10.2.33}$$

the uniformly a priori estimate of the solution u_ε with respect to ε was established and the existence and uniqueness of the solution for the initial value problem of the KdV equation

$$\begin{aligned} &u_t + uu_x + u_{xxx} = 0, \quad (-\infty < x < +\infty, t > 0), \\ &u(x,0) = g(x), \quad (-\infty < x < +\infty) \end{aligned} \tag{10.2.34}$$

were proved. For the existence and uniqueness theorem of solution for all kinds of definite solution problems to the more general class of KdV equations, complex KdV equations, and the higher-order KdV equations, we refer the reader to [31, 71, 104, 110, 161, 294, 320].

10.3 Periodic initial value problem for a class of nonlinear Schrödinger equations

We consider the following periodic initial value problem for a class of nonlinear Schrödinger equations:

$$\begin{aligned} &iu_{jt} - u_{jx} + \beta(x)q(\sigma_{21}|u_1|^2 + \sigma_{31}|u_2|^2)u_j \\ &\quad + k_j(x)u_j = 0, \quad (j = 1, 2,\ 0 < x < 2\pi,\ t > 0), \end{aligned} \tag{10.3.1}$$

$$u_j|_{t=0} = u_0^j(x), \quad 0 \leq x \leq 2\pi,\ j = 1, 2, \tag{10.3.2}$$

$$u_j(x,t) = u_j(x + 2\pi, t), \quad \forall x, t \geq 0,\ j = 1, 2, \tag{10.3.3}$$

where $i = \sqrt{-1}$, σ_{21}, and σ_{31} are positive constants and $\beta(x)$ is a bounded real function with period 2π. $q(s) \geq 0$, $s \in [0, +\infty)$, and $k_j(x)$ $(j = 1, 2)$ are bounded real functions with period 2π. $u_j(x,t)$ are the unknown functions. $u_0^j(x)$ $(j = 1, 2)$ are the given complex-valued functions with period 2π. We define the inner

$$(f, g) = \int_0^{2\pi} f\bar{g}dx, \quad a(u, v) = \int_0^{2\pi} \frac{\partial u}{\partial x}\frac{\partial \bar{v}}{\partial x}dx.$$

We first make an a priori estimate for the solution of the problem (10.3.1)–(10.3.3).

Lemma 10.3.1. *If the following conditions are satisfied:* (i) $\beta(x), q(s), k_j(x)$ *are real functions and* (ii) $u_0^j(x) \in L^2$, *then the solutions* $u_j(x,t)$ *of problem* (10.3.1)–(10.3.3) *satisfy the following equality:*

$$\|u_j(t)\|_{L^2}^2 = \|u_0^j\|_{L^2}^2 \quad (j = 1, 2). \tag{10.3.4}$$

Proof. Multiplying (10.3.1) by $\bar{u}_j$ and integrating with respect to x, we get

$$i(u_{jt}, u_j) + a(u_j, u_j) + (\beta(x)q(\sigma_{21}|u_1|^2 + \sigma_{31}|u_2|^2)u_j, u_j) + (k_j(x)u_j, u_j) = 0. \tag{10.3.5}$$

Since $a(u_j, u_j) \geq 0$, $\beta(x)$ is a real function and

$$(\beta q u_j, u_j) = \int_0^{2\pi} \beta(x)q|u_j|^2 dx,$$
$$(k_j u_j, u_j) = \int_0^{2\pi} k_j(x)q|u_j|^2 dx.$$

Then, taking the imaginary part of (10.3.5), we immediately get (10.3.4). □

Lemma 10.3.2. *If the following conditions hold:* (i) σ_{21} *and* σ_{31} *are real numbers and* $\beta(x)$, $k_j(x)$, *and* $q(s)$ *are all real functions and* (ii) $u_0^j(x) \in L^2$, $\beta(x)$ *and* $Q(\sigma_{21}|u_0^1|^2 + \sigma_{31}|u_0^2|^2) \in L^1$, *where* $Q(s) = \int_0^s q(z)dz$, *then the solution of problem* (10.3.1)–(10.3.3) *satisfies*

$$\begin{aligned}
&\sigma_{21}\|u_{1x}\|_{L^2}^2 + \sigma_{31}\|u_{2x}\|_{L^2}^2 + \int_0^{2\pi} \beta(x)Q(\sigma_{21}|u_1|^2 + \sigma_{31}|u_2|^2)dx \\
&\quad + \int_0^{2\pi} [k_1(x)\sigma_{21}|u_1|^2 + k_2(x)\sigma_{31}|u_2|^2]dx \\
&= \sigma_{21}\|u_{0x}^1\|_{L^2}^2 + \sigma_{31}\|u_{0x}^2\|_{L^2}^2 + \int_0^{2\pi} \beta(x)Q(\sigma_{21}|u_0^1|^2 + \sigma_{31}|u_0^2|^2)dx \\
&\quad + \int_0^{2\pi} [k_1(x)\sigma_{21}|u_0^1|^2 + k_2(x)\sigma_{31}|u_0^2|^2]dx.
\end{aligned} \tag{10.3.6}$$

Proof. Multiplying (10.3.1) by $\bar{u}_{jt}$ and integrating with respect to x, we have

$$i(u_{jt}, u_{jt}) + (u_{jx}, u_{jxt}) + (\beta(x)q(\sigma_{21}|u_1|^2 + \sigma_{31}|u_2|^2)u_j, u_{jt}) + (k_j u_j, u_{jt}) = 0. \tag{10.3.7}$$

Because

$$\operatorname{Re}(u_{jx}, u_{jxt}) = \frac{1}{2}\frac{d}{dt}\|u_{jx}\|_{L^2}^2,$$
$$\operatorname{Re}(k_j u_j, u_{jt}) = \frac{d}{dt}\frac{1}{2}\int_0^{2\pi} k_j(x)|u_j|^2 dx,$$

$$\begin{aligned}
&\mathrm{Re}(\beta(x)q(\sigma_{21}|u_1|^2+\sigma_{31}|u_2|^2)u_1,\sigma_{21}u_{1t})+\mathrm{Re}(\beta(x)q(\sigma_{21}|u_1|^2+\sigma_{31}|u_2|^2)u_2,\sigma_{31}u_{2t})\\
&\quad=\frac{1}{2}\int_0^{2\pi}\beta(x)q(\sigma_{21}|u_1|^2+\sigma_{31}|u_2|^2)\frac{\partial}{\partial t}(\sigma_{21}|u_1|^2+\sigma_{31}|u_2|^2)dx\\
&\quad=\frac{d}{dt}\frac{1}{2}\int_0^{2\pi}\beta(x)Q(\sigma_{21}|u_1|^2+\sigma_{31}|u_2|^2)dx.
\end{aligned}$$

Taking the imaginary part of (10.3.7), multiplying for $j=1$ by σ_{21}, multiplying for $j=2$ by σ_{31}, and adding the resultants, we obtain

$$\begin{aligned}
&\frac{1}{2}\frac{d}{dt}(\sigma_{21}\|u_1\|_{L^2}^2+\sigma_{31}\|u_2\|_{L^2}^2)+\frac{1}{2}\frac{d}{dt}\int_0^{2\pi}\beta(x)Q(\sigma_{21}|u_1|^2+\sigma_{31}|u_2|^2)dx\\
&\quad+\frac{1}{2}\frac{d}{dt}\int_0^{2\pi}[k_1(x)\sigma_{21}|u_1|^2+k_2(x)\sigma_{31}|u_2|^2]dx=0.
\end{aligned}$$

That is, we obtain (10.3.6). □

Lemma 10.3.3. *If the conditions of Lemma* 10.3.2 *hold,* $k_j(x)$ $(j=1,2)$ *are bounded real functions,* $q(s)\ge 0$, $\beta(x)\ge 0$, *and they are bounded,* $\sigma_{21}>0$, *and* $\sigma_{31}>0$, *then, for the solution of problem* (10.3.1)–(10.3.3), *we have the following estimates:*

$$\begin{aligned}
&\|u_{1x}\|_{L^2}^2\le c_1,\quad \|u_{2x}\|_{L^2}^2\le c_2,\\
&\int_0^{2\pi}\beta(x)Q(\sigma_{21}|u_1|^2+\sigma_{31}|u_2|^2)dx\le c_3,
\end{aligned}\tag{10.3.8}$$

where the constants c_1, c_2 *only depend on the initial functions and their derivatives.*

Proof. By (10.3.6) and the lemma conditions, we immediately get the results. □

Corollary 10.3.4. *We have*

$$\sum_{j=1}^{2}\|u_j\|_{L^\infty}\le c_4,\tag{10.3.9}$$

where the constant c_4 *only depends on the initial functions and their derivatives.*

Proof. From the conclusion of the lemma and the Sobolev inequality, (10.3.9) immediately follows. □

Lemma 10.3.5. *If the conditions of Lemma* 10.3.3 *hold and we assume that* $u_0^j(x)\in H^2$ $(j=1,2)$, *then the solution of problem* (10.3.1)–(10.3.3) *satisfies the following estimates:*

$$\sup_{0\le t\le T}\|u_{1t}\|_{L^2}^2\le c_5,\quad \sup_{0\le t\le T}\|u_{2t}\|_{L^2}^2\le c_6,\tag{10.3.10}$$

where the constants c_5, c_6 *only depend on the initial functions and their derivatives up to the second order.*

Proof. Differentiating equation (10.3.1) with respect to t, multiplying by $\bar{u}_{jt}$, and integrating, we get

$$i(u_{jtt}, u_{jt}) + (u_{jxt}, u_{jxt}) + \left(\beta(x)\frac{d}{dt}u_j q(\sigma_{21}|u_1|^2 + \sigma_{31}|u_2|^2), u_{jt}\right) + (k_j u_{jt}, u_{jt}) = 0. \quad (10.3.11)$$

Because

$$\begin{aligned}
&\left(\beta(x)\frac{d}{dt}u_j q(\sigma_{21}|u_1|^2 + \sigma_{31}|u_2|^2), u_{jt}\right) \\
&\quad = \int_0^{2\pi} \beta(x) q(\sigma_{21}|u_1|^2 + \sigma_{31}|u_2|^2)|u_{jt}|^2 dx + \int_0^{2\pi} \beta(x) q' \frac{\partial}{\partial t}(\sigma_{21}|u_1|^2 + \sigma_{31}|u_2|^2) u_j \bar{u}_{jt} dx, \\
&\frac{\partial}{\partial t}|u_j(t)|^2 = u_{jt}\bar{u}_j + u_j \bar{u}_{jt}.
\end{aligned}$$

Therefore,

$$\begin{aligned}
&\left|\int_0^{2\pi} \beta(x) q' \frac{\partial}{\partial t}(\sigma_{21}|u_1|^2 + \sigma_{31}|u_2|^2) u_j \bar{u}_{jt} dx\right| \\
&\quad \le c_1 \|q'(\sigma_{21}|u_1|^2 + \sigma_{31}|u_2|^2)\|_{L^\infty} \int_0^{2\pi} dx [|u_{1t}| \cdot |\bar{u}_1| + |u_2| \cdot |u_{1t}|]|u_j||u_{jt}| \\
&\quad \le c_2 [\|u_{1t}\|_{L^2}^2 + \|u_{2t}\|_{L^2}^2].
\end{aligned}$$

Taking the imaginary part of (10.3.11), we obtain

$$\frac{d}{dt}[\|u_{1t}(t)\|_{L^2}^2 + \|u_{2t}(t)\|_{L^2}^2] \le c_3 [\|u_{1t}(t)\|_{L^2}^2 + \|u_{2t}(t)\|_{L^2}^2].$$

By the Grönwall inequality and the lemma conditions, we immediately get (10.3.10). □

Now we define the generalized solution of the definite solution problem (10.3.1)–(10.3.3). The space 2π-periodic functions $u_j(x,t) \in L^\infty(0,T;H^1)$, $u_{jt} \in L^\infty(0,T;L^2)$ $(j = 1,2)$, and $Q(\sigma_{21}|u_0^1|^2 + \sigma_{31}|u_0^2|^2) \in L^1$ (where $Q(s) = \int_0^s q(z)dz$) are called the generalized solutions of the definite solution problem (10.3.1)–(10.3.3), if the integral equality is satisfied, so we have

$$i(u_{jt}, v_j) + (u_{jx}, v_{jx}) + (\beta(x) u_j q(\sigma_{21}|u_1|^2 + \sigma_{31}|u_2|^2), v_j) + (k_j u_j, v_j) = 0,$$
$$\forall v_j(x) \in H^1,\ t \ge 0,\ j = 1,2, \quad (10.3.12)$$
$$(u_j|_{t=0}, v_j) = (u_0^j(x), v_j), \quad (j = 1,2). \quad (10.3.13)$$

By the Galerkin approximate method or writing (10.3.1) as an integral equation, we have

$$\begin{aligned}
u^j(x,t) = S(t)u_0^j(x) + \int_0^t S(t-\tau)[q(\sigma_{21}|u_1(x,\tau)|^2 \\
+ \sigma_{31}|u_2(x,\tau)|^2) u_j + k_j(x) \cdot u_j(x,\tau)] d\tau,
\end{aligned}$$

where $S(t) = \frac{1}{\sqrt{2\pi i t}} e^{\frac{-x^2 t}{4\pi i}}$. Using the contraction mapping principle, we can easily get the local existence of solution to problem (10.3.1)–(10.3.3). Based on above a priori estimates, we get the following theorem.

Theorem 10.3.6. *If the following conditions are satisfied:* (i) $\sigma_{21} \geq 0$, $\sigma_{31} \geq 0$, *and the* 2π*-period real function* $\beta \geq 0$, (ii) $q(s)$ *is a real function,* $q(s) \in C^1$, $q(s) \geq 0$, $s \in [0, \infty)$, *and* $k_j(x)$ *are bounded real functions with period* 2π, *and* (iii) u_0^j *are periodic complex-valued functions and* $u_0^j(x) \in H^2$, *then the generalized solution of the definite solution problem* (10.3.1)–(10.3.3) *exists.*

Theorem 10.3.7. *If* $q(s) \in C^1$, $s \in [0, \infty)$, *and* $k_j(x)$ *are bounded functions, then the generalized solution of the definite solution problem* (10.3.1)–(10.3.3) *is unique.*

Proof. Assume that there are two pairs of generalized solutions, u_j, z_j $(j = 1, 2)$, of (10.3.1)–(10.3.3). Let $w_j = u_j - z_j$, $j = 1, 2$. Then we obtain from (10.3.12)

$$\begin{aligned} & i(w_{jt}, v_j) + (w_{jx}, w_{jx}) + (\beta(x) u_j q(\sigma_{21}|u_1|^2 + \sigma_{31}|u_2|^2) - \beta(x) z_j q(\sigma_{21}|z_1|^2 \\ & \quad + \sigma_{31}|z_2|^2), v_j) + (k_j w_j, v_j) = 0, \quad v_j \in H^1,\ t > 0, \qquad (10.3.14) \\ & w_j|_{t=0} = 0. \qquad (10.3.15) \end{aligned}$$

In particular, choose $v_i = w_i$ in (10.3.14). Because

$$\begin{aligned} & q(\sigma_{21}|u_1|^2 + \sigma_{31}|u_2|^2) u_j - q(\sigma_{21}|z_1|^2 + \sigma_{31}|z_2|^2) z_j \\ & \quad = q'(\tilde{z})[\sigma_{21}(|u_1|^2 - |z_1|^2) + \sigma_{31}(|u_2|^2 - |z_2|^2)] u_j \\ & \quad\quad + q(\sigma_{21}|z_1|^2 + \sigma_{31}|z_2|^2)(u_j - z_j), \end{aligned}$$

where $\tilde{z}$ is between $|u_1|^2 + |u_2|^2$ and $|z_1|^2 + |z_2|^2$, we have

$$\begin{aligned} & |(\beta(x) q(\sigma_{21}|u_1|^2 + \sigma_{31}|u_2|^2) u_j - \beta(x) q(\sigma_{21}|z_1|^2 + \sigma_{31}|z_2|^2) z_j, w_j)| \\ & \quad \leq \max|\beta(x)| \Bigg[(|\sigma_{21}| + |\sigma_{31}|) \|q'(\tilde{z})\|_{L^\infty} \|u_j\|_{L^\infty} \sum_{j=1}^{2} (\|u_j\|_{L^\infty} + \|z_j\|_{L^\infty}) \\ & \quad\quad + \|q(\sigma_{21}|z_1|^2 + \sigma_{31}|z_2|^2)\|_{L^\infty} \Bigg] \left(\sum_{k=1}^{2} |w_k|, |w_j| \right). \end{aligned}$$

Taking the real part of (10.3.14) and summing with respect to j, we get

$$\frac{1}{2} \frac{d}{dt} \sum_{j=1}^{2} \|w_j\|_{L^2}^2 \leq c \sum_{j=1}^{2} \|w_j\|_{L^2}^2.$$

By the Gronwall inequality and $w_j(0) = 0$, we immediately get $u_j = z_j$. □

Remark 10.3.8. For the initial value and boundary value problems of nonlinear Schrödinger equations (10.3.1), where its solution in addition to satisfying (10.3.1) also satisfies the initial condition (10.3.2) and

$$u_j|_{x=0} = u_j|_{x=1} = 0, \tag{10.3.16}$$

the conclusion of the above Theorem 10.3.6 and Theorem 10.3.7 are still valid.

Remark 10.3.9. If the initial functions and the coefficients of equation have a higher smoothness, the classical global solution of the problem (10.3.1)–(10.3.3) can be obtained by using the method of taking the difference quotient of the equation.

The well-posedness of the global solution for a more generalized class of nonlinear Schrödinger equations, the multi-dimensional nonlinear Schrödinger equation, and the system of nonlinear Schrödinger equations of the integral type is described in [90, 103, 105, 243, 282, 296, 322, 323].

10.4 Initial value problem for the nonlinear Klein–Gordon equation

We now consider the initial value problem for the following nonlinear Klein–Gordon equation:

$$\begin{cases} \dfrac{\partial^2 u}{\partial t^2} - \Delta u + m^2 u = -\lambda |u|^2 u, & x \in R^3,\ t > 0, \\ u(x,0) = f(x), & x \in R^3, \\ \dfrac{\partial u}{\partial t}(x,0) = g(x), & x \in R^3, \end{cases} \tag{10.4.1}$$

where $m > 0$, $\lambda > 0$, $\Delta \equiv \frac{\partial^2}{\partial x_1^2} + \frac{\partial^2}{\partial x_2^2} + \frac{\partial^2}{\partial x_3^2}$. We will use the functional method of the abstract differential operator to prove the existence of the global solution of the initial value problem (10.4.1).

We first turn problem (10.4.1) into the first-order system of the variable t. We have

$$\begin{cases} \dfrac{\partial v}{\partial t} - \Delta u + m^2 u = -\lambda |u|^2 u, \\ \dfrac{\partial u}{\partial t} = v, \\ u(x,0) = f(x), \quad v(x,0) = g(x), \end{cases}$$

or

$$\begin{cases} \dfrac{d\varphi(t)}{dt} - \begin{pmatrix} 0 & I \\ \Delta - m^2 & 0 \end{pmatrix} \varphi(t) = J(\varphi(t)), \\ \varphi(x,0) = \begin{pmatrix} f(x) \\ g(x) \end{pmatrix}, \end{cases} \tag{10.4.2}$$

where

$$\varphi = \begin{pmatrix} u \\ v \end{pmatrix}, \quad J(\varphi) = \begin{pmatrix} 0 \\ -\lambda|u|^2 u \end{pmatrix}$$

and I is the unit matrix.

We will use the general Hilbert space theorem to prove the existence and uniqueness of the global solution for problem (10.4.2). We first choose the Hilbert space $\mathcal{H} = L^2(R^3)$ and let $B^2 = -\Delta + m^2$. It is easy to know that B^2 is closed. We use $\mathcal{H}_B$ to denote the direct sum $\mathcal{H}_B = D(B) \oplus D(\mathcal{H})$, which has the following inner product:

$$(\langle u, v\rangle, \langle u, v\rangle)_B \equiv (Bu, Bu) + (v, v).$$

Letting

$$A = i\begin{pmatrix} 0 & I \\ -B^2 & 0 \end{pmatrix}, \quad i = \sqrt{-1}, \tag{10.4.3}$$

it is easy to verify that A is a symmetric operator in $\mathcal{H}_B$ and has the domain of definition $D \equiv D(B^2) \oplus D(B)$. A is also closed. We further write (10.4.2) into the following operator equation form:

$$\begin{cases} \dfrac{d\varphi}{dt} = -iA\varphi + J(\varphi), \\ \varphi(0) = \varphi_0 = \langle f(x), g(x)\rangle. \end{cases} \tag{10.4.4}$$

Now we first estimate the solution of (10.4.1).

Lemma 10.4.1. *Assume that $u \in C_0^\infty(R^3)$. Then we have*

$$\|u\|_{L^6} \le k\|Bu\|_{L^2}. \tag{10.4.5}$$

Proof. Denote $\frac{\partial u}{\partial x_i}$ as u_{x_i}. Then we have

$$|u(x)|^4 \le 4\int |u_{x_i} u^3| dx_i.$$

Integrating with respect to x_j for fixed $j \ne i$, we get

$$|u(x)|^6 \le K\Big(\int |u_{x_1} u^3| dx_1\Big)^{1/2}\Big(\int |u_{x_2} u^3| dx_2\Big)^{1/2}\Big(\int |u_{x_3} u^3| dx_3\Big)^{1/2}.$$

Integrating the above inequality and applying the Schwartz inequality, we get

$$\int_{R^3} |u|^6 dx \le K\Big(\int_{R^3} |u_{x_1} u^3| dx\Big)^{1/2}\Big(\int_{R^3} |u_{x_2} u^3| dx\Big)^{1/2}\Big(\int_{R^3} |u_{x_3} u^3| dx\Big)^{1/2}$$

$$\le K\left(\int_{R^3}|u|^6dx\right)^{3/4}\left(\int_{R^3}|u_{x_1}|^2dx\right)^{1/4}\left(\int_{R^3}|u_{x_2}|^2dx\right)^{1/4}\left(\int_{R^3}|u_{x_3}|^2dx\right)^{1/4}.$$

Then it is easy to deduce

$$\begin{aligned}\left(\int_{R^3}|u|^6dx\right)^{1/6} &\le K(\|u_{x_1}\|_{L^2}+\|u_{x_2}\|_{L^2}+\|u_{x_3}\|_{L^2})\\ &= K(\|k_1\hat{u}\|_{L^2}+\|k_2\hat{u}\|_{L^2}+\|k_3\hat{u}\|_{L^2})\\ &\le K\|(\Sigma k_i^2+m^2)^{1/2}\hat{u}\|_{L^2}=K\|Bu\|_{L^2},\end{aligned}$$

where

$$\hat{u}(t,k)=\frac{1}{2\pi^{3/2}}\int_{R^3}e^{-ix\cdot k}u(x,t)dx,$$
$$\left(x\cdot k=\sum_{i=1}^{3}x_i\cdot k_i\right).$$

□

Lemma 10.4.2. *Assume that $u_1,u_2,u_3\in D(B)$. Then*

$$\|u_1u_2u_3\|_{L^2}\le K\|Bu_1\|_{L^2}\|Bu_2\|_{L^2}\|Bu_3\|_{L^2}. \tag{10.4.6}$$

Proof. Let $u\in D(B)$. Because B is essentially self-conjugate in $C_0^\infty(R^3)$, we choose a sequence of functions $u_n\in C_0^\infty(R^3)$ such that $u_n\xrightarrow{L^2}u$ and $Bu_n\xrightarrow{L^2}Bu$. Choosing a subsequence of u_n, still denoted by u_n, u_n converges to u point-wise. Since

$$\begin{aligned}\|u_n^3-u_m^3\|_{L^2} &= \|(u_n-u_m)(u_n^2+u_nu_m+u_m^2)\|_{L^2}\\ &\le K\|u_n-u_m\|_{L^6}\|(u_n^2+u_nu_m+u_m^2)\|_{L^3}\\ &\le K\|u_n-u_m\|_{L^6}(\|u_n\|_{L^6}^2+\|u_n\|_6\cdot\|u_m\|_{L^6}\\ &\quad+\|u_m\|_{L^6}^2)\le K\|Bu_n-Bu_m\|_{L^2}(\|Bu_m\|_{L^2}^2\\ &\quad+\|Bu_n\|_{L^2}\|Bu_m\|_{L^2}+\|Bu_m\|_{L^2}^2),\end{aligned}$$

$\{u_n^3\}$ is a Cauchy sequence in L^2 and since it converges to u^3 point-wise, $u^3\in L^2$. Passing to the limit in the above inequality, we have

$$\|u\|_{L^6}^3=\|u^3\|_{L^2}\le K\|Bu\|_{L^2}^3.$$

Applying the Hölder inequality twice, the conclusion of the lemma immediately follows. □

Lemma 10.4.3. *For all $\varphi_1,\varphi_2\in\mathcal{H}$, J satisfies*

$$\|J(\varphi_1)\|_{L^2}\le K\|\varphi_1\|_{L^2}^3,$$
$$\|J(\varphi_1)-J(\varphi_2)\|_{L^2}\le C(\|\varphi_1\|_{L^2},\|\varphi_2\|_{L^2})\|\varphi_1-\varphi_2\|_{L^2}.$$

Proof. Letting $\varphi_i = \langle u_i, v_i\rangle$, from Lemma 10.4.2 we deduce

$$\begin{aligned}
\|J(\varphi_1)\|_{L^2} &= \|\lambda u_1^2\bar{u}_1\|_{L^2} \le K\|Bu_1\|_{L^2}^3 \le K\|\varphi_1\|_{L^2}^3,\\
\|J(\varphi_1) - J(\varphi_2)\|_{L^2} &= \|\lambda(u_1^2\bar{u}_1 - u_2^2\bar{u}_2)\|_{L^2}\\
&\le K\|B(u_1-u_2)\|_{L^2}(\|Bu_1\|_{L^2}^2 + \|Bu_1\|_{L^2}\|Bu_2\|_{L^2} + \|Bu_2\|_{L^2}^2)\\
&\le K\|\varphi_1-\varphi_2\|_{L^2}(\|\varphi_1\|_{L^2}^2 + \|\varphi_1\|_{L^2}\|\varphi_2\|_{L^2} + \|\varphi_2\|_{L^2}^2).
\end{aligned}$$

The lemma follows. □

Lemma 10.4.4. *Assume that $\varphi_1, \varphi_2 \in D(A)$. Then*

$$\begin{aligned}
&\|AJ(\varphi_1)\|_{L^2} \le K\|\varphi_1\|_{L^2}^2\|A\varphi_1\|_{L^2},\\
&\|A(J(\varphi_1) - J(\varphi_2))\|_{L^2}\\
&\quad \le C(\|\varphi_1\|_{L^2}, \|\varphi_2\|_{L^2}, \|A\varphi_1\|_{L^2}, \|A\varphi_2\|_{L^2})\|A\varphi_1 - A\varphi_2\|_{L^2}.
\end{aligned}$$

Proof. Let $\varphi_i = \langle u_i, v_i\rangle$. Then $u_i \in D(B^2)$, $v_i \in D(B)$. We calculate

$$\begin{aligned}
\|Bu_{x_i}\|_{L^2}^2 &= \|(\Sigma k_i^2 + m^2)^{1/2}k_i\hat{u}\|_{L^2}^2 \le \|(\Sigma k_i^2 + m^2)\hat{u}\|_{L^2}^2\\
&= \|B^2u\|_{L^2}^2.
\end{aligned}$$

Thus, by Lemma 10.4.2, we have

$$\begin{aligned}
\|(u^2\bar{u})_{x_i}\|_{L^2} &= \|2uu_{x_i}\bar{u} + u^2\bar{u}_{x_i}\|_{L^2} \le K\|Bu\|_{L^2}^2 \cdot \|Bu_{x_i}\|_{L^2}\\
&\le K\|Bu\|_{L^2}^2\|B^2u\|_{L^2},
\end{aligned}$$

so

$$\begin{aligned}
\|AJ(\varphi_1)\|_{L^2}^2 &= \lambda^2\|Bu_1^2\bar{u}_1\|_{L^2}^2\\
&= \lambda^2\sum_{i=1}^{3}\|(u_1^2\bar{u}_1)_{x_i}\|_{L^2}^2 + \lambda^2m^2\|u_1^2\bar{u}_1\|_{L^2}^2\\
&\le K(\|Bu_1\|_{L^2}^4\|B^2u_1\|_{L^2}^2 + m^2\|Bu_1\|_{L^2}^6)\\
&\le K\|Bu_1\|_{L^2}^4\|B^2u_1\|_{L^2}^2 \le K\|\varphi_1\|_{L^2}^4\|A\varphi_1\|_{L^2}^2.
\end{aligned}$$

Thus, we have proved the first equality. In order to prove the second equality, by Lemma 10.4.2 and the above equality, we have

$$\begin{aligned}
\frac{1}{4}\|(u^2\bar{u}_1 - u_1^2\bar{u}_2)_{x_i}\|_{L^2}^2 &\le \|u_1^2(\bar{u}_1 - \bar{u}_2)_{x_i}\|_{L^2}^2 + \|(u_1^2 - u_2^2)(\bar{u}_2)_{x_i}\|_{L^2}^2\\
&\quad + \|2(u_1)_{x_i}(|u_1|^2 - |u_2|^2)\|_{L^2}^2 + \|2(u_1-u_2)_{x_i}|u_2|^2\|_{L^2}^2
\end{aligned}$$

$$\begin{aligned}&\le K(\|Bu_1\|_{L^2}^4\|B^2(u_1-u_2)\|_{L^2}^2+\|B^2u_2\|_{L^2}^2\|B(u_1+u_2)\|_{L^2}^2\|B^2(u_1-u_2)\|_{L^2}^2)\\&\le K(\|\varphi_1\|_{L^2}^4\|A(\varphi_1-\varphi_2)\|_{L^2}^2\\&\quad+\|A\varphi_2\|_{L^2}^2(\|\varphi_1\|_{L^2}+\|\varphi_2\|_{L^2})\|A(\varphi_1-\varphi_2)\|_{L^2}^2).\end{aligned}$$

Thus,

$$\begin{aligned}\|A(J(\varphi_1)-J(\varphi_2))\|_{L^2}^2&=\lambda^2\|B(u_1^2\bar{u}_1-u_2^2\bar{u}_2)\|_{L^2}^2\\&=\lambda^2\sum_{i=1}^{3}\|(u_1^2\bar{u}_1-u_2^2\bar{u}_2)_{x_i}\|_{L^2}^2+m^2\lambda^2\|u_1^2\bar{u}_1-u_2^2\bar{u}_2\|_{L^2}^2\\&\le C(\|\varphi_1\|_{L^2},\|\varphi_2\|_{L^2},\|A\varphi_2\|_{L^2})\|A(\varphi_1-\varphi_2)\|_{L^2}^2\\&\quad+C(\|\varphi_1\|_{L^2},\|\varphi_2\|_{L^2})\|A(\varphi_1-\varphi_2)\|_{L^2}^2,\end{aligned}$$

where we have used the inequality $\|Bu\|_{L^2}\le K\|B^2u\|_{L^2}$ many times. The lemma follows. □

Lemma 10.4.5. *Suppose $u(x,t)$ is the solution of* (10.4.1) *over* $[0,T]$, *where* $u(x,0)=f(x)\in D(B^2)$ *and* $u_t(x,0)=g(x)\in D(B)$. *Then we have*

$$E(t)=\frac{1}{2}\int\left[|Bu(x,t)|^2+|u_t(x,t)|^2+\frac{\lambda}{2}|u(x,t)|^4\right]dx^3,$$

independent of t.

Proof. Let $\varphi(t)=\langle u(x,t),u_t(x,t)\rangle$. Since, for each $t\in[0,T]$, $\varphi(t)\in D(A)$, we have $u(\cdot,t)\in D(B^2)$, $u_t(\cdot,t)\in D(B)$ $\forall t\in[0,T]$. Since $\varphi(t)$ is strongly differentiable, u and u_t, as the functions of $L^2(R^3)$, are strong differentiable and

$$\begin{cases}\left\|B\left(\dfrac{u(t+h)-u(t)}{h}-u_t(t)\right)\right\|_{L^2}\to 0,\quad h\to 0,\\ \left\|\dfrac{u_t(t+h)-u_t(t)}{h}-u_{tt}(t)\right\|_{L^2}\to 0,\quad h\to 0.\end{cases}\tag{10.4.7}$$

We deduce that the first two terms of $E(t)$ are differentiable. In order to prove its third term is also differentiable, we deduce from Lemma 10.4.1 and the Hölder inequality that

$$\begin{aligned}&\left\|u\left(\frac{u(t+h)-u(t)}{h}-u_t(t)\right)\right\|_{L^2}\\&\quad\le\|u\|_{L^2}^{1/2}\|Bu\|_{L^2}^{1/2}\left\|B\left(\frac{u(t+h)-u(t)}{h}-u_t(t)\right)\right\|_{L^2}.\end{aligned}$$

We deduce from this and (10.4.7) that $u^2(x,t)$ is strong differentiable, so

$$\int |u(x,t)|^4 dx = (u^2(t), u^2(t))$$

is strong differentiable. Also, $E(t)$ is strong differentiable and

$$\begin{aligned} E'(t) &= \frac{1}{2}(Bu_t, Bu) + \frac{1}{2}(u_{tt}, u_t) + \frac{\lambda}{2}(uu_t, u^2) + \frac{1}{2}(Bu, Bu_t) + \frac{1}{2}(u_t, u_{tt}) + \frac{\lambda}{2}(u^2, uu_t) \\ &= \frac{1}{2}(u_t, B^2u + u_{tt} + \lambda|u|^2u) + \frac{1}{2}(B^2u + u_{tt} + \lambda|u|^2u, u_t) = 0, \end{aligned}$$

where we have used the differential equation which u satisfies. □

With the above basic estimation of the solution of problem (10.4.1) and applying the following existence theorem for the solution of the abstract differential operator, we can get the existence, uniqueness, and smoothness of the solution of (10.4.1).

We now consider the operator equation (10.4.4). Assume A is a self-conjugate operator in some Hilbert space $\mathcal{H}$. Suppose that J is a nonlinear map from $D(A)$ to $\mathcal{H}$. Our problem is to find what conditions J should satisfy to ensure that, for any $\varphi_0 \in D(A)$, there exists a unique function $\varphi(t)$, $t \in [0,\infty)$, which belongs to $\mathcal{H}$ and satisfies

$$\begin{cases} \dfrac{d\varphi}{dt} = -iA\varphi + J(\varphi), \\ \varphi(0) = \varphi_0. \end{cases} \tag{10.4.8}$$

We have the following theorems.

Theorem 10.4.6 (Local existence). *Suppose that A is a self-conjugate operator in the Hilbert space $\mathcal{H}$ and J is a map from $D(A)$ to $D(A)$, satisfying*

(H_0) $\|J(\varphi)\|_{L^2} \le C(\|\varphi\|_{L^2})\|\varphi\|_{L^2}$;

(H_1) $\|AJ(\varphi)\|_{L^2} \le C(\|\varphi\|_{L^2}, \|A\varphi\|_{L^2})\|A\varphi\|_{L^2}$;

(H_0^2) $\|J(\varphi) - J(\psi)\|_{L^2} \le C(\|\varphi\|_{L^2}, \|\psi\|_{L^2})\|\varphi - \psi\|_{L^2}$;

(H_1^2) $\|A(J(\varphi) - J(\psi))\|_{L^2} \le C(\|\varphi\|_{L^2}, \|A\varphi\|_{L^2}, \|\psi\|_{L^2}, \|A\psi\|_{L^2})\|A\varphi - A\psi\|_{L^2}, \forall \varphi, \psi \in D(A)$;

where each constant C is a monotone increasing function of the signified norm. Then, for all $\varphi_0 \in D(A)$, there exists $T > 0$ such that (10.4.8) *has a unique continuously differentiable solution on $[0,T)$ and, for all $\varphi_0 \in \{\varphi \mid \|\varphi\|_{L^2} \le a, \|A\varphi\|_{L^2} \le b\}$, T can be selected to hold uniformly.*

Theorem 10.4.7 (Local smoothness). (a) *Assume A is a self-conjugate operator in the Hilbert space $\mathcal{H}$ and J is a map of*

$$D(A^j) \to D(A^j), \quad (1 \le j \le n)$$

and satisfies (for $j - 0, 1, \dots, n$)

(H_j) $\|A^j J(\varphi)\|_{L^2} \le C(\|\varphi\|_{L^2}, \dots, \|A^j\varphi\|_{L^2})\|A^j\varphi\|_{L^2}$;

(H_j^2) $\|A^j(J(\varphi) - J(\psi))\|_{L^2} \le C(\|\varphi\|_{L^2}, \|\psi\|_{L^2}, \dots, \|A^j\varphi\|_{L^2}, \|A^j\psi\|_{L^2})\|A^j\varphi - A^j\psi\|_{L^2}$, $\forall \varphi, \psi \in D(A^j)$;

where each constant C is a monotone increasing function of its variables. Then, for all $\varphi_0 \in D(A^n)$, $n \ge 1$, there exists $T_n > 0$ such that (10.4.8) *on $[0, T_n)$ has a unique solution $\varphi(t) \in D(A^n)$, $t \in [0, T_n)$. If φ_0 belongs to*

$$\{\varphi \mid \|A^j\varphi\|_{L^2} \le a_j,\ j = 0, 1, \dots, n\},$$

then it is possible to select T to hold uniformly.

(b) *If one increases the hypotheses from (a), for $j < n$, J has the following properties. If φ is j times strong continuously differentiable, $\varphi^{(k)}(t) \in D(A^{n-k})$, and $A^{n-k}\varphi^{(k)}(t)$ is continuous (for all $k \le j$), then $J(\varphi(t))$ is j times differentiable. As*

$$\frac{d^j J(\varphi(t))}{dt^j} \in D(A^{n-j-1}), \quad A^{n-j-1} d^j J(\varphi(t))/dt^j$$

is continuous, the solution obtained by (a) is n times strongly differentiable about t and

$$\frac{d^j\varphi(t)}{dt^j} \in D(A^{n-j}).$$

Theorem 10.4.8 (Global existence and smoothness). *Assume that A is a self-conjugate operator in the Hilbert space $\mathcal{H}$, n is a positive integer, and J is a map of $D(A^i) \to D(A^j)$ $(1 \le j \le n)$ and satisfies (for $1 \le j \le n$):*

(H_0) $\|J(\varphi)\|_{L^2} \le C(\|\varphi\|_{L^2})\|\varphi\|_{L^2}$;

($H_j^!$) $\|A^j J(\varphi)\|_{L^2} \le C(\|\varphi\|_{L^2}, \dots, \|A^{j-1}\varphi\|_{L^2})\|A^j\varphi\|_{L^2}$, $j = 1, 2, \dots, n$;

(H_j^L) $\|A^j(J(\varphi) - J(\psi))\|_{L^2} \le C(\|\varphi\|_{L^2}, \dots, \|A^{j-1}\varphi\|_{L^2}, \|A^{j-1}\psi\|_{L^2})\|A^j\varphi - A^j\psi\|_{L^2}$, $\forall \varphi, \psi \in D(A^j)$, $j = 1, 2, \dots, n$;

where C is a monotone increasing function of all its variables. Suppose $\varphi_0 \in D(A^n)$ and $\|\varphi(t)\|$ is bounded on any finite interval, guaranteed by Theorem 10.4.7 (a) *that the solution exists. Then there exists the strongly differentiable function $\varphi(t)$ in $D(A^n)$, which, in $[0, \infty)$, satisfies*

$$\begin{cases} \varphi'(t) = -iA\varphi(t) + J(\varphi(t)), \\ \varphi(0) = \varphi_0. \end{cases} \tag{10.4.9}$$

Furthermore, if J satisfies the assumption of Theorem 10.4.7 (b), *then $\varphi(t)$ is n times strongly differentiable and $\frac{d^j\varphi(t)}{dt^j} \in D(A^{n-j})$.*

Using the a priori estimates of Lemmas 10.4.1–10.4.5 and Theorems 10.4.6–10.4.8, we get the existence theorem of problem (10.4.1).

Theorem 10.4.9. *Suppose $\lambda > 0$, $m > 0$, and*

$$f \in D(-\Delta + m^2), \quad g \in D((-\Delta + m^2)^{1/2}).$$

Then there exists a unique function $u(x,t)$, $t \in R$, $x \in R^3$, such that $t \to u(\cdot,t)$ is a twice strongly differentiable function of t in $L_2(R^3)$. For all t, $u(\cdot,t) \in D(-\Delta+m^2)$, $u(x,0) = f(x)$, $u_t(x,0) = g(x)$, which satisfies

$$u_{tt} - \Delta u + m^2 u = -\lambda |u|^2 u. \tag{10.4.10}$$

For all t, the map $\langle f,g\rangle \mapsto \langle u(\cdot,t), u_t(\cdot,t)\rangle$ is continuous.

Proof. By Lemma 10.4.3 and Lemma 10.4.4, we know that J satisfies the conditions (H_0^2), (H_1^2) of Theorem 10.4.6 as well as (H_0), (H_1'). Thus, the unique local solution $\varphi(t)$ exists in $[0,T)$. By Lemma 10.4.5, $E(t)$ is a constant for all $t \in [0,T)$. Since

$$\begin{aligned} \frac{1}{2}\|\varphi(t)\|_{L^2}^2 &\le \frac{1}{2}\|\varphi(t)\|_{L^2}^2 + \frac{\lambda}{4}\int_{R^3} |u(x,t)|^4 dx^3 \\ &= E(t) = E(0), \end{aligned}$$

we know $\|\varphi(t)\|$ is bounded in $[0,T)$. From Theorem 10.4.8, we know that the solution exists for all $t \ge 0$. The proof is complete. □

For some other nonlinear evolution equations and their systems, using the method of abstract differential operators to prove their existence and uniqueness, we refer the reader to [44, 101, 260, 275].

10.5 RLW equation and Galerkin method

We use the Galerkin method to prove the existence of solution for the RLW equation and discuss the smoothness of solution.

Consider the following initial boundary value problem for the general RLW equation:

$$u_t + f(u)_x - u_{xxt} = g(x,t), \tag{10.5.1}$$

$$u|_{t=0} = u_0(x), \tag{10.5.2}$$

$$u|_{x=0} = u|_{x=1} = 0. \tag{10.5.3}$$

In the following proof of the existence of solution for the problem (10.5.1)–(10.5.3), we require the following two lemmas of the Sobolev space.

Lemma 10.5.1. *If $u \in H^1(0,1)$, then there exists a constant $C > 0$, which is independent of u, such that*

$$\sup_{0\le x\le 1} |u(x)| \le C\|u\|_{L^2(0,1)}^{1/2} \left(\|u\|_{L^2(0,1)} + \|u_x\|_{L^2(0,1)}\right)^{1/2}. \tag{10.5.4}$$

Lemma 10.5.2. *Assume $f \in C^k(R)$, $k \geq 1$, and $f(0) = 0$. If $u(x,t) \in L^\infty(0,T;H^k(0,1))$, then $f(u(x,t)) \in L^\infty(0,T;H^k(0,1))$ and the following inequalities hold:*

$$\|f(u(t))\|_{H^1(0,1)} \leq M\|u(t)\|_{H^1(0,1)}$$

and

$$\|f(u(t))\|_{H^k(0,1)} \leq C_k(1 + \|u(t)\|_{H^{k-1}}^{k-1})\|u(t)\|_{H^k}, \quad (k \geq 2), \tag{10.5.5}$$

where M and C_k are constants.

We let $(0,1) = \Omega$, $Q = \Omega \times [0,T]$, $T > 0$.
We have the following theorem.

Theorem 10.5.3. *Assume $T > 0$ is a real number,*

$$g(x,t) \in L^\infty(0,T;L^2(\Omega)), \quad u_0(x) \in H_0^1(\Omega),$$

and $f(s) \in C^1(R)$. Then there exists a unique function $u(x,t)$, $(x,t) \in Q$, which satisfies

$$u \in L^\infty(0,T;H_0^1(\Omega)), \tag{10.5.6}$$
$$u_t \in L^\infty(0,T;H_0^1(\Omega)), \tag{10.5.7}$$
$$u_t + (f(u))_x - u_{xxt} = g(x,t), \quad \text{in } Q, \tag{10.5.8}$$
$$u(x,0) = u_0(x). \tag{10.5.9}$$

Remark. Without loss of generality, we assume $f(0) = 0$. In fact, if $f(0) \neq 0$, then we have

$$u_t + (h(u))_x - u_{xxt} = g(x,t),$$

where $h(s) = f(s) - f(0)$, which is equivalent to (10.5.8).

Proof. First of all, we see that (10.5.9) is meaningful. In fact, we know from (10.5.6) and (10.5.7) that $u(x,t)$ can be determined at $t = 0$. We prove this theorem by the Galerkin method, which involves the following three major steps: (i) construct the approximate solution of equation (10.5.8), (ii) make a priori estimates for the approximate solution, and (iii) take the limit for the approximate solution.

As a first step, we construct the approximate solution. Suppose $\{w_\nu\}$ are the basis functions of the space H_0^1. For any $m \in N$, $w_1, w_2, \ldots, w_n$ are linearly independent. Construct the approximate solution of equation (10.5.8), $u^m = u^m(x,t) = \sum_{\nu=1}^m g_{\nu m}(t)w_\mu(x)$, where the coefficients $g_{\nu m}$ can be determined by the following equation:

$$(u_t^m, w_\nu) + a(u_t^m, w_\nu) + ((f(u^m))_x, w_\nu) = (g, w_\nu), \quad \nu = 1,2,\ldots,m, \tag{10.5.10}$$

where $a(u,v) = \int_0^1 \frac{\partial u}{\partial x} \cdot \frac{\partial v}{\partial x} dx$. Since $u_0 \in H_0^1(\Omega)$, there exist constants $C_{vm} (v = 1, 2, \ldots, m)$ such that

$$u_{0m} \to u_0, \quad \text{strongly in } H_0^1(\Omega),\ m \to \infty, \tag{10.5.11}$$

while $u_{0m} = \sum_{v=1}^m C_{vm} w_v$.

If we add the initial condition to equation (10.5.10), $u^m(0) = u_{0m}$, then we obtain a system of ordinary differential equations with respect to the unknown function g_{vm}, with the initial condition $g_{vm}(0) = C_{vm}$. Because the basis functions $\{w_v\}_{v \in N}$ are linearly independent, the coefficient matrix of g_{vm} is invertible. Therefore, the local solution of this equation exists. Thus, the solution $\{g_{vm}(t)\}_{1 \le v \le m}$ of (10.5.10) exists in $[0, t_m)$. Since we establish the a priori estimates in the following step, its solution can be extended from $[0, t_m)$ to $[0, T]$ for arbitrary finite positive constant T.

As a second step, we make a priori estimates. Multiplying both sides of equations (10.5.10), satisfied by the approximate solution by $g_{vm}(t)$, and summing with respect to v from 1 to m, we get

$$\frac{1}{2}\frac{d}{dt}(\|u^m\|^2 + a(u^m, u^m)) - \left(f(u^m), \frac{\partial u^m}{\partial x}\right) = (g, u^m). \tag{10.5.12}$$

If we let

$$h(x,t) = \int_0^{u^m(x,t)} f(s) ds,$$

then

$$\frac{\partial h}{\partial x} = f(u^m) \frac{\partial u^m}{\partial x}.$$

Since $w_v \in H_0^1(\Omega)$, $h(0,t) = h(1,t) = 0$, we have

$$\left(f(u^m), \frac{\partial u^m}{\partial x}\right) = h(1,t) - h(0,t), \quad \forall t.$$

From (10.5.12), we obtain

$$\frac{d}{dt}(\|u^m\|^2 + a(u^m, u^m)) \le \|g(t)\|^2 + \|u^m\|^2.$$

By Grönwall's lemma, when $u_{0m} \xrightarrow{H_0^1(\Omega)} u_0$, we obtain

$$\|u^m\|_{H_0^1(\Omega)} \le c, \quad \forall t \in [0, T], \tag{10.5.13}$$

where c is independent of m. Thus, it can be seen that u^m belongs to the bounded set of $L^\infty(0, T; H_0^1(\Omega))$. Therefore, $u^m \to u$ weakly star in $L^\infty(0, T; H_0^1(\Omega))$.

We now try to get the estimates of u_t^m. Multiplying both sides of equations (10.5.10) by $g'_{vm}(t)$ and summing with respect to v from 1 to m, we get

$$\begin{aligned}\|u_t^m\|^2 + a(u_t^m, u_t^m) &\le |(f(u^m)_x, u_t^m)| + |(g, u_t^m)| \\ &\le \|f(u^m)_x\|\|u_t^m\| + \|g(t)\|\|u_t^m\|.\end{aligned}$$

Then, by Lemma 10.5.1 and estimate (10.5.13), we obtain

$$\|u_t^m\|_{H_0^1(\Omega)} \le c, \quad \forall t \in [0, T), \tag{10.5.14}$$

where c is independent of m. Thus, u_t^m belongs to the bounded set of $L^\infty(0, T; H_0^1(\Omega))$. Therefore, there exists a subsequence of $\{u_t^m\}$ such that

$$u^m \to u, \quad \text{weakly star in } L^\infty(0, T; H_0^1(\Omega)). \tag{10.5.15}$$

Thus, u^m belongs to the bounded set of $H^1(Q)$ for all m. By Rellich's embedding theorem, we have $u^m \to u$ strongly in $L^2(Q)$ and there exists a subsequence of u^m almost everywhere converging to u.

As the third step, we take the limit for the approximate solution. We first consider the case of the nonlinear terms. From Lemma 10.5.2 and (10.5.13), we have

$$\|f(u^m)\|_{H^1(\Omega)} \le c, \quad t \in [0, T), \tag{10.5.16}$$

where the constant c is independent of m. Therefore,

$$f(u^m)_x \to \chi, \quad \text{weakly star in } L^\infty(0, T; L^2(\Omega)). \tag{10.5.17}$$

Because u^m belongs to the bounded set of $H^1(\Omega)$ and f is continuous in R, we have

$$f(u^m) \to f(u) \quad \text{almost everywhere in } Q. \tag{10.5.18}$$

By (10.5.13), Lemma 10.5.1, and Lemma 10.5.2, we have

$$\|f(u^m)\|_{L^2(Q)} \le c, \quad \forall m, \ t \in [0, T]. \tag{10.5.19}$$

We deduce from (10.5.18) and (10.5.19) that

$$f(u^m) \to f(u), \quad \text{weakly star in } L^2(Q). \tag{10.5.20}$$

Thus, we obtain

$$f(u^m)_x \to f(u)_x, \quad \text{according to the distribution of sense in } Q. \tag{10.5.21}$$

From (10.5.17) and (10.5.21), we have $f(u)_x = \chi$, so

$$f(u^m)_x \to f(u)_x, \quad \text{weakly star in } L^\infty(0, T; L^2(\Omega)). \tag{10.5.22}$$

Letting $m > \nu$, from (10.5.6), we have

$$(u_t^m, w_\nu) + a(u_t^m, w_\nu) + (f(u^m)_x, w_\nu) = (g, w_\nu). \tag{10.5.23}$$

By (10.5.15) and (10.5.22), we get

$$\begin{aligned}
&(u_t^m, w_\nu) \to (u_t, w_\nu), \quad \text{weakly star in } L^\infty(0,T),\\
&a(u_t^m, w_\nu) \to a(u_t, w_\nu), \quad \text{weakly star in } L^\infty(0,T),\\
&(f(u^m)_x, w_\nu) \to (f(u)_x, w_\nu), \quad \text{weakly star in } L^\infty(0,T).
\end{aligned} \tag{10.5.24}$$

Thus, letting $m \to \infty$ in (10.5.23), we get

$$(u_t, w_\nu) + a(u_t, w_\nu) + (f(u)_x, w_\nu) = (g, w_\nu) \quad \text{for all } \nu. \tag{10.5.25}$$

Since $\{w_\nu\}$ is dense in $H_0^1(\Omega)$, we have

$$(u_t, v) + a(u_t, v) + (f(u)_x, v) = (g, v), \quad \forall v \in H_0^1(\Omega),$$

and u satisfies the conditions (10.5.6)-(10.5.8) of Theorem 10.5.3.

Now we verify that u satisfies the initial condition. In fact, since $u^m \to u$ weakly star in $L^\infty(0,T;L^2(\Omega))$, we have

$$\int_0^T (u^m, v)dt \to \int_0^T (u, v)dt, \quad \forall v \in L^1(0,T;L_2(\Omega)). \tag{10.5.26}$$

By (10.5.15), we have

$$\int_0^T (u_t^m, v)dt \to \int_0^T (u_t, v)dt, \quad \forall v \in L^1(0,T;L^2(\Omega)). \tag{10.5.27}$$

We now consider $v(x,t) = \theta(t)w(x)$, $w(x) \in L^2(\Omega)$, and $\theta \in C^1(0,T)$ such that $\theta(0) = 1$, $\theta(T) = 0$. If we choose $v = \theta' w$ in (10.5.26) and $v = \theta w$ in (10.5.27), we obtain

$$\lim_{m\to\infty} (u^m(0), w) = (u(0), w), \quad \forall w \in L^2(\Omega),$$

so $u^m(0) \to u(0)$ weakly in $L^2(\Omega)$. Therefore, $u(0) = u_0$.

The uniqueness remains to be proved. Suppose that there are two solutions u, v corresponding to the same initial condition. If we let $w = u - v$, then we have

$$\begin{cases} w_t - w_{xxt} + f(u)_x - f(v)_x = 0,\\ w(x,0) = 0, \quad w(0,t) = w(1,t) = 0.\end{cases}$$

Then w satisfies

$$\frac{1}{2}\frac{d}{dt}(\|w\|^2 + \|w_x\|^2) = (f(u) - f(v), w_x). \tag{10.5.28}$$

By Lemma 10.5.1, Lemma 10.5.2, and (10.5.28), we get

$$\frac{d}{dt}(\|w\|^2 + \|w_x\|^2) \le C\|w_x\|^2,$$

where $C > 0$ is independent of t. Thus $w \equiv 0$. □

In the following, we discuss the regularity of the weak solution. We choose the base functions of the Galerkin method to be the eigenfunctions of the one-dimensional Laplacian operator, so we can obtain a higher regularity of the solution. Suppose that $\{\psi_v\}_{v\in N}$ are the eigenfunctions of the one-dimensional Laplacian operator in $L^2(\Omega)$. Then we know from [164] that $\{\psi_v\}$ is the complete orthonormal system in $L^2(\Omega)$ and $H_0^1(\Omega)$. We know $\psi_v \in H^k(\Omega)$, where k is any positive integer, and we let V^k be the closure of the linear combination of ψ_v in $H^k(\Omega)$.

Theorem 10.5.4. *Assume $g(x,t) \in L^\infty(0,T;H^k(\Omega))$ such that*

$$D^{2v}g \in L^\infty(0,T;H_0^1(\Omega)), \quad v = 0,1,2,\dots,j,$$

and $k-2j \geq 1, f(s) = \frac{s^2}{2}$, and $u_0 \in V^{k+1}$. Then, for every nonnegative integer k, there only exists a function $u(x,t)$, which is defined in Q and satisfies the following conditions:

$$u \in L^\infty(0,T;H^{k+1}(\Omega)), \tag{10.5.29}$$
$$D^{2v}u \in L^\infty(0,T;H_0^1(\Omega)), \quad v = 0,1,2,\dots,j,\ k+1-2j \geq 1, \tag{10.5.30}$$
$$u_t \in L^\infty(0,T;H^{k+2}(\Omega)), \tag{10.5.31}$$
$$D^{2v}u_t \in L^\infty(0,T;H_0^1(\Omega)), \quad v = 0,1,2,\dots,j,\ k+2-2j \geq 1, \tag{10.5.32}$$
$$u_t + uu_x - u_{xxt} = g, \quad in\ L^\infty(0,T;L^2(\Omega)), \tag{10.5.33}$$
$$u(x,0) = u_0(x). \tag{10.5.34}$$

Proof. Assume $u^m(x,t)$ is the approximate solution defined by Theorem 10.5.3. We consider the eigenfunctions ψ_v instead of w_v and we have

$$(u_t^m,\psi_v) + (f(u^m)_x,\psi_v) - (u_{xxt}^m,\psi_v) = (g,\psi_v), \quad v = 1,\dots,m, \tag{10.5.35}$$
$$u^m(0) = u_{0m}, \tag{10.5.36}$$

where $u_{0m} \to u_0$ strongly in $H^{k+1}(\Omega)$. We note that

$$\Delta^p\psi_v = (-\lambda_v)^p\psi_v$$

for all p, where p is a nonnegative integer.

We will use the induction to prove (under the assumptions of the theorem)

$$\|u^m(t)\|_{H^{k+1}} \leq C, \quad \forall t \in [0,T], \tag{10.5.37}$$
$$\|u_t^m(t)\|_{H^{k+1}} \leq C, \quad \forall t \in [0,T], \tag{10.5.38}$$

where the constant $C > 0$ is independent of m, t.

First, similar to the proof of Theorem 10.5.3, we choose the base functions to be the eigenfunctions of the one-dimensional Laplacian operator. From the assumptions of Theorem 10.5.3, (10.5.13), and (10.5.14), we have

$$\|u_{xxt}^m(t)\| \leq C, \tag{10.5.39}$$

where the constant $C > 0$ is independent of m, t. In fact, multiplying (10.5.10) on both sides by $(-\lambda_\nu)g'_{\nu m}$ and summing with respect to ν, we get

$$\begin{aligned}\|u^m_{xt}\|^2 + \|u^m_{xxt}\|^2 &= -(g, u^m_{xxt}) + (f(u^m)_x, u^m_{xxt})\\ &\le \|g\|\|u^m_{xxt}\| + \|f(u^m)_x\|\|u^m_{xxt}\|.\end{aligned}$$

By Lemma 10.5.2 and (10.5.13), we immediately get (10.5.39). Thus, for $k = 0$, (10.5.37) and (10.5.38) follow. □

Remark. Replacing the arbitrary base $\{w_\nu\}$, considering the special base $\{\psi_\nu\}$, and using the assumptions of Theorem 10.5.3, we improve the results and obtain

$$\begin{aligned}&u_t \in L^\infty(0, T; H^2(\Omega) \cap H^1_0(\Omega)),\\ &u_t + (f(u))_x - u_{xxt} = g(x,t), \quad \text{weakly star in } L^\infty(0, T; L^2(\Omega)),\end{aligned}$$

instead of (10.5.7) and (10.5.8), respectively.

Suppose that (10.5.37) and (10.5.38) hold for $k \ge 0$. We prove these also hold for $k+1$. It is noticed that while $q > 0$ is an odd integer, we have

$$D^q f(u^m) = c_0 u^m D^q u^m + c_1 Du^m D^{q-1}u^m + \cdots + c_{\frac{q-1}{2}} D^{\frac{q-1}{2}} u^m.$$

Therefore,

$$(Df(u^m), D^{2k+2}u^m) = \pm(D^{k+1}f(u^m), D^{k+2}u^m), \tag{10.5.40}$$

$$(Df(u^m), D^{2k+4}u^m) = \pm(D^{k+2}f(u^m), D^{k+3}u^m). \tag{10.5.41}$$

Multiplying (10.5.35) on both sides by $(-\lambda_\nu)^{k+1}g_{\nu m}$ and summing with respect to ν, we get

$$(u^m_t, D^{2k+2}u^m) + (f(u^m)_x, D^{2k+2}u^m) - (D^2u^m_t, D^{2k+2}u^m) = (g, D^{2k+2}u^m).$$

By (10.5.40), we have

$$(D^{k+1}u^m_t, D^{k+1}u^m) + (D^{k+2}u^m_t, D^{k+2}u^m) = -(D^k g, D^{k+2}u^m) \pm (D^{k+1}f(u^m), D^{k+2}u^m),$$

or

$$\begin{aligned}&\frac{1}{2}\frac{d}{dt}(\|D^{k+1}u^m\|^2 + \|D^{k+2}u^m\|^2)\\ &\quad\le \frac{1}{2}\|D^k g\|^2 + \frac{1}{2}\|D^{k+1}f(u^m)\| + \|D^{k+2}u^m\|^2.\end{aligned}$$

By the induction hypothesis and Lemma 10.5.2, we have

$$\|D^{k+2}u^m(t)\| \le C, \quad \forall t \in [0, T], \tag{10.5.42}$$

where the constant $C > 0$ is independent of m, t. Similarly, multiplying both sides by $(-\lambda_v)^{k+2} g'_{vm}$ and summing with respect to v, we get

$$\begin{aligned}
&\|D^{k+2}u_t^m\|^2 + \|D^{k+3}u_t^m\| \\
&= -(D^{k+1}g, D^{k+3}u_t^m) \pm (D^{k+2}f(u^m), D^{k+3}u_t^m) \\
&\leq 2\|D^{k+1}g\|^2 + 2\|D^{k+2}f(u^m)\|^2 + \frac{1}{4}\|D^{k+3}u_t^m\|^2.
\end{aligned}$$

By the induction hypothesis, Lemma 10.5.2, and (10.5.42), we have

$$\|D^{k+2}u_t^m(t)\| \leq C, \quad \forall t \in [0, T], \tag{10.5.43}$$

where the constant $C > 0$ is independent of m, t. By (10.5.42) and (10.5.43), we immediately obtain (10.5.37) and (10.5.38).

Remark. Theorem 10.5.4 also holds for $f(s) = Cs^n$, where C is a constant and $n > 0$ is an even integer.

For other nonlinear evolution equations and the use of the Galerkin method and the vanishing viscosity method to study these equations, we refer the reader to [106–108, 110, 111, 117].

10.6 Asymptotic behavior of solutions as $t \to \infty$ and the blow up problem

For a class of nonlinear evolution equations, the local L^2 norms of their smooth solutions may tend to zero as $t \to \infty$. We will illustrate this with a simple method. In addition, for some nonlinear evolution equations, although local solutions exist, global solutions do not exist. In fact, when $t \to t_1$ (finite), the L^2 norm of its solution will tend to infinity. This phenomenon is called blow up of the solution. It has been found that many nonlinear evolution equations have this property.

Now we consider the initial value problem of the following generalized KdV equation:

$$u_t + (u_{xx} - f(u) - u)_x = 0, \quad (x \in R, t > 0) \tag{10.6.1}$$

$$u|_{t=0} = \varphi(x), \quad x \in \mathbb{R}. \tag{10.6.2}$$

Lemma 10.6.1. *Assume that $u(x,t)$ is the classical solution of problem* (10.6.1), (10.6.2) *and satisfies:*

(i) $\lim_{|x|\to\infty}(|u| + |u_x| + |u_{xx}|)(x,t) = 0$, $\forall t \geq 0$;

(ii) *$f(s)$ is a real-valued continuous function and $f(s)s \geq 0$.*

If we let $F(u) = \int_0^u f(s)ds$, *then* $F \geq 0$ *and we have*

$$\|u(t)\| = \|\varphi\|, \quad \forall t \geq 0.$$

Proof. Multiplying (10.6.1) by u, we get

$$(u^2)_t + (u^2)_{xxx} - (3(u_x)^2)_x - (2f(u)u)_x + (2F(u))_x - (u^2)_x = 0.$$

Integrating with respect to x over $(-\infty, \infty)$, we immediately get the conclusion. □

Theorem 10.6.2. *Assume that* $u(x,t)$ *is the classical solution of problem* (10.6.1), (10.6.2) *and satisfies* (i), (ii). *We also suppose*

$$f(u)u \geq F(u). \tag{10.6.3}$$

Then we have

$$\int_0^\infty \int_{-r}^r (|u|^2 + |u_x|^2)dxdt < \infty, \quad \forall r > 0. \tag{10.6.4}$$

If we further assume that there exists a positive constant α *such that*

$$(1-\alpha)f(u)u \geq F(u), \quad 0 < \alpha < 1, \tag{10.6.5}$$

then we have

$$\int_{-r}^r |u|^2 \to 0, \quad t \to \infty, \forall r > 0. \tag{10.6.6}$$

Proof. Suppose A is the function of x and $A \in C^3$. Multiplying (10.6.2) by A, we get

$$\begin{aligned}(Au^2)_t &+ \{A(u^2)_{xx} - A_x(u^2)_x + A_{xx}u^2 - 3A(u_x)^2 - 2Af(u)u + 2AF(u) - Au^2\}_x \\ &+ (-A_{xx} + A_x)u^2 + 3A_x(u_x)^2 + 2A_x(f(u)u - F(u)) = 0.\end{aligned} \tag{10.6.7}$$

Further assume that A satisfies $A_x > 0$, $-A_{xxx} + A_x > 0$, and $|A|$, $|A_x|$, and $|A_{xx}$ are bounded. Obviously, it is easy to find this kind of A. Integrating (10.6.7) over $(-\infty, \infty) \times [0, T]$, by (10.6.3) and Lemma 10.6.1, we have

$$\int_0^\infty \int_{-r}^r (u^2 + u_x^2)dxdt < \infty. \tag{10.6.8}$$

If (10.6.5) is also satisfied, then we have

$$\int_0^\infty \int_{-r}^r f(u)u(x,t)dxdt < \infty. \tag{10.6.9}$$

We now use the concept of Morawetz [182] to show

$$\int_{-r}^r u^2 dx \to 0, \quad t \to \infty, \forall r \geq 0.$$

Assume $B(x) \in C_0^\infty(\mathbb{R})$, $0 \le B(x) \le 1$, and

$$B(x) = \begin{cases} 1, & |x| \le r, \\ 0, & |x| \ge 2r. \end{cases}$$

Then, from (10.6.7), we have

$$\left| \int_{-2r}^{2r} Bu \cdot u_t dx \right| \le C \int_{-2r}^{2r} [u^2 + u_x^2 + f(u)u] dx.$$

Let $0 < t_1 < t$. Then we have

$$\begin{aligned}(t - t_1) \int_{-r}^{r} u^2(x,t)dx &\le (t - t_1) \int_{-2r}^{2r} Bu^2(x,t)dx \\ &\le \int_0^t \int_{-2r}^{2r} Bu^2(x,\tau)dxd\tau + 2\int_{t_1}^{t} (\tau - t_1) \left| \int_{-2r}^{2r} Bu \cdot u_t(x,\tau)dx \right| d\tau.\end{aligned}$$

Let $t_1 = t - 1$. Then

$$\int_{-r}^{r} u^2(x,t)dx \le C \int_{t-1}^{t} \int_{-2r}^{2r} [u^2 + u_x^2 + f(u)u] dxd\tau.$$

From (10.6.8) and (10.6.9), we obtain

$$\int_{-r}^{r} u^2(x,t)dx \to 0, \quad t \to \infty, \forall r > 0.$$

□

Remark. If $f(u) = u^p$, $p \ge 3$, p is an odd integer. Then it is easy to verify that condition (ii) of Lemma 10.6.1, (10.6.3), and (10.6.5) are all satisfied.

We now consider two examples of the blow up of the solution.

Example 10.6.3. Consider the following initial value problem:

$$u_{tt} - u_{xx} = u^n \quad (n > 1), \tag{10.6.10}$$

$$u(x,0) = u_0(x) \quad (x \in \mathbb{R}), \tag{10.6.11}$$

$$u_t(x,0) = v_0(x) \quad (x \in \mathbb{R}). \tag{10.6.12}$$

It is easy to prove, if $u_0, v_0 \in C_0^\infty(\mathbb{R})$, there exists a local solution u of problem (10.6.10)–(10.6.12). We will show that, if we suitably choose u_0 and v_0, $F(t) = \int_{\mathbb{R}} u^2(x,t)dx$ will tend to infinity in a finite time. Now suppose that we can find $\alpha > 0$ and initial values u_0, v_0 such that

(A): $(F(t)^{-\alpha})'' \le 0, \forall t \ge 0$;

(B): $(F(t)^{-\alpha})' < 0, t < 0$.

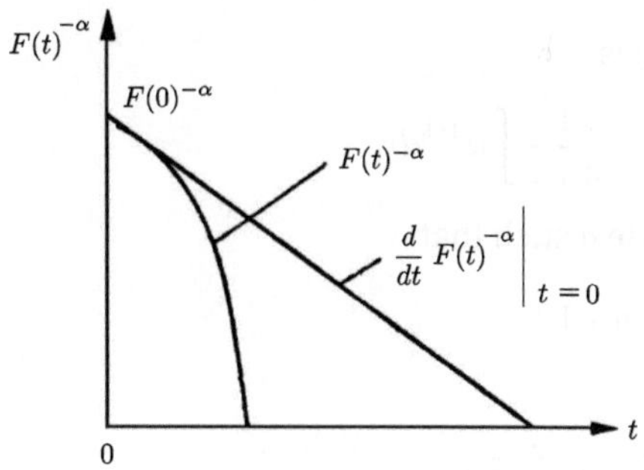

Figure 10.1: The graph of $F(t)^{-\alpha}$.

Then $F(t)^{-\alpha}$ obviously tends to zero in a finite time. Thus $F(t) \to \infty$, as shown in Figure 10.1.

For condition (B), just choose u_0, v_0 with the same symbol over $(-\infty, +\infty)$. Then it is satisfied automatically, because

$$\begin{aligned}(F(0)^{-\alpha})' &= -\alpha F(0)^{-1-\alpha}F'(0)\\ &= -2\alpha F(0)^{-1-\alpha}\int u_0 v_0 dx < 0.\end{aligned}$$

Thus, we only need to check condition (A). Since $F(t) \geq 0$, in order to prove (A), we only need to show $Q(t) \geq 0$, where

$$Q(t) = (-\alpha)^{-1}F^{\alpha+2}(F^{-\alpha})'' = F''F - (\alpha+1)(F')^2.$$

Because

$$\begin{aligned}F'(t) &= 2\int uu_t dx,\\ F''(t) &= 2\int (uu_{tt} + u_t^2)dx\\ &= 4(\alpha+1)\int u_t^2 dx + 2\int (uu_{tt} - (2\alpha+1)u_t^2)dx,\\ Q(t) &= 4(\alpha+1)\left\{\left(\int u^2 dx\right)\left(\int u_t^2 dx\right) - \left(\int uu_t dx\right)^2\right\}\\ &\quad + 2F(t)\left\{\int uu_{tt}dx - \int (2\alpha+1)u_t^2 dx\right\}.\end{aligned}$$

The right-hand side of the above equation is positive by the Schwartz inequality, so we only need to make $H(t) \geq 0$, where

$$\begin{aligned}H(t) &= \int uu_{tt}dx - (2\alpha+1)\int u_t^2 dx\\ &= \int u^{n+1}dx + \int uu_{xx}dx - (2\alpha+1)\int u_t^2 dx\\ &= \int u^{n+1}dx - \int u_x^2 dx - (2\alpha+1)\int u_t^2 dx.\end{aligned}$$

The energy conservation of (10.6.10)–(10.6.12) is

$$E(t) = \frac{1}{2}\int (u_t^2 + u_x^2)dx - \frac{1}{n+1}\int u^{n+1}dx,$$

where, $E(t)$ is independent of t, so if we choose α such that

$$2(2\alpha + 1) = n + 1,$$

we have

$$\begin{aligned} H(t) &= -(n+1)E(t) + 2\alpha\int u_x^2 dx \\ &= -(n+1)E(0) + 2\alpha\int u_x^2 dx. \end{aligned} \tag{10.6.13}$$

Therefore, if $E(0) < 0$ and $\alpha = \frac{1}{4}(n-1) > 0$, then H is always strictly positive. Now, choosing $u_0 \geq 0$, $v_0 \geq 0$, condition (B) is satisfied. We multiply u_0 by a positive constant such that $E(0) < 0$ (when $n + 1 > 2$, this is possible). Then, for any initial value, $F(t)$ will tend to infinity in a finite time.

If we consider the following equation:

$$u_{tt} - u_{xx} = -u^n, \tag{10.6.14}$$

then it is easy to see that $H(t)$ still satisfies (10.6.13). If n is an even number and we choose $u_0(x) \leq 0$, $v_0(x) \leq 0$, and u_0 to be sufficiently large such that $E(0) \leq 0$, then condition (B) is satisfied. Its solution also blows up in a finite time. On the other hand, if n is an odd number, because $E(t) \geq 0$, the conclusion is unclear, which is not surprising. For example, for the case of $-u^3$, we know that it has a global solution.

Example 10.6.4. We consider the initial value problem of the following high-dimensional nonlinear Schrödinger equation:

$$\begin{cases} iu_t = \Delta u + |u|^{p-1}u, & x \in \mathbb{R}^n,\ t > 0, \\ u|_{t=0} = \varphi(x), & x \in \mathbb{R}^n. \end{cases} \tag{10.6.15}$$

We have the following result.

Theorem 10.6.5. *If the following conditions are satisfied:*

(i) $E(0) = \int_{\mathbb{R}^n} (|\nabla\varphi|^2 - \frac{2}{p+1}|\varphi|^{p+1})dx \leq 0$;

(ii) $\operatorname{Im}\int_{\mathbb{R}^n} r\bar{\varphi}\varphi_r dx > 0$, *where* $r^2 = |x|^2$;

(iii) $P > 1 + \frac{4}{n}$;

then $\|\nabla u(t)\|_{L^2}$ *and* $\|u(t)\|_{L^\infty}$ *tend to infinity in a finite time.*

Remark. Condition (ii) is easy to verify. For example, choosing

$$\varphi(x) = e^{i|x|^2}\psi(x),$$

where $\psi(x)$ is an arbitrary real-valued function, direct calculations yield

$$\operatorname{Im}\int_{\mathbb{R}^n} r\bar{\varphi}\varphi_r dx = 2\int r^2|\psi|^2 dx > 0.$$

For the blow up phenomena of solutions to other nonlinear evolution equations and the asymptotic properties of solutions as $t \to \infty$, we refer the reader to [87, 109, 137, 143, 160, 177, 182, 321, 324].

10.7 Definite solutions problem for the Zakharov equations and some other coupling nonlinear evolution equations

In the soliton study of plasma physics, for the interaction between lasers and plasma, Zakharov gave a class of important equations, known as Zakharov equations. Namely,

$$\frac{\partial^2 n}{\partial t^2} - \frac{\partial^2 n}{\partial x^2} - \frac{\partial^2 |\varepsilon|}{\partial x^2} = 0, \tag{10.7.1}$$

$$i\varepsilon_t + \varepsilon_{xx} - n\varepsilon = 0, \tag{10.7.2}$$

where n indicates the perturbation (fluctuation) of the ion density, which is a real-valued function of variables x, t. ε represents the electric field, which is a complex-valued function of variables x, t. Zakharov discovered the soliton solutions of (10.7.1) and (10.7.2) and investigated the features of these solitons. We now study them from the perspective of differential equations. For this purpose, introducing the potential function φ, (10.7.1) will transform into the following equations:

$$\frac{\partial n}{\partial t} - \frac{\partial^2 \varphi}{\partial x^2} = 0, \tag{10.7.3}$$

$$\frac{\partial \varphi}{\partial t} - (n + |\varepsilon|^2) = 0. \tag{10.7.4}$$

We discuss the periodic initial value problem of (10.7.2)–(10.7.4), i.e., we aim to obtain the solutions $\varepsilon(x,t)$, $\varphi(x,t)$ with period 2π with respect to x, such that they satisfy the following initial conditions:

$$\begin{aligned} &n(x,0) = n_0(x), \varphi(x,0) = \varphi_0(x), \\ &\varepsilon(x,0) = \varepsilon_0(x), \quad (-\infty < x < \infty), \end{aligned} \tag{10.7.5}$$

where we suppose that $n_0(x)$, $\varphi_0(x)$, and $\varepsilon_0(x)$ are all functions with period 2π.

We use the Galerkin method to construct the approximate solution for problem (10.7.2)–(10.7.5) and make an a priori estimate about the approximate solution, to obtain the following theorems.

Theorem 10.7.1. *If $\varepsilon_0(x) \in H^6$, $\varphi_0(x) \in H^4$, $n_0(x) \in H^4$, and they are functions with period 2π, then the local classical solution of problem* (10.7.2)–(10.7.5) *exists.*

According to the a priori estimates, we can extend the local solution to a wide range and obtain its global solution. We have the following theorem.

Theorem 10.7.2. *If the conditions of Theorem* 10.7.1 *are satisfied, then the global classical solution of problem* (10.7.2)–(10.7.5) *exists.*

If we further improve the smoothness of initial functions, we get the following smooth solution.

Theorem 10.7.3. *If* $\varepsilon_0(x) \in H^8$, $\varphi_0(x) \in H^6$, $n_0(x) \in H^6$, *and they are functions with period* 2π, *then the globally smooth solution (i.e., the solution is a second-order derivative in* t*) of problem* (10.7.2)–(10.7.5) *exists and is unique.*

With the above results, it is easy to study the periodic initial problem of Zakharov equations, namely, to get the solutions $n(x,t)$, $\varepsilon(x,t)$ with period 2π with respect to x for (10.7.1), (10.7.2), such that they satisfy the following initial conditions:

$$\begin{aligned} &n(x,0) = n_0(x), \quad \frac{\partial n}{\partial t}(x,0) = n_1(x), \\ &\varepsilon(x,0) = \varepsilon_0(x), \end{aligned} \tag{10.7.6}$$

where $n_0(x)$, $n_1(x)$, and $\varepsilon_0(x)$ are all functions with period 2π.

Theorem 10.7.4. *If* $n_0 \in H^6$, $n_1 \in H^4$, $\varepsilon_0 \in H^8$, *and they are functions with period* 2π, *then the globally classical solution of problem* (10.7.1), (10.7.2), (10.7.6) *exists and is unique.*

For the following Cauchy problem of a kind of coupled KdV and nonlinear Schrödinger equations:

$$i\varepsilon_t + a\varepsilon_{xx} - bn\varepsilon = 0, \tag{10.7.7}$$

$$n_t + \frac{1}{2}[\beta n_{xx} + n^2 + |\varepsilon|^2]_x = 0, \tag{10.7.8}$$

$$\varepsilon|_{t=0} = \varepsilon_0(x), \quad n|_{t=0} = n_0(x), \quad (-\infty < x < \infty), \tag{10.7.9}$$

we have the following result.

Theorem 10.7.5. *If* (i) $\varepsilon_0(x), n_0(x) \in H^s$ $(s \geq 3)$ *and* (ii) *the constant coefficients* a *and* $b\beta$ *have opposite signs, then the global solutions for the Cauchy problem and the periodic problem of* (10.7.7), (10.7.8) *exist and are unique, and the solutions satisfy* $n(x,t)$, $\varepsilon(x,t) \in L^\infty(0,T;H^s)$.

For the results of the global solutions to the coupled equations of some other nonlinear evolution equations, we refer the reader to [109, 112, 319, 321].

11 The soliton movements of elementary particles in nonlinear quantum field

11.1 The elementary particles and solitons in nonlinear quantum field

What is a soliton? What are the differences between a soliton and elementary particles? These problems are worthy of being investigated. Although the concept of the soliton was discovered and proposed approximately 200 years ago and it has been used extensively in physics and other fields, its real significance and wide applications in many areas in science and engineering are currently not well known. Therefore, the soliton is worthy of extensive investigation. Obviously, the soliton is in essence different from the microscopic particles, such as the elementary particles.

It is well known that the concept of the soliton comes from water waves, which were first observed by Russell in 1834 in the movement of surface water waves in water channels with suitable widths. Russell, who rode on a horse, observed that the solitary water waves were formed in the channels and propagated several kilometers along the water channels maintaining their amplitude and outlines. These properties are very interesting because the transport features of the soliton are completely different from the features of elementary particles. This means that significant differences between solitons and elementary particles exist. It is necessary to research these differences.

Evidently, the soliton is not an elementary particle. The soliton exists in water waves, so some people thought the soliton cannot be used to describe the corpuscle features of particles. In order to solve this problem, we have to recall and elucidate the process of confirmation of the features of wave-corpuscle duality of the elementary particles.

As is well known, around the year 1900 humans discovered that the microscopic particles, such as the photon, electron, and proton, possess a wave-corpuscle duality. This feature was first verified for the photon, which was found studying optical phenomena. Thus, we thought that light is merely an electromagnetic wave. In accordance with this theory, the energy of the light should be distributed continuously in the light wave. However, the results obtained from the experiments of the “light-electron effect” and “Compton scattering” of the light rejected the above conclusion and indicated further that the energy of light does not follow a continuous but a discrete distribution; it has a quantum feature. As far as monochromatic light is concerned, the unit of the smallest energy is $\hbar\nu$, where $\hbar$ is the Planck constant and ν is the frequency of the light wave [239]. Thus, the concept of “light quantum” or “photon” was introduced and confirmed. Subsequently, on the basis of the concept of “quantum”, proposed by Planck and Einstein, De Broglie also postulated that quantum features exist for microscopic particles, such as electrons and protons. Scientists affirmed that all microscopic particles have not only corpuscle features, but also wave features, as is the case for photons,

https://doi.org/10.1515/9783110549638-011

whose wave feature is clearly exhibited and embodied in the process of propagation. At the same time, De Broglie further gave the representation of wave-corpuscle duality, which is $P = \hbar/\lambda$ and $E = \hbar\omega$, where P and E are the momentum and energy of the microscopic particle and λ and ω are its wavelength and frequency, respectively. The relation was called the De Broglie formula. De Broglie's idea was soon confirmed by experimental physicists by means of experiments showing the stripes of fraction of an electron beam passing through crystal foils, which are similar to the fraction experiments of X-rays through crystal [239].

Soon afterwards, similar results were obtained from experiments with molecular beams and atoms. These results indicated obviously that the microscopic particles all have a wave-corpuscle duality. The theory of quantum fields, which is a branch of physics still in development, was established just to describe these features.

The solitons that appeared in the above phenomena are also thought to possess wave-corpuscle duality, because the soliton has not only wave features, but also corpuscle features, because it maintains both a constant size and a constant outline in its propagations, which are properties that reveal analogy to particles. From this perspective, it would be natural to think that the soliton belongs to the microscopic particles mentioned above. In practice, Born, in the metaphase of 1900, tried to add a nonlinear term to the Maxwell equation, serving as a revision. He proposed and used further the localized singular point of the nonlinear equation to describe the electrons in this system. However, his theory did not include these successful conclusions of wave mechanics, so his new idea was not developed much further soon afterwards. Subsequently, De Broglie did further develop this idea, trying to make the theory of his ideal of dual solutions correspond to the description of the localization of microscopic particles. This theory itself had many flaws or weak points and thereupon, this theory has also not been developed [239].

On the other hand, a lot of experimental results were presented that showed that the elementary particles have an inner structure. For example, hadrons are composed of quarks or stratons, which were accepted widely. However, this experimental evidence does not suffice and direct evidence for the existence of free quarks has not been found up to this day, although many studies and experiments have been carried out. Thanks to the experimental results of deep inelastic scattering, we know that the mass of quarks, which are inside the hadrons, is very small (about several decades to several hundreds MeV). If they do really exist, then these particles with the above masses should have been easily discovered in the accelerator experiments, but this is not the case. The contradiction between these results compels us to think that these quarks are bound inside the hadrons and cannot be separated. This hypothesis is called "quark bound" or "quark imprisonment" [116, 222].

In the research of "quark imprisonment" some people thought the mass of the quark is very great, greater than that of the hadrons, so huge binding energies should be released when the hadrons are formed. Then we must expend huge energies to separate the quarks from the hadrons. In this case the quarks are "partly bound or imprisoned". However, the "permanent imprisonment of the quark" can also occur. Therefore, the phenomenon theories of the "string model" and the "pocket model" were successively proposed to describe the "quark imprisonment". In "the pocket model", the "MTT pocket" and "SLAC pocket" are proposed and included, but their starting points are different although the quarks are all bound in the matter structure resembling the pockets. These pockets may be thought of as some conformations of the solitons. In these models, the quarks are proposed to act as small insects. The solitons provide suitable bunkers or structures for the insects (quarks), or the insects can only be present on the edges of the pockets. Then huge energies (> 10 MeV) must be used, in order for these insects (quarks) to be separated. Clearly, this cannot be done, so if this theory is correct, we cannot observe the existence of free quarks [116, 222].

The string model indicates that the hadron is a string and the quarks are attached on the edges of the strings. This is analogous to the superconductive effect; the vacuums in the external part of the hadrons serve as a superconductive phase and the superconductive strings in the inner part of the hadrons resemble or mount the "magnetic lines". However, in type-II superconductors, the magnetic fields cannot penetrate into them, but they can be bound in the magnetic line tubes. In this case, we can assume that the dumpling field between the quarks can also be bound on a string. However, others thought that a lot of magnetic monopole dipoles are considered in the vacuums, so the vacuum phases can occur as a special "imprisonment phase", different from the normal phase, which provides a vacuum pressure, which impels the dumpling fields in the hadrons to bind into the pocket, so the latter cannot escape again. This phenomenon resembles the effect of extrusion of the bubbles in the liquid solutions.

The magnetic monopoles were proposed by Dirac in 1931. Subsequently, 't Hooft [289] discovered in 1974 that the magnetic monopoles are a solution of a nonlinear equation, which is also analogous to the features of the soliton. Therefore, we call it a soliton.

As a matter of fact, we can easily explain and elucidate plenty of basic problems for the magnetic monopole in the elementary particle physics using the concept of the soliton, in which some nonlinear equations are serving as the classical or approximate equations of the particles in the localized quantum field. We can find the soliton solutions of these equations and investigate their properties further. Because theories of elementary particles should possess the relativity covariant feature and Lorentz invariance, the sine-Gordon equation and φ^4 equation are quite suitable to describe the dynamic properties of the elementary particles in elementary particle physics.

Tsung-Dao Lee et al. [49, 154] researched deeply the movement properties of solitons in elementary particle systems. Their investigations indicated clearly that quark fields are basic fields, but they do not represent the lowest energy state, as their soliton states are the lowest energy state. Therefore, the hadrons observed in the natural world are just a kind of soliton. Using this theory and its conclusion, Lee et al. explained the problem of "quark imprisonment". Subsequently, they elucidated that only if the vacuum is an ideal dielectric medium with a certain resistance to the color, then the hadrons are believed to be solitons. In the explanation of "quark imprisonment", it is not necessary to introduce again the concept of the Higgs boson. At the same time, they subdivided the solitons into the topological and nontopological solitons [116, 222].

What are topological and nontopological solitons? Their investigations indicate that, in the renormalization theory of relativity localized fields, all solitons, especially stable solitons, must meet the Euler–Lagrange (EL) field equation obtained from the principle of minimum action for the Hamitonian $\delta S = 0$ and the stability condition of $\delta^2 S > 0$. In order to meet these requirements, two methods and lines are used. The differences between the two methods are only that their groups should be divided into topological and nontopological solitons. A necessary condition of stable existence of the topological solitons is the existence of a degeneracy vacuum state (basic state). Therefore, there are different degeneracy vacuum states (basic states) in the space at infinity. This implies that the systems could have different border conditions. If we consider the topological soliton, then its border conditions of the space at infinity should have different forms lacking the soliton solution. The different border conditions can be expressed by the distinctions of the topological feature. If the quantum numbers of topological charges are introduced again, then we can judge the stability of the soliton in accordance with the conservation feature of topological charge [116, 222].

Considering nontopological solitons, the case is different from topological solitons. Concretely speaking, it does not demand the condition of existence of degeneracy vacuum states (basic states). The soliton solutions of the corresponding equations all have the same border conditions in the space at infinity, no matter the soliton solution. This implies that the bell soliton belongs to this kind of solution. However, the nonlinear systems having nontopological solitons must satisfy the added condition of conservation of topological charge and there must exist a scalar field. Therefore, the solutions of nontopological solitons have a generality and are widespread in the spaces having any dimensions.

In fact, the theory of nontopological solitons is mainly applied in quantum color dynamics (QCD). QCD is the study of a theory of strong interactions. Its foundation is based on the nonabelian gauge theory. Its field equation should also be nonlinear, but its solutions are very difficult to find. At present, small solutions of nontopological solitons can be found in special cases. For example, the Knot shape solution in a one-dimensional space, the vortex solution in a two-dimensional space, the 't Hooft magnetic unijunction soliton in a three-dimensional space, and the instanton solution in Eu's four-dimensional space were only found. These problems will be investigated in detail in the following sections.

As far as the topological soliton is concerned, it may help us to solve the questions of $U_A(1)$ symmetry of localized gauge invariance in the electromagnetic theory, so we could construct the images of "quark imprisonment of topology" and the topological method of "mechanical color blindness" (the hadron states which are observed are all colorless) in QCD [116, 289]. At the same time, the static solutions with the nonsources in the nonabelian gauge field equations can provide an instance that contains not the singular points and is an organic unity possessing self-constant stability constructed and formed by the mutual actions and dependence between the field and the source. Otherwise, it can also give reasonably dynamic explanations for the quantum numbers, such as the baryon number and the singular number. On the other band, because the soliton is a general solution of corresponding classical fields, we can think that the solution gives the main and predominant contributions in the functional integral. Therefore, the above results clearly show that the movements of the solitons in the field equations play quite important rules for revealing the properties of microscopic particles in elementary particle physics.

On the other hand, the above soliton theory also promotes the development of mathematical science in the theories of quantum fields. This is due to the fact that the behavior of the amplitude of the solitons varies approximately inversely proportionally to the changes of their coupling coefficients. We could supply a new theoretical method, which is beyond the perturbation theory, in elementary particle physics. At the same time, the topological stability of the soliton comes from the dynamic mechanism and topologic features of field configurations, so the stability of nontopological solitons is reached thanks to the dynamic variation of the soliton amplitude with varying time [313]. In this case, the Noether theorem gives the conservation quantity – Noether charges. In practice, the stability of he topological solitons is related to the topological conservation quantity, which is not related to the invariance of Lagrange function existed in non-Abel theory and topogical features. However, the stability of he nontopological solitons is determinated by dynamic features of its amplitude and nöether charges in nöether theory. Therefore, the investigations of these equations promote the development of homotopy and fiber bundles in topological science [129, 131, 313]. On the contrary, in order to better know and understand the features of topological and nontopological solitons, we should research topology and homotopy [129, 131, 313].

What is topology? Topology is a concept in mathematics, related to the changing properties of collection and assembly of a series of quantities and vectors, which can be defined by the features of open sets in the measurement space. We now assume that S is a subset cluster and τ is a subset family, in which the members are called an open set. This open set is defined as follows. If we assume that A is a subset in the measurement space X and each point of set A has a spherical neighboring domain $\in A$, then A is called the open set of X. If τ meets the following conditions: (1) s and the empty set φ exist and they are also the open sets, (2) the cross set of two open sets is one open set, and (3) the parallel set of several open sets is one open set, then we say that τ is one topology space of the assembly s. The assembly s and the topology τ are together called one topologic space, which is represented by (s, t).

Homotopy is a very important concept in topology and relates to the concept of mapping. The latter is a transformation between two topological spaces and can be represented by $X \to Y$, which indicates the correspondence between the topological spaces of X and Y [129, 131, 313]. Therefore, the homotopy delineates the feature of the correspondence or mapping. Thereby, if the homotopy is equivalent to or describes the features of the mapping and it is expressed by $f_0, f_1 : X \to Y$, then we say that f_0 has homotopy with f_1, or, in other words, we can change to f_1 from f_0 between f_0 and f_1 in space Y through one continuous deformation F. The change can be expressed as $f_0 \cong f_1 : X \to Y$. Thus, F is called one homotopy from f_0 to f_1. Therefore, the homotopy relation is one equivalent relation [79, 189, 279]. It has very important significances and is used widely in topology.

11.2 The movements of topological solitons in one-dimensional space

We now use Derrick's theorem to investigate the movements of elementary particles in one-dimensional space. In this case, the Lagrange function $\mathcal{L}$ [65, 116, 222] of the system is defined as

$$L'(x) = \frac{1}{2}\partial_\mu \varphi \partial^\mu \varphi - U(\varphi), \tag{11.2.1}$$

where $L' = \mathcal{L}$. If $U(\varphi) = 0$, it corresponds to the vacuum state of the system. $U(\varphi) > 0$ in equation (11.2.1) describes a scalar field, in which the statically nonsingular soliton solution does not exist, except for the case where the dimension of the space is $D = 1$. The reasons are described as follows.

If $\varphi_s(x)$ is the soliton solution having an energy of $H = V_1 + V_2$, where

$$V_1 = \int (\Delta\varphi_s(x))^2 d^l x, \quad V_2 = \int U\varphi_s(x) d^l x,$$

then the energy corresponding to the field configuration $\varphi_s(x/a)$ can be represented by

$$H(a) = a^{D-2}V_1 + a^D V_2.$$

However, H must be stable for any variation. Especially, for the variation of one scalar field, we must have

$$\left.\frac{\delta H(a)}{\delta a}\right|_{a=1} = (D-2)V_1 + DV_2 = 0.$$

Because V_1 and V_2 are all positive, this equation has a solution in the case where $D = 1$, so the above result can be confirmed. Obviously, this conclusion can be generalized to the scalar field cases with many dimensions, rather than only the scalar field with $D = 1$.

For the pure Yang–Mills theory, which possesses a compact specification group, we can verify that one variant of Derrick's theorem, in a D-dimensional space (except for $D = 4$), is a transformation of $A_\mu = 0$, only for the static state solution with the limit energy for the equation of a pure specification field. This solution was obtained by Coleman using the scale-free property, which is similar to the scalar case. The above description is known as the theorem of "no-go" in quantum field theory. This means that the pure scalar field has only the one-dimensional soliton solution. Then the pure Yang–Mills field has only the four-dimensional soliton solution other than this solution; it has no other soliton solution [66].

Because the dynamic nontopological solitons are related to time, they are not controlled and limited in this theorem. At the same time, this theory neither limits the static scalar solitons in two- and three-dimensional spaces. In practice, the spiral and magnetic monopole soliton solutions, which will be described in the following sections, also belong to this case. Otherwise, we can affirm that the spurious particle or the instanton that appears when $D = 4$ is an exception to Coleman's and Deser's theorems.

We now investigate the features of topological solitons in a one-dimensional space [82]. Because the topological solitons demand a degeneracy vacuum, the minimum of V is not one. We now assume the minimum of V is zero, as shown in Figure 11.1.

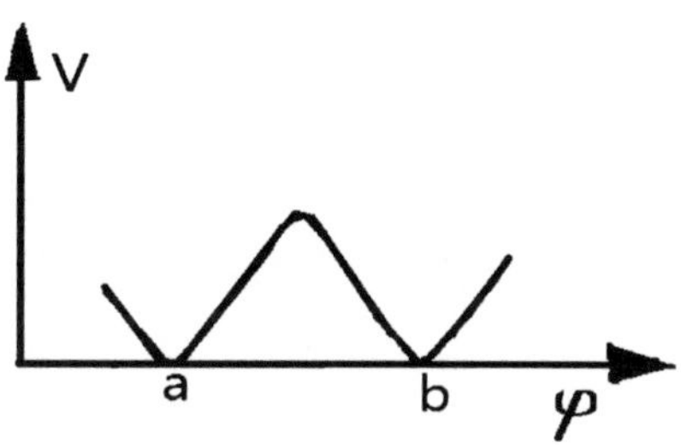

Figure 11.1: The form of $V(\varphi)$, where $V(a) = V(b) = V \cdots = 0$.

From equation (11.2.1), we obtain the dynamic equation, which is represented by

$$\frac{\partial^2\varphi}{\partial x^2} - \frac{dV}{d\varphi} = 0.$$

Let $\varphi = \varphi(x)$ be a real field, unrelated to time. Then we obtain

$$\frac{1}{2}\left(\frac{\partial\varphi}{\partial x}\right)^2 - V(\varphi) = \text{constant}. \qquad (11.2.2)$$

This equation corresponds to the dynamic equation of the particle in the nonrelativity theory, in which the space coordinate is φ, the time coordinate is x, and the mass of the particle is 1. Therefore, equation (11.2.2) is the dynamic equation of the particle

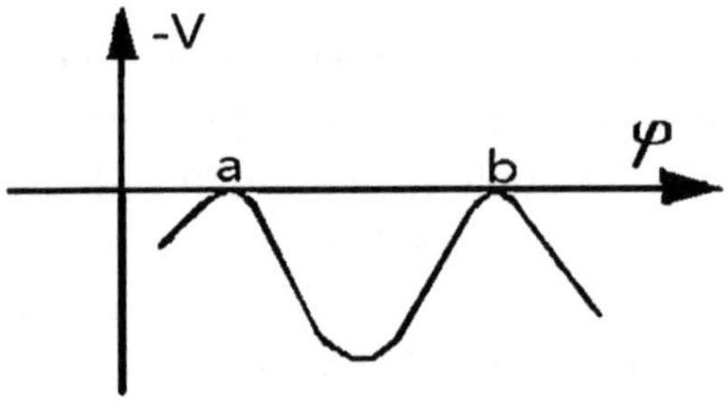

Figure 11.2: The form of $-V$.

with a mass of 1. Equation (11.2.2) shows the energy conservation of the particle, where its potential is $-V$, which is exhibited in Figure 11.2.

In this case, the kinetic energy of the particle is $\frac{1}{2}(\frac{d\varphi}{dx})^2$. We now assume that the particle is at a point at $x = -\infty$. If the particle is promoted to the right direction, then it will move, following the direction of the curve. At $x = +\infty$, the particle can reach point b. In this case, the corresponding energy is still limited. This means that the particle is not dispersed. The soliton solution is shown in Figure 11.3.

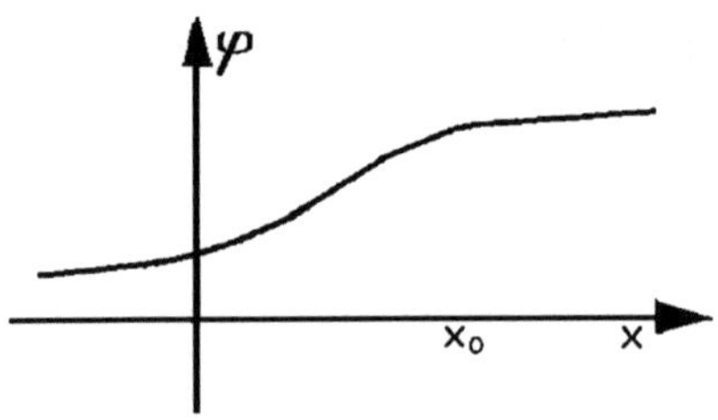

Figure 11.3: The solution of topological solitons.

Clearly, the energy of the soliton is bound between a and b, where $d\varphi/dx \to 0$ and $V \to 0$. However, its boundary conditions at $x = \pm\infty$ are different from the above results, so the solution is called the topological soliton. It may be thought to be the solution having the lowest energy and meeting the boundary conditions of $\varphi = a$ at $x \to -\infty$ and $\varphi = b$ at $x \to +\infty$. Therefore, it is stable. The soliton formed in this case is called a positive soliton. If the soliton is at point b at $x = -\infty$ and at point a at $x = +\infty$, then the soliton formed in this case is called the solution of a negative soliton. Therefore, the positive and negative solitons both exist in this case using classical field theories.

Therefore, the soliton solutions formed from equation (11.2.2) can in this case be presented together by

$$x - x_0 = \int_{\varphi_0}^{\varphi} \frac{d\varphi'}{\sqrt{2V(\varphi')}}, \tag{11.2.3}$$

where x_0 is a constant. On the other hand, we can obtain the dynamic solution of the soliton, if only the Lorentz transformation corresponding to the above soliton solutions is finished, because equation (11.2.1) possesses "/".

In practice, the most general dynamic equations used are φ^4-field equations with the potential $U(\varphi) = (\beta/4)(\varphi^2 - m^2/\beta)^2$ and the sine-Gordon equation with the potential $U(\varphi) = \sin e\varphi$ in the one-dimensional case. They all have knot soliton solutions. Their dynamic equations all have the Lorentz invariant features.

For example, the Lagrange function corresponding to the φ^4-field equation theory can be expressed by

$$L = \int L' dx = \int \left[\frac{1}{2} \partial_\alpha \varphi \partial^\alpha \varphi - \frac{\beta}{4} (\varphi^2 - m^2/\beta)^2 \right] dx \quad (\alpha = 0, 1), \tag{11.2.4}$$

where $L' = \mathcal{L}$. We transform $\varphi' = \sqrt{\beta}\varphi/m$, $x' = mx$, so equation (11.2.4) becomes

$$L = \frac{m^3}{\beta} \int \left[\frac{1}{2} \partial_\alpha \varphi' \partial^\alpha \varphi' - \frac{1}{4} (\varphi'^2 - 1)^2 \right] dx. \tag{11.2.5}$$

The corresponding equation can be presented as

$$(\partial_x^2 - \partial_t^2)\varphi' + \varphi'(1 - \varphi') = 0. \tag{11.2.6}$$

In order to find the static solutions for equation (11.2.6), it is easily written as

$$\varphi'_{xx} + \varphi'(1 - \varphi'^2) = 0. \tag{11.2.7}$$

The solutions of equation (11.2.7) are easily found. We have

$$\varphi' = \pm 1 \quad \text{(vacuum state)}, \tag{11.2.8}$$

$$\varphi' = \pm \tanh\left(\frac{x' - x'_0}{\sqrt{2}} \right), \tag{11.2.9}$$

where "+" expresses the knot soliton and "–" expresses the anti-knot soliton.

Thus, the soliton solutions of the φ^4-field equation in equation (11.2.2) should be presented as

$$\varphi = \pm (m/\sqrt{\beta}) \quad \text{(vacuum state)}, \tag{11.2.10}$$

$$\varphi = \pm \left(\frac{m}{\sqrt{\beta}} \operatorname{th}\left(\frac{m(x - x_0)}{\sqrt{2}} \right) \right), \tag{11.2.11}$$

where "+" expresses the knot soliton and "–" expresses the anti-knot soliton.

If $V(\varphi) = \frac{\beta}{4}(\varphi^2 - m^2/\beta)^2$ is substituted into equation (11.2.3), then its solution can be obtained immediately, which is shown in Figure 11.4, in which the following energy difference between the knot soliton solution and the vacuum state can be found:

$$E_{\text{knot}} - E_{\text{vacuum}} = 2\sqrt{2}m^3/3\beta. \tag{11.2.12}$$

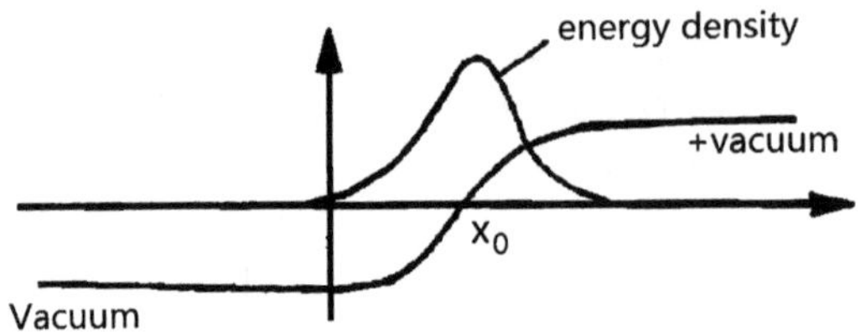

Figure 11.4: The images of a knot soliton localized at x_0 and its distribution of energy density.

The nongeneral topology charge K' associated with the knot soliton is called the twisting number. It is due to the conservation of the flow J_μ that the latter is expressed by

$$J_\mu = \frac{1}{2}\varepsilon_{\mu L}\partial^\nu \varphi \quad (\varepsilon_{01} = \varepsilon_{10} = 1,\ \varepsilon_{00} = \varepsilon_{11} = 0).$$

Then

$$K' = \int_{-\infty}^{\infty} J_0 dx = \frac{1}{2}\int_{-\infty}^{\infty} \frac{d\varphi}{dx} dx = \frac{1}{2}\varphi(x)\bigg|_{-\infty}^{\infty}. \tag{11.2.13}$$

Clearly, $K' = 1$ and -1 correspond to the knot and anti-knot solitons, respectively. They represent the states of nongeneral mapping, while $K' = 0$ corresponds to the vacuum state. Because the topologic charges are absolutely conserved in this case, it is, in general, difficult to dissipate and apply the field configurations to the vacuum state. This means that the action of the field configurations seems to be one infinite potential barrier, which prevents its variation and decay to the vacuum state. However, this change can be made if a large amount of energy is added. This indicates that the knot soliton is quite stable. This stability of the soliton can be thought to be the results produced by the degenerate vacuum. In this case, the two vacuum states may be transformed to each other in the case of variation of $\phi \to -\varphi$. The knot state is inserted just between two vacuum states at $x = \pm\infty$. They approach each vacuum state at the infinitely large position. However, they are only evidently different at $x = x_0$, where the energy density is at its maximum. In this case, the homotopy image is a general image which maps the field energy $\varphi = \pm 1$ to the points $x = \pm\infty$. Therefore, it is necessary to obtain infinite energy, in order to change and distort the images of the knot state to any vacuum state.

As is described above, the knot soliton can be used to describe the hadrons. The image of the "SLAC pocket" of the quark model for the hadrons [18, 82] is a theory based on the above theory, in which the quarks are distributed on the edges of the pockets, as shown in Figure 11.4, and cannot escape from the pocket. Therefore, this theory can explain why we have not observed the free quarks up to now, so it is natural to deduce the existence of the phenomenon of "quark imprisonment" in this case.

11.3 The movements of topological solitons in two-dimensional space

The investigations indicated that the properties of topological solitons in a two-dimensional space resemble those of vortex structures formed by the magnetic flux quantization in the type-II superconductors. Thus, we first must know the properties of magnetic flux quantization formed in the type-II superconductors, in order to elucidate the features of the topological soliton [116, 222].

As is well known, superconductors can be made from some specific elements, compounds or alloys, when their temperature T is lower than the critical temperature T_C, i.e., $T \leq T_C$, in which the electric current in the superconductor will flow forever without being damped, or, in other words, without any resistance. Such a phenomenon is referred to as perfect conductivity. This has been observed in some experiments, when the materials are in the superconducting state. In this case, all magnetic fluxes in the materials repel each other completely. This will result in the presence of zero magnetic fields inside the superconducting material. Similarly, the magnetic fluxes induced by an external magnetic field cannot penetrate into the superconducting materials. This phenomenon is called perfect anti-magnetism or the Maissner effect [241] and is illustrated in Figure 11.5.

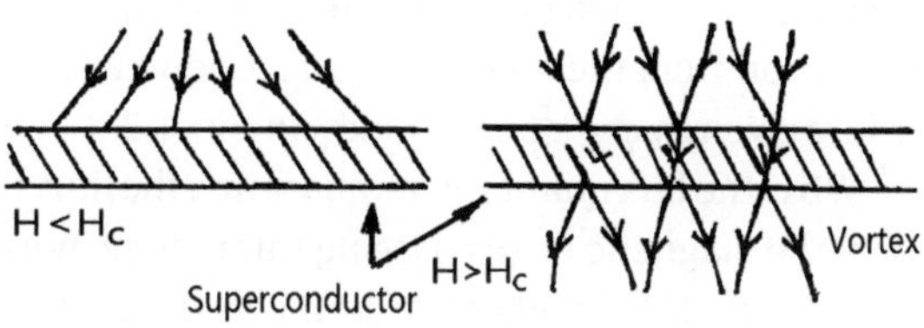

Figure 11.5: Maissner effect in the superconductor.

How can this phenomenon be explained? After more than 40 years of research, Bardeen, Cooper, and Schreiffier proposed the new idea of Cooper pairs of electrons and established the microscopic theory of superconductivity at low temperature to explain and elucidate the superconductive phenomenon. This is called the Bardeen–Cooper–Schreiffier (BCS) theory of superconductivity [14, 17, 51, 272, 298], which was established in 1957 on the basis of the mechanism of electron-phonon interaction proposed by Frohlich.

According to this theory, the electrons with opposite momenta and anti-parallel spins form the pairs, when their attraction due to the interaction between the electron and phonon in these materials overcomes and exceeds the Coulomb repulsion between them. The Cooper pairs condense to form a minimum energy state, resulting in some quantum states, which are highly ordered and coherent over a long range, in which there is essentially no energy exchange between the electron pairs and the lattice. Thus, the electron pairs are no longer scattered by the lattice and flow freely without electric resistance. Then the superconductivity appears. The electron pair in a superconductive state is somewhat similar to a diatomic molecule, but it is not as tightly bound as a molecule. The size of an electron pair, which gives the coherent length, is approximately 10^{-4} cm. A simple calculation shows that there can be up to 10^6 electron pairs in a sphere of 10^{-4} cm in diameter. Therefore, perturbation to any of the electron pairs would certainly affect all others. Thus, various macroscopic quantum effects can be expected to occur in the materials, which are some coherent and long range ordered states. Magnetic flux quantization and a vortex structure in the type-II superconductors occur in this case.

As far as quantization of magnetic flux is concerned, we consider mainly the superconductive rings. Assume that a magnetic field is applied at $T > T_c$. Then the magnetic flux lines ϕ_0 produced by the external field pass through and penetrate into the body of the ring. When the temperature of the superconductive material is lowered to a value below T_c, if the external magnetic field is removed, then the magnetic inductions inside the body of the circular ring are equal to zero ($\overline{B} = 0$) because the ring is in the superconductive state and the magnetic field produced by the superconductive current cancels the magnetic field in the ring. However, a part of the magnetic fluxes in the hole of the ring remains because the induced current in the ring vanishes. These residual magnetic fluxes are referred to as frozen magnetic fluxes. It was observed experimentally that the frozen magnetic fluxes are discrete or quantized. Using the macroscopic quantum wave function from the theory of superconductivity, it can be shown that the magnetic fluxes are given by $\varphi = \oiint L' ds = n\phi_0$ $(n = 0, 1, 2, 3 \ldots)$, where $L' = L$, $\phi_0 = hc/2e = 2.07 \times 10^{-15}$ Wb is the flux quantum, representing the flux of one magnetic flux line. This means that the magnetic fluxes passing through the hole of the ring can only be a multiple of ϕ_0 [84, 271, 272]. In other words, the magnetic field lines are discrete. What does this imply? If the magnetic fluxes of an applied magnetic field are exactly n, then the magnetic fluxes through the hole are $n\phi_0$. However, what are the magnetic fluxes through the hole if the fluxes of an applied magnetic field are $(n + 1/4)\phi_0$? According to the results, the magnetic fluxes cannot be $(n + 1/4)\phi_0$. As a matter of fact, it should only be $n\phi_0$. Similarly, if the fluxes of an applied magnetic field are $(n+3/4)\phi_0$, the magnetic fluxes passing through the hole are not $(n+3/4)\phi_0$, but rather $(n+1)\phi_0$. Therefore, the magnetic fluxes passing through the hole of the circular ring are always quantized. An experiment conducted in 1961 surely proved this. It indicated that magnetic flux does exhibit discrete or quantized characteristics on a macroscopic scale. The above experiment was the first demonstration of a macroscopic quantum effect. Based on the quantization of magnetic flux, we can build a "quantum magnetometer" which can be used to measure weak magnetic fields with a sensitivity of 3×10^{-7} Oersted. A slight modification of this device would allow us to measure electric currents with strengths as low as 2.5×10^{-9} A.

We now research the structure of vortex lines in type-II superconductors. The superconductors discussed above are referred to as type-I superconductors. This type of

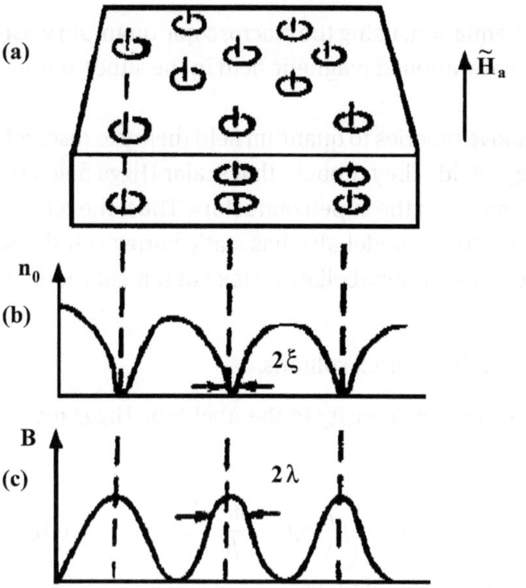

Figure 11.6: Current and vortex lines distributions in a type-II superconductor.

superconductor exhibits the Maissner effect perfectly when the external applied field is higher than a critical magnetic value $\overline{H}^C$, as shown in Figure 11.6.

There exists another type of materials, such as the NbTi alloy and the Nb_3Sn compound, in which the magnetic field partially penetrates inside the material when the external field $\overline{H}$ is greater than the lower critical magnetic field $\overline{H}_c$, but less than the upper critical field $\overline{H}_c$ [49, 129, 131, 154, 222, 239, 289, 313]. This kind of superconductor is classified as type-II superconductors and is characterized by a Ginzburg–Landau (GL) parameter, K, greater than $1/\sqrt{2}$, i.e., $K > 1/\sqrt{2}$. Studies using the Bitter method showed that the penetration of a magnetic field results in some small regions changing from the superconductive to the normal state. These small regions in the normal state are of cylindrical shape and regularly arranged in the superconductor, as shown in Figure 11.6. Each cylindrical region is called a vortex (or magnetic field line) [84, 271, 272]. The vortex lines are similar to the vortex structure formed in a turbulent flow of fluid. Both theoretical analysis and experimental measurements have shown that the magnetic flux associated with one vortex is exactly equal to one magnetic flux quantum, ϕ_0. When the applied field $\overline{H} \geq \overline{H}_{c1}$, the magnetic field penetrates into the superconductor in the form of vortex lines, increasing one by one; the vortex lines or magnetic lines within the cylindrical structure are inserted one by one with the unit of $\phi_0 = hc/2e$ into type-II superconductors in an order forming a mixed phase. These vortex line structures of quantization are called the Abrikosov structure of the superconductor. In ideal type-II superconductors and stable states, these vortex lines are constructed and arranged as triangular structures. The structures can be obtained from the solutions of the nonlinear GL equation, using the macroscopic quantum wave function under the action of an externally applied magnetic field in the superconductors. Therefore, it is correct.

Nielsen and Olesen applied the above theories to quantum field theory to research the properties of vortex lines in a Higgs field. They thought that scalar Higgs fields resemble the functions of ordered parameters in the superconductors. Thus, the relativity field theory of the similar Abel-type Higgs model also has static vortex solutions. We investigate the properties of Abelian and nonabelian vortex soliton solutions in the following section.

(1) The properties of solutions of Abelian vortex solitons.

In the two-dimensional case, the Lagrange density in the Abel-type Higgs model can be represented by

$$L' = -\frac{1}{4}F_{\mu\nu}F^{\mu\nu} + \frac{1}{2}(D_\mu\varphi)^* D^\mu\varphi - \frac{1}{4}\beta\left(\varphi\varphi^* - \frac{m^2}{\beta}\right)^2, \tag{11.3.1}$$

$$F_{\mu\nu} = \partial_\mu A_\nu - \partial_\nu A_\mu, \quad D_\mu\varphi = (\partial_\mu - ieA_\mu)\varphi, \tag{11.3.2}$$

where $L' = \mathcal{L}$, indicating the interaction between the electromagnetic field $A_\mu(x)$ and the complex scalar Higgs field $\varphi(x)$. From the EL equation and equation (11.3.1), we obtain the corresponding dynamic equation [84, 159, 190, 227], which is as follows:

$$\partial^\mu F_{\mu\nu} = j_\mu = -\frac{1}{2}ie(\varphi^*\partial_\mu\varphi - \varphi\partial_\mu\varphi^*) + e^2 A_\mu\varphi\varphi^*, \tag{11.3.3}$$

$$D_\mu D^\mu\varphi = -\beta\varphi(\varphi\varphi^* - m^2/\beta). \tag{11.3.4}$$

Equations (11.3.3)–(11.3.4) are just GL equations in this case, but their solutions are very difficult to find. Thus, we have to fit them into the cylindrical coordinates to find the asymptotic solutions in this ansatz [21, 189].

We now assume

$$A_0 = 0, \quad A = \hat{\theta}A(r), \quad \varphi = f(r)e^{in\theta}, \quad r^2 = x^2 + y^2. \tag{11.3.5}$$

Then equations (11.3.3)–(11.3.4) become

$$-\frac{1}{r}\frac{d}{dr}\left(r\frac{d}{dr}f\right) + \left[\left(\frac{n}{r} - eA\right)^2 + (\beta^2 - m^2/\beta)\right]f = 0, \tag{11.3.6}$$

$$-\frac{d}{dr}\left(\frac{1}{r}\frac{d}{dr}(rA)\right) + \left(Ae^2 - \frac{ne}{r}\right)f^2 = 0. \tag{11.3.7}$$

In order to find the asymptotic solutions of equations (11.3.6)–(11.3.7), we require that the energy of the vortex lines of unit length is limited. This implies that the vortex fields must have the following asymptotic solutions:

$$f(r) \to 1 - \text{const}\exp[1 - r/\xi]m/\beta^{1/2} \quad (r \to \infty), \tag{11.3.8}$$

$$A(r) \to (n/er) + \text{const}\exp[1 - r/\delta] \quad (r \to \infty), \tag{11.3.9}$$

where the particle mass in the Higgs scalar field is $m_s = \sqrt{2}m$, but the particle mass in the directive field $A_\mu(x)$ is $m_A = m_e/\sqrt{\beta}$. The spontaneous break of symmetry of the field occurs when the conditions cause self-interactions to form of the φ^4 scalar field and the imaginary mass of the particles. In this case, the standard particles – photons – "eat off" the Goldstone particles generated by it, so it increases its own mass. The corresponding coherent length $\xi = \sqrt{2}/m_s$ gives the scale of the space change of the Higgs field. The penetration depth of the electromagnetic field $\delta = 1/m_A$ describes the amplitude of space variation of the field. These space variations of the solution of the vortex soliton are shown in Figure 11.7, where the GL parameter is represented by

$$K = \delta/\xi = m_s/\sqrt{2}m_A = \sqrt{\beta}/e. \tag{11.3.10}$$

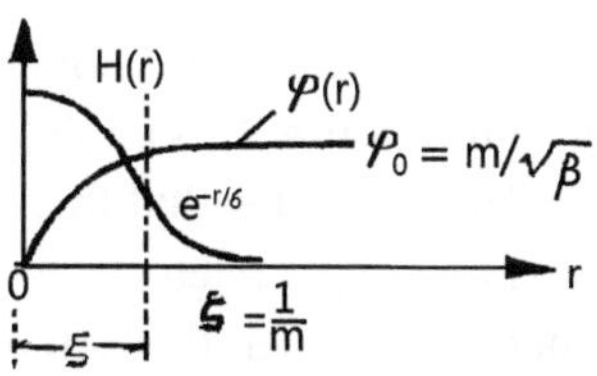

Figure 11.7: The variations of $\varphi(r)$ and $H(r)$ of the vortex lines.

Clearly, the superconductors are separated into two kinds when using this parameter, i.e., if $K < 1/\sqrt{2}$, then it is a type-I superconductor, but it is a type-II superconductor if $K > 1/\sqrt{2}$, in which case a Nielsen–Olesen vortex soliton appears in this system. However, there is possibly a potential Nielsen–Olesen solution of the vortex soliton for any K in this system.

Now, we discuss again the topologic features of these vortex solitons [14, 37, 298].

We now focus mainly on F_{12} in equation (11.3.2) for finding the features of the magnetic vortex along the Z direction, or, in other words, we should find the sizes of the flux through the unit area in the (x, y)-plane by virtue of the above theory. In this case, we use the parameter $\varphi = |\varphi|e^{i\alpha}$ to change the quantity of the Higgs field. Then the fluxes through the area formed by the closed loop P in Figure 11.8 can be represented by

$$\varphi(r) = \int F_{12}dxdy = \int_P A_i dx^i = -\frac{1}{e}\int_P \partial_i \alpha dx^i,$$

where $j_\mu = 0$ along the line P is used. From the demand for a single value feature of $\varphi(x)$, we obtain

$$\varphi(r) = \frac{1}{e}[\partial(2\pi) - \partial(0)] = \frac{2\pi}{e}n = n\varphi_0, \quad (n = 0, \pm 1, \pm 2, \ldots). \tag{11.3.11}$$

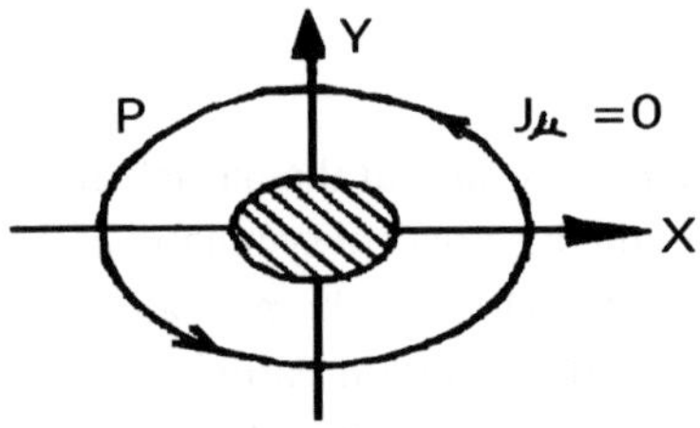

Figure 11.8: The integration along path P surrounding the vortices.

This indicates that the fluxes formed in this case are quantized. The quantum fluxes are just total topologic charges of the vortices surrounded by the path P.

In order to explain the topology features of the magnetic fluxes, we here notice that the vacuum states can be determined by the condition $|\varphi| = \varphi_0 = \sqrt{m^2/2\beta}$.

In this model, because we require only that the energy of the vortex line of unit length is limited, there is necessarily a plural scalar field. Its asymptotic form should be represented by $\varphi(\theta) = e^{in\theta}\varphi_0$, where θ is the polar angle in the two-dimensional plane and n is an integer due to the requirement of a single value feature. This implies that we can only determinate φ for a phase factor $\alpha = n\theta$. This indicates that there is one degenerate round vacuum in the plural plane, marked by the parameter α. We now assume that the round in the (x, y)-plane is represented by R. When the round is shifted, the phase factor $\alpha(x, y) = \chi(\theta)$ can be changed from zero to $2\pi n$. Thus, $\chi(\theta)$ gives just one image, which is formed by means of the mapping from the real round to the round in one plural φ inner space, as shown in Figure 11.9. This mapping can be represented by $U(1) \to S^1$. Therefore, the class of the image is characterized by

$$\pi_1(U(1)) = Z \quad \text{(the integer set).} \tag{11.3.12}$$

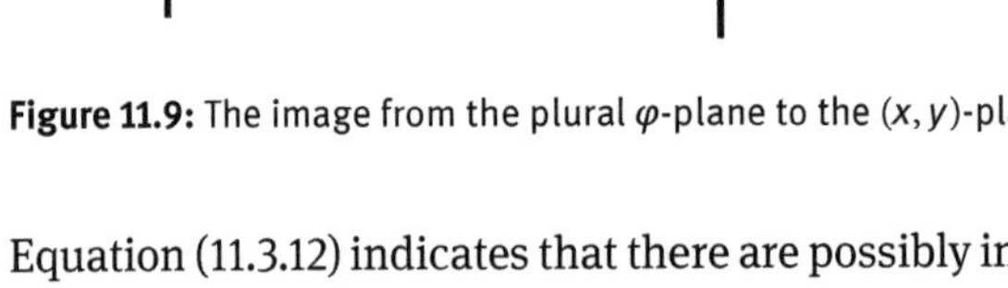

Figure 11.9: The image from the plural φ-plane to the (x, y)-plane.

Equation (11.3.12) indicates that there are possibly infinite vortices with discrete fluxes $\varphi = n\varphi_0$ ($n = 0, \pm 1, \pm 2, \ldots$) in this case, where the integer, which marks the class of each homotopy, is called the winding number. Equation (11.3.12) expresses also the rotation number in the φ-plane corresponding to a rotation of 2π in the (x, y)-plane. Therefore, the pure fluxes of the vortex are directly proportional to the winding number. Meanwhile, if the winding numbers constructed by the field are nonzero, then the corresponding fluxes are also nonzero. Owing to the conservation of topological charges, the configuration of the field having certain winding numbers cannot be changed as the configuration without winding numbers, i.e., it cannot become and deform continuously to the case of $\alpha(\theta) = $ constant, because it would require infinite amounts of energy. This means that the solutions of the vortex soliton obtained as mentioned above are stable, so the shapes of the vortices are constant in the process of topologic variations and it is a nongeneral mapping.

On the other hand, we can also find the energy of each vortex line. For the vortex with n unit magnetic flux, its energy of unit length can be represented by [298]

$$\begin{aligned}
&\varepsilon_n > n\pi m^2\sqrt{2}k/\beta, \quad (k > 1/\sqrt{2}),\\
&\varepsilon_n = n\pi^2 m^2/\beta = n\pi m^2 A/e^2, \quad (k = 1\sqrt{2}),\\
&\varepsilon_n > n\pi m^2/\beta, \quad (k < 1/\sqrt{2}).
\end{aligned} \tag{11.3.13}$$

As far as the interaction between the vortex lines is concerned, they are possibly attracting each other when $k > 1/\sqrt{2}$, they are mutually repulsive when $k < 1/\sqrt{2}$, and the nature of the interaction is uncertain when $k = 1\sqrt{2}$, or they may have any interaction. Matrion et al. found that $\varepsilon_2 > 2\varepsilon_1$ for all $k < 1/\sqrt{2}$ using the numerical simulation method. The investigations affirmed that the presence of two types of quantized vortices is not advantageous for the energy in a type-II superconductor. Bogomolny obtained the same conclusion for the unit fluxes at $n \geq 2$ using the general analytic method of energy functionals. Thus, one vortex, which has n unit magnetic fluxes, can split into n unit vortices having the same topology and equivalent configurations at $k > 1/\sqrt{2}$.

(2) The properties of solutions of nonabelian vortex solitons.

Tze and Ezawn extended and promoted the above Higgs model to research the properties of nonabelian vortex solitons. They obtained the following results.

For a specifical group G, if we demand that the energy of unit length of a static vortex solution having axial symmetry is limited, then this implies that the Higgs scalar must be a covariance constant, when the radius tends to infinity. In this case we obtain the following relation:

$$l_\mu\varphi = (\partial_\mu - iet^\alpha A_\mu)\varphi \to 0 \quad (r \to \infty) \text{ and } (r^2 = x^2 + y^2), \tag{11.3.14}$$

where t^α is a matrix expression generating an element group acting in the φ-plane and A_μ^α is the specific field, which must be accompanied by an expression belonging to the G group. Considering these conditions, we found that the values of the φ-field, at any two points P_1 and P_2 along the path P at the position of a great r, are represented by

$$\varphi(P_2) = s(P_2, P_1)\varphi(P_1), \tag{11.3.15}$$

where $s(P_2, P_1) = T\exp[-ie\int_{P_1}^{P_2} t^\alpha A_\mu^\alpha(x)dx^\mu]$ is the nonintegrable factor and T is the operator arranging the time of the matrix along the path. For the ring path marked by the angle parameter θ, the group decided by the phase factor $s(\theta)$ determines the sort of allowed vortices in this model. For example, the Abel model

$$s(\theta) = e^{in\theta} \in U(1) \tag{11.3.16}$$

leads to the vortex solutions having the allowed fluxes $\varphi_n = n\varphi_0$ $(n = 0, \pm 1, \ldots)$.

However, for the case where $G = \mathrm{SU}(2)$, the results are different from the above results. If the gauge invariance of the double state Higgs scalar is broken, then we obtain

$$s(\theta) = \exp(2in\theta\tau_3) \in \mathrm{SU}(2). \tag{11.3.17}$$

In this case, from the total symmetry of SU(2), we confirm that this theory cannot have a vortex solution. Because $\pi_1(\mathrm{SU}(2)) = 0$, this indicates clearly that any simple vortex solution can deform continuously due to the vacuum, so this solution has no topologic stability.

However, when the scalar triplet state is used to break the symmetry, this will lead to the following relation:

$$s(\theta) = \exp(2in\theta\tau_3) \in \mathrm{SO}(3). \tag{11.3.18}$$

In this case, we can obtain the following relation from the homotopy theory:

$$\pi_1(\mathrm{SO}(3)) = \pi_1(\mathrm{SU}(2)/Z_2) = Z_2. \tag{11.3.19}$$

This implies that this theory possibly has the flux unit solutions of 0, +1, and −1, instead of other vortices. This indicates that the Abelian theory and the nonabelian theory are different; the latter has only some limited solutions.

11.4 The magnetic unipolar solutions in three-dimensional spaces

The so-called magnetic unipolar is a minimum magnet element that has only one magnetic pole. This concept was introduced into the quantum field by Dirac in 1931 [116, 222]. Its appearance and properties are described below.

As is known from quantum field theory, the wave function $\psi(x_1, x_2, x_3, t)$ can be multiplied by a phase factor e^{ir} to give another wave function $\Psi = e^{ir}\psi$. In this case, the phase factor r is a function of (x_1, x_2, x_3, t), i.e., $\Psi(x_1, x_2, x_3, t) = \exp[ir(x_1, x_2, x_3, t)]\psi(x_1, x_2, x_3, t)$.

In this case, we have the relation

$$\frac{\partial\Psi}{\partial x_i} = e^{ir}\left(\frac{\partial}{\partial x_i} + i\Pi_i\right)\psi \quad \left(\text{here } \Pi_i = \frac{\partial r}{\partial x_i}\right) \quad (i = 1, 2, 3).$$

Therefore, when r is related to (x_1, x_2, x_3, t), this operator has the transformation relation of $\frac{\partial}{\partial x_i} \to \frac{\partial r}{\partial x_i} + i\Pi_i$. This is similar to the change in relationship of the momentum operator of the electron in the electric field A, i.e., $\hat{p} \to \hat{p} + eA_i$ or $\frac{\partial}{\partial x_i} \to \frac{\partial}{\partial x_i} + ieA_i$.

If $\Pi_i = eA_i$, the two relations mentioned above are identical. This means that the nonintegrable factor $r(x_1, x_2, x_3, t)$ introduced is identical to the electromagnetic potential A, introduced here. When we follow a closed loop once, the total change of the phase r can be represented by

$$(\Delta r)_{\text{loop}} + 2\pi n = \oint_{\text{loop}} \Pi_i dx_i = e \oint A_i dx_i = e \iint_{\substack{\text{look face}\\ \text{of loop}}} H' dS,$$

where n is the integer and $H' = \mathcal{H} = \nabla \times A$, $\iint H' dS$ is the magnetic flux through the loop hook face surrounding the closed curve, which is related closely to the variation of the phase.

We now consider the domain in which ψ approaches zero. If $\psi = 0$, then r is completely uncertain, but if ψ approaches zero, then its small variation will result in evident changes of r. If the two conditions are satisfied along one line, then this line is called the node line. Thus, several node lines can occur at the position of $\psi = 0$. We now assume that some wave functions contain only one node line that has only one end point. In this case, this end point is the singular point of the field. If we take a closed hook face, which surrounds the singular point (where $(\Delta r)_{\text{loop}} = 0$), then e is multiplied by the total of magnetic fluxes in the closed loop to obtain $2\pi n$, i.e., $e \oiint H' dS = 2\pi n$, where $H' = \mathcal{H}$.

If the magnetic fluxes traversing the closed hook face are not zero, this implies that there is only one magnetic unipolar in this closed hook face. If its strength is expressed by q_n, then the result, $\oiint H' dS = 4\pi q_m$, is true. This conclusion is consistent with the Gauss theorem in the study of electricity. Thus, it may be called the Gauss theorem in magnetism. This indicates that the magnetic fluxes traversing any closed face which surrounds the magnetic unipolar are equal to 4π times q_n, where $q_m = ne/2$. This indicates clearly that the particle charges are related closely to the strength of the magnetic unipolar. Therefore, we conclude that the charges of all charged particles must be quantized in nature.

Otherwise, if the magnetic unipolars appear in quantum theory, then it is quite necessary to obtain this conclusion from quantum mechanics. Similarly, if the magnetic unipolar really exists, then the Maxwell electromagnetic equations in electromagnetism can be express in a symmetric form. Just so, plenty of experimental and theoretical research has been carried out after Dirac's magnetic unipolar idea was proposed, seeking to demonstrate the existence of the magnetic unipolar in nature and in laboratory settings. Blas and Cabrera, from Stanford University, have measured accurately the changes of magnetic fluxes in superconductive niobium coils. After 151 days, they observed and measured a sudden increase of magnetic flux in one experiment, which they thought was induced by magnetic unipolars. Thus, they claimed to have demonstrated the existence of the magnetic unipolar.

However, here we should point out that there are some bothersome node lines and singular strings in Dirac's magnetic unipolar, which are not natural and easy to understand. 't Hooft in the Netherlands in 1974 [289] and Polyakov in the Soviet Union in 1975–1977 [251] both pointed out that the mass of the magnetic unipolar exceeds that of the proton approximately 5 000 times (while others thought that the mass of the magnetic unipolar is 10^{16} times greater than that of the proton), so they thought that magnetic unipolars could not have singular strings, but represent a soliton solution of the nonlinear partial differential equation if the Dirac electromagnetic $U(1)$ specification group is inserted into the nonabelian compact specification group. Just so, we here investigate further the properties of the magnetic unipolar using Dirac's model.

First, we research the 't Hooft structure [289] of compact electric dynamics, as shown in Figure 11.10, in which Φ is the magnetic flux coming into the ball and P_0 is a path surrounding the magnetic line, where the potential along P_0 must be a pure specific.

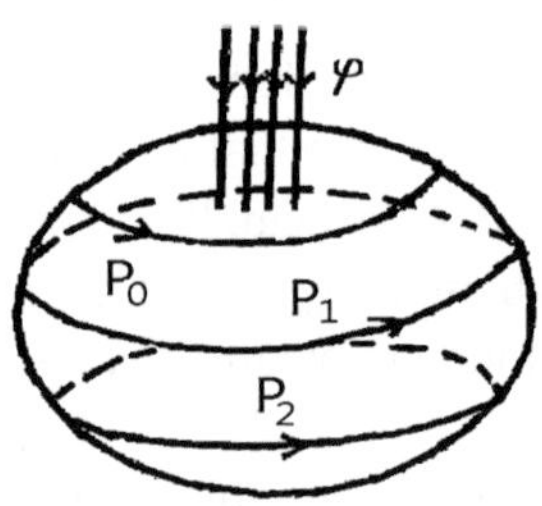

Figure 11.10: The compact electric dynamics structure by 't Hooft.

Because of the single value feature for the whole charge field, we obtain

$$\Phi = \oint P_0 A_i dx^i = \frac{2\pi n}{e}.$$

For the Abel theory, the fluxes must flow fully out from the ball. This means that the outside line P_0 cannot shift and change continuously to a constant ($P_0 \to P_1 \to P_2 \to \cdots \to$ the south pole). This implies that one Dirac spring is required in the case of an Abelian magnetic unipolar. On the other hand, if the electromagnetic $U(1)$ specification group is put into one nonabelian compact group, then the magnetic unipolar is not required to attach to the singular spring. For example, in the SO(3) specification field, the rotation of 4π can vary as a constant in the south pole because its degrees of freedom are increased in the specification transform. Therefore, this theory includes the magnetic unipolar without the spring. Then the magnetic unipolar without the spring and with the magnetic charge of $q_m = n/2e$ ($n = \pm 1, \pm 2, \ldots$) also exists in this case. In order to carry out this idea, 't Hooft et al. [289] investigated the properties of the interaction between the specification fields A_μ and the Higgs isospin vector field φ^u

having SO(3) specification invariance using the following Lagrange density function possessing SO(3) specification invariance in the south pole:

$$\begin{aligned} L' &= -\frac{1}{4}F^{\alpha}_{\mu\nu}F^{\alpha\mu\nu} + \frac{1}{2}D_{\mu}\varphi^{\alpha}D^{\mu}\varphi^{\alpha} - \frac{\beta}{4}(\varphi^{\alpha}\varphi^{\alpha} - m^2/\beta)^2, \\ F^{\alpha}_{\mu\nu} &= \partial_{\mu}A^{\alpha}_{\nu} - \partial_{\nu}A^{\alpha}_{\mu} + e\varepsilon^{abc}A^{b}_{\mu}A^{c}_{\nu}, \\ D_{\mu}\varphi^{\alpha} &= \partial_{\mu}\varphi^{\alpha} + e\varepsilon^{abc}A^{b}_{\mu}\varphi^{c} \quad (a = 1,2,3). \end{aligned} \tag{11.4.1}$$

Obviously, this is the Georgi–Glashow model of electromagnetic interactions having SO(3) specification invariance, namely, the Georgi–Glashow model of weak electromagnetic interaction. It describes the features of the photon without mass and having a dual and charging middle vector boson. The latter acquired the masses from the Higgs mechanism. In this case, the existence of the degenerated vacuum of $\varphi_0^2 = m^2/\beta$ results in the spontaneous break of symmetry of the SO(3) specification, so the $U(1)$ specification symmetry remains the invariance.

By means of the EL equation and from equation (11.4.1) we obtain the classic dynamic equations:

$$\begin{aligned} D^{\mu}F^{\alpha}_{\mu\nu} &= -e\varepsilon^{abc}\varphi^{b}D_{r}\varphi^{c}, \\ D^{\mu}D_{\mu}\varphi^{\alpha} &= -\beta\varphi^{a}(\varphi^{\alpha} - m^2/\beta). \end{aligned} \tag{11.4.2}$$

Meanwhile, 't Hooft and Polyakov found the solution of the magnetic unipolar for the case of static spherical symmetry [49], which is expressed by

$$\begin{aligned} &A^{a}_{0} = 0, \quad A^{\alpha}_{i} = \varepsilon_{aij}x_i[1 - K(r)]er^2, \\ &\varphi^{a} = -x_a H(r)/er^2, \quad r^2 = x^2 + y^2 + z^2. \end{aligned} \tag{11.4.3}$$

They represent the simple forms of the radial equations of equation (11.4.2) as

$$\begin{aligned} r^2K'' &= K(K^2 - 1) + KH^2, \\ r^2H'' &= 2HK^2 + \beta/e^2(H^2 - c^2r^2H) \quad (c = M_c/\beta^{1/2}). \end{aligned} \tag{11.4.4}$$

Here, we are interested only in the solutions having the same solutions of asymptotic forms along the outward direction of the radius of the spheroid, which is significant because it approaches one pure specification ($F^{a}_{\mu\nu} = 0$) ($K(0) = \pm 1$), which is shown in Figure 11.11. This solution is called the magnetic unipolar. The energy or mass of one magnetic unipolar was obtained using the numerical calculation, which is represented by

$$M = 4z\frac{m}{e\beta}f(\beta/e^2) = \frac{M}{\alpha}f(\beta/e^2),$$

where $\alpha = e^2/4z$, f is a monotonous, slowly rising function, $f(0) = 1$, and M_{ω} is the mass of the vector boson. Because $1/\alpha = 137$, we estimate $M_{\omega} \approx 50$ GeV, so we know that the mass of the magnetic unipolar is very great.

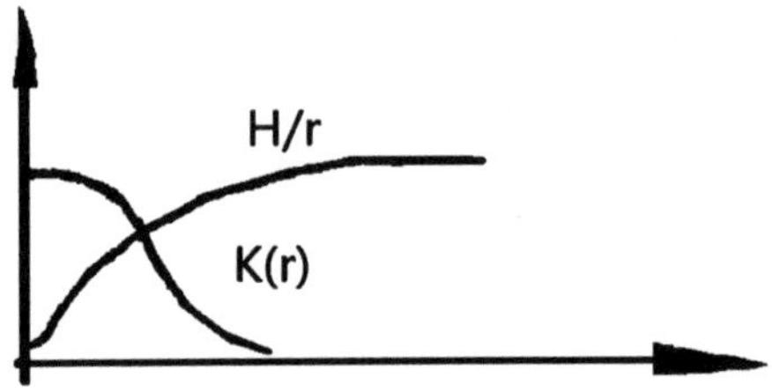

Figure 11.11: The magnetic unipolar solution.

As is well known, the quantities of topologic invariants are the magnetic charges in a three-dimensional space. In order to elucidate the topologic nongenerality or stability of soliton solutions, 't Hooft structured electromagnetic tensors having the specification invariant [116, 222], which are represented by

$$F_{\mu\nu} = \overline{\varphi}^a F^\alpha_{\mu\nu} - \frac{1}{e}\varepsilon^{abc}\overline{\varphi}^a D_\mu \overline{\varphi}^c,$$
$$\overline{\varphi}^a = \varphi^a/|\varphi|. \tag{11.4.5}$$

They have rewritten the above equations as

$$F_{\mu\nu} = \partial_\mu B_\nu - \partial_\nu B_\mu - \frac{1}{e}\varepsilon^{abc}\overline{\varphi}^a \partial_\mu \overline{\varphi}^b \partial_\nu \overline{\varphi}^c,$$
$$B_\mu = \overline{\varphi}^a A^a_\mu. \tag{11.4.6}$$

Substituting equation (11.4.3) into equation (11.4.6), we obtain

$$F_{ij} = \varepsilon_{ijk} \chi_k / er^3. \tag{11.4.7}$$

Based on the topological concept and its significance [141], F_{ij} should correspond to the magnetic field of a point magnetic unipolar with a magnetic charge of $q_m = 1/2e$. If the symbols in equation (11.4.3) are all changed, then we obtain the values of an anti-magnetic unipolar [21, 227]. At present, we elucidate the sources of the topology of the magnetic charge.

If there is no singularity of the spring in B_μ, then the magnetic flow that appears in this case can be expressed as

$$^*j_\mu = \partial^{\nu^*} F_{\mu\nu} = \frac{1}{2e}\varepsilon_{\mu\nu\alpha\beta}\varepsilon^{abc}\partial^\nu(\overline{\varphi}^a \partial^\alpha \overline{\varphi}^b \partial^\beta \overline{\varphi}^c).$$

We can verify that j_μ is the conservation, i.e., $\partial^*_\mu j_\mu = 0$. However, this topologic flow is not the Noether flow. The charge associated with this topologic flow cannot form the symmetry of the Lagrange quantity. In this case, the magnetic flux or magnetic charge is denoted

$$\Phi = 4zq_m = \int d^3x^* j_0 = \partial^{\nu^*} F_{\mu\nu} = \frac{1}{2e}\oint_{s_k^2} \varepsilon_{ijk}\bar{\varphi}^a \partial_j \bar{\varphi}^b \partial_k \bar{\varphi}^c (d^2\sigma)_i, \tag{11.4.8}$$

where s_k^2 is a ball with radius R (in the limit case, $R \to \infty$). Because the ball can be expressed by two parameter coordinates ξ_α ($\alpha = 1, 2$), the above representation can be expressed by

$$4zq_m = \frac{1}{e} \oint_{s_k^2} d^2\xi \left[\frac{2}{2} \varepsilon_{\alpha\beta} \varepsilon^{abc} \overline{\varphi}^a \partial_\alpha \overline{\varphi}^b \partial_\beta \overline{\varphi}^c \right] = \frac{1}{e} \int d^2\xi \sqrt{g} \quad (g = \det(\partial_\alpha \overline{\varphi}^a \partial_\beta \overline{\varphi}^a)). \tag{11.4.9}$$

We know that the integration is 4π times the Kronecker indicator for the mapping of $s_k^2 \to s_\varphi^2$. The Kronecker indicator is certainly an integer, so we obtain $q_m = n/2e$.

To sum up, the appearance of the magnetic unipolar soliton led to the following results:

(1) The model may appear or result in the dyon solution from Julia's and Zee's consistent assumptions $A_0^a = \chi_a J(r)/er^2$ by introducing an electric field [293]. These dyons have limited energies and continuous electric charges and magnetic charges of $q_m = 1/2e$. In quantum theory, the permission values of electric charges become the discrete values of $q = ne$ [293].

(2) This theory can be extended to the specification group with high ranks, such as SU(3). In this case, some new magnetic unipolars having different charges can be formed.

(3) In the limited case of $\beta \to 0$, in which the condition of $H(r) \to C(r)$ (here $r \to \infty$) can be kept, Prasad and Sommerfied [253] obtained serious solutions of equation (11.4.4), which are represented by

$$K(r) = Cr/sh(Cr), \quad H(r) = C(r)_{\text{coth}}(Cr) - 1.$$

(4) Hasenfratz, 't Hooft, and Jackiw [125] and Rebbi pointed out that, if SU(2) is used to add the Lorentz scalar having the isospin double state to the unipolar model, the magnetic unipolar can be added into the isospin state, so its angular momentum becomes 1/2. In this case, the complex system formed meets the rule of Dirac–Fermi statics. Therefore, we conclude that the spin comes from the isospin in this case.

(5) From the above research, we affirm that, in the SU(2) model, the magnetic unipolar exists with $q_m = 1/2e$. However, we cannot affirm whether solutions exist with many magnetic charges and limited energy. We can now think by the above investigations that the model having suited and decided spherical symmetry does not have a SU(2) spherically symmetric magnetic unipolar solution with $|q_m| > 1/2e$. However, in the SU(3) model, the magnetic unipolar solution having many magnetic charges and spherical symmetry exists [136].

(6) The solutions having topological stability and static limited energy cannot be formed or structured in a three-dimensional system in which the long range field cannot be formed. This is due to the fact that the long range specification field requires an infinite barrier in the three-dimensional space, which can provide the topologic stability to stop the damping of the soliton to a general vacuum state [305].

11.5 The topological soliton – instanton – in four-dimensional space

As is well known, the topological soliton – instanton – was first researched by Belavin, Polyakov, Schwarz, and Tyupkin (BPST) in a four-dimensional space [21]. They first obtained the solution of the dynamic equations of a nonabelian specification field in the case of SU(2), which is a solution without the source and is called the instanton. In this case, the specification potential is analytical in the total Euler space. However, the strength of the field exists only in the localized time-space region, where the Euler energy and momentum are zero. In this section, we first describe BPST's work.

In the four-dimensional case, the Lagrange density function of a specification field in the BPST model is denoted

$$L' = -\frac{1}{4}F^a_{\mu\nu}F^a_{\mu\nu} \quad (\mu,\nu = 1,2,3,4), \tag{11.5.1}$$

where $L' = \mathcal{L}$ and $F^a_{\mu\nu} = \partial_\mu A^a_\nu - \partial_\nu A^a_\mu + gC^{abc}A^b_\mu A^c_\nu \ldots$ (C^{abc} = structure constant).

We now consider the specific group SU(2) and use the following matrix representations:

$$A_\mu = A^a_\mu \tau^a/2, \quad F_{\mu\nu} = F^a_{\mu\nu}\tau^2/2,$$

where τ^a is a 2×2 polly matrix. Utilizing again the following relations:

$$\begin{aligned} &[\tau^a, \tau^b] = 2i\varepsilon^{abc}\tau^c, \quad \operatorname{Tr}\tau^a\tau^b = 2\delta^{ab} \\ &\operatorname{Tr}\tau^a\tau^b\tau^c = 2i\varepsilon^{abc} \quad (\operatorname{Tr} = \text{find trace}), \end{aligned} \tag{11.5.2}$$

we can obtain the following field equation from the EL equation:

$$D_\mu F_{\mu\nu} = \partial_\mu F^a_{\mu\nu} - ig[A_\mu, F_{\mu\nu}] = 0. \tag{11.5.3}$$

From this equation, we can determine that there is only one topological invariant in this theory, which is called the Pontryagin indicator, or second Chen's number, and corresponds to $\pi_3(\mathrm{SU}(2)) = Z$. In this case, the topologic charge Q is defined as

$$Q = \frac{g^2}{32\pi^2}\int dx^4 \operatorname{Tr}(\varepsilon_{\mu\nu\alpha\beta}F_{\alpha\beta}F_{\mu\nu}) = 0,$$

where $n = 0, \pm1, \pm2, \pm3, \ldots$ and $Q = n$ is the number covered by SU(2) in the case of topologic reflection, which is expressed by the homotopy class of $g(x)$. Here, $g(x)$ is the matrix of the specification group. In the specification of $A_0 = 0$, the topological charge equals the changes of the winding number between $t = -\infty$ and $t = +\infty$.

In this physical system, we introduce again the following inequality:

$$\int d^4x \operatorname{Tr}\left(F_{\mu\nu} - \frac{1}{2}\varepsilon_{\mu\nu\alpha\beta}F_{\alpha\beta}\right)^2 \geq 0. \tag{11.5.4}$$

This condition is equal to adding one limit for the following quasi-energy:

$$E = \frac{1}{2}\int d^4x\,\mathrm{Tr}(F_{\mu\nu}F_{\mu\nu}) \geq 8\pi^2|Q|, \tag{11.5.5}$$

where Q may be simplified further, i.e., if A_μ has no singularity, so $\partial_\mu\partial_\nu A_\lambda = \partial_\nu\partial_\mu A_\lambda$, then

$$\mathrm{Tr}\left(\frac{1}{2}\varepsilon_{\mu\nu\alpha\beta}F_{\alpha\beta}F_{\mu\nu}\right) = \frac{\partial}{\partial\mu}\left(2\varepsilon_{\mu\nu\alpha\beta}\,\mathrm{Tr}\left(A_\nu\partial_\alpha A_\beta + \frac{2g}{3i}A_\nu A_\alpha A_\beta\right)\right), \tag{11.5.6}$$

where the representation on the right side of equation (11.5.6) is not related to the specification characterization, but $2\varepsilon_{\mu\nu\alpha\beta}\,\mathrm{Tr}(A_\nu\partial_\alpha A_\beta + \frac{2g}{3i}A_\nu A_\alpha A_\beta)$ is related to the specification features. If equation (11.5.3) is integrable and $g = 1$, then Q can also be expressed as

$$Q = \frac{1}{16\pi^2}\iint_{s^3}\left[2\varepsilon_{\mu\nu\alpha\beta}\,\mathrm{Tr}\left(A_\nu\partial_\alpha A_\beta + \frac{2g}{3i}A_\nu A_\alpha A_\beta\right)\right]d^3\sigma^\mu. \tag{11.5.7}$$

In order to make A_μ approach zero, we demand A_μ to have a pure specification. We have

$$A_\mu(x) = ig^{-1}(x)\partial_\mu g(x), \quad x \in s^3 \tag{11.5.8}$$

at $r \to \infty$. Clearly, equation (11.5.5) gives the limitation for the energy E, which indicates that the energy E of the soliton solution having a nongeneral charge has certainly a lower limit. If $F_{\mu\nu} = \varepsilon_{\mu\nu\alpha\beta}F_{\alpha\beta}/2$ in this case, then we should obtain or use the lower limit of the energy in equation (11.5.5). When this condition holds, the above field equations are satisfied automatically because $D_\mu F_{\mu\nu} = D_\mu(\varepsilon_{\mu\nu\alpha\beta}F_{\alpha\beta}/2) = 0$ (this is one identity for the field without singular string).

From the assumptions of spherical symmetry of the specification field, we obtain

$$A_\mu(x) = if(r)g^{-1}(x)\partial_\mu g(x), \quad r^2 = x_1^2 + x_2^2 + x_3^2 + x_4^2, \quad g(x) = \frac{x_4 - ix_a\tau^a}{r}. \tag{11.5.9}$$

Substituting equation (11.5.9) into equation (11.5.8), we find that the following self-dual condition is met:

$$rf' \mp 2f(1-f). \tag{11.5.10}$$

Finding the solutions of equation (11.5.10), we obtain the nongeneral solution, which is

$$f(r) = r^2/(r^2 + \beta'^2). \tag{11.5.11}$$

This solution is called the instanton, where β' expresses the size of the instanton in the length dimension. The instanton is localized at any position and has any volume. From this solution and equation (11.5.9), we find

$$A_\mu = ir^2g^{-1}\partial_\mu g/(r^2 + \beta'^2), \qquad F_{\mu\nu} = 4\beta'^2\sigma_{\mu\nu}/(r^2 + \beta'^2), \tag{11.5.12}$$

where $\sigma_{ij} = [\tau^i, \tau^j]/4i$, $\sigma_{i4} = \tau^i/2 = -\sigma_{ij}$.

In this case, the instanton's Pontryagin indicator is $Q = 1$. Because $F_{\mu\nu} \to 0$, when $r \to \infty$, A_μ also approaches one pure specification. In this case, if the parallel shift of $x_\mu \to x_\mu - a_\mu$ is made, another solution of the instanton is also obtained at position $x_\mu = a_\mu$. At the same time, there is also the anti-instanton at this position, which represents the change of g to $g+$ in the first equation in equation (11.5.12). This solution should meet the following equation [242]:

$$F_{\mu\nu} = -\left(\frac{1}{2}\varepsilon_{\mu\nu\alpha\beta}F_{\alpha\beta}\right) = \frac{4\beta^2}{(y^2+\beta^2)^2}\overline{\sigma}_{\mu\nu},$$

where $\overline{\sigma}_{ij} = \sigma_{ij}$, $\overline{\sigma}_{i4} = -\sigma_{i4}$. Its topologic charge is β'. For $F_{\mu\nu} = \pm\frac{1}{2}\varepsilon_{\mu\nu\alpha\beta}F_{\alpha\beta}$ and $|Q| = 1$, other solutions do not exist, except for the above specification transformation. At the same time, the single instanton formed is represented by five parameters, in which four parameters determine its position and the remainder determines its size.

However, we point out that the instanton is not a real physical particle. It is only a solution of the field equation in the four-dimensional physical space, which does not exist in the real world. We should also point out that the instanton process that appears indicates only the dynamic features of the solution of the field equation in the process of imaginary time, which is quite similar to the tunnel effect in quantum mechanics. Thus, it can be thought of as a classical solution (or the path) linking two classical vacuums of $\Delta n = 1$ in imaginary time. This relation can be expressed by

$$|i\rangle \xleftarrow{\text{anti-instanton}} |i+1\rangle \quad \text{and} \quad |i\rangle \xrightarrow[\text{instanton}]{} |i+1\rangle.$$

Just so, the solution is called "instanton", but there is no interaction between the two instantons. If the quantum fluctuation is considered in this process, then the interaction between the instanton and anti-instanton also exists. Although the interaction exists in this process, its features are different from those in the Euclidean space. In the latter, the instanton is a real physical particle and the instanton's tunnel effect can be observed experimentally. Then the effects can be explained by the perturbation method. Hence, the instanton is also called a quasi-particle. On the other hand, the instantons are related to strong interaction theory (QCD). Therefore, instanton physics is an important subject in modern physics.

Several researchers [49, 82, 154] have introduced the concept of the instanton into the general theory of relativity on the basis of the similarity between the strength of the specification field $F_{\mu\nu}$ and the metric tensor $R_{\mu\nu}$ in the Reimann space. Thus, they are referred to as attractive instantons. The importance of attractive instantons can be envisioned by the process in which they are inserted between different vacuums ($R_{\mu\nu\alpha\beta} = 0$) of Minkowski spaces to improve the method of investigation of difficult problems. On the other hand, in so-called superspecification theory, superattractive instantons also exist. These investigations on instantons are also quite significant.

11.6 The nontopological soliton and its properties

The nontopological soliton is different from the topological soliton. It does not require the existence of the degeneracy vacuum, but its border conditions at infinity are the same as those of field equations without soliton solutions. Meanwhile, it again demands the condition of existence of conservation rules for the addition for charges and scalar fields [116, 222]. Therefore, a simplified method, which is used to produce one nontopological soliton, is to introduce one complex number field [318]. We have

$$\varphi = \varphi_1 + i\varphi_2, \quad \varphi = \varphi_1 - i\varphi_2, \tag{11.6.1}$$

where φ_1 and φ_2 are the Fermi field. Here we only discuss the features of solutions in a one-dimensional space. The corresponding Lagrange density function is represented by

$$L' = \frac{1}{2}\frac{\partial\varphi^*}{\partial x_\mu}\frac{\partial\varphi}{\partial x_\mu} - U(\varphi^*\varphi), \tag{11.6.2}$$

where $L' = \mathcal{L}$. Using the EL equation and from equation (11.6.2), we obtain the corresponding dynamic equation:

$$\frac{\partial^2\varphi}{\partial x_\mu^2} - \varphi\frac{\partial\varphi}{\partial(\varphi^*\varphi)}U(\varphi^*\varphi) = 0. \tag{11.6.3}$$

Utilizing the above equations, we obtain

$$N = i\int(\varphi^*\dot{\varphi} - \dot{\varphi}^*\varphi)dx, \tag{11.6.4}$$

which is a conservation quantity. We here assume

$$\varphi^*\frac{\partial\varphi}{\partial x} - \frac{\partial\varphi^*}{\partial x}\varphi = 0 \quad \text{at } x = \pm\infty.$$

In this system, because $L' = \mathcal{L}$ is not changed under the transformation of $\varphi \to \varphi e^{i\theta}$, the Hamilton function $H' = \mathcal{H}$ is also invariable, so $\mathcal{H}$ is not related to θ. Therefore, we can hypothesize that N is the conjugate momentum of θ. From the Hamilton equation, we obtain $\dot{N} = \partial H'/\partial\theta = 0$, where $H' = \mathcal{H}$. Therefore, N is a conserved quantity. In classical field theory, N may be any real number. Because θ is, in essence, a phase variable, if $\theta \to \theta + 2\pi$, then $\varphi \to \varphi$. Here, N resembles the momentum and must be an integer. This means that $\mathcal{H}$ does not certainly relate to its conjugate coordinates. In quantum field theory, if N is an integer, then θ is thought to be a cyclical variable, so θ may be thought of as a phase variable. This is the exact reason behind the introduction of the complex field mentioned above.

When $N \neq 0, \varphi$ is certainly related to time. In this case, we can verify that, if N is fixed, the solution having minimum energy, which relates to time, must meet the following formula of the harmonic oscillator:

$$\varphi = \sigma(x)e^{-i\omega t}. \tag{11.6.5}$$

Inserting equation (11.6.5) into equation (11.6.3), we obtain

$$\frac{d^2\sigma}{dx^2} + \omega^2\sigma - \sigma\frac{d}{d\sigma^2} - U = 0.$$

Integrating the above equation, we obtain

$$\frac{1}{2}\left(\frac{d\sigma}{dx}\right)^2 - \sigma V(\sigma) = \text{constant}, \tag{11.6.6}$$

where

$$V = \frac{1}{2}(U - \omega^2\sigma^2), \quad U = U(\sigma^2). \tag{11.6.7}$$

Because nontopological solitons do not need degeneracy vacuums, we may choose $U(0) = 0$, where the form of $U(\sigma)$ is shown in Figure 11.12.

In order to acquire the nontopologic soliton solution, we here choose $V = \frac{1}{2}(U - \omega^2\sigma^2)$, which is shown in Figure 11.13, where $U(\varphi^*\varphi) - \omega^2\varphi^*\varphi = 0$. Except for the solution of $\varphi = 0$ in this equation, a solution $\varphi \neq 0$ also exists.

In the case of the mechanical simulation in the nonrelativity, the potential of the particle should be $-V$. Its form is shown in Figure 11.14. When $x = -\infty$, the particle at point O will move to point A along the curve. Subsequently, it will return to the original position from point A. However, the particle will return again to point O when $x = +\infty$. This feature of movement of the particle can be easily obtained.

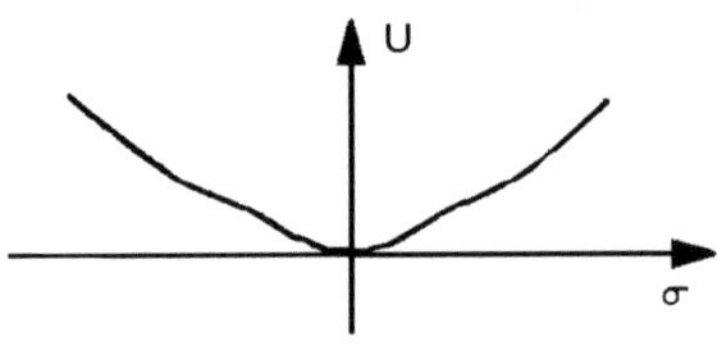

Figure 11.12: The potential curve in equation (11.6.7).

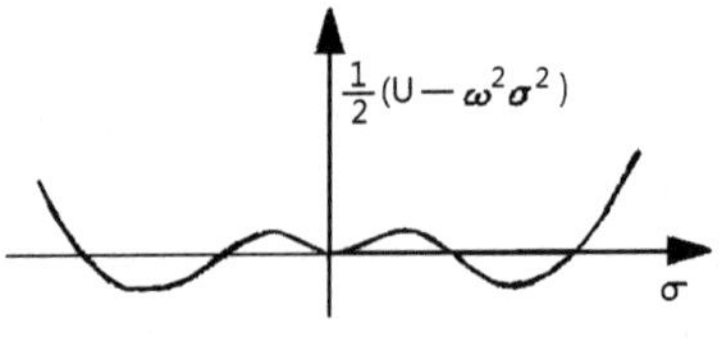

Figure 11.13: The curve of $\frac{1}{2}(U - \omega^2\sigma^2)$.

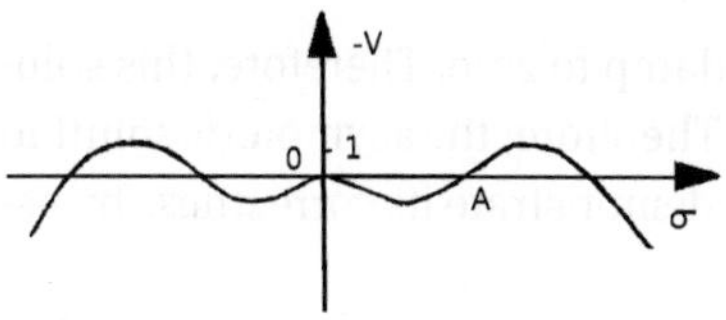

Figure 11.14: The form of $V(\sigma)$.

From equation (11.6.7), we obtain the general solution, denoted

$$x - x_0 = \int_A^\sigma \frac{d\sigma}{\sqrt{2V(\sigma)}}, \quad (x = x_0, \text{ at } \sigma = A). \tag{11.6.8}$$

Evidently, the energy of the field without dispersion is bound in a limited region in space. Therefore, we affirm that this solution is a soliton. Obviously, the solution approaches zero when $x = \pm\infty$. Therefore, it is one nontopological soliton. Its outline is shown in Figure 11.15.

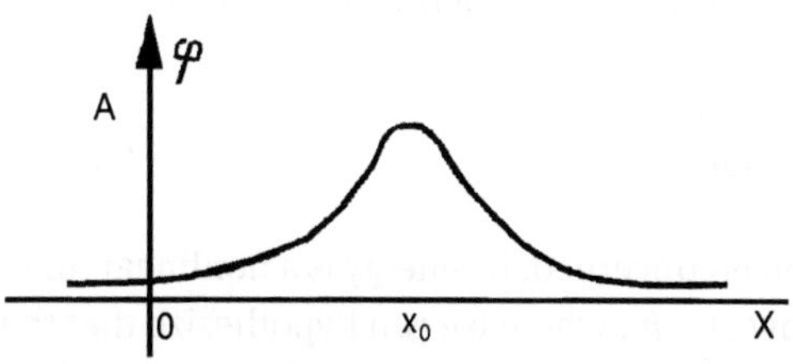

Figure 11.15: The outline of soliton solution $\varphi(x)$.

We see in Figure 11.12 that $U \to m^2\sigma^2$, at $\sigma \to 0$, where m^2 is a constant. However, if we use function V in Figure 11.14, then we verify

$$\omega < m. \tag{11.6.9}$$

If we now choose $U = \frac{m^2\varphi^*\varphi}{1+\varepsilon^2}[(1 - g^2\varphi^*\varphi)^2 + \varepsilon^2]$, then we obtain the following solution from its movement equation:

$$\varphi = \frac{1}{g}\left[\frac{a}{1 + \sqrt{1 - a\cosh y}}\right]^{1/2} e^{-i\omega t}, \tag{11.6.10}$$

where

$$a = (1 + \varepsilon^2)(m^2 - \omega^2) \quad y = 2\sqrt{m^2 - \omega^2}(x - x_0).$$

In equation (11.6.10), if we choose $|x| \to \infty$, then we obtain the asymptotic solution, denoted

$$\varphi \propto \frac{1}{g}\sqrt{m^2 - \omega^2}e^{-\sqrt{m^2-\omega^2}(x-x_0)}. \tag{11.6.11}$$

This is a damping solution. At $|x| \to \infty$, it will damp to zero. Therefore, this solution satisfies the definition of nontopologic soliton. Therefore, the asymptotic solution in equation (11.6.11) has generality, because we can demonstrate its correctness by using other methods.

In fact, we can explain and elucidate the stability of the above soliton [116, 222] as follows.

As is well known, all nonlinear field equations have solutions of plane waves, which can be represented by

$$\varphi = \sqrt{\frac{N}{2\omega Q}} e^{i\vec{K}\vec{x}-i\omega t} \quad (\omega = \sqrt{m^2 + \vec{K}^2}). \tag{11.6.12}$$

Clearly, this is because the variation of amplitudes of the plane waves approaches zero or infinity when the volume of the system approaches infinity, i.e., $\Omega \to \infty$. Thus, we can ignore the higher-order terms of the field in the above Lagrange function. If we retain only the first-order term, then it is changed to equation (11.6.12). In this case, we obtain the following straight line relation between the energy and the conservation quantity N:

$$E_{\text{plan}} = N\omega \geq Nm. \tag{11.6.13}$$

However, as far as the soliton solution is concerned, its energy is a nonlinear function of N. From equation (11.6.9), we know $\omega < m$, where we can hypothesize that the solution of the nontopologic soliton is an analytical continuation and extension of the above plane wave, i.e., it is the result of the extension from $\omega \geq m$ to $\omega < m$. Therefore, we deduce that the minimum energy E_{soliton} of the nontopologic soliton solution has the following relation:

$$E_{\text{soliton}} < Nm \tag{11.6.14}$$

for any conservation quantity N and any coupling content g. Otherwise, we know that the conservation quantity N and the phase angle of complex number field $\theta = \omega t$ are conjugate variables, so we acquire the relations of $\dot{N} = -\partial H'/\partial N$ and $\dot{\theta} = \partial H'/\partial N$, where $H' = \mathcal{H}$. Because $\dot{\theta} = \omega$ and the eigenvalue of $\mathcal{H}$ for any solution is the energy E, the above relation can be written

$$\omega = \frac{dE}{dN}. \tag{11.6.15}$$

For the plane wave, ω is not related to N, so $E = N\omega$. However, this relation does not exist for the nontopologic soliton because we should consider the relation between ω and N. If we consider the limit case of $\omega \to m^-$, from equation (11.6.11) we obtain

$$N \to 2m \int |\varphi|^2 dx \approx \sqrt{m^2 - \omega^2} \to 0.$$

From the integration of equation (11.6.15) and the relation of $\omega < m$, we finally obtain

$$E_{\text{soliton}} = \int_0^N \omega dN < m \int_0^N dN = mN, \tag{11.6.16}$$

from which we conclude $E_{\text{soliton}} < E_{\text{plane}}$. This means that a nontopologic soliton solution exists in the one-dimensional space and the minimum energy state of the system is not the plane wave, but the soliton state. Therefore, we determine that the soliton solution is always stable in this system. At the same time, we can demonstrate that the soliton solution is the minimum energy state and very stable in a two-dimensional space.

The above investigations exhibit clearly that the static soliton solution exists in the one-dimensional space. If the result is further extended to a (3 + 1)-dimensional space, then we can deduce by Derrick's theorem that the pure scalar fields do not have a static soliton, but if one inner symmetry group exists in the nonlinear scalar field, then there is possibly a soliton solution, such as $\{e^{i\alpha_i(t)T_i}\}\varphi_c(x)$, where T_i is the generator of the expression of group G in the scalar field φ and $\alpha_i(t)$ is the parameter of the group relating to time t. When the G group is a nonabelian group, Tsung-Dao Lee et al. [49, 154] researched its properties. Others have also demonstrated that the nontopologic soliton is stable only if the $I_3 = \pm I_{3\,\text{max}}$ component of $\varphi_c(x)$ is nonzero [49, 82, 318], while Zhou Guang Zhao et al. [318] researched the soliton solutions of scalar fields having the nonabelian inner symmetry using a $G = \text{SU}(2)$ isospin group. These investigations are very significant to understand the essence of soliton solutions in quantum field theory.

12 The theory of soliton movement of superconductive features

12.1 The macroscopic quantum effects in the superconductor and its soliton movement properties

So-called macroscopic quantum effects refer to a quantum phenomenon that occurs on a macroscopic scale. Such effects are obviously different from the microscopic quantum effects as described by quantum mechanics. It has been experimentally demonstrated [10, 93–95, 157–159, 196, 221, 226, 227, 232, 233, 241, 263–265] that macroscopic quantum effects are the phenomena that occur in superconductors. Superconductivity is a physical phenomenon in which the resistance of a material suddenly vanishes when its temperature is lower than a certain value, T_c, which is referred to as the critical temperature of superconducting materials. Modern theories [17, 51, 271, 272] tell us that superconductivity arises from the irresistible motion of superconductive electrons. As such, we want to answer, amongst others, the following questions. How is the macroscopic quantum effect formed? What are its essential features? What are the properties and rules of motion of superconductive electrons in a superconductor? Up to now, these problems have not been studied systematically. We will study these problems in this chapter. The experimental observations of properties of macroscopic quantum effects in superconductors are described as follows.

(1) Superconductivity of material. As is well known, superconductors can be pure elements, compounds, or alloys. To date, more than 30 single elements and up to a hundred alloys and compounds have been found to possess the characteristics [10, 93–95, 157–159, 196, 221, 226, 227, 232, 233, 241, 263–265] of superconductors. When $T \leq T_c$, any electric current in a superconductor will flow forever without being damped. Such a phenomenon is referred to as perfect conductivity. Moreover, it has been observed experimentally that, when a material is in the superconducting state, any magnetic flux in the material is completely repelled, resulting in zero magnetic fields inside the superconducting material. Similarly, a magnetic flux applied by an external magnetic field cannot penetrate into superconducting materials. Such a phenomenon is called perfect anti-magnetism or the Maissner effect. Meanwhile, there are also other features associated with superconductivity, which are not presented here.

How can this phenomenon be explained? After more than 40 years of research, Bardeen, Cooper, and Schreiffier (BCS) proposed the new idea of Cooper pairs of electrons and established the microscopic theory of superconductivity at low temperatures, the BCS theory [17, 51, 271, 272], in 1957, on the basis of the mechanism of electron-phonon interactions proposed by Frohlich [84, 85]. According to this theory, electrons with opposite momenta and antiparallel spins form pairs when the attraction between the electron and phonon in these materials overcomes the Coulomb repulsion between them. The so-called Cooper pairs condense to a minimum energy

https://doi.org/10.1515/9783110549638-012

state, resulting in quantum states, which are highly ordered and coherent over the long range and in which there is essentially no energy exchange between the electron pairs and the lattice. Thus, the electron pairs are no longer scattered by the lattice but flow freely, resulting in superconductivity. The electron pairs in a superconductive state are somewhat similar to a diatomic molecule, but are not as tightly bound as a molecule. The size of an electron pair, which gives the coherent length, is approximately 10^{-4} cm. A simple calculation shows that there can be up to 10^6 electron pairs in a sphere of 10^{-4} cm in diameter. There must be mutual overlap and correlation when so many electron pairs are brought together. Therefore, perturbation to any of the electron pairs would certainly affect all others. Thus, various macroscopic quantum effects can be expected in a material with such coherence and long range ordered states. Magnetic flux quantization, vortex structures in type-II superconductors, and the Josephson effect [138–140] in superconductive junctions are only some examples of the phenomena of macroscopic quantum mechanics.

(2) Quantizated effect of magnetic flux. Consider a superconductive ring. Assume that a magnetic field is applied at $T > T_c$. Then the magnetic flux lines ϕ_0 produced by the external field pass through and penetrate into the body of the ring. We now lower the temperature to a value below T_c and then remove the external magnetic field. The magnetic induction inside the body of the circular ring equals zero ($\vec{B} = 0$), because the ring is in the superconductive state and the magnetic field produced by the superconductive current cancels the magnetic field, which was within the ring. However, part of the magnetic fluxes in the hole of the ring remain, because the induced current in the ring vanishes. This residual magnetic flux is referred to as "the frozen magnetic flux". It has been observed experimentally that the frozen magnetic flux is discrete or quantized. Using the macroscopic quantum wave function in the theory of superconductivity, it can be shown that the magnetic flux is established by $\Phi' = n\phi_0$ ($n = 0, 1, 2, \dots$), where $\phi_0 = hc/2e = 2.07 \times 10^{-15}$ Wb is the flux quantum, representing the flux of one magnetic flux line. This means that the magnetic fluxes passing through the hole of the ring can only be multiples of ϕ_0 [10, 93–95, 157–159, 227, 241, 263–265]. In other words, the magnetic field lines are discrete. What does this imply? If the magnetic flux of the applied magnetic field is exactly n, then the magnetic flux through the hole is $n\phi_0$, which is not difficult to understand. However, what is the magnetic flux through the hole if the applied magnetic field is $(n + 1/4)\phi_0$? According to the above, the magnetic flux cannot be $(n + 1/4)\phi_0$. In fact, it should be $n\phi_0$. Similarly, if the applied magnetic field is $(n + 3/4)\phi_0$, then the magnetic flux passing through the hole is not $(n + 3/4)\phi_0$, but rather $(n + 1)\phi_0$. Therefore, the magnetic fluxes passing through the hole of the circular ring are always quantized.

An experiment conducted in 1961 surely proves this to be so, indicating that the magnetic flux does exhibit discrete or quantized characteristics on a macroscopic scale. The above experiment was the first demonstration of the macroscopic quantum effect. Based on quantization of the magnetic flux, we can build a "quantum magnetometer", which can be used to measure weak magnetic fields with a sensitivity of 3×10^{-7} Oersted. A slight modification of this device would allow us to measure electric currents with strengths as low as 2.5×10^{-9} A.

(3) Quantization of magnetic flux lines in type-II superconductors. The superconductors discussed above are referred to as type-I superconductors. This type of superconductor exhibits a perfect Maissner effect when the external applied field is higher than a critical magnetic value $\vec{H}_c$. There exist other types of materials, such as the NbTi alloy and Nb_3Sn compounds, in which the magnetic field partially penetrates inside the material when the external field $\vec{H}$ is greater than the lower critical magnetic field $\vec{H}_{c1}$, but less than the upper critical field $\vec{H}_{c2}$ [10, 93, 94, 241, 263–265]. This kind of superconductor is classified as type-II superconductors and is characterized by a Ginzburg-Landau (GL) parameter greater than 1/2. Studies using the Bitter method showed that the penetration of a magnetic field results in some small regions changing from the superconductive to the normal state. These small regions in the normal state are of cylindrical shape and regularly arranged in the superconductor, as shown in Figure 12.1. Each cylindrical region is called a vortex (or magnetic field line) [10, 93–95, 157–159, 227, 241, 263–265]. The vortex lines are similar to the vortex structure formed in a turbulent flow of fluid. Both theoretical analysis and experimental measurements have shown that the magnetic flux associated with one vortex is exactly equal to one magnetic flux quantum ϕ_0, when the applied field $\vec{H} \geq \vec{H}_{c1}$, the magnetic field penetrating into the superconductor in the form of vortex lines, increases step-wise. For an ideal type-II superconductor, stable vortices are distributed in a triangular pattern. The superconducting current and magnetic field distributions are shown in Figure 11.1 in Chapter 11. For other, nonideal type-II superconductors, the triangular pattern of distribution can be observed in small local regions, even though its overall distribution is disordered. It is evident that the vortex line structure is quantized. This has been verified by many experiments and can be considered a result of the quantization of the magnetic flux. Furthermore, it is possible to determine the en-

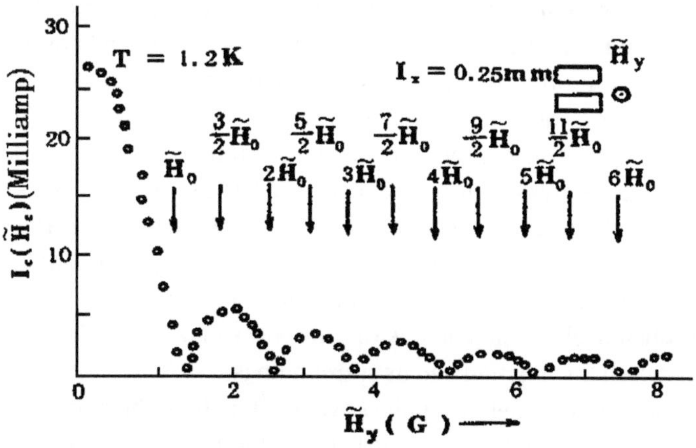

Figure 12.1: Quantum diffraction effect in the superconductor junction.

ergy of each vortex line and the interaction energy between the vortex lines. Parallel magnetic field lines are found to repel each other while anti-parallel magnetic lines attract each other.

(4) The Josephson effect of superconductivity junctions [138–140]. As is well known in quantum mechanics, microscopic particles, such as electrons, have a wave property and can penetrate through a potential barrier. For example, if two pieces of metal are separated by an insulator of a width of tens of Ångströms, an electron can tunnel through the insulator and travel from one metal to the other. If a voltage is applied across the insulator, a tunnel current can be produced. This phenomenon is referred to as a tunneling effect. If two superconductors replace the two pieces of metal in the above experiment, a tunneling current can also occur when the thickness of the dielectric is reduced to about 30 Å. However, this effect is fundamentally different from the tunneling effect discussed above in quantum mechanics and is referred to as the Josephson effect.

Evidently, this is due to the long range coherent effect of the superconductive electron pairs. Experimentally, it was demonstrated that such an effect could be produced via many types of junctions involving a superconductor, such as superconductor-metal-superconductor junctions, superconductor-insulator-superconductor junctions, and superconductor bridges. These junctions can be considered as superconductors with a weak link. On the one hand, they have properties of bulk superconductors. For example, they are capable of carrying certain superconducting currents. On the other hand, these junctions possess unique properties, which a bulk superconductor does not. Some of these properties are summarized in the following.

(A) When a direct current (dc) passing through a superconductive junction is smaller than a critical value I_c, the voltage across the junction does not change with the current. The critical current I_c can range from a few tens of μA to a few tens of mA.

(B) If a constant voltage is applied across the junction and the current passing through the junction is greater than I_c, a high-frequency sinusoidal superconducting current occurs in the junction. The frequency is given by $\upsilon = 2\,\text{eV/h}$ in the microwave and far-infrared regions of $(5–1000)\times10^9$ Hz. The junction radiates a coherent electromagnetic wave with the same frequency. This phenomenon can be explained as follows. The constant voltage applied across the junction produces an alternating Josephson current that, in turn, generates an electromagnetic wave with frequency υ. The wave propagates along the planes of the junction. When the wave reaches the surface of the junction (the interface between the junction and its surroundings), part of the electromagnetic wave is reflected from the interface and the rest is radiated, resulting in the radiation of the coherent electromagnetic wave. The power of radiation depends on the compatibility between the junction and its surroundings.

(C) When an external magnetic field is applied over the junction, the maximum dc, I_{ce}, is reduced due to the effect of the magnetic field. Furthermore, I_c changes periodically as the magnetic field increases. The T_c–H curve resembles the distribution of light intensity in the Fraunhofer diffraction experiment, shown in Figure 12.1. This phenomenon is called quantum diffraction of the superconductivity junction.

(D) When a junction is exposed to a microwave of frequency υ and if the voltage applied across the junction is varied, then it can be seen that the dc passing through the junction increases suddenly at certain discrete values of the electric potential. Thus, a series of steps appear on the dc I–V curve and the voltage at a given step is related to the frequency of the microwave radiation by $n\upsilon = 2\mathrm{eV}\, n/\mathrm{h}$ ($n = 0, 1, 2, 3 \ldots$). More than 500 steps have been observed in experiments.

Josephson first derived these phenomena theoretically and each was experimentally verified subsequently. All these phenomena are, therefore, called Josephson effects [138–140]. In particular,

(1) and (3) are referred to as dc Josephson effects while (2) and (4) are referred to as alternating current (ac) Josephson effects. Evidently, Josephson effects are macroscopic quantum effects, which can be explained well by the macroscopic quantum wave function. If we consider a superconducting junction as a weakly linked superconductor, the wave functions of the superconducting electron pairs in the superconductors on both sides of the junction are correlated due to a definite difference in their phase angles. This results in a preferred direction for the drifting of the superconducting electron pairs and a Josephson dc is developed in this direction. If a magnetic field is applied to the plane of the junction, the magnetic field produces a gradient of phase difference, which makes the maximum current oscillate along with the magnetic field and radiation of the electromagnetic wave occurs. If a voltage is applied across the junction, the phase difference will vary with time and result in the Josephson effect. In view of this, the change in the phase difference of wave functions of superconducting electrons plays an important role in the Josephson effect, which will be discussed in more detail in Section 12.5.

The discovery of the Josephson effect opened the door for a wide range of applications of superconductor theory. Properties of superconductors have been explored to produce the superconducting quantum interferometer-magnetometer, sensitive ammeter, voltmeter, electromagnetic wave generator, detector, frequency mixer, etc.

12.2 The properties of Boson condensation and spontaneous coherence of macroscopic quantum effects in the superconductor

12.2.1 The nonlinear model of theoretical description for the macroscopic quantum effects

From the above studies, we know that the macroscopic quantum effect is obviously different from the microscopic quantum effect, the former having been observed for physical quantities, such as resistance, magnetic flux, vortex line, and voltage.

In the latter, the physical quantities, characteristics of microscopic particles, such as energy, momentum, and angular momentum, are quantized. Thus, it is reasonable to believe that the fundamental nature and the rules governing these effects are different.

We know that the microscopic quantum effect is described by quantum mechanics. However, the question what the mechanisms of macroscopic quantum effects are remains. How can these effects be properly described?

And what are the states of microscopic particles in the systems where macroscopic quantum effects occur? In other words, what are the essential features of macroscopic quantum states? These questions need to be addressed.

We know that materials are composed of a great number of microscopic particles, such as atoms, electrons, and nuclei, which exhibit quantum features. We then assume that the macroscopic quantum effects result from the collective motion and excitation of these particles under certain conditions, such as extremely low temperatures, high pressure, and high density. Under such conditions, a huge number of microscopic particles paired with each other condense in a low-energy state, resulting in a high order and long range coherence. In such a highly ordered state, the collective motion of a large number of particles is the same as the motion of single particles. Since the latter is quantized, the collective motion of the many particle system gives rise to a macroscopic quantum effect. Thus, the condensation of the particles and their coherent state play an essential role in the macroscopic quantum effect.

What is the concept of condensation? On a macroscopic scale, the process of transforming gas into liquid, as well as that of changing vapor into water, is called condensation. This, however, represents a change in the state of molecular positions and is referred to as condensation of positions. The phase transition from a gaseous state to a liquid state is a first-order transition in which the volume of the system changes and the latent heat is produced, but the thermodynamic quantities of the systems are continuous and have no singularities. The word condensation in the context of macroscopic quantum effects has a special meaning. The condensation concept discussed here is similar to the phase transition from gas to liquid, in the sense that the pressure depends only on the temperature, not on the volume. Thus, it is essentially different from the first-order phase transition, such as that from vapor to water. It is not the condensation of particles into a high-density material in normal space. On the contrary, it is the condensation of particles to a single-energy state or to a low-energy state with a constant or nonexistent momentum. It is thus also called a condensation of momentum. This differs from a first-order phase transition and theoretically it should be classified as a third-order phase transition, even though it is really a second-order phase transition, because it is related to the discontinuity of the third derivative of a thermodynamic function. Discontinuities can be clearly observed when measuring specific heat or magnetic susceptibility of certain systems when condensation occurs. The phenomenon results from a spontaneous breakdown of symmetry of the system due to nonlinear interactions within the system under some special conditions, such as extremely low temperatures and high pressure. Different systems have different critical temperatures of condensation. For example, the condensation temperature of a superconductor is its critical temperature T_c (see also [197, 198, 209, 213, 245, 259]).

From the above discussions on the properties of superconductors, we know, even though the microscopic particles involved can be either bosons or fermions, that the ones that actually condense are either bosons or quasi-bosons, since fermions are bound as pairs. Bosons obey the Bose–Einstein statistics, so the condensation is referred to as Bose–Einstein condensation [19, 29, 30, 166]. Properties of bosons are different from those of fermions, as they do not follow the Pauli exclusion principle and there is no limit to the number of particles occupying the same energy levels. At finite temperatures, bosons can distribute in many energy states and each state can be occupied by one or more particles, while some states may not be occupied at all. Due to the statistical attractions between bosons in the phase space (consisting of generalized coordinates and momenta), groups of bosons tend to occupy one quantum energy state under certain conditions. Then, when the temperature of the system falls below a critical value, the majority or all bosons condense to the same energy level (e.g., the ground state), resulting in Bose condensation and a series of interesting macroscopic quantum effects. Different macroscopic quantum phenomena are observed because of differences in the fundamental properties of the constituting particles and their interactions in different systems.

In the highly ordered state of these phenomena, the behavior of each condensed particle is closely related to the properties of the system. In this case, the wave function $\phi = fe^{i\theta}$ or $\phi = \sqrt{\rho}e^{i\theta}$ of the macroscopic state [19, 29, 30] is also the wave function of an individual condensed particle. The macroscopic wave function is also called the order parameter of the condensed state. This term was used to describe the superconductive states in the study of these macroscopic quantum effects. The essential features and fundamental properties of macroscopic quantum effects are given by the macroscopic wave function ϕ and it can be further shown that the macroscopic quantum states, such as the superconductive states, are coherent and are Bose condensed states formed through second-order phase transitions after the symmetry of the system is broken due to nonlinear interactions in the system.

In the absence of any externally applied field, the Hamiltonian of a given macroscopic quantum system can be represented by the macroscopic wave function ϕ and written as

$$H = \int dxH' = \int dx\left[-\frac{1}{2}|\nabla\phi|^2 - \alpha|\phi|^2 + \lambda|\phi|^4\right], \tag{12.2.1}$$

where $H' = \mathcal{H}$ represents the Hamiltonian density function of the system. The unit system in which $m = \hbar = c = 1$ is used here for convenience. If an externally applied electromagnetic field does exist, the Hamiltonian given above should be replaced by

$$H = \int dxH' = \int dx\left[-\frac{1}{2}|\nabla - ie^*\vec{A}\phi|^2 - \alpha|\phi|^2 + \lambda|\phi|^4 + \frac{\vec{H}^2}{8\pi}\right] \tag{12.2.2}$$

or, equivalently,

$$H = \int dxH' = \int dx\left[-\frac{1}{2}|(\partial_j - ie^* A_j)\phi|^2 - \alpha|\phi|^2 + \lambda|\phi|^4 + \frac{1}{4}F_{ji}.F^{ji}\right],$$

where $F_{ji} = \partial_j A_i - \partial_i A_j$ is the covariant field intensity, $\vec{H} = \nabla \times \vec{A}$ is the magnetic field intensity, e is the charge of an electron, $e^* = 2e$, $\vec{A}$ is the vector potential of the electromagnetic field, and α and λ can be said to be some of the interaction constants. The above Hamiltonians in equations (12.2.1) and (12.2.2) have been used in the study of superconductivity by many scientists, including de Gennes [57], Saint-James et al. [285], Kivshar [145, 146], Bullough [38, 39], Huepe [130], Sonin [281], and Davydov et al. [56]. They can also be derived from the free energy expression of a superconductive system given by Landau et al. [150, 152]. As a matter of fact, the Lagrangian function of a superconducting system can be obtained from the well-known GL equation [8, 9, 91, 92, 96–98, 150] using the Lagrangian method. The Hamiltonian function of a system can then be derived using the Lagrangian approach. The results, of course, are the same as equations (12.2.1) and (12.2.2). Evidently, the Hamiltonian operator corresponding to equations (12.2.1) and (12.2.2) represents a nonlinear function of the wave function of a particle, where the nonlinear interaction is caused by the electron-phonon interaction and by the vibration of the lattice in the BCS theory in the superconductors. Therefore, it truly exists. Evidently, the Hamiltonians of the systems are different from those in quantum mechanics and a nonlinear interaction related to the state of the particles is involved in equations (12.2.1) and (12.2.2). Hence, we can expect that the states of particles depicted by the Hamiltonian also differ from those in quantum mechanics and the Hamiltonian can describe the features of macroscopic quantum states including superconducting states. These problems are treated in the following pages. Evidently, the Hamiltonians in equations (12.2.1) and (12.2.2) possess $U(1)$ symmetry, that is, they remain unchanged while undergoing the following transformation:

$$\phi(\vec{r}, t) \to \phi'(\vec{r}, t) = e^{-iQ_j\theta}\phi(\vec{r}, t),$$

where Q_j is the charge of the particle, θ is a phase, and, in the case of one dimension, each term in the Hamiltonian in equation (12.2.1) or equation (12.2.2) contains the product of $\phi_j(x, t)$. From this, we obtain

$$\phi_1'(x, t)\phi_2'(x, t)\dots\phi_n'(x, t) = e^{-i(Q_1+Q_2+\dots+Q_n)\theta}\phi_1(x, t)\phi_2(x, t)\dots\phi_n(x, t).$$

Since charge is invariant under the transformation and neutrality is required for the Hamiltonian, $(Q_1 + Q_2 + \dots + Q_n) = 0$ in such a case. Furthermore, since θ is independent of x, $\nabla\phi_j \to e^{-i\theta Q_j}\nabla\phi_j$, so each term in the Hamiltonian in equation (12.2.1) is invariant under the above transformation or it possesses $U(1)$ symmetry.

If we rewrite equation (12.2.1) as follows:

$$H' = -\frac{1}{2}(\nabla\phi)^2 + U_{\text{eff}}(\phi), \quad U_{\text{eff}}(\phi) = -\alpha\phi^2 + \lambda\phi^4, \tag{12.2.3}$$

then we see that the effective potential energy $U_{\text{eff}}(\phi)$ in equation (12.2.3) has two sets of extrema and $\phi_0 = 0$, but the minimum is located at

$$\phi_0 = \pm\sqrt{\alpha/2\lambda} = \langle 0|\phi|0\rangle, \tag{12.2.4}$$

rather than $\phi_0 = 0$. This means that the energy at $\phi_0 = \pm\sqrt{\alpha/2\lambda}$ is lower than that at $\phi_0 = 0$. Therefore, $\phi_0 = 0$ corresponds to the normal ground state, while $\phi_0 = \pm\sqrt{\alpha/2\lambda}$ is the ground state of the macroscopic quantum systems.

In this case, the macroscopic quantum state is the stable state of the system. This shows that the Hamiltonian of a normal state differs from that of the macroscopic quantum state, in which the two ground states satisfy $\langle 0|\phi|0\rangle \neq -\langle 0|\phi|0\rangle$ under the transformation $\phi \to -\phi$. In other words, they no longer have $U(1)$ symmetry, so the symmetry of the ground states has been destroyed. The reason for this is evidently the nonlinear term $\lambda\phi^4$ in the Hamiltonian of the system. Therefore, this phenomenon is referred to as a spontaneous breakdown of symmetry. According to Landau's theory of phase transition, the system undergoes a second-order phase transition in such a case and the normal ground state $\phi_0 = 0$ is changed to the macroscopic quantum ground state $\phi_0 = \pm\sqrt{\alpha/2\lambda}$. Proof will be presented in the following example.

In order to make the expectation value in a new ground state zero in the macroscopic quantum state, the following transformation [221, 232] is made:

$$\phi' = \phi + \phi_0, \tag{12.2.5}$$

so that

$$\langle 0|\phi'|0\rangle = 0. \tag{12.2.6}$$

After this transformation, the Hamiltonian density of the system becomes

$$H'(\phi + \phi_0) = \frac{1}{2}|\nabla\phi|^2 + (6\lambda\phi^2 - \alpha)\phi^2 + 4\lambda\phi_0\phi^3 + (4\lambda\phi_0^3 - 2\alpha\phi_0)\phi + \lambda\phi^4 - \alpha\phi_0^2 + \lambda\phi_0^4. \tag{12.2.7}$$

Inserting equation (12.2.4) into equation (12.2.7), we have $\langle\phi_0|(4\lambda\phi_0^2 - 2\alpha)|\phi_0\rangle = 0$. Consider now the expectation value of the variation $\delta H'/\delta\phi$ in the ground state, i.e., $\langle 0|(\delta H'/\delta\phi)|0\rangle = 0$. Then, from equation (12.2.1), we get

$$\langle 0|\frac{\delta H'}{\delta\phi}|0\rangle = \langle 0| - \nabla^2\phi + 2\alpha\phi - 4\lambda\phi^3|0\rangle = 0. \tag{12.2.8}$$

After the transformation of equation (12.2.6), this becomes

$$\nabla^2\phi_0 + (4\lambda\phi_0^2 - 2\alpha)\phi_0 + 12\lambda\phi_0\langle 0|\phi^2|0\rangle + 4\lambda\langle 0|\phi^3|0\rangle - (2\alpha - 12\lambda\phi_0^2)\langle 0|\phi|0\rangle = 0, \tag{12.2.9}$$

where the terms $\langle 0|\phi^3|0\rangle$ and $\langle 0|\phi|0\rangle$ are both zero, but the fluctuation $12\lambda\phi_0\langle 0|\phi^2|0\rangle$ of the ground state is not zero. However, for a homogeneous system, at $T = 0\,\mathrm{K}$, the term $\langle 0|\phi^2|0\rangle$ is very small and can be neglected.

Then equation (12.2.9) can be written as

$$-\nabla^2\phi_0 - (4\lambda\phi_0^2 - 2\alpha)\phi_0 = 0. \tag{12.2.10}$$

Obviously, two sets of solutions, $\phi_0 = 0$ and $\phi_0 = \pm\sqrt{\alpha/2\lambda}$, can be obtained from the above equation. We can demonstrate that the former is unstable and that the latter is stable.

If the displacement is very small, i.e., $\phi_0 \to \phi_0 + \delta\phi_0 = \phi_0'$, then the equation satisfied by the fluctuation $\delta\phi_0$ is relative to the normal ground state $\phi_0 = 0$ and is

$$\nabla^2\delta\phi_0 - 2\alpha\delta\phi_0 = 0. \tag{12.2.11}$$

Its solution attenuates exponentially, indicating that the ground state $\phi_0 = 0$ is unstable. On the other hand, the equation satisfied by the fluctuation $\delta\phi_0$ relative to the ground state $\phi_0 = \pm\sqrt{\alpha/2\lambda}$ is $\nabla^2\delta\phi_0 - 2\alpha\delta\phi_0 = 0$. Its solution, $\delta\phi_0$, is an oscillatory function and thus the macroscopic quantum state ground state $\phi_0 = \pm\sqrt{\alpha/2\lambda}$ is stable. Further calculations show that the energy of the macroscopic quantum ground state is lower than that of the normal state by $\varepsilon_0 = -\alpha^2/4\lambda < 0$. Therefore, the ground state of the normal phase and that of the macroscopic quantum phase are separated by an energy gap of $\alpha^2/4\lambda$, so, at $T = 0\,\mathrm{K}$, all particles can condense to the ground state of the macroscopic quantum phase rather than filling the ground state of the normal phase. Based on this energy gap, we conclude that the specific heat of the macroscopic quantum systems has an exponential dependence on the temperature and the critical temperature is given by $T_c = 1.11\omega_p \exp[-1/(3\lambda/\alpha)N(0)]$ [221, 232]. This is a feature of the second-order phase transition. The results are in agreement with those of the BCS theory of superconductivity.

Therefore, the transition from the state $\phi_0 = 0$ to the state $\phi_0 = \pm\sqrt{\alpha/2\lambda}$ and the corresponding condensation of particles are second-order phase transitions. This is obviously the result of a spontaneous breakdown of symmetry due to the nonlinear interaction $\lambda\phi^4$.

In the presence of an electromagnetic field with a vector potential $\vec{A}$, the Hamiltonian of the system is given by equation (12.2.2). It still possesses $U(1)$ symmetry. Since the existence of the nonlinear terms in equation (12.2.2) has been demonstrated, a spontaneous breakdown of symmetry can be expected. Now consider the following transformation:

$$\phi(x) = \frac{1}{\sqrt{2}}[\phi_1(x) + i\phi_2(x)] \to \frac{1}{\sqrt{2}}[\phi_1(x) + \phi_0 + i\phi_2(x)]. \tag{12.2.12}$$

Since $\langle 0|\phi_1|0\rangle = 0$ under this transformation, equation (12.2.2) becomes

$$H' = \frac{1}{4}(\partial_i A_j - \partial_j A_i)^2 - \frac{1}{2}(\nabla\phi_2)^2 - \frac{1}{2}(\nabla\phi_1)^2 + \frac{(e^*)^2}{2}[(\phi_1 + \phi_0)^2 + \phi_2^2]A_i^2 - e^*\phi_0 A_i \nabla\phi_2$$

$$+e^*(\phi_2\nabla\phi_1-\phi_1\nabla\phi_2)A_i-\frac{1}{2}(-12\lambda\phi_0^2+2\alpha)\phi_1^2-\frac{1}{2}(12\lambda\phi_0^2+2\alpha)\phi_2^2$$
$$+4\lambda\phi_0\phi_1(\phi_1^2+\phi_2^2)+4\lambda(\phi_1^2+\phi_2^2)^2-\phi_0(4\lambda\phi_0^2+2\alpha)\phi_1-\alpha\phi_0^2+\lambda\phi_0^2. \qquad (12.2.13)$$

We see that the effective interaction energy of ϕ_0 is still given by

$$U_{\text{eff}}(\phi_0)=-\alpha\phi_0^2+\lambda\phi_0^4 \qquad (12.2.14)$$

and is in agreement with that given in equation (12.2.4). Therefore, using the same argument, we conclude that the spontaneous symmetry breakdown and the second-order phase transition also occur in the system. The system changes from the ground state of the normal phase $\phi_0 = 0$ to the ground state $\phi_0 = \pm\sqrt{\alpha/2\lambda}$ of the condensed phase in such a case. The above result can also be used to explain the Meissner effect and to determine its critical temperature in the superconductor. Thus, quantum states are formed through a second-order phase transition following a spontaneous symmetry breakdown due to nonlinear interaction in the system, regardless of the existence of any external field macroscopic quantum states, such as the superconducting state.

12.2.2 The features of the coherent state of macroscopic quantum effects

Proof that the macroscopic quantum state described by equations (12.2.1) and (12.2.2) is a coherent state, using either the second quantization theory or the solid state quantum field theory, is presented in the following paragraphs.

As discussed above, when $\delta H'/\delta\phi = 0$, from equation (12.2.1), we have

$$\nabla^2\phi-2\alpha\phi+4\lambda|\phi|^2\phi=0. \qquad (12.2.15)$$

It is a time-independent nonlinear Schrödinger equation (NLSE), which is similar to the GL equation. Expanding ϕ in terms of the creation and annihilation operators, b_p^+ and b_p, we have

$$\phi=\frac{1}{\sqrt{\bar{V}}}\sum_p\frac{1}{\sqrt{2\varepsilon_p}}(b_pe^{-ipx}+b_p^+e^{ip.x}), \qquad (12.2.16)$$

where $\bar{V}$ is the volume of the system. After a spontaneous breakdown of symmetry, ϕ_0, the ground-state of ϕ, is no longer zero, but $\phi_0 = \pm\sqrt{\alpha/2\lambda}$. The operation of the annihilation operator on $|\phi_0\rangle$ no longer gives zero, i.e.,

$$b_p|\phi_0\rangle \neq 0. \qquad (12.2.17)$$

A new field ϕ' can then be defined in accordance with the transformation equation (12.2.5), where ϕ_0 is a scalar field and satisfies equation (12.2.10) in such a case. Evidently, ϕ_0 can also be expanded into

$$\phi_0=-\frac{1}{\sqrt{\bar{V}}}\sum_p\frac{1}{\sqrt{2\varepsilon_p}}(\zeta_pe^{-ip_x}+\zeta_p^+e^{ip.x}). \qquad (12.2.18)$$

The transformation between the fields ϕ and ϕ' is obviously a unitary transformation. We have

$$\phi' = U\phi U^{-1} = e^{-s}\phi e^{s} = \phi + \phi_0, \tag{12.2.19}$$

where

$$S = i\int dx'[\phi'(x',t)\phi_0(x',t) - \phi_0(x',t)\phi(x',t)]. \tag{12.2.20}$$

ϕ and ϕ' satisfy the following commutation relation:

$$[\phi'(x',t),\phi(x,t)] = i\delta(x'-x). \tag{12.2.21}$$

From equation (12.2.6), we now have $\langle 0|\phi'|0\rangle = \phi_0' = 0$. The ground state $|\phi_0'\rangle$ of the field ϕ' thus satisfies

$$b_p|\phi_0'\rangle = 0. \tag{12.2.22}$$

From equation (12.2.6), we obtain the following relationship between the annihilation operator a_p of the new field ϕ' and the annihilation operator b_p of the ϕ field:

$$a_p = e^{-S}b_p e^{S} = b_p + \zeta_p, \tag{12.2.23}$$

where

$$\zeta_p = \frac{1}{(2\pi)^{3/2}}\int \frac{dx}{\sqrt{\varepsilon_p}}[\phi_0(x,t)e^{ip.x} + i\phi_0^*(x,t)e^{-ip.x}]. \tag{12.2.24}$$

Therefore, the new ground state $|\phi_0'\rangle$ and the old ground state $|\phi_0\rangle$ are related through $|\phi_0'\rangle = e^S|\phi_0\rangle$.

Thus, we have

$$a_p|\phi_0'\rangle = (b_p + \zeta_p)|\phi_0'\rangle = \zeta_p|\phi_0'\rangle. \tag{12.2.25}$$

According to the definition of the coherent state, from equation (12.2.25) we see that the new ground state $|\phi_0'\rangle$ is a coherent state. Because such a coherent state is formed after the spontaneous breakdown of symmetry of the system, it is referred to as a spontaneous coherent state. When $\phi_0 = 0$, the new ground state is the same as the old state, which is not a coherent state. The same conclusion can be directly derived from the BCS theory [17, 51, 271, 272]. In the BCS theory, the wave function of the ground state of a superconductor is written

$$|\phi_0'\rangle = \prod_k(\mu_k + \upsilon_k\hat{a}_k^+a_{-k}^+)|\phi_0\rangle = \prod_k(\mu_k + \upsilon_k\hat{b}_{k-k}^+)|\phi_0\rangle \sim \eta' \exp\left(\sum_k \frac{\upsilon_k}{\mu_k}\hat{b}_{k-k}^+\right)|\phi_0\rangle, \tag{12.2.26}$$

where $\hat{b}^+_{k-k} = \hat{a}^+_k \hat{a}^+_{-k}$. This equation shows that the superconducting ground state is a coherent state. Hence, we conclude that the spontaneous coherent state in superconductors is formed after the spontaneous breakdown of symmetry.

By reconstructing a quasi-particle operator-free new formulation of the Bogoliubov–Valatin transformation parameter dependence [297], Lin et al. [165] demonstrated that the BCS state is not only a coherent state of single Cooper pairs, but also the squeezed state of the double Cooper pairs, and reconfirmed thus the coherent feature of the BCS superconductive state.

12.2.3 The boson condensed features of macroscopic quantum effects

We will now employ the method used by Bogoliubov in the study of superfluid ^{4}He to prove that the above state is indeed a Bose condensed state. In order to do so, we rewrite equation (12.2.16) in the following form [196, 221, 226, 227, 232, 233]:

$$\phi(x) = \frac{1}{\sqrt{\bar{V}}} \sum_p q_p e^{ipx}, \quad q_p = \frac{1}{\sqrt{2\varepsilon_p}}(b_p + b^+_{-p}). \tag{12.2.27}$$

Since the field ϕ describes a boson, such as the Cooper electron pair in a superconductor, Bose condensation can occur in the system. We will apply the following traditional method in quantum field theory. Consider the following transformation:

$$b_p = \sqrt{N_0}\delta(p) + \gamma_p, \quad b^+_p = \sqrt{N_0}\delta(p) + \beta_p, \tag{12.2.28}$$

where N_0 is the number of bosons in the system and $\delta(p) = \begin{cases} 1, & \text{if } p=0 \\ 0, & \text{if } p\neq 0 \end{cases}$. Substituting equations (12.2.27) and (12.2.28) into equation (12.2.1), we arrive at the Hamiltonian operator of the system as follows:

$$\begin{aligned} H = {} & \left(\frac{4\lambda N_0^2}{\varepsilon_0^2 \bar{V}} - \frac{\alpha}{\varepsilon_0}\right)\sqrt{N_0}(\gamma_0 + \gamma_0^+ + \beta_0 + \beta_0^+) + \sum_P \left(\frac{4\lambda}{\varepsilon_0 \varepsilon_P}\frac{N_0}{\bar{V}} - \varepsilon_p\right)(\gamma_p^+ \beta^+_{-p} + \gamma_p \beta_{-p}) + \frac{4\lambda N_0^2}{\varepsilon_0^2 \bar{V}} \\ & - \frac{2N_0\alpha}{\varepsilon_0} + \frac{\lambda N_0}{\varepsilon_0 \bar{V}} \sum_p \frac{1}{\varepsilon_p}(\beta_p^+\beta^+_{-p} + \beta_p\beta_{-p} + \gamma_p^+\gamma^+_{-p} + \gamma_p\gamma_{-p} + 2\gamma_p^+\beta_p^+ + 2\beta_p^+\gamma_p) \\ & + \sum_p \left(\varepsilon_p - \frac{\alpha}{2\varepsilon_0} + \frac{4\lambda N_0}{\varepsilon_0\varepsilon_p \bar{V}}\right)(\gamma_p^+\gamma_p + \beta_p^+\beta_p) + \sum_P \frac{4\lambda}{\varepsilon_0\varepsilon_P}\frac{N_0}{\bar{V}} + O\left(\frac{\sqrt{N_0}}{\bar{V}}\right) + O\left(\frac{N_0}{\bar{V}^2}\right). \end{aligned} \tag{12.2.29}$$

Because the condensed density $N_0/\bar{V}$ must be finite, it is possible that the higher-order terms $O(\sqrt{N_0}/\bar{V})$ and $O(N_0/\bar{V}^2)$ may be neglected. Next, we perform the following canonical transformation:

$$\gamma_p = u_p^* c_p + \upsilon_p c^+_{-p}, \quad \beta_p = u_p^* d_p + \upsilon_p d^+_{-p}, \tag{12.2.30}$$

where ν_p and μ_p are real and satisfy $(\mu_p^2 - \nu_p^2) = 1$. This introduces another transformation,

$$\varsigma_p = \frac{1}{\sqrt{2}}(u_p\gamma_p^+ - \upsilon_p\gamma_{-p} + u_p\beta_p^+ - \upsilon_p\beta_{-p}), \quad \eta_p^+ = \frac{1}{\sqrt{2}}(u_p\gamma_p^+ - \upsilon_p\gamma_{-p} - u_p\beta_p^+ + \upsilon_p\beta_{-p}), \tag{12.2.31}$$

so the following relations are obtained:

$$[\varsigma_p, H] = g_p\varsigma_p + M_p\varsigma_{-p}^+, \quad [\eta_p, H] = g_p'\eta_p + M_p'\eta_{-p}^+, \tag{12.2.32}$$

where

$$\begin{cases} g_p = G_p(u_p^2 + \upsilon_p^2) + F_p 2u_p\upsilon_p, \ M_p = F_p(u_p^2 + \upsilon_p^2) + G_p 2u_p\upsilon_p, \\ g_p' = G_p'(u_p^2 + \upsilon_p^2) + F_p' 2u_p\upsilon_p, \ M_p' = F_p'(u_p^2 + \upsilon_p^2) + G_p' 2u_p\upsilon_p, \end{cases} \tag{12.2.33}$$

while

$$G_p = \varepsilon_p - \frac{\alpha}{2\varepsilon_p} + 6\xi_p', \quad F_p = -\frac{\alpha}{2\varepsilon_p} + 6\xi_p', \quad G_p' = \varepsilon_p - \frac{\alpha}{2\varepsilon_p} + 2\xi_p', \quad F_p' = \frac{\alpha}{2\varepsilon_p} - 2\xi_p', \tag{12.2.34}$$

where $\xi_p' = \lambda N_o/(\varepsilon_0\varepsilon_p\bar{V})$.

We will now study two cases to illustrate the concepts.

(A) Let $M_p' = 0$. Then it can be seen from equation (12.2.32) that η_p^+ is the creation operator of elementary excitation and its energy is given by

$$g_p' = \sqrt{\varepsilon_p^2 + 4\varepsilon_p\xi_p' - 2\alpha}. \tag{12.2.35}$$

Using this concept, we obtain the following form from equations (12.2.32) and (12.2.34):

$$(u_p')^2 = \frac{1}{2}\left(1 + \frac{G_p'}{g_p'}\right) \quad \text{and} \quad (\upsilon_p')^2 = \frac{1}{2}\left(-1 + \frac{G_p'}{g_p'}\right). \tag{12.2.36}$$

From equation (12.2.32), we know that ξ_p^+ is not a creation operator of the elementary excitation, so another transformation must be made. We have

$$B_p = \chi_p\varsigma_p' + \mu_p\varsigma_p'^+, \quad |\chi_p|^2 - |\mu_p|^2 = 1. \tag{12.2.37}$$

We can then prove that

$$[B_p, [B_p, \bar{H}]] = E_p B_p, \tag{12.2.38}$$

where $E_p = \sqrt{12\varepsilon_p\xi_p' + \xi_p^2 - 2\alpha}$.

Now, inserting equations (12.2.30), (12.2.37), and (12.2.38) and $M_p' = 0$ into equation (12.2.29), after some reorganization, we have

$$\bar{H} = U + E_0 + \sum_{p>0}[E_p(B_p^+B_p + B_{-p}^+B_{-p}) + g_p'(\eta_p^+\eta_p + \eta_{-p}^+\eta_{-p})], \tag{12.2.39}$$

where

$$E_0 = -2\sum_{p>0} E_p|\mu_p|^2 = -\sum_{p>0}(g_p' - E_p). \tag{12.2.40}$$

Both U and E_0 are now independent of the creation and annihilation operators of the bosons. $U + E_0$ gives the energy of the ground state. N_0 can be determined from the condition of $\delta(U + E_0)/\delta N_o$, so we obtain the following formula:

$$\frac{N_0}{\bar{V}} = \frac{\alpha\varepsilon_0}{4\lambda} = \frac{1}{2}\varepsilon_0\phi_0^2. \tag{12.2.41}$$

This is the condensed density of the ground state ϕ_0. From equations (12.2.36), (12.2.37), and (12.2.40), we arrive at

$$g_p' = \sqrt{\varepsilon_p^2 - \alpha}, \quad E_p = \sqrt{\varepsilon_p^2 - \alpha}. \tag{12.2.42}$$

These correspond to the energy spectra of η_p^+ and B_p^+, respectively, and they are similar to the energy spectra of the Cooper pair and phonon in the BCS theory. Substituting equation (12.2.42) into equation (12.2.36), we have

$$u_p'^2 = \frac{1}{2}\left(1 + \frac{2\varepsilon_p^2 - \alpha}{2\sqrt{\varepsilon_p^2 - \alpha e_p}}\right), \quad v_p'^2 = \frac{1}{2}\left(-1 + \frac{2\varepsilon_p^2 - \alpha}{2\sqrt{\varepsilon_p^2 - \alpha\varepsilon_p}}\right). \tag{12.2.43}$$

(B) In the case where $M_p = 0$, a similar approach can be used to arrive at the energy spectrum corresponding to ξ_p^+ as $E_p = \sqrt{\varepsilon_p^2 + \alpha}$, while that corresponding to $A_p^+ = \chi_p\eta_p^+ + \mu_p\eta_{-p}$ is $g_p' = \sqrt{\varepsilon_p^2 + \alpha}$, where

$$u_p^2 = \frac{1}{2}\left(1 + \frac{2\varepsilon_p^2 + \alpha}{2\varepsilon_p\sqrt{\varepsilon_p^2 + \alpha}}\right), \quad v_p^2 = \frac{1}{2}\left(-1 + \frac{2\varepsilon_p^2 + \alpha}{2\varepsilon_p\sqrt{\varepsilon_p^2 + \alpha}}\right). \tag{12.2.44}$$

Based on experiments in quantum statistical physics, we know that the occupation number of the level with an energy of ε_p, for a system in thermal equilibrium at certain temperature ($T \neq 0$), is shown as

$$N_p = \langle b_p^+ b_p \rangle = \frac{1}{e^{\varepsilon_p/K_B T} - 1}, \tag{12.2.45}$$

where $\langle \ldots \rangle$ denotes the Gibbs average, defined as $\langle \ldots \rangle = \frac{\mathrm{sp}[e^{-H/K_B T}\ldots]}{\mathrm{SP}[e^{-H/K_B T}]}$, where SP denotes the trace in a Gibbs statistical description. When $T \to 0$ K, the majority of the bosons or Cooper pairs in a superconductor condense to the ground state with $p \to 0$. Therefore, $\langle b_0^+ b_0 \rangle \approx N_0$, where N_0 is the total number of bosons or Cooper pairs in the system and $N_0 \gg 1$, i.e., $\langle b^+ b \rangle = 1 \ll \langle b_0^+ b_0 \rangle$.

As can be seen from equations (12.2.27) and (12.2.28), the number of particles is extremely large when they lie in the condensed state, that is,

$$\phi_0 = \phi_{p=0} = \frac{1}{\sqrt{2\varepsilon_0 \bar{V}}}(b_0 + b_0^+). \tag{12.2.46}$$

Because $\langle \gamma_0 | \phi_0 \rangle = 0$ and $\langle \beta_0 | \phi_0 \rangle = 0$, b_0 and b_0^+ can be taken to be $\sqrt{N_0}$. The average value of $\phi^* \phi$ in the ground state then becomes

$$\langle \phi_0 | \phi^* \phi | \phi_0 \rangle = \langle \phi^* \phi \rangle_0 = \frac{1}{2\varepsilon_0 \bar{V}} \cdot 4N_0 = \frac{2N_0}{\varepsilon_0 \bar{V}}. \tag{12.2.47}$$

Substituting equation (12.2.41) into equation (12.2.47), we see that

$$\langle \phi^* \phi \rangle_0 = \frac{\alpha}{2\lambda} \quad \text{or} \quad \langle \phi^* \rangle_0 = \pm \sqrt{\frac{\alpha}{2\lambda}},$$

which is the ground state of the condensed phase, or the superconducting phase, that we have seen. Thus, the density $N_0/\bar{V}$ of the condensed phase or the superconducting phase formed after the Bose condensation coincides with the average value of the boson's (or Cooper pair's) field in the ground state. We can then conclude from the above investigation shown in equations (12.2.1) and (12.2.2) that the macroscopic quantum state or the superconducting ground state formed after the spontaneous symmetry breakdown is indeed a Bose–Einstein condensed state. This clearly shows the essence of the nonlinear properties of macroscopic quantum effects.

In the last few decades, Bose–Einstein condensation has been observed in a series of remarkable experiments using weakly interacting atomic gases, such as vapors of rubidium, sodium lithium, and hydrogen. Its formation and properties have been extensively studied. These studies show that Bose–Einstein condensation is a nonlinear phenomenon, analogous to nonlinear optics, and that the state is coherent and can be described by the following NLSE or the Gross–Pitaevskii equation [100, 249, 250]:

$$i\frac{\partial \phi}{\partial t'} = -\frac{\partial^2 \phi}{\partial x'^2} - \lambda |\phi|^3 + V(x)\phi, \tag{12.2.48}$$

where $t' = t/\hbar$, $x' = x\sqrt{2m}/\hbar$. This equation was used to discuss the realization of the Bose–Einstein condensation in the $(D+1)$ dimensions $(D = 1, 2, 3)$ by Bullough et al. [38, 39]. Also, Elyutin et al. [74, 75] gave the corresponding Hamiltonian density of a condensate system as follows:

$$H' = \left| \frac{\partial \phi}{\partial x'} \right|^2 + V(x')|\phi|^2 - \frac{1}{2}\lambda |\phi|^4, \tag{12.2.49}$$

where $H' = \mathcal{H}$, the nonlinear parameters of λ are defined as $\lambda = -2Naa_1/a_0^2$, N is the number of particles trapped in the condensed state, a is the ground state scattering length, and a_0 and a_1 are the transverse (y, z) and the longitudinal (x) condensation sizes (without self-interaction), respectively. Note that integrations over y and z

have been carried out in obtaining the above equation. λ is positive for condensation with self-attraction (negative scattering length). The coherent regime was observed in Bose–Einstein condensation of lithium. The specific form of the trapping potential $V(x')$ depends on the details of the experimental setup. Work on Bose–Einstein condensation based on the above Hamiltonian model was carried out and reported by Barenghi et al. [259].

It is not surprising to see that equation (12.2.48) is exactly the same as equation (12.2.15), corresponding to the Hamiltonian density in equation (12.2.49). As used in this study, it is also equivalent to equation (12.2.1). This prediction confirms the correctness of the above theory for Bose–Einstein condensation. As a matter of fact, immediately after the first experimental observation of this condensation phenomenon, it was realized that the coherent dynamics of the condensed macroscopic wave function could lead to the formation of nonlinear solitary waves. For example, self-localized bright, dark, and vortex solitons, formed by increased (bright) or decreased (dark or vortex) probability densities, respectively, were experimentally observed, particularly for the vortex solution, which has the same form as the vortex lines found in type II-superconductors and superfluids. These experimental results were in concordance with the results of the above theory. In the following sections of this text, we will study the soliton motions of quasi-particles in macroscopic quantum systems, superconductors, and superfluid systems. We will see that the dynamic equations in macroscopic quantum systems do have such soliton solutions.

12.2.4 Differences between macroscopic quantum effects and the microscopic quantum effects and their nonlinear quantum mechanic features

From the above discussion we may understand the nature and characteristics of macroscopic quantum systems. It is interesting to compare the macroscopic and microscopic quantum effects. Here we give a summary of the main differences between them.

(1) Concerning the origins of these quantum effects, the microscopic quantum effect is produced when microscopic particles, which have only a wave feature, are confined in a finite space or are constituted as matter, while the macroscopic quantum effect is due to the collective motion of the microscopic particles in systems with nonlinear interaction. It occurs through second-order phase transition following the spontaneous breakdown of symmetry of the systems.

(2) From the point of view of their characteristics, the microscopic quantum effect is characterized by quantization of physical quantities, such as energy, momentum, and angular momentum, wherein the microscopic particles remain constant. On the other hand, the macroscopic quantum effect is represented by discontinuities in macroscopic quantities, such as resistance, magnetic flux, vortex lines, and voltage.

The macroscopic quantum effects can be directly observed in experiments on a macroscopic scale, while the microscopic quantum effects can only be inferred from other effects related to them.

(3) The macroscopic quantum state is a condensed and coherent state, but the microscopic quantum effect occurs in determinant quantization conditions, which are different for bosons and fermions. So far, only bosons or combinations of fermions have been found in macroscopic quantum effects.

(4) The microscopic quantum effect is a linear effect, in which the microscopic particles are in an expanded state, their motions being described by linear differential equations, such as the Schrödinger equation, the Dirac equation, and the Klein–Gordon equations.

On the other hand, the macroscopic quantum effect is caused by nonlinear interactions. The motions of the particles are described by nonlinear partial differential equations, such as the NLSE (12.2.17).

We conclude that the macroscopic quantum effects are, in essence, a nonlinear quantum phenomenon, in which the properties of microscopic particles, such as the electron in superconductors, are described by an NLSE, such as equations (12.2.15) and (12.2.48). The corresponding Hamiltonians of the systems are also nonlinear. Therefore, we affirm that the macroscopic quantum systems possess the nonlinear quantum mechanical properties, so their effects and properties should be described by nonlinear quantum mechanics [220, 228, 230, 231, 235, 238, 240], whereas microscopic quantum effects are described by traditional quantum mechanics.

12.3 The soliton movements of electrons in superconductors

It is clear from the previous section that the superconductivity of a material is a kind of nonlinear quantum effect, formed after the breakdown of the symmetry of the system due to the electron-phonon interaction, which is a nonlinear interaction.

In this section, we discuss the properties of the motion of superconductive electrons in superconductors and the relation of the solutions of dynamic equations to the above macroscopic quantum effects. The study presented here shows that the superconductive electrons move in the form of a soliton, which results in a series of macroscopic quantum effects in superconductors. Therefore, the properties and motions of the quasi-particles are important for understanding the properties of superconductivity and macroscopic quantum effects.

12.3.1 The soliton features of motion of electrons in steady superconductors

As is well known, in the superconductor the states of the electrons are often represented by a macroscopic wave function. We have

$$\phi(\vec{r},t) = f(\vec{r},t)\phi_0 e^{i\theta(\vec{r},t)}, \quad \text{or} \quad \phi = \sqrt{\rho}e^{i\theta},$$

as mentioned above, where $\phi_0^2 = \alpha/2\lambda$. Landau et al. [56, 152] used the wave function to give the free energy density function f of a superconducting system, which is represented by

$$f_s = f_n - \frac{\hbar^2}{2m}|\nabla\phi|^2 - \alpha|\phi|^2 + \lambda|\phi|^4 \tag{12.3.1}$$

in the absence of any external field. If the system is subjected to an electromagnetic field specified by a vector potential $\vec{A}$, the free energy density of the system is of the following form:

$$f_s = f_n - \frac{\hbar^2}{2m}\left|\left(\nabla - \frac{ie^*}{c\hbar}\vec{A}\right)\phi\right|^2 - \alpha|\phi|^2 + \lambda|\phi|^4 + \frac{1}{8\pi}\vec{H}^2, \tag{12.3.2}$$

where $e^* = 2e$, $\vec{H} = \nabla\times\vec{A}$, α and λ are some interaction constants related to the features of the superconductor, m is the mass of an electron, e^* is the charge of a superconductive electron, c is the velocity of light, h is the Planck constant, $\hbar = h/2\pi$, and f_n is the free energy of the normal state. The free energy of the system is $F_s = \int f_s d_3x$. In terms of the conventional field, $F_{ij} = \partial_j A_i - \partial_i A^j$ $(j, i = 1, 2, 3)$, the term $\vec{H}^2/8\pi$ can be written as $F_{ji}F^{ji}/4$. Equations (12.3.1) and (12.3.2) show the nonlinear features of the free energy of the systems because $\phi(\vec{r}, t)$ is the nonlinear function of the wave function of the particles. Thus, we predict that the superconductive electrons have many new properties compared to the normal electrons. From $\delta F_s/\phi = 0$, we get

$$\frac{\hbar^2}{2m}\nabla^2\phi - \alpha\phi + 2\lambda\phi^3 = 0 \tag{12.3.3}$$

in the absence of external fields and

$$\frac{\hbar^2}{2m}\left(\nabla - \frac{ie^*}{c\hbar}\vec{A}\right)^2\phi - \alpha\phi + 2\lambda\phi^3 = 0 \tag{12.3.4}$$

in the presence of an external field, as well as

$$\vec{J} = +\frac{e^*\hbar}{2mi}(\phi^*\nabla\phi - \phi\nabla\phi^*) - \frac{e^*}{mc}|\phi|^2\vec{A}. \tag{12.3.5}$$

Equations (12.3.3)–(12.3.5) are the well-known GL equations [8, 9, 91, 92, 96–98] in steady state and a time-independent Schrödinger equation. Here, equation (12.3.3) is the GL equation in the absence of external fields. It is the same as equation (12.2.15), which was obtained from equation (12.2.1). Equation (12.3.5) can be obtained from equation (12.2.2). Therefore, equations (12.2.1) and (12.2.2) are the Hamiltonians corresponding to the free energy in equations (12.3.1) and (12.3.2).

From equations (12.3.3) and (12.3.4) we clearly see that superconductors are nonlinear systems. GL equations containing the nonlinear term of $2\lambda\phi^3$ are fundamental equations of superconductors, describing the motion of the superconductive electrons. However, the equations contain two unknown functions ϕ and $\vec{A}$, which make them extremely difficult to resolve.

We first study the properties of motion of superconductive electrons in the case where there is no external field. Then we consider only a one-dimensional pure superconductor [199, 203], where

$$\phi = \phi_0 \varphi(x, t), \quad \xi'^2(T) = \hbar^2/2m|\alpha|, \quad x' = x/\xi'(T), \tag{12.3.6}$$

where $\xi'(T)$ is the coherent length of the superconductor, which depends on the temperature. For a uniform superconductor, $\xi'(T) = 0.94\xi_0[T_c/(T_c - T)]^2$, where T_c is the critical temperature and ξ_0 is the coherent length of superconductive electrons at $T = 0$. Under the boundary conditions $\varphi(x' = 0) = 1$ and $\varphi(x' \to \pm\infty) = 0$, from equations (12.3.3) and (12.3.5), we easily find the following solution:

$$\varphi = \pm\sqrt{2}\sec h\left[\frac{x - x_0}{\xi'(T)}\right]$$

or

$$\phi = \pm\sqrt{\frac{\alpha}{\lambda}}\sec h\left[\frac{x - x_0}{\xi'(T)}\right] = \pm\sqrt{\frac{\alpha}{\lambda}}\sec h\left[\frac{\sqrt{2m\alpha}}{\hbar}(x - x_0)\right]. \tag{12.3.7}$$

This is a well-known wave packet type soliton solution. It can be used to represent the bright soliton occurring in the Bose–Einstein condensate found by Perez-Garcia et al. [244]. If the signs of α and λ in equation (12.3.3) are reversed, we get the following kink soliton solution under the boundary conditions of $\varphi(x' = 0) = 0$ and $\varphi(x' \to \pm\infty) = \pm 1$:

$$\phi = \pm(\alpha/2\lambda)^{1/2}\tanh\{[m\alpha(x - x_0/\hbar^2)]^{1/2}\}. \tag{12.3.8}$$

The energy of the soliton described by (12.3.7) is given by

$$E_{\text{sol}} = \int_{-\infty}^{\infty}\left[\frac{\hbar^2}{2m}\left(\frac{d\phi}{dx}\right)^2 - \alpha\phi^2 - \lambda\phi^4\right]dx = \frac{4\hbar\alpha^{3/2}}{3\lambda\sqrt{2m}}. \tag{12.3.9}$$

We assume here that the lattice constant $r_0 = 1$. The energy of the above soliton can be compared with the ground state energy of the superconducting state, $E_{\text{ground}} = -\alpha^2/4\lambda$. Their difference can be represented by

$$E_{\text{sol}} - E_{\text{ground}} = \alpha^{3/2}\left(\sqrt{\alpha} + \frac{16\hbar}{3\sqrt{2m}}\right)\Big/2\lambda > 0.$$

This indicates clearly that the soliton is not in the ground state, but in an excited state of the system. Therefore, the soliton is a quasi-particle.

From the above discussion, we see that, in the absence of external fields, the superconductive electrons move in the form of solitons in a uniform system. These solitons are formed by a nonlinear interaction among the superconductive electrons, which suppresses the dispersive behavior of electrons. A soliton can carry a certain amount of energy while moving in superconductors. It can be demonstrated that these soliton states are very stable.

12.3.2 The features of soliton motion of electrons in superconductors under the action of an electromagnetic field

We now consider the motion of superconductive electrons in the presence of an electromagnetic field $\vec{A}$. Its equation of motion is denoted by equations (12.3.4) and (12.3.5). Assume now that the field $\vec{A}$ satisfies the London gauge $\nabla.\vec{A} = 0$ [167] and that the substitution of $\phi(\vec{r},t) = \varphi(\vec{r},t)e^{i\theta(\vec{r})}$ into equations (12.3.4) and (12.3.5) [199, 203] yields

$$J = \frac{e^*\phi_0^2}{m}\left(\hbar\nabla\theta - \frac{e^*}{c}\vec{A}\right)\varphi^2 \tag{12.3.10}$$

and

$$\nabla^2\varphi - \left[\left(\nabla\theta - \frac{e^*}{\hbar c}\vec{\boldsymbol{A}}\right)^2\varphi\right] - \frac{2m}{\hbar^2}(\alpha - 2\lambda\phi_0^2\varphi^2)\varphi = 0. \tag{12.3.11}$$

For bulk superconductors, J is a constant (permanent current) for a certain value of $\vec{A}$ and it can thus be taken as a parameter.

Let $B^2 = m^2J^2/\hbar^2(e^*)^2\phi_0^4$, $b = 2m\alpha/\hbar^2 = \xi^{-2}$. From equations (12.3.10) and (12.3.11) [235, 238], we obtain

$$\left(\hbar\nabla\theta - \frac{e^*}{c}\vec{\boldsymbol{A}}\right) = \frac{Jm}{e^*\phi_0^2\varphi^2}, \tag{12.3.12}$$

$$\frac{d^2\varphi}{dx^2} = -\frac{d}{d\varphi}U_{\text{eff}}(\varphi), \quad U_{\text{eff}}(\varphi) = \frac{B^2}{2\varphi^2} - \frac{1}{2}b\varphi^2 + \frac{1}{4}b\varphi^4, \tag{12.3.13}$$

where U_{eff} is the effective potential of the superconductive electron, as schematically shown in Figure 12.2. Comparing this case with that in the absence of external fields, we find that the equations have the same form and the electromagnetic field changes only the effective potential of the superconductive electron. When $\vec{A} = 0$, the effective potential well is characterized by double wells. In the presence of an electromagnetic field, there are still two minima in the effective potential, corresponding to the two

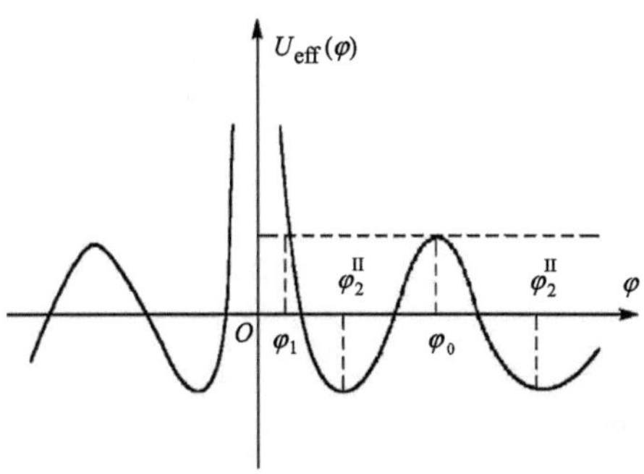

Figure 12.2: The effective potential energy in equation (12.3.18).

ground states of the superconductor in this condition. This shows that the spontaneous breakdown of symmetry still occurs in the superconductor, so the superconductive electrons also move in the form of solitons. To obtain the soliton solution, we integrate equation (12.3.13) and get

$$x = \int_{\varphi_1}^{\varphi} \frac{d\varphi}{\sqrt{2[E - U_{\text{eff}}(\varphi)]}}, \tag{12.3.14}$$

where E is a constant of integration which is equivalent to the energy and φ_1 is the lower limit of the integral, determined by the value of φ at $x = 0$, i.e., $E = U_{\text{eff}}(\varphi_0) = U_{\text{eff}}(\varphi_1)$. We introduce the following dimensionless quantities: $\varphi^2 = u$, $E = b\varepsilon/2$, $2\tilde{d} = 4J^2 m\lambda/[(e^*)^2 d^2]$, $\varphi^2 = u$ and, after performing the transformation $u \to -u$, equation (12.3.14) can be written in the following form:

$$-\sqrt{2b}x = \int_{u_1}^{u} \frac{du}{\sqrt{u^3 - 2u^2 - 3\varepsilon u - 2\tilde{d}^2}}. \tag{12.3.15}$$

It can be seen from Figure 12.3 that the denominator in the integrand in equation (12.3.15) approaches zero linearly when $u = u_1 = \varphi_1^2$, but approaches zero gradually when $u = u_2 = \varphi_0^2$. Thus [205, 206], we have

$$u(x) = \varphi^2(x) = u_0 - g \sec h^2\left(\sqrt{\frac{1}{2}gbx}\right) = u_1 + g \tan h^2\left(\sqrt{\frac{1}{2}gbx}\right), \tag{12.3.16}$$

where $g = u_0 - u_1$ and satisfies

$$(2+g)^2(1-g) = 27\tilde{d}^2, \quad 2u_0 + u_1 = 2, \quad u_0^2 + 2u_0 u_1 = -2\varepsilon, \quad u_1 u_0^2 = 2\tilde{d}^2. \tag{12.3.17}$$

It can be seen from equation (12.3.16) that, for a large part of the sample, u_1 is very small and may be neglected; the solution u is very close to u_0. We then get from

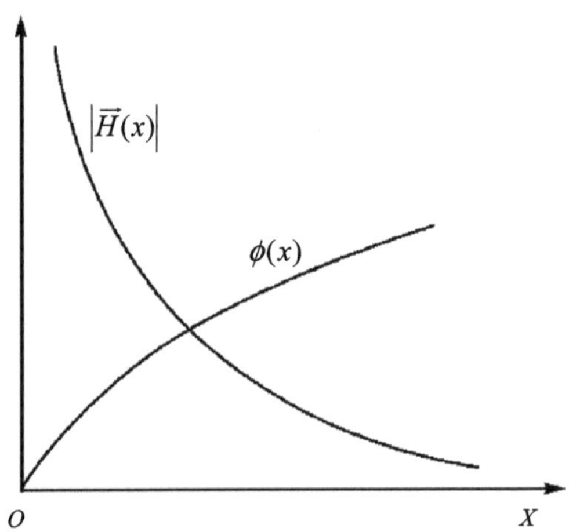

Figure 12.3: Changes of $\phi(x)$ and $|\tilde{H}(x)|$ with x in equations (12.3.18) and (12.3.19).

equation (12.3.16)

$$\varphi(x) = \varphi_0 \tan h\left(\sqrt{\frac{1}{2}}gbx\right). \tag{12.3.18}$$

Substituting the above into equation (12.3.12), the electromagnetic field $\vec{A}$ in the superconductors can be obtained. We have

$$\vec{A} = -\frac{\vec{J}mc}{(e^*)^2\phi_0^2}\frac{1}{\varphi^2} - \frac{\hbar c}{e^*}\nabla\theta = \frac{\vec{J}mc}{(e^*)^2\phi_0^2\varphi_0^2}\cot h(\sqrt{\frac{1}{2}}gbx) - \frac{\hbar c}{e^*}\nabla\theta.$$

For a large portion of the superconductor, the phase change is very small. Using $\vec{H} = \nabla \times \vec{A}$, the magnetic field can be determined and is given [205, 206] by

$$\vec{H} = \frac{\vec{J}mc\sqrt{2gb}}{(e^*)^2\phi_0^2\varphi_0^2}\left[\cot h^3\left(\sqrt{\frac{1}{2}}gbx\right) + \cot h\left(\sqrt{\frac{1}{2}}gbx\right)\right]. \tag{12.3.19}$$

Equations (12.3.18) and (12.3.19) are analytical solutions of the GL equations (12.3.14) and (12.3.15) in the one-dimensional case, which are shown in Figure 12.2 and Figure 12.3, respectively. Equation (12.3.18) or (12.3.16) shows that the superconductive electron in the presence of an electromagnetic field is still a soliton. However, its amplitude, phase, and shape are different from those in a uniform superconductor and in the absence of external fields. The soliton here is obviously influenced by the electromagnetic field, as reflected by the change in the form of the solitary wave. This is why a permanent superconducting current can be established by the motion of superconductive electrons in a certain direction in such a superconductor; solitons have the ability to maintain their shape and velocity while in motion.

It is clear from Figure 12.4 that $\tilde{H}(x)$ is large where $\phi(x)$ is small and vice versa. When $x \to 0$, $\tilde{H}(x)$ reaches a maximum, while $\phi(x)$ approaches zero. On the other hand, when $x \to \infty$, $\phi(x)$ becomes very large, while $\tilde{H}(x)$ approaches zero. This shows that the system is still in the superconductive state. These are exactly the well-known

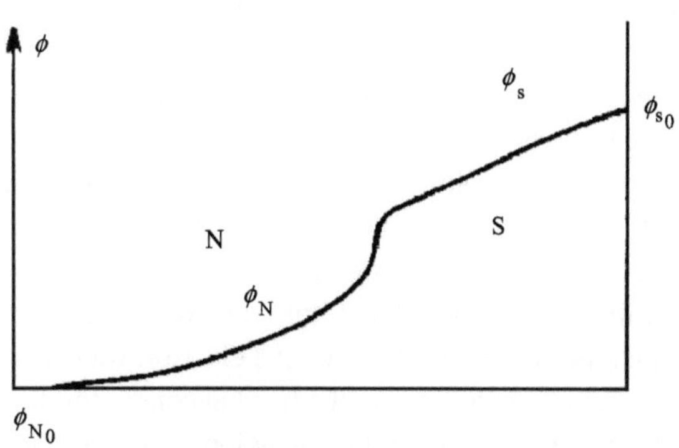

Figure 12.4: Proximity effect in an S-N junction.

behaviors of vortex lines or magnetic flux lines in type-II superconductors [205, 206]. With this, we conclude the explanation of macroscopic quantum effects in type-II superconductors using the GL equation of motion of superconductive electrons under action of an electromagnetic field.

Recently, Caradoc-Davies et al. [43], Matthews et al. [174], and Madison et al. [169] observed vertex solitons in the Bose–Einstein condensates. Tonomure [295] experimentally observed magnetic vortices in superconductors. These vortex lines in the type-II superconductors are quantized. The macroscopic quantum effects are well described by the nonlinear theory discussed above, demonstrating the correctness of the theory.

We now proceed to determine the energy of the soliton given by (12.3.18). From earlier discussions, we know the energy of the soliton is given by

$$E = \int_{-\infty}^{+\infty} \left[\frac{1}{2} \left(\frac{d\varphi}{dx} \right)^2 + \frac{b}{2}\varphi^2 - \frac{b}{4}\varphi^4 - \frac{B^2}{2\varphi^2} \right] dx \approx \varphi_0^2 \left[\frac{2b\varphi_0^2}{3} - 1 + \frac{b}{2}\left(1 - \frac{\varphi_0^2}{2}\right) \right] - \frac{B^2}{2\varphi_0^2},$$

which depends on the interaction between superconductive electrons and the electromagnetic field.

From the above discussion, we understand that, for a bulk superconductor, the superconductive electrons behave as solitons, regardless of the presence of external fields. Thus, the superconductive electrons are a special type of soliton. Obviously, the solitons are formed because the nonlinear interaction $\lambda|\phi|^2\phi$ suppresses the dispersive effect of the kinetic energy in equations (12.3.3) and (12.3.4). They move in the form of solitary waves in the superconducting state. In the presence of external electromagnetic fields, we demonstrate theoretically that a permanent superconductive current is established and that the vortex lines or magnetic flux lines also occur in type-II superconductors.

12.3.3 The properties of soliton movement of the electrons in superconductive junctions and its relation to the macroscopic quantum effects

(1) The features of the motion of electrons in an S-N junction and the proximity effect. The superconductive junction consists of a superconductor (S) which contacts a normal conductor (N), in which the latter can be superconductive. If this is the case, this phenomenon is referred to as the proximity effect. This is obviously the result of the long range coherent properties of superconductive electrons. It can be regarded as the penetration of electron pairs from the superconductor into the normal conductor or a result of diffraction and transmission of superconductive electron waves. In this phenomenon, superconductive electrons can occur in the normal conductor, but their amplitudes are much smaller than in the superconductive region, so the nonlinear term $\lambda|\phi|^2\phi$ in the GL equations (12.3.4) and (12.3.5) can be neglected. Because of

this, GL equations in the normal and superconductive regions have different forms. On the S side of the S-N junction, the GL equation [207] is

$$\frac{\hbar^2}{2m}\left(\nabla - \frac{ie^*}{ch}\vec{A}\right)\phi - \alpha\phi + 2\lambda\phi^3 = 0, \tag{12.3.20}$$

while that on the N side of the junction is

$$\frac{\hbar^2}{2m}\left(\nabla - \frac{ie^*}{ch}\vec{A}\right)\phi - \alpha'\phi = 0. \tag{12.3.21}$$

Thus, the expression for $\vec{J}$ remains the same on both sides. We have

$$\vec{J} = \frac{e^*\hbar}{2mi}(\phi^*\nabla\phi - \phi\nabla\phi^*) - \frac{(e^*)^2}{mc}|\phi|^2\vec{A}. \tag{12.3.22}$$

In the S region, we obtain a solution of equation (12.3.20), which is given by equation (12.3.16) or equations (12.3.18) and (12.3.19). In the N region, from equations (12.3.21) and (12.3.22), we easily obtain

$$\begin{cases} \varphi^2 = \frac{1}{2}\sqrt{(\varepsilon')^2 - 4\tilde{d}^2}\sin(2\sqrt{b'x}) + \frac{\varepsilon'}{2}, \\ \phi_N^2 = \varphi^2\phi_0^2 e^{-2i\theta} = \frac{1}{2}\sqrt{(\varepsilon')^2 - 4\tilde{d}^2}\sin(2\sqrt{b'x})e^{-i2\theta} + \frac{\varepsilon'}{2}\phi_0^2 e^{-i2\theta}, \end{cases} \tag{12.3.23}$$

where

$$b' = \frac{2m\alpha'}{\hbar^2} = \frac{1}{\xi'^2}, \quad 2\tilde{d}^2 = \frac{4\vec{J}^2 m\lambda}{(e^*)^2\alpha'^2}, \quad E' = \frac{b'}{2}\varepsilon'.$$

Here, ε' is an integral constant. A graph of ϕ vs. x in both the S and the N regions, as shown in Figure 12.4, coincides with that obtained by Blackburn [25]. The solution given in equation (12.3.23) is the analytical solution in this case. On the other hand, Blackburns result was obtained by expressing the solution in terms of elliptic integrals and then integrating numerically. From this, we see that the proximity effect is caused by diffraction or transmission of the superconductive electrons.

(2) The Josephson effect in S-insulator (I)-S and S-N-S as well as S-I-N-S junctions. An S-N-S or an S-I-S consists of a normal conductor or an insulator sandwiched between two superconductors, as schematically shown in Figure 12.5a. The thickness of the normal conductor or the insulator layer is referred to as L and we choose the z coordinate such that the normal conductor or the insulator layer is located at $-L/2 \le x \le L/2$. The features of S-I-S junctions were studied by Jacobson et al. [135]. We will treat this problem using the above ideas and methods [200, 214].

The electrons in the superconducting regions ($x \ge L/2$) are depicted by the GL equation (12.3.20). Its solution is given by equation (12.3.18). After eliminating u_1 from

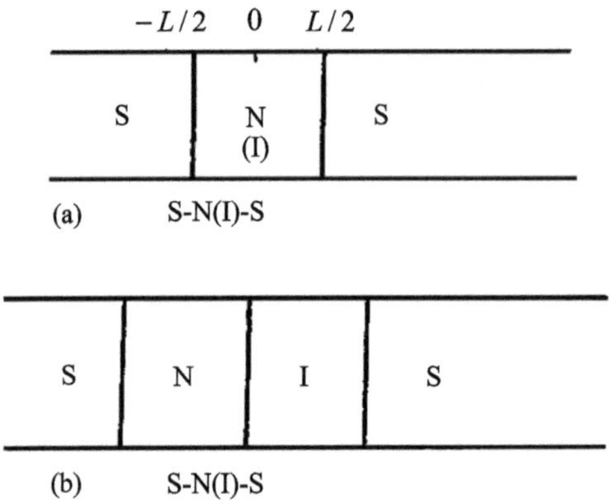

Figure 12.5: Superconductive junction of S-N(I)-S and S-N-I-S.

equation (12.3.17) [25, 135], we have $J = (e^* \alpha u_0/2)\sqrt{(1-u_0)\alpha/(m\lambda)}$. Setting $dJ/du_0 = 0$, we get the maximum current $J_c = (e^*\alpha/3)\sqrt{\alpha/(3m\lambda)}$.

This is the critical current of a perfect superconductor, corresponding to the three-fold degenerate solution of equation (12.3.17), i.e., $u_1 = u_0$.

From equation (12.3.22), we have $\vec{A} = -m\vec{J}_c/[(e^*)^2\phi_0^2\varphi^2]$. Using the London gauge, $\nabla.\vec{A} = 0$ [200, 214], we get $d^2\theta/dx^2 = mJ/(e^*\phi_0^2\hbar)\frac{d}{dx}(1/\varphi^2)$. Integrating the above equation twice, we get the change of the phase to be

$$\Delta\theta = \frac{mJ}{e^*\phi_0^2\hbar}\int\left(\frac{1}{\varphi^2} - \frac{1}{\varphi_\infty^2}\right)dx, \tag{12.3.24}$$

where $\varphi^2 = u$ and $\varphi_\infty^2 = u_0$. Here we have used the following de Gennes boundary conditions in obtaining equation (12.3.24):

$$\left.\frac{d\phi}{dx}\right|_{|x|\to\infty} = 0, \quad \left.\frac{d\theta}{dx}\right|_{|x|\to\infty} = 0, \quad \phi(|x|\to\infty) = \phi_\infty. \tag{12.3.25}$$

If we substitute equations (12.3.15)–(12.3.18) into equation (12.3.24), the phase shift of the wave function from an arbitrary point x to infinity can be obtained directly from the above integral. It takes the following form:

$$\Delta\theta_L(x\to\infty) = -\tan^{-1}\sqrt{\frac{u_1}{u_0-u_1}} + \tan^{-1}\sqrt{\frac{u_1}{u-u_1}}. \tag{12.3.26}$$

For the S-N-S or S-I-S junctions, the superconducting regions are located at $|x| \geq L/2$ and the phase shift in the S region is thus

$$\Delta\theta_s = 2\Delta\theta_L\left(\frac{L}{2}\to\infty\right) \approx 2\tan^{-1}\sqrt{\frac{u_1}{u_s-u_1}}. \tag{12.3.27}$$

According to the results in (12.3.21) and (12.3.22) and the above similar method, the change of the phase in the I or N region of the S-N-S or S-I-S junction may be expressed [200, 214] as

$$\Delta\theta_N = -2\tan^{-1}\left[\frac{2e^*h'}{\dot{J}}\sqrt{\frac{\alpha^2}{8m\lambda}}\tan\left(\frac{\sqrt{b'}L}{2}\right)\right] + \frac{m\dot{J}L}{2e^*h'\mu_0}, \tag{12.3.28}$$

where

$$h' = \sqrt{\frac{8m\lambda}{\alpha^2}\frac{\dot{J}}{2e^*}\frac{\tan(\Delta\theta_N/2)}{\tan(\sqrt{b'}L/2)}} \quad \text{and} \quad \frac{m\dot{J}L}{2e^*h'\mu_0}$$

is an additional term to satisfy the boundary conditions (12.3.25) that may be neglected in the case being studied. Near the critical temperature ($T < T_c$), the current passing through a weakly linked superconductive junction is very small, i.e., $\dot{J} \ll 1$, so we obtain

$$\mu_1' = 4\dot{J}^2 m\lambda/[(e^*)^2\alpha^2] = 2\bar{A}^2 \quad \text{and} \quad g' = 1.$$

Since $\eta\varphi^2$ and $d\varphi^2/dx$ are continuous at the boundary $x = L/2$, we have

$$\left.\frac{d\mu_s}{dx}\right|_{x=L/2} = \left.\frac{d\mu_N}{dx}\right|_{x=L/2}, \quad \eta_s\mu_s|_{x=L/2} = \eta_N\mu_N|_{x=L/2},$$

where η_s and η_N are the constants related to features of superconductive and normal phases in the junction, respectively. These [200, 214] give

$$2\sqrt{b'}\bar{A}\sin(2\Delta\theta_N) = \varepsilon_1[1-\cos(2\Delta\theta_s)]\sin(\sqrt{b'}L),$$
$$\cos(\sqrt{b'}L)\sin(2\Delta\theta_s) = \varepsilon\sin(2\Delta\theta_N) + \sin(2\Delta\theta_s + \Delta\theta_N),$$

where $\varepsilon_1 = \eta_N/\eta_S$. From the two equations, we obtain

$$\sin(\Delta\theta_s + \Delta\theta_N) = \frac{2\sqrt{2m\lambda}\dot{J}}{e^*\alpha}\sqrt{b'}\sin(\sqrt{b'}L).$$

Thus,

$$\dot{J} = \dot{J}_{\max}\sin(\Delta\theta_s + \Delta\theta_N) = \dot{J}_{\max}\sin(\Delta\theta), \tag{12.3.29}$$

where

$$\dot{J}_{\max} = \frac{e^*\alpha_s}{2\sqrt{2m\lambda b'}}\cdot\frac{1}{\sin(\sqrt{b'}L)}, \quad \Delta\theta = \Delta\theta_s + \Delta\theta_N. \tag{12.3.30}$$

Equation (12.4.5) is the well-known example of the Josephson current. From Section 12.1 we know that the Josephson effect is a macroscopic quantum effect. We have seen that this effect can be explained by nonlinear quantum theory. This again shows that the macroscopic quantum effect is just a nonlinear quantum phenomenon.

From equation (12.4.6), we see that the Josephson critical current is inversely proportional to $\sin(\sqrt{b'}L)$, which means that the current increases suddenly whenever $\sqrt{b'}L$ approaches $n\pi$, suggesting some resonant phenomena occur in the system. This has not been observed before. Moreover, $e^*\alpha/2\sqrt{2m\lambda b'} = e^*\hbar\alpha_s/4m\sqrt{\lambda\alpha_N}$, which is related to $(T - T_c)^2$.

Finally, it is worthwhile to mention that no explicit assumption was made in the above on whether the junction is a potential well ($\alpha < 0$) or a potential barrier ($\alpha > 0$). The results are thus valid and the Josephson effect in equation (12.3.29) occurs both with potential wells and with potential barriers.

We now study the Josephson effect in the S-N-I-S junction as shown schematically in Figure 12.5b. It can be regarded as a multi-layer junction consisting of an S-N-S and S-I-S junction. If appropriate thicknesses for the N and I layers are used (approximately 20–30 Å), the Josephson effect similar to that discussed above can occur. Since the derivations are similar to those in the previous sections, we will skip much of the details and give the results in the following. The Josephson current in the S-N-I-S junction is still given by

$$\dot{J} = \dot{J}_{\max}\sin(\Delta\theta),$$

but, where

$$\Delta\theta = \Delta\theta_{s1} + \Delta\theta_N + \Delta\theta_I + \Delta\theta_{s2}$$

and

$$\begin{aligned}\dot{J}_{\max} = {} & \frac{1}{\sqrt{b'_N}}\left\{\frac{\varepsilon_1\sinh(\sqrt{b'_N}L)}{2[\cosh(\sqrt{b'_N}L) - \cos(2\Delta\theta_N)]}\right\}\\ & \times\frac{1}{\sqrt{[1+\cos(2\Delta\theta_N)][1+\cos(2\Delta\theta_I)]} - \sqrt{[1-\cos(2\Delta\theta_N)][1-\cos(2\Delta\theta_I)]}}\\ & - \frac{1}{\sqrt{b'_N}}\left\{\frac{\varepsilon_1\sqrt{[1-\cos^2(2\Delta\theta_N)]}\sinh(\sqrt{b'_N}L)}{2[\cosh(\sqrt{b'_N}L) - \cos(2\Delta\theta_N)]^2 - 1 + \cos^2(2\Delta\theta_N)}\right\}\\ & \times\frac{1}{\sqrt{[1-\cos(2\Delta\theta_N)][1-\cos(2\Delta\theta_I)]} + \sqrt{[1+\cos(2\Delta\theta_N)][1+\cos(2\Delta\theta_I)]}},\end{aligned}$$

it can be shown that the temperature dependence of $J_{\max}$ is $J_{\max} \propto (T_c - T_0)^2$, which is similar to the results obtained by Blackburn et al. [25] for the S-N-I-S junction and those by Romagnan et al. [93] using the Pb-PbO-Sn-Pb junction. Here, we obtain the same results using a completely different approach. This indicates again that we can theoretically obtain some results which agree with the experimental data.

12.4 The soliton movement features of the electrons with time dependence in the superconductor

12.4.1 The dynamic equations of the superconductive electron and its soliton solutions

We have so far only studied the properties of the motion of superconductive electrons in steady states in superconductors, as described by the time-independent GL equation. In such a case, the superconductive electrons move as solitons. What are the features of time-dependent motion in nonequilibrium states of a superconductor? Naturally, this motion should be described by the time-dependent GL (TDGL) equation [8, 9, 67, 91, 92, 96–98]. Unfortunately, there are many different forms of the TDGL equation under different conditions. The following is commonly used when an electromagnetic field $\vec{A}$ is involved:

$$\Gamma\left[\hbar\frac{\partial}{\partial t} - 2ie\mu(r)\right]\phi = \frac{-1}{2m}\left(\hbar\nabla - \frac{2ie}{c}\vec{A}\right)^2 + \alpha\phi - \lambda|\phi|^2\phi \tag{12.4.1}$$

and

$$\vec{J} = \sigma\left[-\frac{1}{c}\frac{\partial\vec{A}}{\partial t} - \nabla\mu(r)\right] + \frac{ie\hbar}{m}(\phi^*\nabla\phi - \phi\nabla\phi^*) - \frac{4e^2}{mc}\vec{A}|\phi|^2, \tag{12.4.2}$$

where $i = \sqrt{-1}$, $\nabla\times\nabla\times\vec{A} = \frac{1}{c}\frac{\partial}{\partial t}(-\frac{1}{c}\frac{\partial\vec{A}}{\partial t} - \nabla\mu) + \frac{4\pi\vec{J}}{c}$, σ is the conductivity in the normal state, Γ is an arbitrary constant, and μ is the chemical potential of the system. In practice, equation (12.4.1) is simply a time-dependent Schrödinger equation with a damping effect.

In certain situations, the following forms of the TDGL equation are also used:

$$i\hbar\frac{\partial\phi}{\partial t} = -\frac{\hbar^2}{2m}\left(\nabla - \frac{2ie}{\hbar c}\vec{A}\right)^2\phi + \alpha\phi - \lambda|\phi|^2\phi \tag{12.4.3}$$

or

$$i\left(\hbar\frac{\partial}{\partial t} - i2e\mu\right)\phi = \frac{1}{\Gamma}(\alpha - \lambda|\phi|^2)\phi + \frac{\xi'^2}{\Gamma}\left(\nabla - \frac{2ie}{\hbar c}\vec{A}\right)^2\phi, \tag{12.4.4}$$

where $\xi' = \hbar/\sqrt{2m}$ and equation (12.4.3) is a NLSE under an electromagnetic field having soliton solutions. However, these solutions are very difficult to find and no analytic solutions have been obtained. An approximate solution was obtained by Kusayanage et al. [149] by neglecting the φ^3 term in equation (12.4.1) or equation (12.4.3) in the case where $\vec{A} = (0, \tilde{H}x, 0)$, $\mu = -K\tilde{E}x$, $\vec{H} = (0, 0, \tilde{H})$, and $\vec{E} = (\tilde{E}, 0, 0)$, where $\vec{H}$ is the magnetic field and $\vec{E}$ is the electric field. We will solve the TDGL equation in the case of weak fields in the following. The TDGL equation (12.4.4) can be written in the following form [116, 224] when $\vec{A}$ is very small:

$$i\hbar\frac{\partial\phi}{\partial t}+\frac{\hbar^2}{2m\Gamma}\nabla^2\phi+\frac{\lambda}{\Gamma}|\phi|^2\phi=\left(\frac{\alpha}{\Gamma}-2e\mu\right)\phi, \tag{12.4.5}$$

where α and Γ are material-dependent parameters, λ is the nonlinear coefficient, and m is the mass of the superconductive electron. Equation (12.4.5) is actually an NLSE in a potential field $\alpha/\Gamma - e\mu$. Cai and Bhattacharjee [42] and Davydov [56] used it in their studies of superconductivity. However, this equation is also difficult to solve. In the following, we show how Pang solved the equation in the one-dimensional case.

For convenience, let $t' = t/\hbar$, $x' = x\sqrt{2m\Gamma}/\hbar$. Then equation (12.4.5) becomes

$$i\frac{\partial\phi}{\partial t'}+\frac{\partial^2\phi}{\partial x'^2}+\frac{\lambda}{\Gamma}|\phi|^2\phi=\left[\frac{\alpha}{\Gamma}-2e\mu(x')\right]\phi. \tag{12.4.6}$$

If we let $\alpha/\Gamma - 2e\mu = 0$, then equation (12.4.6) is the usual NLSE, whose solution [116, 224] is of the following form:

$$\phi_s^0=\varphi_0(x',t')e^{i\theta_0(x',t')}, \tag{12.4.7}$$

$$\varphi_0(x',t')=\sqrt{\frac{\Gamma(v_e^2-2v_cv_e)}{2\lambda}}\times\operatorname{sec}h\left[\sqrt{\frac{(v_e^2-2v_cv_e)}{4}}(x'-v_et')\right], \tag{12.4.8}$$

where $\theta_0(x',t') = v_s(x' - v_ct')/2$. In the case where $\alpha/\Gamma - 2e\mu \neq 0$, let $\mu = K\tilde{E}x'$, where K is a constant, and assume that the solution [135, 200] is of the following form:

$$\phi=\varphi'(x',t')e^{i\theta(x',t')}. \tag{12.4.9}$$

Substituting equation (12.4.9) into equation (12.4.7), we get

$$-\varphi'\frac{\partial\theta}{\partial t'}-\varphi'\left(\frac{\partial\theta}{\partial t'}\right)^2+\frac{\partial^2\varphi'}{\partial(x')^2}+\frac{\lambda}{\Gamma}(\varphi')^3=\left(2Ke\tilde{E}x'+\frac{\alpha}{\Gamma}\right)\varphi', \tag{12.4.10}$$

$$\frac{\partial\varphi'}{\partial t'}+2\frac{\partial\varphi'}{\partial x'}\frac{\partial\theta}{\partial x'}+\varphi'\frac{\partial^2\theta}{\partial(x')^2}=0. \tag{12.4.11}$$

Now let $\varphi'(x',t') = \varphi(\xi)$, $\xi = x' - u(t')$, $u(t') = -2\tilde{E}Ke(t')^2 + vt' + d$, where $u(t')$ describes the accelerated motion of $\phi'(t')$. The boundary condition at $\xi' \to \infty$ requires $\phi(\xi)$ to approach zero rapidly. When $2\partial\theta/\partial\xi - \dot{u} \neq 0$, equation (12.4.11) can be written as $\varphi^2 = \frac{g(t')}{(\partial\theta/\partial\xi - \dot{u}/2)}$, or

$$\frac{\partial\theta}{\partial x'}=\frac{g(t')}{\varphi^2}+\frac{\dot{u}}{2}, \tag{12.4.12}$$

where $\dot{u} = du/dt'$. Integration of equation (12.4.12) yields

$$\theta(x',t') = g(t')\int_0^{x'} \frac{dx''}{\varphi^2} + \frac{\dot{u}}{2}x' + h(t'), \tag{12.4.13}$$

where $h(t')$ is an undetermined constant of integration. From equation (12.4.13), we get

$$\frac{\partial\theta}{\partial t'} = \dot{g}(t')\int_0^{x'} \frac{dx''}{\varphi^2} - \frac{g\dot{u}}{\varphi^2} + \left.\frac{g\dot{u}}{\varphi^2}\right|_{x'=0} + \frac{\ddot{u}}{2}x' + \dot{h}(t'). \tag{12.4.14}$$

Substituting equations (12.4.13) and (12.4.14) into equation (12.4.10), we have

$$\frac{\partial^2\varphi}{\partial(x')^2} = \left[\left(2K\tilde{E}ex' + \frac{\alpha}{\Gamma}\right) + \frac{\ddot{u}}{2}x' + \dot{h}(t') + \frac{\dot{u}^2}{4} + \dot{g}\int_0^{x'} \frac{dx''}{\varphi^2} + \left.\frac{g\dot{u}}{\varphi^2}\right|_{x'=0}\right]\varphi - \frac{\lambda}{\Gamma}\varphi^3 + \frac{g^2}{\varphi^3}. \tag{12.4.15}$$

Since $\partial^2\varphi/\partial(x')^2 = d^2\varphi/d\xi^2$, which is a function of ξ only, the right-hand side of equation (12.4.15) is also a function of ξ only, so it is necessary that $g(t') = g_0 =$ constant and

$$\left(2K\tilde{E}ex' + \frac{\alpha}{\Gamma}\right) + \frac{\ddot{u}}{2}x' + \dot{h}(t') + \frac{\dot{u}^2}{4} + \left.\frac{g\dot{u}}{f^2}\right|_{x'=0} = \bar{V}(\xi).$$

Next, we assume that $V_0(\xi) = \bar{V}(\xi) - \beta$, where β is real and arbitrary. Then

$$2K\tilde{E}ex' + \frac{\alpha}{\Gamma} = V_0(\xi) - \frac{\ddot{u}}{2}x' + \left[\beta - \left.\frac{g\dot{u}}{\varphi^2}\right|_{x'=0} - \dot{h}(t') - \frac{\dot{u}^2}{4}\right]. \tag{12.4.16}$$

Obviously, in this case, $V_0(\xi) = 0$ and the function in the brackets in equation (12.4.16) is a function of t'. Substituting equation (12.4.16) into equation (12.4.15) [116, 224], we get

$$\frac{\partial^2\tilde{\varphi}}{\partial\xi^2} = \beta\tilde{\varphi} - \frac{\lambda}{\Gamma}\tilde{\varphi}^3 + \frac{g_0^2}{\tilde{\varphi}^3}. \tag{12.4.17}$$

This shows that $\tilde{\varphi}$ is the solution of equation (12.4.17) when β and g are constant. For large $|\xi|$, we assume that $|\tilde{\varphi}| \le \beta/|\xi|^{1+\Delta}$, when Δ is a small constant. To ensure that $\tilde{\varphi}$ and $d^2\tilde{\phi}/d\xi^2$ approach zero when $|\xi| \to \infty$, only the solution corresponding to $g_0 = 0$ in equation (12.4.17) is kept. It can be shown that this soliton solution is stable in such a case. Therefore, we choose $g_0 = 0$ and obtain the following from equation (12.4.12):

$$\partial\theta/\partial x' = \dot{u}/2. \tag{12.4.18}$$

Thus, we obtain from equation (12.4.16)

$$\begin{aligned} 2K\tilde{E}ex' + \frac{\alpha}{\Gamma} &= -\frac{\ddot{u}}{2}x' + \beta - \dot{h}(t') - \frac{\dot{u}^2}{4}, \\ h(t') &= \left(\beta - \frac{\alpha}{\Gamma} - \frac{1}{4}v^2\right)t' - \frac{4}{3}(K\tilde{E}e)^2(t')^3 + evK\tilde{E}(t')^2. \end{aligned} \tag{12.4.19}$$

Substituting equation (12.4.19) into equations (12.4.13) and (12.4.14), we obtain

$$\theta = \left(-2K\tilde{E}et' + \frac{1}{2}\upsilon\right)x' + \left(\beta - \frac{\alpha}{\Gamma} - \frac{1}{4}\upsilon^2\right)t' - \frac{4}{3}(K\tilde{E}e)^2(t')^3 + e\upsilon K\tilde{E}(t')^2. \qquad (12.4.20)$$

Finally, substituting equation (12.4.20) into equation (12.4.17), we get

$$\frac{\partial^2\tilde{\varphi}}{\partial\xi^2} - \beta\tilde{\varphi} + \frac{\lambda}{\Gamma}\tilde{\varphi}^3 = 0. \qquad (12.4.21)$$

When $\beta > 0$, the solution of equation (12.4.21) is of the following form:

$$\tilde{\varphi} = \sqrt{2\beta\Gamma/\lambda}\,\sec h(\sqrt{\beta}\xi). \qquad (12.4.22)$$

Thus [135, 200],

$$\phi = \sqrt{\frac{2\beta\Gamma}{\lambda}}\sec h\left[\sqrt{\beta}\left(\sqrt{\frac{2m\Gamma}{\hbar^2}}x + \frac{2eKEt^2 - \upsilon t - d}{\hbar}\right)\right]$$
$$\times \exp\left\{i\left[\left(\frac{-2eKEt}{\hbar} + \frac{\upsilon}{2}\right)\sqrt{\frac{2m\Gamma}{\hbar^2}}x + \left(\beta - \frac{\alpha}{\Gamma} - \frac{1}{4}\upsilon^2\right)\frac{t}{\hbar} - \frac{4(eK\tilde{E})^2t^3}{3\hbar^3} + \frac{\upsilon Ke\tilde{E}t^2}{\hbar}\right]\right\}. \qquad (12.4.23)$$

This is also a soliton solution, but its shape, amplitude, and velocity have been changed relative to that of equation (12.4.8). It can be shown that equation (12.4.15) does indeed satisfy equation (12.4.6), so equation (12.4.6) has a soliton solution. It can also be shown that this soliton solution is stable.

12.4.2 The properties of soliton motion of the electrons in superconductors

For the solution of equation (12.4.23), we define a generalized time-dependent wave number, $k = \frac{\partial\theta}{\partial x'} = \frac{v}{2} - 2K\tilde{E}et'$, and a frequency

$$\omega = -\frac{\partial\theta}{\partial t'} = 2K\tilde{E}ex' - (\beta - \frac{\alpha}{\Gamma} - \frac{1}{4}\upsilon^2) + e(K\tilde{E}e)^2(t')^2$$
$$- 2K\tilde{E}e\upsilon t' = 2K\tilde{E}ex' - \beta - \frac{\alpha}{\Gamma} + k^2. \qquad (12.4.24)$$

The usual Hamilton equations for the superconductive electron (soliton) in the macroscopic quantum systems are still valid. They can be written [116, 224] as

$$\frac{dk}{dt'} = -\left.\frac{\partial\omega}{\partial x'}\right|_k = -2K\tilde{E}e.$$

Then the group velocity of the superconductive electrons can be denoted by

$$\upsilon_g = \frac{dx'}{dt'} = \left.\frac{\partial\omega}{\partial k}\right|_{x'} = 2\left(\frac{\upsilon}{2} - 2K\tilde{E}et'\right) = \upsilon - 4K\tilde{E}et'. \qquad (12.4.25)$$

This means that the frequency ω of the soliton still serves as Hamiltonian in the case of nonlinear quantum systems. Hence, the following relation still exists:

$$\frac{d\omega}{dt'} = \frac{d\omega}{dk}\bigg|_{x'} \frac{dk}{dt'} + \frac{\partial\omega}{\partial x'}\bigg|_{k} \frac{dx'}{dt'} = 0,$$

which is similar to that in the usual stationary linear medium [167, 205, 206].

The relations in equations (12.4.24) and (12.4.25) show that the superconductive electrons move as if they were classical particles with a constant acceleration in the invariant electric field, for which the acceleration is given by $-4K\bar{E}e$. If $v > 0$, the soliton initially travels toward the overdense region. It then suffers a deceleration and its velocity changes sign. The soliton is then reflected and accelerated toward the underdense region. The penetration distance into the overdense region depends on the initial velocity v.

From the above studies, we see that the time-dependent motion of a superconductive electron still resembles that of a soliton in the nonequilibrium state of a superconductor. Therefore, we conclude that electrons in superconductors are essentially solitons in both time-independent steady state and time-dependent dynamic state systems. This means that the soliton motion of the superconductive electrons causes the superconductivity of the material. Then the superconductors have a complete conductivity and nonresistance property because the solitons can move over macroscopic distances, retaining their amplitude, velocity, energy, and other quasi-particle features. In such a case, the motions of the electrons in the superconductors are described by the NLSEs (12.3.3), (12.3.4), (12.4.1), (12.4.3), or (12.4.5). According to the soliton theory, the electrons in the superconductors are localized and have a wave-corpuscle duality due to the nonlinear interaction, which is completely different from electrons as described by quantum mechanics. Therefore, the electrons in superconductors should be described in terms of nonlinear quantum mechanics [221, 232].

12.5 The transmission features of magnetic flux lines along the Josephson junctions

12.5.1 The transmission equation of magnetic flux lines

We have learned that, in a homogeneous bulk superconductor, the phase $\theta(\vec{r},t)$ of the electron wave function $\phi(\vec{r},t) = f(\vec{r},t)e^{i\theta(\vec{r},t)}$ is constant, independent of position and time. However, in an inhomogeneous superconductor, such as a superconductive junction discussed above, $\theta(\vec{r},t)$ becomes dependent on $\vec{r}$ and t. In the previous section, we discussed the Josephson effects in the S-N-S or S-I-S and S-N-I-S junctions, starting from the Hamiltonian and the GL equations satisfied by $\phi(\vec{r},t)$, and showed that the Josephson current, whether dc or ac, is a function of the phase change $\varphi = \Delta\theta = \theta_1 - \theta_2$. The dependence of the Josephson current on "'" is clearly seen in equation (12.3.29). This clearly indicates that the Josephson current is caused by the phase change of the superconductive electrons. Josephson himself derived the equations satisfied by the phase difference φ, known as the Josephson relations, through his studies on both the dc and ac Josephson effects. The Josephson relations for the Josephson effects in superconductor junctions can be summarized as follows:

$$J_s = J_m \sin\varphi, \quad \hbar\frac{\partial\varphi}{\partial t} = 2eV, \quad \hbar\frac{\partial\varphi}{\partial x'} = 2ed'\bar{H}_y/c, \quad \hbar\frac{\partial\varphi}{\partial y} = 2ed'\bar{H}_{x'}/c, \tag{12.5.1}$$

where *d'* is the thickness of the junction. Because the voltage V and magnetic field $\vec{H}$ are not determined, equation (12.5.1) is not a complete set of equations. Generally, these equations are solved simultaneously with the Maxwell equation $\nabla\times\vec{H} = (4\pi/c)\vec{J}$. Assuming that the magnetic field is applied in the (x, y)-plane, i.e., $\vec{H} = (\bar{H}_x, \bar{H}_y, 0)$, the above Maxwell equation becomes

$$\frac{\partial}{\partial x}\bar{H}_y(x,y,t) - \frac{\partial}{\partial y}\bar{H}_x(x,y,t) = \frac{4\pi}{c}J(x,y,t). \tag{12.5.2}$$

In this case, the total current in the junction is given by

$$J = J_s(x,y,t) + J_n(x,y,t) + J_d(x,y,t) + J_0.$$

In the above equation, $J_s(x,y,t)$ is the superconductive current density, $J_n(x,y,t)$ is the normal current density in the junction ($J_n = V/R(V)$ if the resistance in the junction is $R(V)$ and a voltage V is applied at two ends of the junction), $J_d(x,y,t)$ is called the displacement current, given by $J_d = CdV(t)/dt$, where C is the capacity of the junction, and J_0 is a constant current density. Solving the equations in equations (12.4.23) and (12.5.2) simultaneously, we get

$$\nabla^2\varphi - \frac{1}{v_0^2}\left(\frac{\partial^2\varphi}{\partial t^2} - \gamma_0\frac{\partial\varphi}{\partial t}\right) = \frac{1}{\lambda_J^2}\sin\varphi + I_0, \tag{12.5.3}$$

where

$$v_0 = \sqrt{c^2/4\pi Cd'}, \quad \gamma_0 = 1/RC, \quad \lambda_J = \sqrt{c^2/4\pi Cd'}, \quad I_0 = 4e^*\pi J_0/c^2\hbar, \quad e^* = 2e.$$

Equation (12.5.3) is the equation satisfied by the phase difference. It is a sine-Gordon equation with a dissipative term. From equation (12.5.1), we see that the phase difference ϕ depends on the external magnetic field $\vec{H}$, so the magnetic flux in the junction $\Phi' = \int \bar{H}ds = \oint \vec{A}d\vec{l} = \frac{c\hbar}{e^*}\oint \varphi dl$ can be specified in terms of φ, where $\vec{A}$ is the vector potential of the electromagnetic field and $d\vec{l}$ is the line element of vortex lines. The nonlinear equation (12.5.3) represents the transmission of superconductive vortex lines. Therefore, we know clearly that the Josephson effect and the related transmission of the vortex line, or magnetic flux, along the junctions are also nonlinear problems. The sine-Gordon equation given above has been extensively studied by many scientists, including Kivshar and Malomed. We will solve it here using different approaches.

12.5.2 The transmission features of magnetic flux lines

Assuming that the resistance R in the junction is very high, so that $J_n \to 0$, or, equivalently, $\gamma_0 \to 0$, and setting $I_0 = 0$, equation (12.5.3) reduces to

$$\nabla^2\varphi - \frac{1}{v_0^2}\frac{\partial^2\varphi}{\partial t^2} = \frac{1}{\lambda_J^2}\sin\varphi. \tag{12.5.4}$$

We define $X = x/\lambda_J$, $T = v_0 t/\lambda_J$. Then, in the one-dimensional case, the above equation becomes

$$\frac{\partial^2\varphi}{\partial X^2} - \frac{\partial^2\varphi}{\partial T^2} = \sin\varphi.$$

This is the one-dimensional sine-Gordon equation. If we assume that $\varphi = \varphi(X,T) = \varphi(\theta')$, where

$$\theta' = X' - X_0' - vT', \quad X' = x\sqrt{hc/2eLI_0}, \quad T' = T\sqrt{2eI_0/hc},$$

then the above equation becomes

$$(1 - v^2)\varphi_{\theta'}^2(\theta') = 2(A' - \cos\varphi),$$

where A' is a constant of integration. Thus [116, 224], we have

$$\int_{\varphi_0}^{\varphi(\theta)} [(A' - \cos\varphi)]^{-1/2} d\varphi = \sqrt{2}\delta v'\theta',$$

where $v' = 1/\sqrt{1-v^2}$, $\delta = \pm 1$. Choosing $A = 1$, we have

$$\int_{\pi}^{\varphi(\theta')} [\sin(\varphi/2)]^{-1/2} d\varphi = 2v\theta'.$$

Then a kink soliton solution can be obtained from the above equation, denoted by

$$\pm v\theta' = \ln[\tan(\varphi/2)], \quad \text{or} \quad \varphi(\theta') = 4\tan^{-1}[\exp(\pm v\theta')].$$

Thus, we obtain

$$\varphi(X',T') = 4\tan^{-1}\{\exp[\delta v(X' - X_0' - vT')]\}. \tag{12.5.5}$$

From the Josephson relations, we obtain the electric potential difference across the junction, which is represented by

$$V = \frac{\hbar}{2e}\frac{d\varphi}{dT'} = \frac{\varphi_0}{2c\pi}\frac{d\varphi}{dT'} = 2\delta vv\sqrt{\frac{2I_0 e}{\hbar c^2}\frac{\varphi_0}{2c\pi}}\sec h[v(X' - X_0' - vT')],$$

where $\varphi_0 \pi\hbar c = 2 \times 10^{-7}$ Gauss/cm^{-2} is a quantum fluxon and c is the speed of light. In this case, a similar expression for the magnetic field can be derived from the above results. We have

$$\bar{H}_z = \frac{\hbar}{2e}\frac{d\varphi}{dX'} = \frac{\varphi_0}{2c\pi}\frac{d\varphi}{dX'} = \pm 2\delta v \sqrt{\frac{2I_0 e}{\hbar c^2}\frac{\varphi_0}{2c\pi}} \sec h[v(X' - X_0' - vT')].$$

We can then determine the magnetic flux through a junction with length L and a cross section of 1 cm^2. The result is denoted

$$\Phi' = \int_{-\infty}^{\infty} H_x(x,t)dx = B_0' \int_{-\infty}^{\infty} H_x(X',T')dX' = \delta\varphi_0.$$

Therefore, the kink ($\delta = \pm 1$) carries a single quantum of magnetic flux in the extended Josephson junction. Such an excitation is often called a fluxon and the sine-Gordon equation or equation (12.5.3) is often referred to as the transmission equation of the quantum flux or fluxon. The excitation corresponding to $\delta = -1$ is called an anti-fluxon. A fluxon is an extremely stable formation, which can be easily controlled with the help of external effects. It may be used as a basic unit of information.

This result shows clearly that magnetic flux in superconductors is quantized and this is a macroscopic quantum effect as mentioned in Section 12.1. The transmission of the quantum magnetic flux through the superconductive junctions is described by the above nonlinear dynamic equations (12.5.3) and (12.5.4). The energy of the soliton can be determined and it is given by $E = 8m^2/\beta$, where $m^2/\beta = 1/\lambda_J^2$.

However, the boundary conditions must be considered for real superconductors. Various boundary conditions have been considered and studied. For example, we can assume the following boundary conditions for a one-dimensional superconductor: $\varphi_x(0,t) = \varphi_x(L,t) = 0$. Lamb obtained the following soliton solution for the sine-Gordon equation (12.5.4):

$$\varphi(x,t) = 4\tan^{-1}[h(x)g(t)], \tag{12.5.6}$$

where h and g are the general Jacobian elliptical functions and satisfy the following equations:

$$[h(x)]^2 = a'h^4 + (1+b')h^2 - c', \quad [g(x)]^2 = c'h^4 + b'h^2 - a',$$

where a', b', and c' are arbitrary constants. Coustabile et al. also gave the plasma oscillation, breathing oscillation, and vortex line oscillation solutions for the sine-Gordon equation under certain boundary conditions. All of these can be regarded as the soliton solution under the given conditions. The solutions of equation (12.5.4) in the two- and three-dimensional cases can also be found. In the two-dimensional case, the solution is given by

$$\varphi(X,Y,T) = 4\tan^{-1}\left[\frac{g(X,Y,T)}{f(X,Y,T)}\right], \tag{12.5.7}$$

where

$$X = x/\lambda_J, \quad Y = y/\lambda_J, \quad T = v_0 t/\lambda_J,$$
$$f = 1 + a(1,2)e^{y_1+y_2} = a(2,3)e^{y_2+y_3}a(3,1)e^{y_1+y_3},$$
$$g = e^{y_1} + e^{y_2} + a(1,2)a(2,3)a(3,1)e^{y_1+y_2+y_3},$$
$$y_1 = p_i X + q_i Y - \Omega_i \tau - y_i^0, p_i^2 + q_i^2 - \Omega_i^2 = 1, \quad (i = 1,2,3),$$
$$a(i,j) = \frac{(p_i - p_j)^2 + (q_i - q_j)^2 + (\Omega_i - \Omega_j)^2}{(p_i + p_j)^2 + (q_i + q_j)^2 + (\Omega_i + \Omega_j)^2}, \quad (1 \le i \le j \le 3),$$

where p_i, q_i, and Ω_i satisfy the following formula:

$$\det \begin{vmatrix} p_1 & q_1 & \Omega_1 \\ p_2 & q_2 & \Omega_2 \\ p_3 & q_3 & \Omega 3 \end{vmatrix} = 0.$$

In the three-dimensional case, the solution can also be found. We have

$$\varphi(X,Y,Z,T) = 4\tan^{-1}\left[\frac{g(X,Y,Z,T)}{f(X,Y,Z,T)}\right], \tag{12.5.8}$$

where X, Y, and T are similarly defined as in the two-dimensional case given above and $Z = z/\lambda_J$. The functions f and g are defined as

$$f = dX_2 e^{y_1+y_2} + dY_3 e^{y_2+y_3} + dZ_3 e^{y_1+y_3} + 1, \quad g = e^{y_1} + e^{y_2} + e^{y_3} + dX_2 dY_3 dZ_3 e^{y_1+y_2+y_3},$$
$$y_i = a_{i1}X + a_{i2}Y + a_{i3} + b_i T - C_i, \quad a_{i1}^2 + a_{i2}^2 + a_{i3}^2 - b_i^2 = 1, \quad (i = X, Y, Z),$$

with

$$d(i,j) = \frac{\sum_{k=1}^{3}[(a_{ik} - a_{jk})^2 - (b_i - b_j)^2]}{\sum_{k}^{3}[(a_{ik} + a_{jk})^2 - (b_i + b_j)^2]}, \quad (1 \le j \le 3),$$

where y_3 is a linear combination of y_1 and y_2, i.e., $y_3 = \alpha y_1 + \beta y_2$.

We now discuss the sine-Gordon equation with a dissipative term $y_0 \partial\varphi/\partial t$. First we make the following substitutions to simplify the equation:

$$X = x/\lambda_J, \quad T = v_0 t/\lambda_J = t/\omega_J, \quad a = y_0 \lambda_J^2/v_0, \quad B' = I_0 \lambda_J^2.$$

In terms of these new parameters, the one-dimensional sine-Gordon equation (12.5.3) can be rewritten as

$$\frac{\partial^2\varphi}{\partial X^2} - \frac{\partial^2\varphi}{\partial T^2} - a\frac{\partial\varphi}{\partial T} = \sin\varphi + B'. \tag{12.5.9}$$

The analytical solution of equation (12.5.6) is not easily found. Now let

$$\alpha = \frac{1 - v_0^2}{a^2 v_0^2}, \quad \eta = \frac{1}{\sqrt{\alpha}}\frac{X - v_0 T}{a v_0}, \quad q' = \frac{a v_0}{\sqrt{1 - v_0^2}}, \quad \varphi = \pi + \varphi'. \tag{12.5.10}$$

Equation (12.5.6) then becomes

$$\frac{\partial^2\varphi}{\partial\eta^2} + q'\frac{\partial\varphi}{\partial\eta} + \sin\varphi - B' = 0. \tag{12.5.11}$$

This equation is the same as that of a pendulum being driven by a constant external moment and a frictional force which is proportional to the angular displacement. The solution of the latter is well known; generally there exists a stable soliton solution. Let $Y = d\varphi'/d\eta$. Then equation (12.5.11) can be written as

$$\frac{\partial Y}{\partial\eta} + q'Y + \sin\varphi' - B' = 0. \tag{12.5.12}$$

For $0 < B' < 1$, we let $B' = \sin\varphi_0$ $(0 < \varphi_0 < \pi/2)$ and $\varphi' = \pi - \varphi_0 + \varphi_1$. Then equation (12.5.12) becomes

$$Y\frac{\partial Y}{\partial\eta} = -q'Y + \sin\varphi_0 + \sin(\varphi_1 - \varphi_0). \tag{12.5.13}$$

Expanding Y as a power series of φ_1, i.e., $Y = \sum_n C_n\varphi_1^n$, inserting it into equation (12.5.13), and comparing the coefficients of terms of the same power of φ_1 on both sides, we get

$$\begin{aligned} c_1 &= -\frac{q'}{2} \pm \sqrt{\frac{q'^2}{4} + \cos\varphi_0}, \quad c_2 = -\frac{1}{q' + 3c_1}\frac{\sin\varphi_0}{2}, \\ c_3 &= \frac{1}{q' + 4c_1}\left(-2c_2^2 - \frac{\cos\varphi_0}{6}\right), \quad c_4 = \frac{1}{q' + 5c_1}\left(-5c_2c_3 - \frac{\sin\varphi_0}{24}\right), \end{aligned} \tag{12.5.14}$$

etc. Substituting these c_n into $Y = d\varphi'/d\eta = \sum_n c_n\varphi_1^2$, the solution of φ_1 can be found by integrating $\eta = \int d\varphi_1/\sum_n c_n\varphi_1^2$. In general, this equation has a soliton solution or elliptical wave solution. For example, when $d\varphi'/d\eta = c_1\varphi_1 + c_2\varphi_1^2 + c_3\kappa_1^3$, it can be found that

$$\eta = \frac{2}{\sqrt{A - C}}F\left(\sqrt{\frac{A - B}{A - C}}, \sin^{-1}\left(\sqrt{\frac{A - \varphi_1}{A - B}}\right)\right),$$

where $F(k, \varphi_1)$ is the first Legendre elliptical integral and A, B, and C are constants. The inverse function φ_1 of $F(k, \varphi_1)$ is the Jacobian amplitude $\varphi_1 = \mathrm{am}\,F$. Thus,

$$\sin^{-1}\left(\sqrt{\frac{A - \varphi_1}{A - B}}\right) = \mathrm{am}\sqrt{\frac{A - C}{A - B}}\eta \quad \text{or} \quad \sqrt{\frac{A - \varphi_1}{A - B}} = \mathrm{sn}\left(\sqrt{\frac{A - C}{A - B}}\eta\right),$$

where $\mathrm{sn}\,F$ is the Jacobian sine function. Introducing the symbol $\csc F = 1/\mathrm{sn}\,F$, the solution can be written as

$$\varphi_1 = A - (A - B)\left[\csc\left(\sqrt{\frac{A - C}{A - B}}\eta\right)\right]^2. \tag{12.5.15}$$

This is an elliptic function. It can be shown that the corresponding solution at $|\eta| \to \infty$ is a solitary wave.

It can be seen from the above discussion that the quantum magnetic flux lines (vortex lines) move along a superconductive junction in the form of solitons. The transmission velocity v_0 can be obtained from $h = \alpha v_0 \sqrt{(1 - v_0^2)}$ and c_n in equation (12.5.14). It is given by

$$v_0 = 1/\sqrt{1 + [\alpha/h(\varphi_0)]^2}.$$

That is, the transmission velocity of the vortex lines depends on the current I_0 injected and the characteristic decaying constant α of the Josephson junction. When α is finite, the greater the injection current I_0 is, the faster the transmission velocity will be; when I_0 is finite, the greater α is, the smaller v_0 will be. These are realistic conclusions.

12.6 Conclusions

We here first reviewed the properties of superconductivity and macroscopic quantum effects, which are different from the microscopic quantum effects, obtained from some experiments. The macroscopic quantum effects are caused by the collective motion of microscopic particles, such as electrons in superconductors, after the symmetry of the system is broken due to nonlinear interactions. Such interactions result in Bose condensation and self-coherence of particles in these systems. Meanwhile, we also studied the properties of the motion of superconductive electrons and arrived at the soliton solutions of the time-independent and time-dependent Ginzburg-Landau equation in superconductors, which are, in essence, a kind of nonlinear Schrödinger equation. These solitons, with wave-corpuscle duality, come to being due to the nonlinear interactions arising from the electron-phonon interactions in superconductors, in which the nonlinear interaction suppresses the dispersive effect of the kinetic energy in these dynamic equations. Thus, soliton states of the superconductive electrons, which can move over macroscopic distances retaining the energy, momentum, and other quasi-particle properties, are formed. Meanwhile, we used these dynamic equations and their soliton solutions to obtain and explain these macroscopic quantum effects and superconductivity of the systems. Effects such as quantization of magnetic flux in superconductors and the Josephson effect of superconductivity junctions prompted us to conclude that the superconductivity and macroscopic quantum effects are a kind of nonlinear quantum effects that arise from the soliton motions of superconductive electrons. This shows clearly that the study of the features of macroscopic quantum effects and of the properties of motion of microscopic particles in the superconductor has significant importance in physics.

13 The soliton movements in condensed state systems

13.1 Soliton movements of helium atoms in superfluid systems

13.1.1 The macroscopic quantum effects in helium superfluid

As is well known, helium is a common inert gas. It is also the most difficult gas to liquefy. There are two isotopes of helium, ^{4}He and ^{3}He, with the former being the major constituent in a normal helium gas. The boiling temperatures of ^{4}He and ^{3}He are 4.2 K and 3.19 K, respectively. The critical pressure for ^{4}He is 1.15 atm. Because of their light mass, both ^{4}He and ^{3}He have extremely high zero-point energies and remain in gaseous form from room temperature down to a temperature near the absolute zero. Helium attains the solid state due to cohesive forces only when the interatomic distances become sufficiently small under high pressure. For example, a pressure of 25–34 atm is required in order to solidify ^{3}He. For ^{4}He, when it is crystallized at a temperature below 4 K, it neither absorbs nor releases heat, i.e., the entropies of the crystalline and liquid phases are the same and only its volume is changed in the crystallization process. However, ^{3}He absorbs heat when it is crystallized at a temperature $T < 3.19$ K under pressure. In other words, the temperature of ^{3}He rises during crystallization under pressure. This endothermic crystallization process is called the Pomeranchuk effect. This indicates that the entropy of liquid ^{3}He is lower than that of ^{3}He in its crystalline phase. That is, the liquid phase represents a more ordered state [29, 30, 142, 246]. These peculiar characteristics are the result of the unique internal structures of ^{4}He and ^{3}He. Both ^{4}He and ^{3}He can crystallize in a body-centered cubic (bcc) or hexagonal close-stacked structure.

A phase transition for ^{4}He will occur at a pressure of 1 atm and a temperature of 2.17 K. Above this temperature, ^{4}He has no difference from a normal liquid and this liquid phase is referred to as He-I. However, when the temperature is below 2.17 K, the liquid phase, referred to as He-II, is completely different from He-I and attains a superfluid state. This superfluid state can pass through capillaries with a diameter of less than 10^{-6} cm, without experiencing any resistance. The superfluid state has a low viscosity ($< 10^{-11}$ P) and its fluid velocity is independent of the pressure difference over the capillary and its length. If a test tube is inserted into liquid He-II in a container, the level of liquid He-II inside the test tube is the same as that in the container. If the test tube is pulled up, the He-II inside the test tube will rise along the inner wall of the tube, climb over the mouth of the tube and then flow back to the container along the outer wall of the tube, until the liquid level inside the test tube reaches the same level as that in the container. On the other hand, if the test tube is lifted up above the container, the liquid in the test tube drips directly into the container until the tube becomes empty. This property is called the superfluidity of ^{4}He [19, 29, 30, 70, 142, 166, 202, 246].

https://doi.org/10.1515/9783110549638-013

Experiments [142, 246] have shown that the quantization of current circulation and vortex structures, similar to those of magnetic flux in a superconductor, can exist in ^{4}He-II and in superfluid ^{3}He. In terms of the phase θ of the macroscopic wave function of a superfluid helium atom, its velocity v_s is given by $v_s = \hbar\nabla\theta/m$, where m is the mass of the helium atom. v_s satisfies the following quantization condition:

$$\oint v_s dr = n\hbar/m, \quad (n = 1, 2, 3, \ldots).$$

The above suggests that the circulation velocity of the superfluid helium atom is quantized with a quantum of n/m. In other words, as long as the superfluid is rotating, a new whirl in the superfluid is developed whenever the circulation of the current is increased by n/m, i.e., the circulation of the whirl (or energy of the vortex lines) is quantized. An experiment was done in 1963 to measure the energy of the vortex lines and the results obtained were consistent with the theoretical prediction. The quantization of the circulation was thus proved, which is a macroscopic quantum effect, completely different from a normal fluid.

If the helium superfluid flows, without rotation, through a tube with a varying diameter, then $\nabla \times \vec{v}_s = 0$ and it can be shown, based on the above quantization condition, that the pressure is the same everywhere inside the tube, even though the fluid flows faster at a point where the diameter of the tube is smaller and slower where the diameter is larger.

Firbake and Maston in the U.S.A. observed the macroscopic quantum effect of ^{4}He-II experimentally once again. When the superfluid ^{4}He-II was set into rotational motion in a cup, a whirl would be formed when the temperature of the liquid was reduced to below the critical temperature. In this case, an effective viscosity develops between the fluid and the cup, which is very similar to normal fluid in a cup being stirred. The surface of the superfluid becomes inclined at a certain angle and the cross section of the liquid surface has the shape of a parabola, due to the combined effects of gravitational and centrifugal forces. Fluid distant from the center has a tendency to converge toward its center, which is balanced by the centrifugal force, and a dynamic equilibrium is reached. The angular momentum of such a whirl is very small and consists of only a small number of discrete quantum packets. The angular momentum of the quantum packets can be obtained via quantum theory. In other words, the whirl can exist only in discrete form over a certain range in certain materials, such as superfluid helium. Firbake and Maston managed to obtain a sufficiently large angular momentum in their experiment and were able to observe the whirl's surface shape using visible light. They used a thin layer of rotating helium superfluid in their experiment. While the rotating superfluid was illuminated from both the top and the bottom by laser beams with a wavelength of 6328 Å from a He–Ne laser, the whirls formed were observed. Alternate bright and dark interference fringes were formed when the reflected beams were focused on an observing screen. The analysis of the interference pattern showed that the surface was indeed inclined at an angle. Based on this, the shape of the surface can be correctly constructed. The observed interference pattern was found in excellent agreement with those predicted by the theory. This experiment further confirmed the existence of quantized vortex rings in the ^{4}He superfluid.

The mechanisms of superfluidity and vortex structure in ^{4}He and ^{3}He have been extensively studied and the reader is encouraged to read Barenghi et al. [19] for a review of recent work.

13.1.2 The nonlinear theory of macroscopic quantum effects in a superfluid system

How to theoretically explain the superfluidity and the macroscopic quantum effects of ^{4}He is still a subject of current research. In the 1940s, Bogoliubov calculated the critical temperature of Bose–Einstein condensation in ^{4}He based on an ideal boson gas model [29, 30, 166]. The value of the critical temperature Bogoliubov obtained was 3.3 K, quite close to the experimental value of 2.17 K. In Bogoliubov's model, some ^{4}He atoms condense to the state with minimal energy at a temperature below T_c and a Bose–Einstein condensation state is formed [19, 29, 30, 70, 166, 202]. Pang [202, 221, 232] believed that the macroscopic quantum phenomena occurring in superfluid helium could be attributed to Bose–Einstein condensation of the ^{4}He atoms. When the temperature of liquid ^{4}He is below T_c, the symmetry of the system breaks due to nonlinear interactions within the system. Thus, some of the ^{4}He atoms spontaneously condense to the state of lower energy. When the temperature approaches the absolute zero, all the ^{4}He atoms will condense to the state with zero momentum. According to the relation $\lambda = \hbar/p$, the wavelength of each ^{4}He atom is infinite and an ordered state over the entire space can be formed in this case, which leads to a highly ordered and long range coherent state. Thus, the macroscopic quantum effect could appear in the systems. In the following, we study the nature of the macroscopic quantum effect in helium superfluid using a nonlinear theory.

(1) The soliton movements of helium atoms in the superfluid and nonrelativistic case.

As mentioned above, the helium atoms form a quantum liquid without viscosity in the superfluid state at temperatures below 2.17 K, which is very similar to the superconducting state and thus also described by a macroscopic wave function $\phi(\vec{r}, t)$ similar to that in equation (11.2.1) in Chapter 11. Here, $\phi(\vec{r}, t)$ is also called an order parameter of the helium superfluid or an effective wave function of helium atoms.

It is known that the effective wave function of helium atoms, $\phi(\vec{r}, t)$, satisfies the following Gross–Pitaevskii (GP) equation, which was derived by Gross and Pitaevskii in 1950 [100, 248, 249]:

$$i\hbar\frac{\partial \phi}{\partial t} = -\frac{\hbar^2}{2m}\nabla^2\phi + \lambda|\phi|^2\phi - \mu'\phi, \tag{13.1.1}$$

where λ is a nonlinear interaction coefficient and μ' is a constant related to the ground state energy of a helium atom. This equation is similar to the above GL equation for superconducting electrons. This is understandable because the superfluid, similar to a superconducting system, is also a nonlinear system due to the existence of nonlinear interaction $\lambda|\phi|^2\phi(\vec{r},t)$. Equation (13.1.1) is a nonlinear Schrödinger equation and was extensively used by some researchers in their studies of superfluidity. A similar equation was derived by Dewitt in 1966. According to Dewitt investigation, λ in Eq. (13.1.1) should be a negative value.

The corresponding Lagrangian density function of the system can be obtained and is given by

$$L' = \frac{i\hbar}{2}\left(\phi\frac{\partial\phi^*}{\partial t} - \phi^*\frac{\partial\phi}{\partial t}\right) - \frac{\hbar^2}{2m}|\nabla\phi|^2 - \frac{\lambda}{2}|\phi|^4 + \mu'|\phi|^2, \tag{13.1.2}$$

where $L' = \mathcal{L}$. In the one-dimensional case where $\mu = 0$, equation (13.1.1) is the usual nonlinear Schrödinger equation whose solution is of the following form [194, 198, 204, 208, 209]:

$$\phi_0 = \varphi_0(x,t)e^{i\theta_0(x,t)}, \tag{13.1.3}$$

where

$$\varphi_0(x,t) = \sqrt{\frac{(v_e^2 - 2v_c v_e)}{2|\lambda|}} \times \sec h\left[\sqrt{\frac{(v_e^2 - 2v_c v_e)}{4}}(x - v_e t)\right] \tag{13.1.4}$$

and $\theta_0(x,t) = v_e(x - v_c t)/2$, v_e and v_c are the group and phase velocities of the superfluid helium atom, respectively.

In the one-dimensional case where $\mu \neq 0$, if we let $t' = t/\hbar$ and $x' = x\sqrt{(2m/\hbar^2)}$, then equation (13.1.1) becomes

$$i\frac{\partial\phi}{\partial t} = -\frac{\partial^2\phi}{\partial^2 x'} + \lambda|\phi|^2\phi - \mu'\phi. \tag{13.1.5}$$

Assume that the solution is of the following form [15, 162]:

$$\phi = \varphi(x',t')e^{i\theta(x',t')}. \tag{13.1.6}$$

Substituting equation (13.1.6) into equation (13.1.5), we get

$$-\varphi\frac{\partial\theta}{\partial t'} - \varphi\left(\frac{\partial\theta}{\partial t'}\right)^2 + \frac{\partial^2\varphi}{\partial(x')^2} + \lambda(\varphi)^3 = \mu'\varphi \tag{13.1.7}$$

and

$$\frac{\partial\varphi}{\partial t'} + 2\frac{\partial\varphi}{\partial x'}\frac{\partial\theta}{\partial x'} + \varphi\frac{\partial^2\theta}{\partial(x')^2} = 0. \tag{13.1.8}$$

Now let $\varphi(x',t') = \varphi(\xi)$, $\xi = x' - u(t')$, $u(t') = vt' + d$, where $u(t')$ describes the accelerated motion of $\varphi(x',t')$. The boundary condition at $\xi' \to \infty$ requires $\varphi(\xi)$ to approach zero rapidly. When $2\partial\theta/\partial\xi - \dot{u} \neq 0$, equation (13.1.8) can be written as

$$\varphi^2 = \frac{g(t')}{(\partial\theta/\partial\xi - \dot{u}/2)}$$

or

$$\frac{\partial\theta}{\partial x'} = \frac{g(t')}{\varphi^2} + \frac{\dot{u}}{2}, \tag{13.1.9}$$

where $g(t')$ is an undetermined constant of integration and $\dot{u} = du/dt'$. Integration of (13.1.9) yields

$$\theta(x',t') = g(t')\int_0^{x'} \frac{dx''}{\varphi^2} + \frac{\dot{u}}{2}x' + h(t'), \tag{13.1.10}$$

where $h(t')$ is an undetermined constant of integration. From equation (13.1.10), we get

$$\frac{\partial\theta}{\partial t'} = \dot{g}(t')\int_0^{x'} \frac{dx''}{\varphi^2} - \frac{g\dot{u}}{\varphi^2} + \left.\frac{g\dot{u}}{\varphi^2}\right|_{x'=0} + \frac{\ddot{u}}{2}x' + \dot{h}(t'). \tag{13.1.11}$$

Substituting equations (13.1.10) and (13.1.11) into equation (13.1.7), we have

$$\frac{\partial^2\varphi}{\partial(x')^2} = \left[-\mu' + \frac{\ddot{u}}{2}x' + \dot{h}(t') + \frac{\dot{u}^2}{4} + \dot{g}\int_0^{x'} \frac{dx''}{\varphi^2} + \left.\frac{g\dot{u}}{\varphi^2}\right|_{x'=0}\right]\varphi - \lambda\varphi^3 + \frac{g^2}{\varphi^3}. \tag{13.1.12}$$

Since $\partial^2\varphi/\partial(x')^2 = d^2\varphi/d\xi^2$, which is a function of ξ only, the right-hand side of equation (13.1.12) is also a function of ξ only, so it is necessary that $g(t') = g_0 =$ constant and

$$-\mu' + \frac{\ddot{u}}{2}x' + \dot{h}(t') + \frac{\dot{u}^2}{4} + \left.\frac{g\dot{u}}{f^2}\right|_{x'=0} = \overline{V}(\xi).$$

Next, we assume that $V_0(\xi) = \overline{V}(\xi) - \beta$, where β is real and arbitrary. Then

$$-\mu' = V_0(\xi) - \frac{\ddot{u}}{2}x' + \left[\beta - \left.\frac{g\dot{u}}{\varphi^2}\right|_{x'=0} - \dot{h}(t') - \frac{\dot{u}^2}{4}\right]. \tag{13.1.13}$$

Obviously, $V_0(\xi) = 0$ and the function in the brackets in equation (13.1.13) is a function of t'. Substituting equation (13.1.12) into equation (13.1.11) [198, 209], we get

$$\frac{\partial^2\overline{\varphi}}{\partial\xi^2} = \beta\overline{\varphi} - \lambda\overline{\varphi}^3 + \frac{g_0^2}{\overline{\varphi}^3}. \tag{13.1.14}$$

This shows that $\overline{\varphi}$ is the solution of equation (13.1.14) when β and g are constant. For large $|\xi|$, we may assume that $|\overline{\varphi}| \leq \beta/|\xi|^{1+\Delta}$ when Δ is a small constant. To ensure that φ and $d^2\overline{\varphi}/d\xi^2$ approach zero when $|\xi| \to \infty$, only the solution corresponding to $g_0 = 0$ in equation (13.1.14) is kept. It can be shown that this soliton solution is stable. Therefore, we choose $g_0 = 0$ and obtain the following from equation (13.1.9):

$$\frac{\partial\theta}{\partial x'} = \frac{\dot{u}}{2}. \tag{13.1.15}$$

Thus, we obtain from equation (13.1.11)

$$-\mu' = \frac{\ddot{u}}{2}x' + \beta - \dot{h}(t') - \frac{\dot{u}^2}{4}, \tag{13.1.16}$$

where

$$h(t') = \left(\beta - \frac{1}{4}v^2 + \mu'\right)t'.$$

Substituting equation (13.1.16) into equations (13.1.10) and (13.1.11), we obtain

$$\theta = \frac{1}{2}vx' + \left(\beta + \mu' - \frac{1}{4}v^2\right)t'. \tag{13.1.17}$$

Finally, substituting equation (13.1.17) into equation (13.1.14), we get

$$\frac{\partial^2\overline{\varphi}}{\partial\xi^2} - \beta\overline{\varphi} + \lambda\overline{\varphi}^3 = 0. \tag{13.1.18}$$

When $\beta > 0$, the solution of equation (13.1.18) [116, 224] is of the following form:

$$\overline{\varphi} = \sqrt{\frac{2\beta}{|\lambda|}}\operatorname{sec}h(\sqrt{\beta}\xi). \tag{13.1.19}$$

Thus, the solution of equation (13.1.14) can be obtained [204, 208] and represented by

$$\phi = \sqrt{\frac{2\beta}{|\lambda|}}\operatorname{sec}h\{\sqrt{\beta}[x' - v(t' - t_0')]\}\exp\{i[vx'/2 - (\beta + v^2/4 - \mu')t]\} \tag{13.1.20a}$$

or

$$\phi = \sqrt{\frac{2\beta}{|\lambda|}}\operatorname{sec}h\left\{\frac{\sqrt{2m\beta}}{\hbar}[(x-x_0)-v't]\right\}\exp\{i[mv'x/\hbar-(\beta+v^2/4-\mu')t/\hbar]\}, \tag{13.1.20b}$$

where β is an arbitrary constant, v and v' are the group velocity of the helium atom in the (x',t')-coordinate and (x,t)-coordinate, respectively, and $x_0 = v't_0$. This is a bell-type soliton solution, but its shape, amplitude, and velocity have been changed as compared to that of equation (13.1.4). It can be shown that equation (13.1.20) does indeed satisfy equation (13.1.1), so equation (13.1.1) has a soliton solution. It can also be shown that this soliton solution is stable.

This soliton in equation (13.1.20) consists of an envelop and a carrier wave, the former being denoted

$$\overline{\varphi}(x,t) = \sqrt{2\beta/|\lambda|}\,\mathrm{sec}\,h\{\sqrt{2m\beta}[(x-x_0)-v't]/\hbar\},$$

which is a bell-type soliton with an amplitude of $\sqrt{2\beta/|\lambda|}$, and the latter being denoted

$$\exp\{i[mv'x/\hbar-(\beta+v'^2/4-\mu')t/\hbar]\}.$$

The envelop $\overline{\varphi}(x,t)$ denotes the dynamic feature of the mass centre of the helium atom. Its position is at x_0, its amplitude is $\sqrt{2\beta/|\lambda|}$, and its width is $W' = 2\pi\hbar\sqrt{2m\beta}$. Thus, the size of the soliton is $W'\sqrt{2\beta|\lambda|} = 2\pi\hbar\sqrt{m|\lambda|}$, which is a constant. This result shows clearly that the helium atom has a well-determined size and is localized at x_0. Therefore, we conclude that the helium atom in a superfluid system possesses wave and corpuscle features.

In the three-dimensional case, equation (13.1.1) becomes

$$i\hbar\frac{\partial\phi}{\partial t} = -\frac{\hbar^2}{2m}\left[\frac{\partial^2\phi}{\partial x^2}+\frac{\partial^2\phi}{\partial y^2}+\frac{\partial^2\phi}{\partial z^2}\right]-\mu'\phi+\lambda|\phi|^2\phi. \tag{13.1.21}$$

If we assume that $X = a_1x + a_2y + a_3z$, where a_1, a_2, and a_3 are some constants, then the above dynamic equation becomes

$$i\hbar\frac{\partial\phi}{\partial t} = \frac{\hbar^2}{2m}(a_1^2+a_2^2+a_3^2)\frac{\partial^2\phi}{\partial X^2}-\mu'\phi+2\lambda|\phi^2|\phi.$$

Its solution can be obtained using the same method mentioned above. The soliton solution is represented [116, 224] by

$$\phi = \sqrt{\frac{2\beta}{|\lambda|B}}\,\mathrm{sec}\,h\{\sqrt{\beta}[(X'-X_0')-v_et']\}\times\exp\{i[v_eX'/2-(\beta+v_e^2/4-\mu')t']\}, \tag{13.1.22}$$

where

$$\begin{aligned}
&X'-X_0' = a_1(x'-x_0')+a_2(y'-y_0')+a_3(z'-z_0'), \quad t' = t/\hbar,\\
&x' = x\sqrt{(2m/\hbar^2)}, \quad x_0' = x_0\sqrt{(2m/\hbar^2)}, \quad y' = y\sqrt{(2m/\hbar^2)}, \quad y_0' = y_0\sqrt{(2m/\hbar^2)},\\
&z' = z\sqrt{(2m/\hbar^2)}, \quad z_0' = z_0\sqrt{(2m/\hbar^2)},\\
&B = \sqrt{a_1^2+a_2^2+a_3^2}.
\end{aligned}$$

This solution is still a bell-type soliton in a three-dimensional space.

From these results, we know that the motion of helium atoms in a superfluid system are described by the nonlinear Schrödinger equation (13.1.1) or (13.1.22). Its solution is a soliton, which is expressed by equation (13.1.20) or equation (13.1.22). This exhibits clearly that the motion of the helium atoms in a superfluid system has a nonlinear quantum feature. Therefore, the system undergoes a second-order phase transition and changes from the normal He-I state to the superfluid He-II state at 2.17 K. When this happens, the nonlinear interactions generated in the system suppress the dispersion effect of helium atoms, so the superfluid helium atoms behave as solitons due to the spontaneous Bose condensation. Because solitons can preserve their energy, momentum, wave form, and other properties as quasi-particles throughout their motion, the superfluidity occurs naturally when the liquid helium moves as solitons.

In this case, we can prove that the superfluid helium atom (soliton) moves with a uniform speed and we can determine its speed from the solution (13.1.20). For example, from (13.1.20) we find that the wave number of the soliton is $k = \partial\theta/\partial x' = v/2$ and its frequency can be denoted by $\omega = \partial\theta/\partial t' = \beta + v^2/4 - \mu' = \beta - \mu' + k^2$. Thus, the acceleration and group velocity of the helium soliton satisfy the following relations [204, 208]:

$$\frac{dk}{dt'} = -\left.\frac{\partial\omega}{\partial x'}\right|_k = 0$$

and

$$v_g = \frac{dx'}{dt'} = \frac{\partial\omega}{\partial k} = 2k = v,$$

respectively. That is, the group velocity of the helium soliton, v_g, is a constant, v, and the helium atoms move in the form of a soliton with constant velocity in the superfluid state. This is a basic feature of the superfluidity and the above discussion gives a clear physical interpretation of the phenomenon. In such a case, the mass of the helium soliton can be determined from (13.1.20), that is,

$$M = \int_{-\infty}^{\infty} |\phi|^2 dx = \frac{4}{\lambda\hbar}\sqrt{m\beta} = \text{constant}.$$

The energy of the helium soliton is denoted

$$\begin{aligned} E &= \int_{-\infty}^{\infty}\left[\left|\frac{\partial\phi}{\partial t'}\right|^2 + \frac{1}{2}\lambda|\phi|^4 - \mu'|\phi|^2\right]dx' = \frac{4\sqrt{2m\beta}}{3\lambda\hbar}(4+\beta) - \mu' M + \frac{1}{2}Mv^2 \\ &= E_0 + \frac{1}{2}Mv^2, \end{aligned} \tag{13.1.23}$$

where

$$E_0 = \frac{4\sqrt{2m\beta}}{3\lambda\hbar}(4+\beta) - \mu' M.$$

The results exhibit clearly that the helium soliton in the superfluid system has features of a classical particle.

(2) The theoretical explanation of macroscopic quantum mechanics in helium superfluid.

We now discuss the properties of circulation (vortex lines) produced by the motion of superfluid helium atoms, using the above result. The properties are defined by the velocity of superfluid helium atom, that is,

$$Q = \int_r v_s dr. \tag{13.1.24}$$

In terms of the phase, $\theta(x, t)$, of the macroscopic wave function of superfluid helium, the velocity of the superfluid can be written as

$$v_s = \frac{\hbar}{m}\nabla\theta. \tag{13.1.25}$$

Earlier we concluded from equation (13.1.20) that the velocity of the superfluid is equal to 2 times the group velocity of the soliton, i.e., $v_s = 2v$. This indicates that the motion of the soliton is the motion of the superfluid and the vortex lines in superfluid are a result of soliton motion of the superfluid helium atoms. The phase difference along a closed path is given by the line integral corresponding to the circulation of the velocity v_s. We have

$$\Delta\theta = \oint_r \nabla\theta dr = \frac{m}{\hbar}\oint_r v_s dr. \tag{13.1.26}$$

Thus, the circulation is related to the phase difference $\Delta\theta$ of the superfluid helium atoms. If the path of integration lies in a multiply connected domain or it encloses a vortex line, then $\nabla\theta(r) \neq 0$. Furthermore, if $\phi(r) \neq 0$ and it is single-valued, then we have $\Delta\theta = 2\pi$ and

$$Q = \oint_r v_s dr = n\frac{h}{m}, \quad (n = 1, 2, 3, \ldots), \tag{13.1.27}$$

where n is an integer. Equation (13.1.27) implies that, whenever the velocity of the rotating superfluid helium exceeds a critical velocity, the vortex can be produced in the superfluid. The circulation (the vortex) is quantized and is given a value by an integer multiple of h/m. Thus, a macroscopic quantum effect occurs in the superfluid system, consistent with the experimental result mentioned above. Therefore, the nonlinear GP equation indeed gives an adequate description of "quantum vortex" in superfluid helium. Thus, the superfluid can be viewed as a Bose condensate with local interactions [100, 248, 249].

The concept of "quantum vortex" was proposed first by Ginzburg and Pitaevskii in 1960, but quantization of vortices in a superfluid was earlier suggested, by Onsager in

1949 [194], on the basis of classical vortex flows and turbulence. The superfluidity of ^{4}He and its quantum vortex lines and loops, weak turbulence, and dissipative vortex dynamics in the superfluids were studied by Barenghi et al. [19], Roberts et al, Pismeu and Rica, and many others, all using the nonlinear Schrödinger equation or the GP equation given above (see Barenghi et al. [19], Donnely [70], Avenel et al. [15], and Lindensmith et al. [162]). Experimental observations of these vortices were reported early by Yamchuk [312], Packard [195], and Zieve et al. [325, 326]. Numerical simulations were presented by Frish [83], Pomeau and Rica [252], and Schwarz [273, 274]. The energy of the vortex lines was also measured.

If the superfluid liquid does not rotate, then $\nabla \times v_s = (h/m)\nabla \times \nabla\theta = 0$ and the superfluid velocity field is a conservative field without rotation. This suggests that, when the superfluid helium flows through a tube with a graduate decreasing diameter, the pressure inside the tube is equal everywhere, irrespective of the diameter of the tube. This is completely different from that of a normal fluid, but it has been demonstrated experimentally. Moreover, the macroscopic wave function $\phi(x,t)$, given in (13.1.20), approaches zero when x approaches ∞. That is, $\phi(x,t)$ vanishes at the boundary. This implies that the superfluid density $\rho_s(\propto |\phi|^2)$ should also approach zero at the boundary. The value of ρ_s was measured in 1970 and it was found that its value dropped from the value in the bulk to zero over a few atomic layers. This gave a direct verification of the theoretical results.

In the last few decades, Bose–Einstein condensation has been observed in a series of remarkable experiments using weakly interacting atomic gases, such as vapors of rubidium, sodium lithium, and hydrogen. Its properties have been extensively studied. These studies show that Bose–Einstein condensation is a nonlinear phenomenon, analogous to nonlinear optics, and that the state is coherent and can be described by the following nonlinear Schrödinger equation or the GP equation [100, 248, 249]:

$$i\frac{\partial\phi}{\partial t'} = -\frac{\partial^2\phi}{\partial x'^2} - \lambda|\phi|^3 + V(x)\phi, \tag{13.1.28}$$

where $t' = t/\hbar$, $x' = x\sqrt{2m}/\hbar$. This equation was used to discuss the realization of the Bose–Einstein condensation in the $d+1$ dimensions ($d = 1,2,3$) by Bullough et al. [38, 39]. Elyutin et al. [74, 75] gave the corresponding Hamiltonian density, H, of a condensate system as follows:

$$H' = \left|\frac{\partial\phi}{\partial x'}\right|^2 + V(x')|\phi|^2 - \frac{1}{2}\lambda|\phi|^4, \tag{13.1.29}$$

where $H' = H$, the nonlinear parameters of λ are defined as $\lambda = -2Naa_1/a_0^2$, N is the number of particles trapped in the condensed state, a is the ground state scattering length, and a_0 and a_1 are the transverse (y,z) and the longitudinal (x) condensate sizes (without self-interaction), respectively. Note that integrations over y and z have been carried out in obtaining the above equation. λ is positive for condensation with self-attraction (negative scattering length). The coherent regime was observed in Bose–Einstein condensation in lithium. The specific form of the trapping potential $V(x')$ depends on the details of the experimental setup. Work on Bose–Einstein condensation based on the above Hamiltonian model was carried out and reported by Barenghi et al. [15].

It is not surprising to see that equation (13.1.28) is exactly the same as equation (13.1.1). This prediction confirms the correctness of the above nonlinear theory for Bose–Einstein condensation. As a matter of fact, immediately after the first experimental observation of this condensation phenomenon, it was realized that the coherent dynamics of the condensed macroscopic wave function could lead to the formation of nonlinear solitary waves. For example, self-localized bright, dark, and vortex solitons, formed by an increased (bright) or decreased (dark or vortex) probability density, respectively, were experimentally observed, particularly for the vortex solution which has the same form as the vortex lines found in superfluids. These experimental results were in concordance with the results of the above theory.

13.1.3 The soliton motion of superfluid helium atoms in a relativistic case

The GP equation (13.1.1) is not relativistic and neither takes the gravitational field into consideration. Anandan [12] and others extended the theory to include the relativistic effect. The generalized relativistic equation of motion for the quantum superfluid helium is given by

$$\frac{\partial^2\phi}{\partial t^2} - \nabla^2\phi + \alpha^2\phi = -\lambda'|\phi|^2\phi, \tag{13.1.30}$$

where

$$\alpha^2 = \frac{m^2c^2}{\hbar^2}, \quad \lambda' = \frac{2m\lambda}{\hbar^2}.$$

Equation (13.1.30) is called the Gross–Pitaevskii–Anandan equation. It is in essence a type of ϕ^4 equation. Anandan did not find its solutions. Instead, he gave an order of magnitude estimate of $\phi = \varphi e^{i\theta}$ for all types of solutions using the Einstein–Planck law. As will be shown below, an exact solution of equation (13.1.30) is actually possible [210, 215].

Let us assume the following trial solution [234, 236]:

$$\phi(x,y,z,t) = \varphi(Z)e^{i\theta}, \tag{13.1.31}$$

where

$$Z = \vec{p}.\vec{r} - \Omega t, \quad \theta = \vec{k}.\vec{r} - \omega t = k_1x + k_2y + k_3z - \omega t.$$

Substituting equation (13.1.31) into equation (13.1.30), the latter can be written

$$(\Omega^2 - p^2)\frac{d^2\varphi}{dZ^2} + (\alpha^2 + k^2 - \omega^2)\varphi + \lambda'\varphi^3 = 0 \tag{13.1.32}$$

in terms of $\vec{k} = (k_1, k_2, k_3)$ and $\vec{p} = (p_1, p_2, p_3)$, where $\omega\Omega = \vec{k}.\vec{p}$. In the integrate equation (13.1.32), we finally obtain the solution of equation (13.1.30) [210, 215] as follows:

$$\phi(x, y, z, t) = \varphi(Z)e^{i\theta} = \sqrt{\frac{W}{R}}\sec h[\sqrt{w}(\vec{p}.\vec{r} - \Omega t)]e^{i(\vec{k}.\vec{r}-\omega t)}, \tag{13.1.33}$$

where

$$w = \frac{\omega^2 - \alpha^2 - k^2}{\Omega^2 - p^2}, \quad R = \frac{\lambda'}{2(\Omega^2 - p^2)}. \tag{13.1.34}$$

This is a soliton solution of the wave packet type and its group velocity is v.

From the above study, we see that the time-dependent motion of superfluid helium atoms still resembles that of a soliton, so we conclude that the superfluid helium atoms are, in essence, a soliton, because it is the soliton motion of the superfluid helium atoms that causes the superfluidity. Because the solitons can move over macroscopic distances retaining their amplitude, velocity, and energy, the motion of the liquid helium necessarily resembles superfluidity. In such a case, the motions of the superfluid helium atoms are described by the nonlinear Schrödinger equation (13.1.1). According to the soliton theory, the superfluid helium atoms are localized and have a wave-corpuscle duality due to the nonlinear interaction which suppresses the dispersive effect of the kinetic energy in equation (13.1.30). Obviously, the nonlinear interaction is caused by the self-interaction among the helium atoms. We seek the direction of the development of quantum mechanics through this investigation, so the nonlinear theory of helium superfluid can truly serve as the theoretical foundation establishing nonlinear quantum mechanics.

13.2 The soliton movement and macroscopic quantum effects in physical systems

13.2.1 The macroscopic quantum effects in superfluids

From the above discussion, we clearly understand the nature and characteristics of macroscopic quantum systems, which are completely different from microscopic quantum effects on single particles, described by linear quantum mechanics. It is interesting to compare the two, so here we give a summary of the main differences.

(1) From the point of view of their characteristics, the microscopic quantum effect is characterized by the quantization of physical quantities, such as energy, momentum, and angular momentum. On the other hand, the macroscopic quantum effect

is represented by discontinuities in macroscopic quantities, such as resistance, magnetic flux, vortex lines, and voltage. The macroscopic quantum effects can be directly observed in experiments on the macroscopic scale, while the microscopic quantum effects can only be inferred from other effects related to them.

(2) Concerning the origins of these quantum effects, the microscopic quantum effect is produced when microscopic particles, which have only a wave feature, are confined in a finite space, or are constituted as matter, while the macroscopic quantum effect is due to the collective motion of the microscopic particles in systems with nonlinear interaction. It occurs through second-order phase transitions following the spontaneous breakdown of symmetry of the systems.

(3) The macroscopic quantum state is a condensed and coherent state, but the microscopic quantum effect occurs in determined quantization conditions, which are different for the bosons and fermions. So far, only the bosons or combinations of fermions have been found in macroscopic quantum effects.

(4) The microscopic quantum effect is a linear effect, in which the microscopic particles are in an expanded state, their motions being described by linear quantum mechanics and the linear Schrödinger equation, the Dirac equation, and the Klein–Gordon equations. On the other hand, the macroscopic quantum effect is caused by nonlinear interactions. Linear quantum mechanics failed to describe it, but it can be described by nonlinear partial differential equations such as the nonlinear Schrödinger equation (13.1.28) or equation (13.1.1).

We conclude that the macroscopic quantum effects are, in essence, a nonlinear quantum effect or phenomenon. Because its nature and fundamental characteristics are different from those of the microscopic quantum effects, it may be said that the effects should be depicted by a new nonlinear quantum theory, instead of linear quantum mechanics. Therefore, this investigation shows the necessity to develop nonlinear quantum theory [234, 236]. The macroscopic quantum effects mentioned above can serve as the experimental foundation establishing a new approach to nonlinear quantum mechanics.

13.2.2 The relation between soliton movement in superconductors and superfluids

Macroscopic quantum effects are nonlinear phenomena. The Bardeen–Cooper–Schreiffier (BCS) theory of superconductivity and the modern theory of superfluidity are both nonlinear theories and have been well established. A basic feature of the nonlinear theories is that the Hamiltonian and free energy Lagrangian functions of the systems are nonlinear functions of the wave function of microscopic particles as given in equations (13.1.1) and (13.1.2), respectively. The dynamic equation becomes a nonlinear Schrödinger equation as shown in equations (13.1.1) and (13.1.28), due to the nonlinear nature of the samples. These microscopic particles behave differently in such systems from those in linear quantum mechanics and become as some solitons with a wave-corpuscle duality, because the nonlinear interactions balance and suppress the dispersive effect of the kinetic energy in these dynamic equations. In such a case, these particles are in an ordered coherent state or a Bose–Einstein condensed state. These states occur following a second-order phase transition and spontaneous break of symmetry of the systems under the nonlinear interactions. In this process, the nonlinear interactions play a very important role. The BCS theory indicates that the nonlinear interaction is caused by the electron–phonon interaction due to the vibration of the lattice and exists truly in the superconductor, but the nonlinear interaction is not involved in linear quantum mechanics. It also suggests that one should pay particular attention to nonlinear interactions in order to establish a correct and new quantum theory, so the right direction for solving problems encountered by linear quantum mechanics is to establish a nonlinear quantum theory.

Therefore, present theories of superconductivity and superfluids could build the foundation for establishing nonlinear quantum theory. On what foundation should a new theory be based? How can a comprehensive nonlinear quantum theory be established? We will discuss how the superconductive and superfluid theories differ from those of linear quantum mechanics.

(a) The Hamiltonian or Lagrangian function and the free energy of these systems are all dependent on and are nonlinear functions of the wave function $\phi(\vec{r}, t)$ of the microscopic particles, i.e., the superconductive electron or superfluid helium. These go directly against the fundamental hypothesis of linear quantum mechanics, which suggests that the Hamiltonian of the system is independent of the wave functions of the microscopic particle. It was exactly because of this nonlinear feature in the theories of superconductivity and superfluidity that they were able to correctly describe the nonlinear behavior of superconductivity and superfluidity and successfully explain these macroscopic quantum effects. Lacking such a nonlinear feature was also the reason for other theories to fail. For example, although Frohlich's superconducting theory [84] gave the correct superconducting mechanism, electron–phonon coupling, in 1951 he failed to establish a complete superconducting theory because his theory was based on linear perturbation theory of linear quantum mechanics. Therefore, the new nonlinear theory should abandon this hypothesis.

(b) The fundamental dynamic equations in linear quantum mechanics are a linear Schrödinger equation and the Klein–Gordon equations, which are wave equations. They are linear equations of the wave function of the particles. As a result, solutions of these linear equations cannot describe the wave-corpuscle duality of microscopic particles as discussed in Chapter 11. On the other hand, the Ginzburg-Landau equations (13.1.1) and (13.1.28) and the GP equation (13.1.28), satisfied by the quasi-particles (e.g., the superconductive electron and the superfluid helium atom), as well as the ϕ^4 equation (13.1.30) in the superfluid, are nonlinear Schrödinger equations and the ϕ^4 equation of the wave function of the quasi-particles. With these nonlinear equations, the experiments on superconductivity and superfluidity, as well as other macroscopic quantum effects, can be explained. This suggests that, in establishing a new theory, the linear dynamic equation must be replaced with a nonlinear equation. Fortunately, all the nonlinear equations mentioned above are natural generalizations of the linear Schrödinger equation or Klein–Gordon equations and are the dynamic equations in linear quantum mechanics. Therefore, the new nonlinear quantum theory should be developed on the basis of linear quantum mechanics rather than anything else. Certainly, it is still necessary to further examine whether these dynamic equations have the correct space-time symmetries and what physical invariance they might possess. Thus, it seems that only nonlinear dynamic equations or Hamiltonians which satisfy the required symmetries of space-time and invariances of physical quantities can be adopted in the new theory.

(c) It is well known that the nonlinear dynamic equations describing superconductivity and superfluidity states admit stable soliton solutions. This shows that the microscopic particles in linear quantum mechanics evolve into solitons in nonlinear systems due to nonlinear interactions. It is therefore natural to use the concept of solitons in the description of microscopic particles in nonlinear systems. A soliton is a new form of physical entity, one which cannot be described by linear theory. According to modern soliton theory, a soliton, which differs completely from a microscopic particle in linear quantum mechanics, possesses wave-particle duality. Its wave property appears in the form of a traveling solitary wave which has all the essential features of wave motion, including frequency, period, amplitude, group and phase velocities, diffraction, transmission, and reflection. Its corpuscle feature is reflected by the stable shape analogous to a classical particle, even after going through a collision with another particle, a definite energy, momentum and mass, its uniform motion in free space and its motion with a constant acceleration in the presence of a constant external field, etc. This suggests that modern soliton theory should be an integral part of any new theory on nonlinear quantum mechanics.

To summarize, we see clearly that the direction for developing a new quantum theory is only likely through nonlinear quantum mechanics. The superconductivity, superfluidity, macroscopic quantum effects, and soliton theory prepare and provide the sufficient conditions for establishing the nonlinear quantum mechanics, the former being the experimental foundation. Its mathematical basis is formed by the nonlinear partial differential equations and soliton theory. Therefore, the conditions establishing the nonlinear quantum mechanics are already sufficient in present cases.

(d) In this chapter we review the properties of superconductivity, superfluidity, and macroscopic quantum effects, which are different from the microscopic quantum effects obtained from some experiments. The macroscopic quantum effects occurring on the macroscopic scale are caused by the collective motion of microscopic particles after the symmetry of the system is broken due to nonlinear interactions. Such interactions result in Bose condensation of particles in these systems. Meanwhile, we also study the properties of motion of superconductive electrons and superfluid helium atoms and arrived at the soliton solutions of the Ginzburg-Landau equation in superconductors and of the GP equation in superfluidity, which are, in essence, a kind of nonlinear Schrödinger equation. These solitons, with wave-corpuscle duality, are due to the nonlinear interactions arising from the electron–phonon interaction in superconductors or self-interaction of helium atoms in the superfluids in which the nonlinear interaction suppresses the dispersive effect of the kinetic energy in these dynamic equations. Meanwhile, we use these dynamic equations and their soliton solutions to obtain and further explain these macroscopic quantum effects, such as the quantization of magnetic flux in superconductors and the Josephson effect of superconductivity junctions, as well as quantized vortex rings in the superfluid ^{4}He. We conclude that the superconductive electrons and the superfluid helium atoms are some solitons and that the superconductivity, superfluidity, and macroscopic quantum effects are a kind of nonlinear quantum effects that arise from the soliton motions of superconductive electrons and superfluid helium atoms, respectively.

Therefore, studying the essences of macroscopic quantum effects and the properties of motion of microscopic particles in the superconductor and superfluid has important significance. From these studies, we see that the superconductive electrons and the superfluid helium atoms can be described by the nonlinear Schrödinger equations. The superconductivity, superfluidity, and macroscopic quantum effects observed can serve as the experimental foundations establishing nonlinear quantum mechanics and the present superconductive and superfluid theories can serve as the theoretical foundations of nonlinear quantum mechanics.

13.3 The soliton excitations in the anti-ferromagnetic systems

13.3.1 Dynamic features of solitons under the action of magnon–phonon interaction

The collective excitation and motion of magnons due to magnon–phonon interactions or magnon–magnon interactions in Heisenberg anti-ferromagnetic systems have been extensively studied by Pang et al. [201, 211, 212, 216–219] and many other scientists [55, 170, 172, 254]. The results show that the characteristics of the collective excitation in these systems are quite different from those in ferromagnetic systems. In this section, we present some results obtained by Pang et al. based on collective excitation in anisotropic Heisenberg anti-ferromagnets with magnon–phonon and magnon–magnon interactions by using the double-sublattice model [219].

When the double-sublattice model is used, the Hamiltonian of the Heisenberg anti-ferromagnet can be expressed [219] as

$$H = T + V + \frac{1}{2}\sum_{n}^{A}\sum_{\delta}^{A}[\xi_{n,n+\delta}S_n^x S_{n+\delta}^x + \eta_{n,n+\delta}S_n^y S_{n+\delta}^y + J_{n,n+\delta}S_n^z S_{n+\delta}^z]$$
$$+ \frac{1}{2}\sum_{j}^{B}\sum_{\delta}^{B}[\xi_{j,j+\delta}S_j^x S_{j+\delta}^x + \eta_{j,j+\delta}S_j^y S_{j+\delta}^y + J_{j,j+\delta}S_j^z S_{j+\delta}^z], \qquad (13.3.1)$$

where

$$T = \frac{m}{2}\sum_n \dot{u}_n^2, \quad V = \frac{mv_0^2}{r_0}\sum_n (u_{n+1} - u_n - u_0)$$

are the kinetic and potential energies of lattice oscillations, respectively, m is the "mass" of a spin, r_0 is the lattice constant, v_0 is the sound velocity in the crystal which we set equal to unity in subsequent calculations, and $S^k_{n(j)}$ $(k = x, y, z)$ is the spin component at site $n(j)$ in the k-direction. We apply the transformation $S^+_{n(j)} = S^x_{n(j)} + iS^y_{n(j)}$ and change equation (13.3.1) into

$$\begin{aligned} H = T + V + \frac{1}{2}\sum_n^A \sum_\delta^A \Big[& J_{n,n+\delta}S^z_n S^z_{n+\delta} + \frac{1}{2}(\xi_{n,n+\delta} + \eta_{n,n+\delta})(S^+_n S^-_{n+\delta} + S^-_n S^+_{n+\delta}) \\ & + \frac{1}{4}(\xi_{n,n+\delta} - \eta_{n,n+\delta})(S^+_n S^+_{n+\delta} + S^-_n S^-_{n+\delta})\Big] \\ \frac{1}{2}\sum_j^B \sum_\delta^B \Big[& J_{jj+\delta}S^z_j S^z_{j+\delta} + \frac{1}{4}(\xi_{jj+\delta} + \eta_{jj+\delta})(S^+_j S^-_{j+\delta} + S^-_j S^+_{j+\delta}) \\ & + \frac{1}{4}(\xi_{jj+\delta} - \eta_{jj+\delta})(S^+_j S^+_{j+\delta} + S^-_j S^-_{j+\delta})\Big]. \end{aligned} \tag{13.3.2}$$

We again use the Dyson–Maleev representation of the spin operators in virtue of Bose creation and annihilation operators [77, 255–257]. We have

$$\begin{aligned} S^+_a &= \sqrt{2S}\left(1 - \frac{a^+a}{4S}\right)a, \quad S^-_a = \sqrt{2S}a^+\left(1 - \frac{a^+a}{4S}\right), \quad S^z_a = S - a^+a, \\ S^+_b &= \sqrt{2S}b^+\left(1 - \frac{b^+b}{4S}\right), \quad S^-_b = \sqrt{2S}\left(1 - \frac{b^+b}{4S}\right)b, \quad S^z_b = b^+b - S, \end{aligned}$$

where $a^+(a)$ and $b^+(b)$ are the creation (annihilation) operators of the Heisenberg magnon field on the two sublattices. Taking into account the symmetry of the sublattices A and B and the fact that the A and B sublattices are neighbors of each other, the Hamiltonian in equation (13.3.2) can be approximately written [219] as

$$\begin{aligned} H \approx\ & T + V - J_0 NS^2 + S\sum_n^A \sum_\delta^A J_{n,n+\delta}a^+_n a_n + S\sum_j^B \sum_\delta^B J_{J,j+\delta}b^+_j b_j \\ & + \frac{S}{2}\sum_n^A \sum_\delta^A (\xi_{n,n+\delta} - \eta_{n,n+\delta})(a_n b^+_{n+\delta} + b_{n+\delta}a^+_n) \\ & + \frac{S}{2}\sum_n^A \sum_\delta^A (\xi_{n,n+\delta} + \eta_{n,n+\delta})(a_n b_{n+\delta} + a^+_n b^+_{n+\delta}) \\ & - \sum_j^B \sum_\delta^B J_{J,j+\delta}a^+_j a_j b^+_{j+\delta} b_{j+\delta} \end{aligned}$$

$$-\frac{1}{8}\sum_{n}^{A}\sum_{\delta}^{A}(\xi_{n,n+\delta}-\eta_{n,n+\delta})(a_n^+a_na_nb_{n+\delta}^+ + a_nb_{n+\delta}^+b_{n+\delta}b_{n+\delta})$$
$$+ a_n^+a_{n+\delta}a_n^+a_n + b_{n+\delta}^+b_{n+\delta}a_n^+b_{n+\delta}$$
$$-\frac{1}{8}\sum_{n}^{A}\sum_{\delta}^{A}(\xi_{n,n+\delta}+\eta_{n,n+\delta})(a_n^+a_na_nb_{n+\delta} + a_nb_{n+\delta}b_{n+\delta}^+b_{n+\delta})$$
$$+ a_n^+b_{n+\delta}a_n^+a_n + a_nb_{n+\delta}^+b_{n+\delta}^+b_{n+\delta}, \tag{13.3.3}$$

where the last four terms are anomalous terms resulting from magnon–magnon interactions. They can be neglected because they are weak compared to magnon–phonon interactions. The above Hamilltonian can become simple, which will be considered first in the following section.

We can here apply the methods of Makhankov and Fedyanin et al. [170, 172] and Pang [219] to study the properties of the collective excitations of the magnons in a one-dimensional anti-ferromagnetic system. In the Heisenberg representation, the equations of the operators a_f and b_f of sublattices A and B can, after some proper transformations, be written as

$$i\hbar\dot{a}_f = [a_f, H] \approx S\sum_{\delta} J_{ff+\delta}a_f + S/2\sum_{\delta}^{A}(\xi_{ff+\delta}-\eta_{ff+\delta})b_{j+\delta} + S/2\sum_{\delta}(\xi_{ff+\delta}+\eta_{ff+\delta})b_{j+\delta}^+, \tag{13.3.4}$$

$$i\hbar\dot{b}_f = [b_f, H] \approx S\sum_{\delta} J_{ff+\delta}b_f + S/2\sum_{\delta}^{A}(\xi_{ff+\delta}-\eta_{ff+\delta})a_{j+\delta} + S/2\sum_{\delta}(\xi_{ff+\delta}+\eta_{ff+\delta})a_{j+\delta}^+. \tag{13.3.5}$$

We further assume that the wave function of the collective excitation state of the quasi-particles in the system is of the following form:

$$|\varphi(t)\rangle = \frac{1}{\lambda}\left[2 + \sum_{n}^{A}\varphi_{an}(t)a_n^+ + \sum_{j}^{A}\varphi_{bj}(t)a_j^+\right]|0\rangle, \tag{13.3.6}$$

where $|0\rangle$ is the vacuum state (ground state), a_{ai} and b_{ai} are expression coefficients related to the characteristics of the magnons, which obviously are functions of time and space, and λ is a normalization constant.

Note that

$$\langle\varphi(t)|a_f|\varphi(t)\rangle = \alpha_{af}/\lambda^2 = \varphi_{af}, \quad \langle\varphi(t)|b_f|\varphi(t)\rangle = \alpha_{bf}/\lambda^2 = \varphi_{bf}$$

is the Schrödinger probability amplitude of the magnon. Using these representations, from equations (13.3.4) and (13.3.5), we have

$$i\hbar\dot{\varphi}_f = S\sum_{\delta} J_{ff+\delta}\varphi_{af} + S/2\sum_{\delta}(\xi_{ff+\delta}-\eta_{ff+\delta})\varphi_{bj+\delta}(t) + S/2\sum_{\delta}(\xi_{ff+\delta}+\eta_{ff+\delta})\varphi_{bj+\delta}^*, \tag{13.3.7}$$

$$i\hbar\dot{\varphi}_{bf} = S\sum_{\delta} J_{ff+\delta}\varphi_{bf} + S/2\sum_{\delta}(\xi_{ff+\delta} - \eta_{ff+\delta})\varphi_{aj+\delta}(t) + S/2\sum_{\delta}(\xi_{ff+\delta} + \eta_{ff+\delta})\varphi^{*}_{aj+\delta}. \tag{13.3.8}$$

If the oscillation amplitude of the lattice is small and the magnon–phonon interaction is also weak, we proceed to the continuum limit for the coefficients and the probability amplitude. Dropping the terms with higher derivatives as a result, we get

$$J_{f,f+1} \approx J_0 - J_1 r_0 \frac{\partial u_f}{\partial x}, \quad \xi_{f,f+1} \approx \xi_0 - \xi_1 r_0 \frac{\partial u_f}{\partial x}, \quad \eta_{f,f+1} \approx \eta_0 - \eta_1 r_0 \frac{\partial u_f}{\partial x},$$
$$\varphi_{nf\pm 1} \approx \varphi_{nf} \pm r_0 \frac{\partial}{\partial x}\varphi_{n'f} + \frac{1}{2} r_0^2 \frac{\partial^2}{\partial x^2}\varphi_{n'f} + \cdots, \quad (n = a, b, n' = b, a) \text{ etc.,} \tag{13.3.9}$$

where we get the following from the Heisenberg equations for φ_f:

$$i\hbar(\varphi_f)_t = 2SJ_0\varphi_f + S(\xi_0 - \eta_0) + S(\xi_0 + \eta_0)\varphi_f^* + \frac{S}{2}(\xi_0 - \eta_0)r_0^2\varphi_{fxx}$$
$$+ \frac{S}{2}(\xi_0 + \eta_0)r_0^2\varphi^*_{fxx} - S(\xi_1 - \eta_1)r_0 u_{fx}\varphi_f - S(\xi_1 + \eta_1)r_0 u_{fx}\varphi_f^* - 2J_1 S r_0 u_{fx}\varphi_f, \tag{13.3.10}$$

where r_0 is the average distance between neighbouring sites. Using equation (13.3.9) and bearing in mind the symmetry of the sublattices, we easily derive the following approximate equations of motion for φ_{af} and φ_{bf}:

$$i\hbar\dot{\varphi}_{af} \approx 2SJ_0\varphi_{af} + S(\xi_0 - \eta_0)\varphi_{bf} + S(\xi_0 - \eta_0)\varphi^*_{bf} + \frac{S}{2}(\xi_0 - \eta_0)r_0^2\frac{\partial^2}{\partial x^2}\varphi_{af}$$
$$+ \frac{S}{2}(\xi_0 + \eta_0)r_0^2\frac{\partial^2}{\partial x^2}\varphi^*_{af} + S(\xi_1 - \eta_1)r_0\frac{\partial u_f}{\partial x}\varphi_{bf} - S(\xi_1 + \eta_1)r_0\frac{\partial u_f}{\partial x}\varphi^*_{bf}$$
$$- 2SJ_1 r_0\frac{\partial u_f}{\partial x}\varphi_{af}, \tag{13.3.11}$$

$$i\hbar\dot{\varphi}_{bf} \approx 2SJ_0\varphi_{bf} + S(\xi_0 - \eta_0)\varphi_{aj} + S(\xi_0 - \eta_0)\varphi^*_{aj} + \frac{S}{2}(\xi_0 - \eta_0)r_0^2\frac{\partial^2}{\partial x^2}\varphi_{bf}$$
$$+ \frac{S}{2}(\xi_0 + \eta_0)r_0^2\frac{\partial^2}{\partial x^2}\varphi^*_{bf} + S(\xi_1 - \eta_1)r_0\frac{\partial u_f}{\partial x}\varphi_{af} - S(\xi_1 + \eta_1)r_0\frac{\partial u_f}{\partial x}\varphi^*_{af}$$
$$- 2SJ_1 r_0\frac{\partial u_f}{\partial x}\varphi_{bf}. \tag{13.3.12}$$

We define $\varphi_f(t) = \varphi_{af}(t) + \varphi_{bf}(t)$ and obtain

$$i\hbar\dot{\varphi}_f \approx 2SJ_0\varphi_f + S(\xi_0 - \eta_0)\varphi_f + S(\xi_0 + \eta_0)\varphi^*_{bj} + \frac{S}{2}(\xi_0 - \eta_0)r_0^2\frac{\partial^2}{\partial x^2}\varphi_f$$
$$+ \frac{S}{2}(\xi_0 + \eta_0)r_0^2\frac{\partial^2}{\partial x^2}\varphi_f^* + S(\xi_1 - \eta_1)r_0\frac{\partial u_f}{\partial x}\varphi_f - S(\xi_1 + \eta_1)r_0\frac{\partial u_f}{\partial x}\varphi_f^* - 2SJ_1 r_0\frac{\partial u_f}{\partial x}\varphi_f. \tag{13.3.13}$$

Clearly, equation (13.3.11) is, in essence, a nonlinear Schrödinger equation of the magnons.

We [219] have previously discussed the part representing the lattice oscillations. From equation (13.3.3) we have

$$\langle\varphi(t)|H|\varphi(t)\rangle = T + V - \frac{1}{2}S^2\sum_n^A\sum_\delta^A J_{nn+\delta} - \frac{1}{2}S^2\sum_n^B\sum_\delta^B J_{nn+\delta} - \sum_n^A\left(S\sum_\delta^A J_{nn+\delta}\right)|\varphi_{an}|^2 + \sum_j^B\left(S\sum_\delta^B J_{jj+\delta}\right)|\varphi_{bj}|^2 + \frac{1}{2}S\sum_j^A\sum_\delta^A(\xi_{nn+\delta} - \eta_{jj+\delta})(\varphi_{an}\varphi^*_{bj+\delta} + \varphi^*_{an}\varphi_{bj+\delta}). \tag{13.3.14}$$

It should be point out that, with the wave function in the form of equation (13.3.6), the anomalous terms in the original Hamiltonian have already been removed.

We here use the classical Hamiltonian equation

$$-M(u_f)_{tt} = \frac{\partial}{\partial u_f}\langle\varphi(t)|H|\varphi(t)\rangle, \tag{13.3.15}$$

where M is the mass of a lattice point (atom, for example) and u_f is its displacement, a classical quantity. Having in mind the symmetry of the sublattices A and B as well as the fact that the neighbors of sublattice A belong to sublattice B and vice versa and using again

$$\sum_n^A\left(S\sum_\delta^A J_{nn+\delta}\right)|\varphi_{an}|^2 = S\sum_n^A\{2J_0 - J_1(u_{bn+1} - u_{bn-1})\}|\varphi_{an}|^2$$

$$\sum_j^B\left(S\sum_\delta^B J_{jj+\delta}\right)|\varphi_{bj}|^2 = S\sum_j^b\{2J_0 - J_1(u_{aj+1} - u_{aj-1})\}|\varphi_{bj}|^2$$

from equations (13.3.14) and (13.3.15), we obtain

$$\begin{aligned}
-M\ddot{u}_{af} &= \frac{\partial}{\partial u_{af}}\langle\varphi(t)|H|\varphi(t)\rangle = K(2u_{af} - u_{bf+1} - u_{bf-1}) + SJ_1[|\varphi_{bf+1}|^2 - |\varphi_{bf+1}|^2] \\
&\quad + \frac{1}{2}S(\xi_1 - \eta_1)[\varphi_{af}(\varphi^*_{bf+1} - \varphi^*_{bf-1}) + \varphi^*_{af}(\varphi_{bf+1} - \varphi_{bf-1})], \\
-M\ddot{u}_{bf} &= \frac{\partial}{\partial u_{bf}}\langle\varphi(t)|H|\varphi(t)\rangle = K(2u_{bf} - u_{af+1} - u_{af-1}) + SJ_1[|\varphi_{af+1}|^2 - |\varphi_{af+1}|^2] \\
&\quad + \frac{1}{2}S(\xi_1 - \eta_1)[\varphi_{bf}(\varphi^*_{af+1} - \varphi^*_{af-1}) + \varphi^*_{bf}(\varphi_{af+1} - \varphi_{af-1})].
\end{aligned}$$

Assuming $M_a \approx M_b \approx M$, we have

$$\begin{aligned}
-M\ddot{u}_{af} &= K(2u_{af} - u_{bf+1} - u_{bf-1}) + SJ_1[|\varphi_{bf+1}|^2 - |\varphi_{bf+1}|^2] \\
&\quad + \frac{1}{2}S(\xi_1 - \eta_1)[\varphi_{af}(\varphi^*_{bf+1} - \varphi^*_{bf-1}) + \varphi^*_{af}(\varphi_{bf+1} - \varphi_{bf-1})],
\end{aligned}$$

$$-M\ddot{u}_{bf} = K(2u_{bf} - u_{af+1} - u_{af-1}) + SJ_1[|\varphi_{af+1}|^2 - |\varphi_{af+1}|^2]$$
$$+\frac{1}{2}S(\xi_1 - \eta_1)[\varphi_{bf}(\varphi^*_{af+1} - \varphi^*_{af-1}) + \varphi^*_{bf}(\varphi_{af+1} - \varphi_{af-1})],$$

where K is the force coefficient $K = mC_0^2/2r_0$.

Defining $u_f = u_{af} + u_{bf}$, we have no difficulty in obtaining the following equation for the continuum approximation:

$$-Mu_{ftt} \approx -Kr_0^2u_{fxx} + J_1Sr_0\partial/\partial x(|\varphi_f|^2)_x + \frac{1}{4}(\xi_1 - \eta_1)Sr_0\partial/\partial x(|\varphi_f|^2). \tag{13.3.16}$$

Equations (13.3.13) and (13.3.16) form a complete set of equations for the collective excitations in a Heisenberg anti-ferromagnetic system with magnon–phonon interactions. Now we proceed to find the solutions to the equations.

In the case of a quasi-steady state, we assume

$$u(x,t) = u(\varsigma), \quad \varphi(x,t) = \varphi(\varsigma)e^{i\theta} \quad (\varsigma = x - vt). \tag{13.3.17}$$

Substituting equation (13.3.17) into equation (13.3.16) [55, 212, 254], we have

$$(-Kr_0^2 - Mv^2)\frac{\partial^2}{\partial\varsigma^2}u_f = \left[J_1 + \frac{1}{4}(\xi_1 - \eta_1)\right]Sr_0\frac{\partial}{\partial\varsigma}|\varphi_f|^2. \tag{13.3.18}$$

By solving the above equation, we get

$$\frac{\partial u_f}{\partial x} = \frac{\partial u_f}{\partial\varsigma} = \left[J_1 + \frac{1}{4}(\xi_1 - \eta_1)\right]Sr_0(Kr_0^2 - Mv^2)^{-1}|\varphi_f|^2 + C', \tag{13.3.19}$$

where C' is an integration constant to be determined from the boundary conditions. Substituting equation (13.3.19) into equation (13.3.13) and solving this, we immediately find the characteristics of the magnons caused by magnon–phonon coupling in an isotropic anti-ferromagnet.

For simplicity, we consider here only the anisotropic anti-ferromagnet with $\xi = \eta$ (the other cases, of course, can be discussed in the same way). In this case, $J > \xi$ and $J < \xi$ correspond to the easy magnetic axis ($0z$) and the easy magnetic plane ($x0y$) in an anti-ferromagnet, respectively. Equation (13.3.13) reduces to

$$i\hbar\varphi_t = 2J_0S\varphi + 2\xi_0S\varphi^* + S\xi_0r_0^2\varphi^*_{xx} - 2J_1Sr_0u_x\varphi - 2\xi_1Sr_0u_x\varphi^*. \tag{13.3.20}$$

Its conjugate equation reads

$$-i\hbar\varphi^*_t = 2J_0S\varphi^* + 2\xi_0S\varphi + S\xi_0r_0^2\varphi_{xx} - 2J_1Sr_0u_x\varphi^* - 2\xi_1Sr_0u_x\varphi, \tag{13.3.21}$$

where we have dropped the subscripts of $u(t)$ and $\varphi(t)$.

We now perform the transformation $\varphi_\pm = \varphi \pm \varphi^*$. Then the following equations [201] can be obtained from equations (13.3.20) and (13.3.21):

$$\begin{aligned} i\hbar\dot{\varphi}_+ &= 2S(J_0+\xi_0)\varphi_- - S\xi_0 r_0^2\frac{\partial^2}{\partial x^2}\varphi_- - 2(J_1-\xi_1)Sr_0\frac{\partial u}{\partial x}\varphi_-, \\ i\hbar\dot{\varphi}_- &= 2S(J_0+\xi_0)\varphi_+ + S\xi_0 r_0^2\frac{\partial^2}{\partial x^2}\varphi_+ - 2(J_1+\xi_1)Sr_0\frac{\partial u}{\partial x}\varphi_+. \end{aligned} \tag{13.3.22}$$

These are some coupling nonlinear Schrödinger equations of φ_+ and φ_- and their solutions are quite difficult to find out. However, in the case where $v \ll C_0$ or $v^2 \ll Kr_0^2/M$, we obtain the following nonlinear dynamic equation of $\varphi_+(x,t)$ and $\varphi_-(x,t)$:

$$-\hbar^2\ddot{\varphi}_+ = 4S^2(J_0^2-\xi_0^2)\varphi_+ - 4S^2\xi_0^2 r_0^2\frac{\partial^2}{\partial x^2}\varphi_+ - 8(J_1J_0-\xi_1\xi_0)S^2 r_0\frac{\partial u}{\partial x}\varphi_+, \tag{13.3.23}$$

$$-\hbar^2\ddot{\varphi}_- = 4S^2(J_0^2-\xi_0^2)\varphi_- - 4S^2\xi_0^2 r_0^2\frac{\partial^2}{\partial x^2}\varphi_- - 8(J_1J_0-\xi_1\xi_0)S^2 r_0\frac{\partial u}{\partial x}\varphi_-. \tag{13.3.24}$$

Adding equation (13.3.23) to equation (13.3.24) and substituting equation (13.3.19) into them, we finally get

$$\varphi_{tt} - A_0\varphi_{xx} - B_0\varphi - C_0|\varphi|^2\varphi = 0, \tag{13.3.25}$$

where

$$\begin{aligned} A_0 &= \frac{4\xi_0^2S^2r_0^2}{\hbar^2} > 0, \quad C_0 = \frac{16J_1S^3r_0^2(J_0J_1-\xi_0\xi_1)}{\hbar^2(K'r_0^2-Mv^2)}, \\ B_0 &= \frac{1}{\hbar^2}[4S^2(J_0^2-\xi_0^2) - 8S^2r_0(J_0J_1-\xi_0\xi_1)C]. \end{aligned} \tag{13.3.26}$$

Very clearly, equation (13.3.25) is a φ^4-equation, instead of the nonlinear Schrödinger equation, but it is still the same with the dynamic equation (11.2.6) in nonlinear quantum mechanics [223, 225, 229, 237, 238]. Obviously, this is due to the interaction between the magnons or the coupling effect between the double sublattices in the antiferromagnets, although the motion of the magnons in a single ferromagnetic chain is described by a nonlinear Schrödinger equation in equation (13.1.1), which also appears in the case of a single ferromagnetic chain as mentioned above. However, in the antiferromagnets, the features of the magnons are still described as a soliton because the φ^4-equation is a representation or result of the nonlinear Schrödinger equation in the relativistic case [223, 225, 229, 237, 238]. Therefore, they share some features. In order to verify this point, we now find the soliton solutions of equation (13.3.25).

We can prove that, for the anti-ferromagnet magnetized along the z-direction, the magnon–phonon coupling in this direction plays an important part in forming a localized soliton. As a first step, we take equation (13.3.25) [55, 116, 212, 217, 219, 224, 254] and assume

$$\phi = f(x-vt)\exp[i(k'x-\omega t)]. \tag{13.3.27}$$

Inserting this into equation (13.3.25) [55, 212, 216–219, 254], we have

$$K'A_0 = v\omega, \quad (v^2 - A_0)\frac{d^2f}{d\varsigma^2} + (B_0 + A_0K'^2 - \omega^2)f^3 = 0.$$

After integration, this becomes

$$\left(\frac{df}{d\varsigma}\right)^2 = -\frac{C_0}{2(A_0 - v^2)}f^4 + \frac{B_0A_0 + \omega^2(A_0 - V_0)}{A_0(A_0 - v^2)}f^2 = 0,$$

where Pang [217, 219] has set the integration constant equal to zero because of the boundary conditions. Integrating once more, we arrive at

$$x = \int_{f(0,0)}^{f(x,t_0)} \left[\frac{C_0 f^2}{2(A_0 - v^2)}(f_0^2 - f^2)\right]^{1/2} df + \text{constant}, \tag{13.3.28}$$

where

$$f_0^2 = \frac{2}{A_0C_0^2}[B_0A_0 + \omega^2(A_0 - V_0)], \quad K' = v\omega/A_0.$$

In the case where $\omega^2 < (A_0B_0)/(A_0 - v^2)$, there are nontopological bell-type soliton solutions for equation (13.3.25) if the requirements $C_0/(A_0 - v^2) > 0$ and $f_0^2 > 0$ are satisfied. In other words, if $J_0 > \xi_0$ and at the same time either $J_0J_1 > \xi_0\xi_1$, $v < \min[\sqrt{K/mr_0}, 2\xi_0Sr_0/\hbar]$ or $J_0J_1 < \xi_0\xi_1$, $2\xi_0S_0r_0 < \hbar < v < \sqrt{K/mr_0}$, then its normalized solitary wave is

$$\varphi = \sqrt{\frac{r_0}{2W_s}}\, \mathrm{sec}\, h\left(\frac{x - \upsilon t}{W_s}\right)e^{i(k'x - \omega t)}, \tag{13.3.29}$$

where

$$k' = \frac{\upsilon\omega}{A_0}, \quad W_s = \frac{4(A_0 - \upsilon^2)}{C_0r_0}, \quad \omega^2 = \frac{A_0B_0}{A_0 - \upsilon^2} - \frac{d^2A_0C_0^2}{16(A_0 - \upsilon^2)^2},$$

where v is the velocity of the soliton.

Equation (13.3.29) indicates clearly that the magnon in an anti-ferromagnet moves as a bell-type soliton; its outline and features are the same as those described by the nonlinear Schrödinger equation in equation (13.1.31) or equation (13.1.1). It is clear that the magnon is still a soliton in nonlinear anti-ferromagnetic systems, as its essence and basic features have not changed, even though it satisfies the φ^4-equation in nonlinear quantum mechanics [223, 225, 229, 237, 238] because the nonlinear interactions in the anti-ferromagnets are still due to the magnon–phonon or magnon–magnon interactions. Therefore, the nonlinear nature of magnons does not change in the anti-ferromagnets.

However, if and $v^2 - A_0 > 0$, i.e., $C_0/(A_0 - v^2) > 0$, equation (13.3.23) also has another topological soliton solution, which is denoted by

$$\varphi = \sqrt{\frac{r_0}{2W_s}} \tanh\left(\frac{x - vt}{W_s}\right) e^{i(k'x-\omega t)}. \tag{13.3.30}$$

At the same time, it can be seen from equation (13.3.26) that $C_0 = 0$ and $B_0 = 0$ because $J_0 = \xi_0$, $J_1 = \xi_1$ in the isotropic anti-ferromagnets, so there is no solution to equation (13.3.25) in this case.

It can be seen from the above conditions that localized solitons can be excited by the nonlinear magnon–phonon coupling only for the magnetic anti-ferromagnet axis. To the best of our knowledge, this has never been observed before and no parallel observations have been made in ferromagnets. In the case being discussed, the coupling of the longitudinal lattice oscillations with the magnons, which results in a nonlinear interaction, causes a remarkable change in the transverse exchange integral of the anti-ferromagnet. It is the nonlinear interaction caused by the coupling that is vital for the formation of the soliton. In this case, the velocity of the soliton satisfies $v < \min[\sqrt{K/mr_0}, 2\xi_0 S r_0/\hbar]$.

13.3.2 Properties of motion of solitons under the action of magnon–magnon interactions

We have so far only considered the collective excitation caused by magnon–phonon interactions. In fact, when the magnon–magnon interactions in a system become too strong to be neglected, a new nonlinear interaction source will contribute to the collective excitation in anti-ferromagnetic systems. The formation process and the properties of these collective excitations will change accordingly if this interaction is taken into consideration. The Hamiltonian of the system in this case is still given by equation (13.3.1), but the interaction term now includes direct interactions between neighboring magnons and other two-magnon effects, such as influences of a magnon on the transfer of other magnons and on magnon "resonance."

Pang [84, 201, 211, 215, 218, 234, 236] employed the following quasi-average field approximation to treat the effects of the anomalous correlation terms in equation (13.3.1) on the soliton formation and the quasi-particle energy in the collective excitation [219]:

$$\begin{aligned} a_n^+ a_n a_n b_{n+\delta}^+ &= \langle a_n^+ a_n \rangle a_n b_{n+\delta}^+ + \langle a_n b_{n+\delta}^+ \rangle a_n^+ a_n - \langle a_n^+ a_n \rangle \langle a_n b_{n+\delta}^+ \rangle, \\ a_n^+ a_n a_n^+ b_{n+\delta}^+ &= \langle a_n^+ a_n \rangle a_n^+ a_{n+\delta}^+, \\ &\cdots \end{aligned}$$

Then the Hamiltonian of the system (13.3.1) becomes

$$H = E_0 + S \sum_n^A \sum_\delta^A J_{n,n+\delta} a_n^+ a_n + S \sum_j^B \sum_\delta^B J_{j,j+\delta} b_n^+ b_n$$

$$+\frac{1}{2}S\sum_{n}^{A}\sum_{\delta}^{A}(\xi_{n,n+\delta}-\eta_{n,n+\delta})(a_nb_{n+\delta}^{+}+a_n^{+}b_{n+\delta})$$
$$+\frac{1}{2}S\sum_{n}^{A}\sum_{\delta}^{A}(\xi_{n,n+\delta}-\eta_{n,n+\delta})(a_nb_{n+\delta}+a_n^{+}b_{n+\delta}^{+})$$
$$-\sum_{n}^{A}\sum_{\delta}^{A}J_{n,n+\delta}(\langle a_n^{+}a_n\rangle b_{n+\delta}^{+}b_{n+\delta}+\langle b_{n+\delta}^{+}b_{n+\delta}\rangle a_n^{+}a_n)$$
$$-\frac{1}{8}\sum_{n}^{A}\sum_{\delta}^{A}(\xi_{n,n+\delta}-\eta_{n,n+\delta})[(\langle a_n^{+}a_n\rangle+\langle b_{n+\delta}^{+}b_{n+\delta}\rangle)(a_nb_{n+\delta}^{+}+a_n^{+}b_{n+\delta})$$
$$+(\langle a_nb_{n+\delta}^{+}\rangle+\langle a_n^{+}b_{n+\delta}\rangle)(a_n^{+}a_n+b_{n+\delta}^{+}b_{n+\delta})]$$
$$-\frac{1}{8}\sum_{n}^{A}\sum_{\delta}^{A}(\xi_{n,n+\delta}+\eta_{n,n+\delta})[(\langle a_n^{+}a_n\rangle+\langle b_{n+\delta}^{+}b_{n+\delta}\rangle)(a_nb_{n+\delta}+a_n^{+}b_{n+\delta}^{+})],\tag{13.3.31}$$

where

$$E_0=T+V-J_0NS^2+\sum_{n}^{A}\sum_{\delta}^{A}J_{n,n+\delta}\langle a_n^{+}a_n\rangle\langle b_{n+\delta}^{+}b_{n+\delta}\rangle$$
$$-\frac{1}{8}\sum_{n}^{A}\sum_{\delta}^{A}(\xi_{n,n+\delta}-\eta_{n,n+\delta})[(\langle a_n^{+}a_n\rangle+\langle b_{n+\delta}^{+}b_{n+\delta}\rangle)(a_nb_{n+\delta}^{+}+a_n^{+}b_{n+\delta})].\tag{13.3.32}$$

From the second quantum Hamiltonian in equation (13.3.32) and the same method mentioned above, with the aid of Makhankov et al. [170, 172] and Pang's method [201, 211, 216–219], we obtain the following equations of motion for the operators a_f and b_f in the Heisenberg representation:

$$i\hbar\frac{\partial a_f}{\partial t}=[a_f,H],\quad i\hbar\frac{\partial b_f}{\partial t}=[b_f,H].$$

Then, using the Schrödinger probability amplitude defined by equation (13.3.6), we have

$$\varphi_{af}=\langle\varphi(t)|a_f|\varphi(t)\rangle=\alpha_{af}/\lambda^2,\quad \varphi_{bf}=\langle\varphi(t)|b_f|\varphi(t)\rangle=\alpha_{bf}/\lambda^2,$$

from which we finally derive the following nonlinear equation for φ_{af} and φ_{bf}:

$$i\hbar\dot{\varphi}_{af}=S\sum_{\delta}J_{ff+\delta}\varphi_{af}+S/2\sum_{\delta}(\xi_{ff+\delta}-\eta_{ff+\delta})\varphi_{bj+\delta}(t)+S/2\sum_{\delta}(\xi_{ff+\delta}+\eta_{ff+\delta})\varphi_{bj+\delta}^{*}$$
$$-\sum_{\delta}J_{ff+\delta}|\varphi_{bf+\delta}|^2\varphi_{aj}-\frac{1}{8}\sum_{\delta}(\xi_{ff+\delta}-\eta_{ff+\delta})[(|\varphi_{af}|^2+|\varphi_{bf+\delta}|^2)\varphi_{bf+\delta}$$
$$+|\varphi_{af}|^2\varphi_{bf+\delta}^{*}+|\varphi_{af}|^2\varphi_{bf+\delta}]$$
$$-\frac{1}{8}\sum_{\delta}(\xi_{ff+\delta}+\eta_{ff+\delta})(|\varphi_{af}|^2+|\varphi_{bf+\delta}|^2)\varphi_{bf+\delta}^{*}.\tag{13.3.33}$$

Interchanging the symbols a and b in equation (13.3.31), we get the corresponding equation for φ_{bf}. We then proceed to the continuum limit as before, dropping the terms with higher than third derivatives and taking into account the symmetry of sublattices A and B and the distribution characteristics mentioned above, to approximate $\varphi_f(t) = \varphi_{af}(t) + \varphi_{bf}(t)$ as

$$
\begin{aligned}
i\hbar\dot{\varphi} \approx\ & 2J_0S\varphi + (\xi_0 - \eta_0)S\varphi^* + S(\xi_0 - \eta_0)\varphi + \frac{1}{2}Sr_0^2(\xi_0 + \eta_0)\frac{\partial^2}{\partial x^2}\varphi^* + \frac{1}{2}Sr_0^2(\xi_0 - \eta_0)\frac{\partial^2}{\partial x^2}\varphi \\
& - (2J_1 + \xi_1 - \eta_1)Sr_0\frac{\partial u}{\partial x}\varphi - (\xi_1 - \eta_1)Sr_0\frac{\partial u}{\partial x}\varphi^* - [2J_1 + (\xi_0 - \eta_0)/8]|\varphi|^2\varphi \\
& - \nu(\xi_0 + \eta_0)|\varphi|^2\varphi^*,
\end{aligned}
\tag{13.3.34}
$$

where the subscript of the function in equation (13.3.32) has been dropped and a parameter ν has been introduced, which equals 1/16 in the case studied. Equation (13.3.34) differs from equation (13.3.13) only by an additional term and the dispersion effects being neglected. There will be no correlation between magnon–magnon interactions and magnon–phonon interactions and they become two independent nonlinear interaction sources. Meanwhile, equation (13.3.15) still expresses the lattice oscillations. Therefore, we can solve problems of this type by combining equation (13.3.19) with equation (13.3.34).

It can be proved that there is still no soliton solution for isotropic anti-ferromagnets even if the interactions between magnons are taken into account. In the case of isotropy, equation (13.3.34) becomes

$$
i\hbar\dot{\varphi} \approx 2J_0S(\varphi + \varphi^*) + J_0Sr_0^2\frac{\partial^2}{\partial x^2}\varphi^* - 2J_1Sr_0\frac{\partial u}{\partial x}(\varphi + \varphi^*) - 2J_0|\varphi|^2(\varphi - \nu\varphi^*).
$$

Its conjugation equation reads

$$
-i\hbar\dot{\varphi}^* = 2J_0S(\varphi + \varphi^*) + J_0Sr_0^2\frac{\partial^2}{\partial x^2}\varphi - 2J_1Sr_0\frac{\partial u}{\partial x}(\varphi + \varphi^*) - 2J_0|\varphi|^2(\varphi^* + \nu\varphi).
$$

Introducing $\varphi_\pm(t) = \varphi(t) \pm \varphi^*(t)$ and applying the same method mentioned above, we obtain

$$
\hbar^2\ddot{\varphi} = -4J_0S^2r_0^2\frac{\partial^2}{\partial x^2}\varphi - 8J_{01}^2S(1 - \nu)|\varphi|^2\varphi.
\tag{13.3.35}
$$

Comparing this with equation (13.3.25), we now have $B_0 = 0$ and only nonlinear terms exist. Therefore, soliton solutions of a type similar to equation (13.3.25) cannot be found for this equation [116, 201, 216–219, 224], thereby completing the proof. It can then be concluded that, in the case of isotropic Heisenberg anti-ferromagnetic chains, neither magnon–phonon interactions nor interactions between magnons can result in a collective excitation of the type of nontopological solitons, which is similar to the situation in isotropic ferromagnetic chains [178, 179].

For anisotropic anti-ferromagnets, we also limit our discussion to the cases of the magnetic axis (0z) and the magnetic plane (x0y), where equation (13.3.34) reduces to

$$i\hbar\dot{\varphi} \approx 2J_0S\varphi + 2S\xi_0\varphi^* + \xi_0Sr_0^2\frac{\partial^2}{\partial x^2}\varphi^* - 2J_1Sr_0\frac{\partial u}{\partial x}\varphi - 2S\xi_0r_0\frac{\partial u}{\partial x}\varphi^* - 2J_0|\varphi|^2\varphi - v2\xi_0|\varphi|^2\varphi^*. \tag{13.3.36}$$

Again introducing $\varphi_\pm(t) = \varphi(t) \pm \varphi^*(t)$ and adopting the same method applied to equations (13.3.21)–(13.3.23), we arrive at the following equation for $\varphi(x,t)$:

$$-\hbar^2\ddot{\varphi} = 4(J_0^2 - \xi_0^2)S^2\varphi - 4\xi_0^2S^2r_0^2\frac{\partial^2}{\partial x^2}\varphi - 8(J_0J_1 - \xi_0\xi_1)Sr_0\frac{\partial u}{\partial x}\varphi - 8(J_0^3 - v^2\xi_0^2)S|\varphi|^2\varphi. \tag{13.3.37}$$

Similar to the derivations of equations (13.3.4)–(13.3.6), we obtain

$$\varphi_{tt} - A\varphi_{xx} + B\varphi - g'|\varphi|^2\varphi = 0, \tag{13.3.38}$$

where

$$\begin{aligned} &A = A_0 = 4\xi_0S^2r_0^2/\hbar^2, \quad B = B_0 = [4S^2(J_0^2 - \xi_0^2)/\hbar^2] + 8(J_0J_1 - \xi_0\xi_1)Sr_0C_0,\\ &g' = C_0 + 8S(J_0^2 - v^2\xi_0^2)/\hbar^2. \end{aligned} \tag{13.3.39}$$

Therefore, equation (13.3.38) has the same soliton solution as equation (13.3.25) or equation (13.3.29). Simply replacing C_0, A_0, and B_0 in equations (13.3.25) and (13.3.29) by g, A, and B, respectively, we get the solutions. Therefore, taking into account magnon–magnon interaction only changes the amplitude and velocity of the soliton and does not alter the fundamental nature of the magnon solitons. The magnon–magnon interactions enhance the effects of the nonlinear interactions and thereby prompt the formation of more stable solitons. This is because g is always greater than C_0 if the magnon solitons exist. Furthermore, $J_0 \gg J_1$ and $\xi_0 \gg \xi_1$. We see that, in the presence of magnon–magnon interactions, they are the soliton solution of only the type described in equations (13.3.25) or (13.3.29) in the velocity range of $v^2 < A$. In this case, we have

$$\omega^2 = \frac{AB}{A - v^2} - \frac{r_0^2Ag^2}{16(A - v^2)^2} = \frac{AB}{A - v^2} - \frac{r_0^2}{16}\frac{A}{A - v^2}[g + 8S(J_0^2 - v^2\xi_0^2)]^2. \tag{13.3.40}$$

It was found that the nonlinear excitations of magnons can still be described by equation (13.3.25) or equation (13.3.38). The only differences are the coefficients in these equations. Therefore, the magnon solitons in nonlinear anti-ferromagnetic systems obey the laws of nonlinear quantum mechanics.

The above investigations indicate clearly that the properties of the magnons can be described well by nonlinear quantum mechanics [223, 225, 229, 237, 238]. Magnons can become solitons under the action of the nonlinear interactions. Then the magnons have both a wave feature, a solitary wave or a spin wave, and a corpuscle feature, as a localized particle. The nonlinear interactions are generated by virtue of the interaction between moving magnons or between a moving magnon and phonon. These interactions are always existent in the anti-ferromagnets, so the nature of the wave-corpuscle features of the magnons cannot be changed, i.e., the wave-corpuscle feature of the magnons is its inherent nature. Therefore, nonlinear quantum mechanics is a suitable theory to describe the properties of magnons in anti-ferromagnets. Once the magnon–magnon and magnon–phonon interactions are not considered, the magnon has only a spin wave feature.

In practice, the above nonlinear interactions are always existent in all anti-ferromagnets, but the generation mechanism of magnon–phonon coupling and of magnon–magnon interactions are different, although their effects on the nonlinear interaction and the formation of solitons are the same [223, 225, 229, 237, 238]. In the first place, the mechanism of localized nonlinear collective excitation (magnon soliton) caused by the first type is the breaking of kinetic symmetry, that is, it is caused by the interaction between the magnon and lattice oscillation. In a steady state, the magnon soliton and the localized deformation, which depend on lattice oscillation, propagate together with the same speed along the anti-ferromagnetic chain. The mechanism of excitation caused by the second type is the spontaneous breaking of symmetry brought about by the magnon–magnon interactions in the single-axis anisotropic anti-ferromagnet. Both mechanisms result in structural anisotropy and collective excitation. As mentioned, once the two mechanisms cancel each other in isotropic anti-ferromagnetic chains, no soliton can exist, as in the case for ferromagnetic chains. This also indicates that soliton excitation of magnons in such systems is determined by the anisotropy of the system. As long as the anisotropy exists in a given system, the magnon–phonon coupling and nonlinear interactions between the magnons will make the magnons "self-trapping" as a soliton in a range of dimension 2 Ws in the one-dimensional chains and a stable magnon soliton will propagate along the anti-ferromagnetic chain. When the anisotropy changes, the amplitude, the momentum, and the number of the solitons all change accordingly.

On the other hand, the formation of magnon solitons due to nonlinear interactions in anisotropic anti-ferromagnetic chains leads to many interesting physical phenomena. Indeed, anomalies have been observed in experiments. Attempts have been made to explain them using the magnetic soliton model, even though analytical expressions in place of equations (13.3.11) and (13.3.12) have not been obtained. For more detailed descriptions, the reader is referred to the work by Mikeska and Steiner [178, 179]. For example, Boucher et al. [35, 36] used the soliton concept to explain the phenomenon of nuclear spin-lattice relaxation (NSLR) in an anti-ferromagnetic chain of $(CH_3)_4NMnCl_3$, even though a theoretical expression for magnetic solitons has not been obtained. Through measurement, Boucher et al. obtained the ratio T_1^{-1} of NSLR of ^{15}N in the anti-ferromagnet, as a function of the external field H ($2\,\mathrm{kAm}^{-1} < H < 80\,\mathrm{kAm}^{-1}$) and temperature T ($2K \le T \le 4.2K$), and observed that T_1^{-1} diverged exponentially with H/T at a certain temperature. With the analytical results for the soliton given above, we can explain the excitation and the behavior of the soliton in such systems, which in turn validates the correctness of the above theory.

We now find the specific heat of the anti-ferromagnets, at which the temperature effect of the system should be considered. The inner energy of the magnon excited in the anti-ferromagnetic systems can be represented [28, 155] by

$$U(T) = \sum_k \langle a_k^+ a_k \rangle_T \hbar\omega_k + \sum_k \langle b_k^+ b_k \rangle_T \hbar\omega_k \approx 2 \sum_k \hbar\omega_k / [\exp(\hbar\omega_k / K_B T)]. \tag{13.3.41}$$

In the cubic crystal system, we can approximately represent the frequency of the magnon, which is directly proportional to its wavevector k, so its dispersion relation can approximately be denoted by $\hbar\omega \approx 8\hbar S(J_0^2 - v\xi_0^2)kr_0$, by equation (13.3.40) in the long wave approximation of $kr_0 \ll 1$, which is the same as that of the acoustical phonon in the same systems. We assume the anti-ferromagnet is a bcc crystal which is composed of two mutually penetrating simple cubic sublattices, in which the side length of each sublattice is denoted by a. When $\hbar\omega_k \gg K_B T$, the following relation [27, 28, 155] can be obtained:

$$\begin{aligned} U(T) &= \frac{2Nr_0^3}{8\pi^3} .4\pi \int_0^\infty \frac{\hbar\omega_k k^2 dk}{[\exp(\hbar\omega_k / K_B T) - 1]} \\ &= \frac{2N(K_B T)^4}{2\pi^2 [8S\hbar(J_0^2 - v\xi_0^2)]^3} \int_0^\infty x^3 [e^{-x} + e^{-2x} + \cdots] dx. \end{aligned} \tag{13.3.42}$$

Thus, we obtain the specific heat in the form of

$$C_m = \frac{\partial U(T)}{\partial T} = \frac{13{,}71 K_B}{[48S\hbar(J_0^2 - v\xi_0^2)]^3} (K_B T)^3 \propto (K_B T)^3. \tag{13.3.43}$$

This result is basically consistent with the experimental data.

It should also be pointed out that the effects of external fields are not included in our discussion. If any such field is present and it is in the direction along the easy magnetic axis, then its effect, due to the opposite magnetization directions of the two sublattices, will be equivalent to a periodic external field of period $2r_0$ which will strengthen the discreteness of the lattice and thus invalidates the continuum approximation. However, if the direction of the external field is perpendicular to the anti-ferromagnetic spin direction, the continuum approximation will still be valid. For this reason, earlier experimental and theoretical studies were concentrated mainly on transverse fields, rather than longitudinal fields.

13.3.3 The nonlinear properties of magnons in different anti-ferromagnetic systems

Xu and Pang [55, 254–257] further studied nonlinear excitations and properties of magnons in anti-ferromagnetic molecular crystals, such as $Ni(C_2H_8N_2)_2{-}NO_2(ClO_4)$

(NENP) and $Ni(C_3H_{10}N_2)_2NO_2(ClO_4)$ (NINO), using the above model and method. Their molecular structure is represented in Figure 13.1, where NENP's space group is $Pn2_{1a}$ and NINO's space group belongs to $Pbn2_1$. The constant crystal structures have the following dimensions: $a = 15.223$ nm, $b = 8.295$ nm, $c = 10.300$ nm for NENP and $a = 15.384$ nm, $b = 8.507$ nm, $c = 10.590$ nm for NINO. In two molecular crystals, their molecular structures are all composed of Ni^{2+} ionic chains, which are arranged along the c-axis, in which the distance between two Ni^{2+} ions is approximately 5.15 nm in each chain, but between two Ni^{2+} ions from different chains, the distance is approximately 8.295 nm, as shown in Figure 13.1. There are many nonmagnetic perchlorate anions between the two chains and the Ni^{2+} ions are related with a nitrogen atom and an oxygen atom by means of covalent bonds formed by the nitrite base in the same chain.

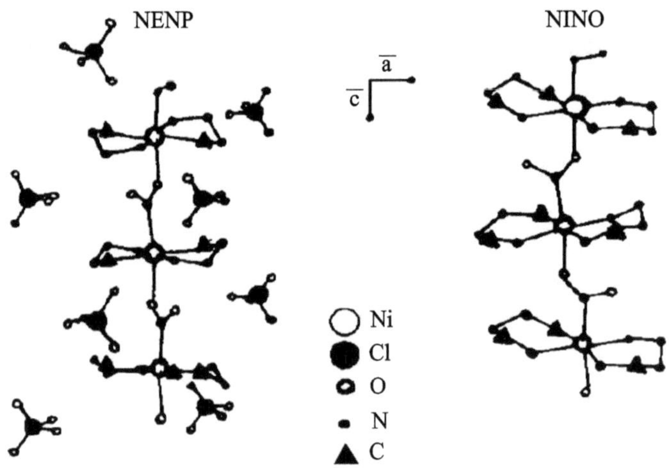

Figure 13.1: Molecular structures of NENP and NINO crystals.

The localized crystal structure of the Ni^{2+} ions is a distorted octahedron. Its fundamental surface is perpendicular to the c-axis and has the space group $Pn2_{1a}$. There is also an inversion center in point near arranged along he c-axis, the ions chains compounds, the distance between two Ni^{2+} ions is approximately the size of one Ni^{2+} ion. This structure has Ni^{2+} ions forming two magnetic tracks, one of which has a high spin state with $S = 1$. The octahedron configuration also makes Ni^{2+} ions form two magnetic tracks. One has d_{xy} symmetry in the fundament surface, the other has d_Z^2 symmetry along the axis of the chain. There is also the strong superposition of the magnetic tracks of d_Z^2 due to the influence of nitrite bases along the chain axis. These structural properties predict that the Ni^{2+} ions in one chain form anti-ferromagnetic-like interactions because of very weak interactions. Renard et al. [54, 262] discovered that the rate of the interactions between the chains relative to that in one chain is about 4×10^{-4}. Therefore, NENP and NIPO are good anisotropic anti-ferromagnetic chains for the Heisenberg model. Its Hamiltonian [308] is of the following form:

$$H = T + V + 2\sum_{nm} J_{nm} S_n^z S_m^z + 2\sum_{nm} \xi m(S_n^x S_m^x + S_n^y S_m^y) + 2D\sum_n (S_n^z)^2 + 2E\sum_n [(S_n^x)^2 - (S_n^y)^2],$$

where D and E are the anisotropic energies of single ions and J_{nm} is the interaction energy of spin ions in one chain.

We studied [35, 36, 307–309] the properties of magnons in anti-ferromagnetic molecular crystals with order parameter conservation (OPCAFMs), such as the magnetic compound CeAs, which was first investigated by Bose [32–34], in which the Hamiltonian is

$$H_{\text{Bose}} = \sum_{nm} J_{nm} S_n^z S_m^z - \frac{1}{2}\sum_{nm} \Delta_{nm}(S_n^+ S_m^+ + S_n^- S_m^-) - h\sum_n S_n^z.$$

For an OPCAFM which contains the magnon–phonon interaction, we gave its Hamiltonian for nonlinear collective excitation by

$$H = T + V + \sum_{nm} J_{nm} S_n^z S_m^z - \frac{1}{2}\sum_{nm} (S_n^+ S_m^+ + S_n^- S_m^-) - h\sum_n S_n^z + h\sum_m S_m^z.$$

On the other hand, we [307–311] researched the properties of magnons and the motion rules in double ferromagnetic anti-ferromagnetic interactions, such as $CsNiF_3$ as one-dimensional Heisenberg anti-ferromagnet. $CsNiF_3$ has, in fact, a hexagonal crystal structure [142], with P63/mmc and an MMC crystal structure. Its lattice constants are $a = b = 62.1$ nm and $c = 52$ nm. In this structure, the public face of the octahedron is composed of NF6, which is homogeneously arranged along the c-axis, with the chains separated by the Cs ions. The coupling interaction between the Ni^{2+} ions along the c-axis possesses the ferromagnetic feature, but we find an anti-ferromagnetic feature along the a and b chains [144, 168, 261]. Therefore, we think that $CsNiF_3$ is a ferromagnet with two chains and a chain–chain interaction. In such a case, its Hamiltonian is represented by

$$H = -J_1\sum_{n\delta} S_{An}^z S_{An+\delta}^z - J_1\sum_{n\delta} S_{Bn} S_{Bn+\delta} + J_2\sum_n S_{An} S_{Bn} + D\sum_n (S_{An}^z)^2 + D\sum_n (B_n)^2.$$

We used the above methods to obtain the dynamic equations of these magnons from the above Hamiltonians for these anti-ferromagnetic systems. These results indicate these dynamic equations are all similar to equation (13.3.25) or (13.3.38). Their differences are only distinctions of the parameters and coefficients. This indicates clearly that the nonlinear interactions in these systems are still due to the interactions between the moving magnons and phonons or between the moving magnons. Therefore, the properties and motion rules of the magnons also have the wave-corpuscle duality, so they can still be described by nonlinear quantum mechanics.

Bibliography

[1] K. Abe and O. Inoue, *J. Comput. Phys.* **34** (1980), 202–210.
[2] M. J. Ablowitz, *Stud. Appl. Math.* **58** (1978), 17–94.
[3] M. J. Ablowitz and A. S. Fokas, *Complex Variables: Introduction and Applications*, Cambridge University Press, Cambridge, 2003.
[4] M. J. Ablowitz and A. C. Newell, *J. Math. Phys.* **14** (1973), 1277–1284.
[5] M. J. Ablowitz and H. Segur, *Stud. Appl. Math.* **57** (1977), 13–44.
[6] M. J. Ablowitz, D. J. Kaup, A. C. Newell, and H. Suger, *Stud. Appl. Math.* **53** (1974), 249–315.
[7] J. Ablowitz, M. D. Kruskal, and J. F. Ladik, *SIAM J. Appl. Math.* **36** (1979), 428–437.
[8] A. A. Abrikosov, *Zh. Èksp. Teor. Fiz.* **32** (1957), 1442.
[9] A. A. Abrikosov and L. P. Gorkov, *Zh. Èksp. Teor. Fiz.* **39** (1960), 781.
[10] A. A. Abrikosov, L. P. Gorkov, and I. V. Dzyaloshinkii, *Quantum Field Theoretical Methods in Statistic Physics*, Pergamon Press, Oxford, 1965.
[11] M. E. Alexander and J. L. Morris, *J. Comput. Phys.* **30** (1979), 428–451.
[12] J. Anandan, *Phys. Rev. Lett.* **47** (1981), 463.
[13] D. L. T. Anderson and G. H. Derrick, *J. Math. Phys.* **11** (1970), 1336–1346.
[14] J. Arafune et al., *J. Math. Phys.* **16** (1975), 432.
[15] O. Avenel et al., *Physica B* **194–196** (1994) 491.
[16] G. A. Baker, V. A. Dougalis, and O. A. Karakashian, *Math. Comput.* **40** (1983), 419–433.
[17] L. N. Bardeen, L. N. Cooper, and J. R. Schrieffer, *Phys. Rev.* **108** (1957), 1175.
[18] W. A. Bardeen et al., *Phys. Rev. D* **13** (1975), 1094.
[19] C. F. Barenghi, R. J. Donnerlly, and W. F. Vinen, *Quantized Vortex Dynamics and Superfluid Turbulence*, Springer, Berlin, 2001.
[20] R. Beals and R. Coifman, *Commun. Pure Appl. Math.* **37** (1984), 39–90.
[21] A. A. Belavin, A. Polgakov, A. Schwartzand, and Y. Tyupkin, *Phys. Lett. B* **59** (1975), 85.
[22] T. B. Benjamin, J. L. Bona, and J. J. Mahony, *Philos. Trans. R. Soc. Lond. A* **272** (1972), 47–78.
[23] Y. A. Berezin and V. I. Karpman, *Sov. Phys. JETP* **24** (1967), 1049–1056.
[24] M. S. Berger, *J. Funct. Anal.* **9** (1972), 249–261.
[25] J. A. Blackburn, H. J. T. Smith, and N. L. Rowell, *Phys. Rev. B* **11** (1975), 1053.
[26] A. Blanco-Redondo, C. Husko, D. Eades, Y. Zhang, J. Li, T. F. Krauss, and B. J. Eggleton, *Nat. Commun.* **5** (2014), 3160.
[27] F. Bloch, *Z. Phys.* **74** (1932), 295.
[28] F. Boch, *Z. Phys.* **61** (1930), 206.
[29] N. N. Bogoliubov, *Quantum Statistics*, Moscow, Nauka, 1949; J. Exp. Theor. Phys. USSR. **34** (1958) 58.
[30] N. N. Bogoliubov, V. V. Toimachev, and D. V. Shirkov, *A New Method in the Theory of Superconductivity*, AN SSSR, Moscow, 1958.
[31] J. L. Bona and R. Smith, *Philos. Trans. R. Soc. Lond. A* **278** (1975), 555–601.
[32] I. Bose, *J. Phys. C* **21** (1988), L841.
[33] I. Bose, *Physics B* **174** (1991).
[34] I. Bose and S. Chatterjee, *Phys. Rev. B* **29** (1984), 2741.
[35] J. P. Boucher, R. Pynn, M. Remoissewet, L. P. Regnault, Y. Endo, and J. P. Renard, *Phys. Rev. Lett.* **65** (1980), 488.
[36] J. P. Boucher, R. Pynn, M. Remoissewet, L. P. Regnault, Y. Endo, and J. P. Renard, *Phys. Rev. Lett.* **64** (1990), 1557.
[37] D. Boulware et al., *Phys. Rev. D* **14** (1976), 2708.
[38] R. K. Bullough and P. T. Caudeey, *Solitons*, Plenum Press, New York, 1980.

https://doi.org/10.1515/9783110549638-014

[39] R. K. Bullough, N. M. Bogolyubov, V. S. Kapitonov, C. Malyshev, J. Timonen, A. V. J. Rybin, G. G. Vazugin and M. Lindberg, *Theor. Math. Phys.* **134** (2003), 47.
[40] V. S. Buslaev, *Mat. Nauk.* **34** (1981), 217–218.
[41] V. S. Buslaev and V. V. Sukhanov, *J. Sov. Math.* **34** (1986), 1905–1920.
[42] S. Y. Cai and A. Bhattacharjee, *Phys. Rev. A* **43** (1991), 6934.
[43] B. M. Caradoc-Davies, R. J. Ballagh, and K. Bumett, *Phys. Rev. Lett.* **83** (1999), 895.
[44] J. Chadam, *J. Math. Phys.* **13** (1972), 597–604.
[45] Q. Chang, *Sci. Bull.* **26** (1981), 1094.
[46] H. H. Chen, *Phys. Rev. Lett.* **33** (1974), 925.
[47] S. S. Chern and C. K. Peng, *Manuscr. Math.* **28** (1979), 207–217.
[48] A. R. Chowdhury and T. Roy, *J. Math. Phys.* **21** (1980), 1416–1417.
[49] N. Christ and T. D. Lee, *Phys. Rev. D* **12** (1975), 1606.
[50] D. B. Cohn and K. R. MacKenzie, *Phys. Rev. Lett.* **30** (1973), 258.
[51] L. N. Cooper, *Phys. Rev.* **104** (1956), 1189.
[52] J. Corones, *J. Math. Phys.* **17** (1976), 756–759.
[53] J. Corones, *J. Math. Phys.* **18** (1977), 163–164.
[54] M. Date and K. Kindo, *Phys. Rev. Lett.* **65** (1990), 1659.
[55] A. S. Davydov, *The Soliton in Molecular Systems*, Dordrecht, Reidel, 1985.
[56] A. S. Davydov and V. N. Ermakov, *Phys. Status Solidi B* **148** (1988), 305.
[57] P. G. de Gennes, *Superconductivity of Metals and Alloys*, W. A. Benjamin, New York, 1966.
[58] L. M. Degtiarev and V. E. Zakharov, *JETP Lett.* **20** (1974), 165.
[59] P. Deift and E. Trubowitz, *Commun. Pure Appl. Math.* **32** (1979), 121–251.
[60] P. Deift and X. Zhou, *Ann. Math.* **137** (1993), 295–368.
[61] P. Deift and X. Zhou, *Long-Time Behavior of the Non-Focusing Nonlinear Schrödinger Equation–A Case Study*. New Series: Lectures in Mathematical Sciences, **5**, University of Tokyo, Tokyo, 1994.
[62] P. Deift and X. Zhou, *Acta Math.* **188** (2002), 163–262.
[63] P. Deift and X. Zhou, *Commun. Pure Appl. Math.* **56** (2003), 1029–1077.
[64] P. Deift, A. R. Its, and X. Zhou, *Ann. Math.* **146** (1997), 149–235.
[65] G. H. Derrick, *J. Math. Phys.* **5** (1964), 1252.
[66] S. Deser, *Phys. Lett. B* **64** (1976), 463.
[67] B. S. Dewitt, *Phys. Rev. Lett.* **16** (1966), 1092.
[68] R. K. Dodd and J. D. Gibbon, *Proc. R. Soc. Lond. A* **358** (1978), 287–296.
[69] R. K. Dodd and J. D. Gibbon, *Proc. R. Soc. Lond. A* **359** (1978), 411–433.
[70] R. J. Donnely, *Quantum Vortices in Helium II*, Cambridge Univ. Press, Cambridge, 1991.
[71] T. E. Dushane, *Trans. Am. Math. Soc.* **188** (1974), 77–96.
[72] J. C. Eilbeck and G. R. McGuire, *J. Comput. Phys.* **19** (1975), 43–57.
[73] J. C. Eilbeck and G. R. McGuire, *J. Comput. Phys.* **23** (1977), 63–73.
[74] P. V. Elyutin and A. N. Rogovenko, *Phys. Rev. E* **63** (2001), 026610.
[75] P. V. Elyutin, A. V. Buryak, V. V. Gubernov, R. A. Sammut, and I. N. Towers, *Phys. Rev. E* **64** (2001), 016607.
[76] F. B. Estabrook and H. D. Wahlquist, *J. Math. Phys.* **17** (1976), 1293–1297.
[77] V. Fedyanin and L. Yakushevich, *Phys. Lett. A* **85** (1981), 100.
[78] E. Fermi, J. Pasta, and S. Ulam, *Studies of Non-Linear Problems, Los Alamos Report LA-1940* **266**, 1955.
[79] D. Finkelstein and J. Rubintein, *J. Math. Phys.* **9** (1968), 1762.
[80] A. S. Fokas and A. R. Its, *SIAM J. Math. Anal.* **27** (1996), 738–764.
[81] A. S. Fokas, A. R. Its, and L. Y. Sung, *Nonlinearity* **18** (2005), 1771–1822.
[82] R. Friedbere, T. D. Lee, and A. Sirlin, *Phys. Rev. D* **13** (1976), 2739–2761.

[83] T. Frisch, Y. Pomeau, and S. Rica, *Phys. Rev. Lett.* **69** (1992), 1644.
[84] H. Frohlich, *Phys. Rev.* **79** (1950), 845.
[85] H. Frohlich, *Proc. R. Soc. A* **223** (1954), 296.
[86] H. Fu, *Math. Numer. Sin.* **1** (1983), 36–50.
[87] H. Fujita, *J. Fac. Sci., Univ. Tokyo, Sect. 1A, Math.* **13** (1966), 109–124.
[88] C. S. Gardner, J. M. Greene, M. D. Kruskal, and R. M. Miura, *Phys. Rev. Lett.* **19** (1967), 1095–1097.
[89] J. Gibbons, S. G. Thornhill, M. J. Wardrop, and D. Ter Haar, *J. Plasma Phys.* **17** (1977), 153–170.
[90] J. Ginibre and G. Velo, *J. Funct. Anal.* **32** (1979), 1–32, 33–71.
[91] V. L. Ginzberg and D. A. Kirahnits, *Problems in High-Temperature Superconductivity*, Moscow, Nauka, 1977.
[92] V. L. Ginzberg and L. D. Landau, *Zh. Èksp. Teor. Fiz.* **20** (1950), 1064.
[93] L. V. Ginzburg, *Superconductivity*, Izd. Moscow, AN SSSR, Moscow-Leningrad, 1946.
[94] L. V. Ginzburg, *Superconductivity, Superdiamagnetism, Superfluidity*, MIR Publ., Moscow, 1987.
[95] V. L. Ginzburg, *Phys. Usp.* **40** (1997), 407.
[96] V. L. Ginzburg, *Phys. Usp.* **47** (2004), 1155–1170.
[97] L. P. Gorkov, *Sov. Phys. JETP* **7** (1958), 505.
[98] L. P. Gorkov, *Sov. Phys. JETP* **9** (1959), 1364.
[99] I. S. Greig and J. L. Morris, *J. Comput. Phys.* **20** (1976), 64–80.
[100] E. F. Gross, *Nuovo Cimento* **20** (1961), 454.
[101] L. Gross, *Commun. Pure Appl. Math.* **19** (1966), 1–15.
[102] B. Guo, *Math. Numer. Sin.* **3** (1981), 211–223.
[103] B. Guo, *Acta Math. Sci.* **1** (1981), 261–274.
[104] B. Guo, *Acta Math. Sin.* **25** (1982), 641–656.
[105] B. Guo, In: *Proceedings of the 1980 Beijing Symposium on Differential Geometry and Differential Equations*, **3**, 1227–1246, Science Press/Gordon and Breach, Science Publishers, Inc., Beijing/New York, 1982.
[106] B. Guo, *Sci. Bull.* **27** (1982), 915–920.
[107] B. Guo, *Acta Math. Sin.* **2** (1982), 641–656.
[108] B. Guo, *Sci. Sin., Ser. A* **XXV** (1982), 897–910.
[109] B. Guo, Some problems for a wide system of Zakharov equations, In: *Proceedings of the 1982 Changchun Symposium on Differential Geometry and Differential Equations*, Science Publishers, Inc. New York, 1982.
[110] B. Guo, *Acta Math. Sin.* **26** (1983), 513–532.
[111] B. Guo, *Sci. Sin., Ser. A* **XXVI** (1984), 561–575.
[112] B. Guo, *Chin. Ann. Math., Ser. B* **6** (1985), 281–287.
[113] B. Guo and Q. Chang, *Sci. Bull.* **16** (1982), 2739.
[114] B. Guo and Q. Chang, *Sci. Bull.* **28** (1983), 310–315.
[115] B. Guo and L. Ling, *Chin. Phys. Lett.* **28** (2011), 110202.
[116] B. Guo and X. Pang, *Solitons*, Chinese Science Press, Beijing, 1987.
[117] B. Guo and L. Shen, *Acta Math. Appl. Sin.* **5** (1982), 310–324.
[118] B. Guo and Y. Wang, *Wave Motion* **67** (2016), 47–54.
[119] B. Guo, L. Xu, and S. Huang, *Sci. China* **6** (1983), 485–494.
[120] B. Guo, L. Ling, and Q. Liu, *Stud. Appl. Math.* **130** (2012), 317–344.
[121] B. Guo, L. Ling, and Q. Liu, *Phys. Rev. E* **85** (2012), 026607.
[122] A. Hasegawa and Y. Kodama, *Solitons in Optical Communications*, Oxford University Press, Oxford, 1995.
[123] A. Hasegawa and F. Tappert, *Appl. Phys. Lett.* **23** (1973), 142–144.

[124] A. Hasegawa and F. Tappert, *Appl. Phys. Lett.* **23** (1973), 171–172.
[125] P. Hasenfratz et al., *Phys. Rev. Lett.* **36** (1976), 1119.
[126] R. Hirota, *J. Phys. Soc. Jpn.* **35** (1973), 1566.
[127] R. Hirota, *The Direct Method in Soliton Theory*, Cambridge University Press, Cambridge, 2004.
[128] R. H. Hobart, *Proc. Phys. Soc.* **85** (1965), 610.
[129] S. T. Hu, *Homotopy, Theory*, Academic Press, New York, 1959.
[130] C. Huepe and M. E. Brachet, *Physica D* **140** (2000), 126.
[131] D. Husenmolber, *Fibre, Bundles*, Springer-Velag, Berlin, Heidelberg, New York, 1966.
[132] H. Ikezi, R. J. Taylor, and D. R. Baker, *Phys. Rev. Lett.* **25** (1970), 11–14.
[133] A. R. Its, *Sov. Math. Dokl.* **24** (1981), 452–456.
[134] A. R. Its and V. Y. Novokshenov, *The Isomonodromic Deformation Method in the Theory of Painlevé Equations*, Springer-Verlag, Berlin, 1986.
[135] D. A. Jacobson, *Phys. Rev. B* **8** (1965), 1066.
[136] R. Jockiw and C. Rebbl, *Phys. Rev. Lett.* **36** (1976), 1116.
[137] F. John, *Manuscr. Math.* **28** (1979), 235–268.
[138] B. D. Josephson, *Phys. Lett.* **1** (1962), 251.
[139] B. D. Josephson, *Thesis*, unpublished, Cambridge University, Cambridge, 1964.
[140] B. D. Josephson, *Adv. Phys.* **14** (1965), 419.
[141] R. Julia and A. Zee, *Phys. Rev. D* **11** (1975), 2227.
[142] G. W. Kayfield and F. Reif, *Phys. Rev. Lett.* **11** (1963), 305.
[143] J. Keller, *Commun. Pure Appl. Math.* **10** (1957), 523–532.
[144] J. K. Kiems and M. Steiner, *Phys. Rev. Lett.* **41** (1978), 1137.
[145] Yu. S. Kivshar and B. A. Malomed, *Rev. Mod. Phys.* **61** (1989) 763.
[146] Yu. S. Kivshar, T. J. Alexander, and S. K. Turitsy, *Phys. Lett. A* **278** (2001), 225.
[147] D. J. Korteweg and G. de Vries, *Philos. Mag.* **39** (1895), 422–443.
[148] M. D. Kruskal, M. Miura, and C. S. Gardner, *J. Math. Phys.* **11** (1970), 952.
[149] E. Kusayanage, T. Kawashima and K. Yamafuji, *J. Phys. Soc. Jpn.* **33** (1972), 551.
[150] G. L. Lamb, *Rev. Mod. Phys.* **43** (1971), 99.
[151] G. L. Lamb, *Phys. Rev. Lett.* **31** (1973), 196–199.
[152] L. D. Landau and E. M. Lifshitz, *Quantum Mechanics*, Pergamon Press, Oxford, 1987.
[153] P. D. Lax, *Commun. Pure Appl. Math.* **21** (1968), 467–490.
[154] T. D. Lee, *Phys. Rep. C* **23** (1976), 254.
[155] Z. Z. Lee, *Solid State Physics*, Advanced Education Press (Chinese), Beijing, 1987.
[156] T. D. Lee and Y. Pang, *Phys. Rep.* **221** (1992), 251–350.
[157] A. J. Leggett, *Phys. Rev. Lett.* **41** (1978), 586.
[158] A. J. Leggett, In: *Percolation, Localization and Superconductivity*, A. M. Goldlinan and S. A. Bvilf (eds.), Plenum Press, New York, 1984, pp. 1–41.
[159] A. J. Leggett, Topics in superfluidity and superconductivity, In: *Low Temperature Physics*, Springer, Berlin, 1991, pp. 1–93.
[160] H. Levine, *Arch. Ration. Mech. Anal.* **51** (1973), 371–386.
[161] J.-E. Lin, *J. Differ. Equ.* **29** (1978), 467–473.
[162] C. A. Lindensmith et al., *Czechoslov. J. Phys.* **46** (1996), 131.
[163] J. L. Lions, *Equations differentielles Operationnelles et Problems aux limites*, Springer, Berlin, 1961.
[164] J. L. Lions, *Quelques Methodes de Resolution des Problems aux limites nonlineaires*, Dunod, Paris, 1969.
[165] W. S. Liu and X. P. Li, *Eur. Phys. J. D* **2** (1998), 1.
[166] F. London, *Superfluids, Vol. 1*, Wiley, New York, 1950.
[167] F. London and H. London, *Proc. R. Soc. Lond. A* **149** (1935), 71.

[168] J. M. Loveluk, T. Schneider, E. Stoll, et al., *Phys. Rev. Lett.* **54** (1980), 150, 5.
[169] K. W. Madison, F. Chevy, W. Wohlleben, and J. Dalibard, *Phys. Rev. Lett.* **84** (2000), 806.
[170] V. G. Makhandov and V. K. Fedyanin, *Phys. Rep.* **104** (1984), 1–80.
[171] V. G. Makhankov, *Phys. Rep.* **35** (1978), 1–128.
[172] V. G. Makhankov, *Phys. Lett. A* **81** (1981), 158.
[173] S. V. Manakov, *Sov. Phys. JETP* **38** (1974), 693–696.
[174] M. R. Matthews, B. P. Anderson, P. C. Haljan, D. S. Hall, C. E. Wieman and E. A. Cornell, *Phys. Rev. Lett.* **83** (1999), 2498.
[175] V. B. Matveev and M. A. Salle, *Darboux Transformations and Solitons*, Springer-Verlag, Berlin, 1991.
[176] S. Maxon, *Lect. Notes Phys.*, **64** (1977).
[177] L. A. Medeiros and M. M. Miranda, *J. Math. Anal. Appl.* **59** (1977), 432–441.
[178] H. J. Mikeska, *J. Phys. C* **11** (1979), l28.
[179] H.-J. Mikeska and M. Steiner, *Adv. Phys.* **40** (1991), 191–356.
[180] M. Miura, C. S. Gardner, and M. D. Kruskal, *J. Math. Phys.* **9** (1968), 1204–1209.
[181] L. F. Mollenauer, R. H. Stolen, and J. P. Gordon, *Phys. Rev. Lett.* **45** (1980), 1095–1098.
[182] C. S. Morawetz, *Proc. R. Soc. Lond. A* **306** (1968), 291–296.
[183] H. Morikawa, I. Matsuda, and S. Hasegawa, *Phys. Rev. B* **70** (2004), 085412.
[184] H. C. Morris, *J. Math. Phys.* **17** (1976), 1867–1869.
[185] H. C. Morris, *J. Math. Phys.* **17** (1976), 1870–1872.
[186] H. C. Morris, *J. Math. Phys.* **18** (1977), 533–536.
[187] H. C. Morris, *J. Math. Phys.* **21** (1980), 327–333.
[188] H. C. Morris, In: *Amee Conference*, Hemann (ed.) (1976).
[189] H. B. Nidsen and P. Olesen, *Nucl. Phys. B* **61** (1971), 45.
[190] I. R. Niek and J. C. Fraser, *J. Low Temp. Phys.* **3** (1970), 225.
[191] V. Y. Novokshenov, *Sov. Math. Dokl.* **21** (1980), 529–533.
[192] V. Y. Novokshenov, *Izv. Acad. Nauk SSSR* **48** (1984), 372–410.
[193] V. Y. Novokshenov, *Differ. Equ.* **21** (1985), 1915–1926.
[194] L. Onsager, *Suppl. Nuovo Cim.* **6** (1949), 2, 249.
[195] R. E. Packard, *Rev. Mod. Phys.* **70** (1998), 641.
[196] X.-f. Pang, *Chin. J. Nature* **5** (1982), 254.
[197] X.-f. Pang, *Chin. J. Low Temp. Supercond.*, **3** (1982) 62.
[198] X.-f. Pang, *J. Low Temp. Phys.* **58** (1985), 333.
[199] X.-f. Pang, *Investig. Met. Mater.*, **12** (1986) 31.
[200] X.-f. Pang, *J. Sci. Explor. Sin.* **1**(4) (1986) 70.
[201] X.-f. Pang, *Chin. J. At. Mol. Phys.* **5** (1987), 383.
[202] X.-f. Pang, The Properties of Soliton Motion in Superfluid He4, Proc. ICNP, Shanghai, 1987, p. 181.
[203] X.-f. Pang, *J. Xinjiang Univ. Nat. Sci.* **5** (1988) 33.
[204] X.-f. Pang, *J. Xinjiang Univ. Nat. Sci.* **3** (1988) 33.
[205] X.-f. Pang, *J. Kunming Tech. Sci. Univ.*, **14** (1989) 78.
[206] X.-f. Pang, *J. Kunming Tech. Sci. Univ.*, **14** (1989) 83.
[207] X.-f. Pang, *J. Chinghai Norm. Univ. Sin.* **1** (1989) 37.
[208] X.-f. Pang, *J. Xinjiang Univ. Nat. Sci.* **3** (1989) 72.
[209] X.-f. Pang, The Properties of Soliton Motion for Superconductivity Electrons, In: Proc. ICNP, Shanghai, 1989, p. 139.
[210] X.-f. Pang, *J. Xuntan Univ. Sin.* **12** (1990) 43 and 65.
[211] X.-f. Pang, *J. Phys. Condens. Matter* **2** (1990) 9541.
[212] X.-f. Pang, *Acta Math. Phys. Sin.* **10** (1990) 47.

[213] X.-f. Pang, *J. Southwest Inst. Nationalities* **17** (1991) 1.
[214] X.-f. Pang, *J. Southwest Inst. Nationalities Sin.* **17** (1991) 18.
[215] X.-f. Pang, *J. Xuntan Univ. Sin.* **13** (1991) 63.
[216] X.-f. Pang, *J. Semicond. Sin.* **15** (1992) 917.
[217] X.-f. Pang, *Chin. J. Light Scattering*, **4** (1992) 126.
[218] X.-f. Pang, *Chin. Phys.* **22** (1993), 282.
[219] X.-f. Pang, *Phys. Status Solidi B* **189** (1993) 237.
[220] X.-f. Pang, *Phys. Bull. Sin.* **5** (1983) 6.
[221] X.-f. Pang, *J. Potent. Sci. Sin.* **6** (1986) 16.
[222] X. Pang, The theory of nonlinear quantum mechanics, In: *Research of New Sciences*, H. Lui (ed.), Chinese Science and Tech. Press, Beijing, China, 1993.
[223] X.-f. Pang, *Nonlinear Quantum Mechanical Theory*, Chongqing Press, Chongqing, 1994.
[224] X.-f. Pang, *Soliton Physics*, Sichuan Sci. and Tech. Press Sin., Chengdu, 2003.
[225] X.-f. Pang, *Quantum Mechanics in Nonlinear Systems*, World Scientific Publishing Co., Singapore, 2005.
[226] X.-f. Pang, *Nature Science* **2**(1) (2007), 42.
[227] X.-f. Pang, *J. Electron. Sci. Technol. China*, **6**(2) (2008) 205–211.
[228] X.-f. Pang, *Res. Devel. World Sci. Tech.*, **28** (2006), 11.
[229] X.-f. Pang, The behaviours and properties of microscopic particles in nonlinear systems, In: *Handbook of Solitons, Research, Technology and Applications*, S. P. Lang and S. H. Bedore (eds.), Nova Science Publishers, Inc., New York, 2009, pp. 185–249.
[230] X.-f. Pang, The soliton theory of bio-energy transport in protein molecules, In: *Handbook of Solitons, Research, Technology and Applications*, S. P. Lang and S. H. Bedore (eds.), Nova Publishers, New York, 2009, pp. 251–298.
[231] X.-f. Pang, *Far East J. Mech. Eng. Phys.*, **2** (2010), 1.
[232] X.-f. Pang, *Nonlinear Quantum Mechanics*, Chinese Electronic Industry Press, Beijing, 2009, pp. 20–63.
[233] X.-f. Pang, Investigation of solutions of a time-dependent Ginzburg-Landau equation in superconductor by nonlinear quantum theory, *IEEE Compendex*, IEEE, New York, 2009, 274–277, doi: 10.1109/ASEMD.2009.5306641(EI).
[234] X.-f. Pang, *Nature Science*, **3**(7) (2011) 978.
[235] X.-f. Pang, *Nonlinear Quantum Mechanics*, LAP Lambert Publishing, Germany, 2012.
[236] X.-f. Pang, *Int. J. Res. Rev. Appl. Sci.*, **10** (3) (2012).
[237] X.-f. Pang, *Int. J. New Innov. Sci. Technol.* **1**(1) (2013), 25–78.
[238] X.-f. Pang, Establishment of nonlinear quantum mechanics and changes of property of microscopic particles as well as their experimental evidences, In: *Contemporary Research in Quantum Systems*, Z. Wzziane (ed.), Nova Publishers, New York, 2014, pp. 207–338.
[239] X. Pang, *Nonlinear Quantum Mechanics and Its Applications*, NAVO publisher, New York, 2015.
[240] X.-f. Pang and Y.-p. Feng, *Quantum Mechanics in Nonlinear Systems*, World Scientific Publishing Co., New Jersey, 2005.
[241] R. D. Parks, *Superconductivity*, Marcel Dekker, New York, 1969.
[242] A. Patrascioiu, *Phys. Rev.* **D12** (1975), 523.
[243] H. Pecher and W. v. Wahl, *Manuscr. Math.* **27** (1979), 125–157.
[244] V. M. Perez-Garcia, M. Michinel, and H. Herrero, *Phys. Rev. A* **57** (1998), 3837.
[245] J. K. Perring and T. H. R. Skyrme, *Nucl. Phys.* **31** (1962), 550–555.
[246] J. K. Perring et al., *Nucl. Phys.* **61** (1963), 1443.
[247] J. Pfeiffer, M. Schuster, A. A. Jr. Abdumalikov, and A. V. Ustinov, *Phys. Rev. Lett.* **96** (2006), 034103.

[248] L. P. Pitaevskii, *Sov. Phys. JETP* **10** (1960), 553.
[249] L. P. Pitaevskii, *Sov. Phys. JETP* **13** (1961), 451 (Zh. Eksp., Theor. Fiz. **40** (1961) 646).
[250] L. P. Pitaevskii and S. Stringari, *Bose–Einstein Condensation*, Clarendon Press, Oxford, 2003.
[251] A. M. Polyakov, *Phys. Lett. B* **59** (1975), 82.
[252] Y. Pomeau and S. Rica, *C. R. Acad. Sci. (Paris)* **316**(II) (1993), 1523.
[253] M. Prasad and C. Sommerfied, *Phys. Rev. Lett.* **35** (1975), 760.
[254] D. Pushkarov and Kh. Pushkarov, *Teor. Mat. Fiz.* **30** (1977), 1.
[255] D. Pushkarov and Kh. Pushkarov, *J. Phys. C* **10** (1977), 3711.
[256] D. Pushkarov and Kh. Pushkarov, *Phys. Status Solidi B* **81** (1979), 703.
[257] D. Pushkarov and Kh. Pushkarov, *Phys. Status Solidi B* **123** (1984), 573.
[258] R. Rajaraman, Soliton and Instantons: An Introduction to Solitons and Instantons in Quantum Field Theory, North-Holland Personal Library, 1st-edition, 1987.
[259] G. W. Rayfield and F. Reif, *Phys. Rev. Lett.* **11** (1963), 305.
[260] M. Reed and B. Simon, *Methods of Modern Mathematical Physics II: Fourier Analysis, Self Adjointness*, Academic Press, New York, San Francisco, London, 1978.
[261] G. Reiter, *Phys. Rev. Lett.* **46** (1981), 202.
[262] J. P. Renard et al., *J. Appl. Phys.* **63** (1988), 3538.
[263] D. Rogovin, *Phys. Rev. B* **11** (1975), 1906–1908.
[264] D. Rogovin, *Phys. Rev. B* **12** (1975), 130–133.
[265] D. Rogovin and M. Scully, *Phys. Rep.* **25** (1976), 178.
[266] A. F. J. Runge, N. G. R. Broderick, and M. Erkintalo, *Optica* **2** (2015), 36–39.
[267] J. C. Russell, Report on waves, *14th Meeting of the British Association for the Advancement of Science*, John Murray, London, 1844 p. 311.
[268] R. Sasaki, *Nucl. Phys. B* **154** (1979), 343–357.
[269] J. Satsuma and D. J. Kaup, *J. Phys. Soc. Jpn.* **43** (1977), 692–697.
[270] S. Sawada and T. Koterz, *Prog. Theor. Phys.* **51** (1974), 1355–1367.
[271] J. R. Schrieffer, *Theory of Superconductivity*, Benjamin, New York, 1964.
[272] J. R. Schrieffer, *Superconductivity*, Benjamin, New York, 1969.
[273] K. W. Schwarz, *Phys. Rev. Lett.* **64** (1990), 1130.
[274] K. W. Schwarz, *Phys. Rev. B* **47** (1993), 12030, *Phys. Rev. B* **71** (1993) 259.
[275] I. Segal, *Ann. Math.* **78** (1963), 339–364.
[276] H. Segur and M. J. Ablowitz, *J. Math. Phys.* **17** (1976), 710.
[277] H. Segur and M. J. Ablowitz, *Physica D* **3** (1981), 165–184.
[278] A. B. Shabat, *Sov. Math. Dokl.* **14** (1973), 1266.
[279] R. Shankar, *Phys. Rev. D* **14** (1976), 1107.
[280] A. Sjöberg, *J. Math. Anal. Appl.* **29** (1970), 569–579.
[281] E. B. Sonin, *Physica B* **210** (1995), 234–250.
[282] W. A. Strauss, *Arch. Ration. Mech. Anal.* **55** (1974), 86–92.
[283] W. A. Strauss, *Commun. Math. Phys.* **55** (1977), 149–162.
[284] C. H. Su and C. S. Gardner, *J. Math. Phys.* **10** (1969), 536.
[285] D. Suint-James et al., *Type-II Superconductivity*, Pergamon, Oxford, 1966.
[286] V. V. Sukhanov, *Sov. Phys. Dokl.* **28** (1983), 325–327.
[287] V. V. Sukhanov, *Zap. Nauč. Semin. LOMI* **161** (1987), 122–138.
[288] V. V. Sukhanov, *Teor. Mat. Fiz.* **84** (1990), 23–27.
[289] G. 't Hooft, *Nucl. Phys. B* **79** (1974), 276.
[290] R. Temam, *J. Math. Pures Appl.* **48** (1969), 159–172.
[291] K. Tenenblat and C. L. Terng, *Ann. Math.* **111** (1980), 477–490.
[292] C. L. Terng, *Ann. Math.* **111** (1980), 491–510.
[293] E. Tomboulis and G. Woo, *Nucl. Phys. B* **107** (1976), 221.

[294] B. A. Ton, *J. Differ. Equ.* **25** (1977), 288–309.
[295] A. Tonomura, H. Kasai, O. Kamimura, T. Matsuda, K. Harada, T. Yoshida, T. Akashi, J. Shimoyama, K. Kishio, T. Hanaguri, K. Kitazawa, T. Masui, S. Tajima, N. Koshizuka, P. L. Gammel, D. Bishop, M. Sasase, and S. Okayasu, *Phys. Rev. Lett.* **88** (2002), 237001.
[296] M. Tsutsumi and N. Hayashi, *Math. Z.* **177** (1981), 217–234.
[297] J. G. Valatin, *Nuovo Cimento* **7** (1958), 843.
[298] H. J. De Vega, F. A. Schaposnik, *Phys. Rev. D* **14** (1976), 1100.
[299] I. Ventura, Solitons in Liquid 4He, In: *Solitons and Applications*, V. G. Makhankov, V. K. Fedyanin and O. K. Pashaev (eds.), World Scientific, Singapore, 1990.
[300] A. C. Vliegenthart, *J. Eng. Math.* **5** (1971), 137–155.
[301] P. K. C. Wang, *J. Math. Phys.* **19** (1978), 1286–1292.
[302] Y. F. Wang, B. Tian, and M. Wang, *Stud. Appl. Math.* **131** (2013), 119–134.
[303] Y. F. Wang, B. Tian, L. C. Liu, M. Li, and B. Qin, *Z. Angew. Math. Phys.* **66** (2015), 2543–2553.
[304] S. Watanabe, H. S. J. van der Zant, S. H. Strogatz, and T. E. Orlando, *Physica D* **97** (1996), 429–470.
[305] E. Weinberg and A. Guth, *Phys. Rev.* **D14** (1976), 1660.
[306] B. Whitham, *Proc. R. Soc. Lond. A* **283** (1965), 238–261.
[307] C.-t. Xu and X.-f. Pang, *Chia. J. At. Mol. Phys.* **13** (1993), 219–226.
[308] C.-t. Xu and X.-f. Pang, *Chia. J. At. Mol. Phys.* **13** (1996), 508–514.
[309] C.-t. Xu and X.-f. Pang, *Chia. J. At. Mol. Phys.* **15** (1999), 506.
[310] C.-t. Xu and X.-f. Pang, *Chia. J. At. Mol. Phys.* **16** (2000), 99.
[311] C.-t. Xu and X.-f. Pang, *Chia. J. At. Mol. Phys.* **16** (2000), 513.
[312] E. J. Yarmchuk, M. J. V. Gordon, and R. E. Packard, *Phys. Rev. Lett.* **43** (1979), 214.
[313] J. Ze hang, *An Introduction to Topology*, Shanghai Science and Technology Press, Shanghai, 1977.
[314] N. J. Zabusky and M. D. Kruskal, *Phys. Rev. Lett.* **15** (1965), 240–243.
[315] V. E. Zakharov, *Sov. Phys. JETP* **34** (1972), 62.
[316] V. E. Zakharov and S. V. Manakov, *Sov. Phys. JETP* **44** (1976), 106–112.
[317] V. E. Zakharov and A. B. Shabat, *Sov. Phys. JETP* **34** (1972), 62–69.
[318] G. Z. Zhou et al., *Sci. China* **11** (1979) 1057.
[319] Y. Zhou and H. Fu, *Acta Math. Sin.* **26** (1983), 234–249.
[320] Y. Zhou and B. Guo, *Acta Math. Sin.* **27** (1984), 154–176.
[321] Y. Zhou and B. Guo, *Sci. Sin., Ser. A* **27** (1984), 799–811.
[322] Y. Zhou and B. Guo, On the solvability of the initial value problem for the quasilinear degenerate parabolic system, In: *Proceedings of the 1982 Changchun Symposium on Differential Geometry and Differential Equations*, Science Publishers, Inc. New York, 1982.
[323] Y. Zhou and B. Guo, The Existence of weak solution of the boundary problem for Systems of Ferro-Magnetic Chain, In: *Proceedings of the 1982 Changchun Symposium on Differential Geometry and Differential Equations*, Science Publishers, Inc. New York, 1982.
[324] Y. Zhou and B. Guo, The Periodic Boundary Problems and the Initial Value Problems for the systems of the Generalized Korteweg-de Vries Type of High order, In: *Proceedings of the 1982 Changchun Symposium on Differential Geometry and Differential Equations*, Science Publishers, Inc. New York, 1982.
[325] R. J. Zieve, Yu. Mukhjarsky, J. D. Close, J. C. Davis and R. E. Packard, *Phys. Rev. Lett.* **68** (1992), 1327.
[326] R. J. Zieve, J. D. Close, J. C. Davis and R. E. Packard, *J. Low Temp. Phys.* **90** (1993), 243.

Bei Fragen zur Produktsicherheit wenden Sie sich bitte an:
If you have any questions regarding product safety,
please contact:

Walter de Gruyter GmbH
Genthiner Straße 13
10785 Berlin
productsafety@degruyterbrill.com